S0-ASQ-562

ELEMENTS OF HEAT TRANSFER

ELEMENTS OF HEAT TRANSFER

Yıldız Bayazıtoğlu

Associate Professor
Mechanical Engineering and
Materials Science
Rice University

M. Necati Özışık

Professor, Mechanical and
Aerospace Engineering
North Carolina State University

McGraw-Hill Book Company

New York St. Louis San Francisco Auckland Bogotá Caracas Colarado Springs
Hamburg Lisbon London Madrid Mexico Milan Montreal New Delhi
Oklahoma City Panama Paris San Juan São Paulo Singapore Sydney Tokyo Toronto

This book was set in Times Roman.
The editors were Anne Duffy, John J. Corrigan, and Steven Tenney;
the designer was John Hite;
the production supervisor was Diane Renda.
Drawings were done by J & R Services, Inc.
R. R. Donnelley & Sons Company was printer and binder.

ELEMENTS OF HEAT TRANSFER

Copyright © 1988 by McGraw-Hill, Inc. All rights reserved.
Printed in the United States of America. Except as permitted under the
United States Copyright Act of 1976, no part of this publication may be
reproduced or distributed in any form or by any means, or stored in a data
base or retrieval system, without the prior written permission of the publisher.

1234567890DOCDOC89321098

ISBN 0-07-004154-7

Library of Congress Cataloging-in-Publication Data

Bayazitoglu, Yildiz.
   Elements of heat transfer / Yildiz Bayazitoglu, M. Necati Özişik.
      p.   cm.
   Bibliography: p.
   Includes index.
   ISBN 0-07-004154-7
   1. Heat-Transmission.   I. Özişik, M. Necati.   II. Title.
TJ260.B38 1988
621.402'2-dc19   87 – 21646
                  CIP

ABOUT THE AUTHORS

Y. Bayazıtoğlu is associate professor of mechanical engineering at Rice University. She received her M.S. and Ph.D. in mechanical engineering from the University of Michigan, where she was a Barbour Fellow. Prior to joining Rice University, she taught at the M. E. Technical University of Ankara and the University of Houston. Dr. Bayazıtoğlu is a member of the fundamentals of heat transfer committee of the American Society of Mechanical Engineers. With her research interest in thermal sciences, she has published extensively in the area of heat transfer, radiation, hydrodynamic and thermal stability, and solar energy. She lives in Houston, Texas, with her husband Yıldırım and her three sons, Ozgur, Mert, and Kunt.

M. N. Özısık is professor of mechanical engineering at North Carolina State University. A graduate of the University of London, Dr. Özısık has been honored here and abroad for his sustained contributions to heat transfer research and education. In London, he received the F. Bernard Hall Price Award from the Institution of Mechanical Engineers; in the United States, he has received the Western Electric Fund Award of ASEE, Alcoa Foundation Research Award, Alcoa Foundation Distinguished Engineering Research Award, Turkish Science Award, O. Max Gardner Award, R. J. R. Nabisco Award, and Heat Transfer Memorial Award. He is the author or coauthor of seven books and over 200 research papers in the areas of heat transfer, mass diffusion, radiation, and heat exchangers. He is a fellow of the American Society of Mechanical Engineers.

ABOUT THE AUTHORS

CONTENTS

PREFACE

This book is intended to serve as a one-semester introductory text for teaching the principles of heat transfer to engineering students at the undergraduate level. It differs from traditional heat transfer books in that all the material in this book can be covered in a one-semester, 3-credit-hour course. To achieve this objective, detailed derivations and complex mathematical analyses of heat transfer are omitted; instead, an exposure to a broader range of topics on the subject is emphasized in order to enable the instructor to cover more heat transfer topics during a single semester. The text is intended to provide a good understanding of the basic concepts and physical principles of heat transfer and to enhance the skills of the reader in the identification of simple problems and the development of solutions for them.

While traditional heat transfer books can well serve the need for a two-semester sequence in heat transfer education at the undergraduate level, there are still a large number of engineering schools where single-semester undergraduate-level heat transfer courses are given, with emphasis on the coverage of a broader range of topics in one semester. Furthermore, the field of heat transfer today is undergoing an unprecedented expansion into areas hitherto unthought of. Almost every branch of science and engineering includes some kind of heat transfer problem, and there is an ever-growing need for scientists and engineers to have some background in heat transfer. Therefore, this book can also serve as a one-semester course for the engineering disciplines that customarily do not have a formal heat transfer course in their curriculum. There are 134 worked-out examples illustrating the application of the basic theory to the solution of simple heat transfer problems and the implications of the mathematical modeling of physical problems of heat transfer. Over 500 problems, arranged in the same order as the material presented in the text, are included at the ends of the chapters, with the answers given for some of them. In Chapter 1, basic concepts on the mechanisms of heat flow by conduction, convection, and radiation are presented, while in Chapter 2 the concept of lumped system analysis is introduced to illustrate the time variation of temperature in solids.

Chapters 3 through 6 are devoted to applications of heat flow by conduction. One-dimensional simple heat conduction problems in bodies having such simple shapes as a plate, a cylinder, and a sphere are studied by utilizing elementary analytic solution techniques and the thermal resistance concepts. The use of finite-difference methods with computer applications is introduced for the solution of one-dimensional steady and transient heat flow in solids. Transient-temperature charts are given for rapid estimation of transient-temperature distribution and heat flow in a semi-infinite body, flat plate, solid cylinder and sphere.

Heat flow by forced convection is covered in Chapters 7 and 8. Emphasis is placed on understanding the concepts of velocity and temperature distribution in flow and their consequences on drag force and heat transfer for forced flow over bodies having simple shapes, as well as for forced flow inside ducts. Chapter 9 deals with free convection, while Chapter 10 covers the topics of condensation and boiling.

Chapter 11 introduces the basic concepts in heat transfer by radiation and utilizes the method to determine radiation heat exchange between surfaces. Finally, Chapter 12 is devoted to heat exchangers and illustrates with representative examples the solution of simple sizing and rating problems.

Heat transfer calculations are commonly performed in engineering using either the English system or the SI (Système International) system of units. In this book the SI system of units is used throughout the main body of the text, in the solution of the examples, and in the physical property tables.

We would like to express our thanks for the many useful comments and suggestions provided by colleagues who reviewed this text during the course of its development, especially to V. S. Arpaci, University of Michigan; I. S. Habib, University of Michigan; Sadik Kakac, University of Miami; Amir Karimi, University of Texas; Vishwanath Prasad, Columbia University; and E. M. Sparrow, University of Minnesota.

Y. Bayazıtoğlu   M. N. Özışık

ELEMENTS OF HEAT TRANSFER

CHAPTER

1

CONCEPTS OF THE MECHANISM OF HEAT FLOW

The earliest evidence of human use of fire for warmth and light comes from caves occupied by Peking man about half a million years ago. In spite of this use, it has been only in relatively recent times that people have understood that heat is energy and that temperature is a measure of the amount of that energy present in a body.

As a form of energy, heat can be converted into work. Although it is customary to speak of sensible heat and latent heat, it is incorrect to speak of the heat accumulation in a body, since the word *heat* is restricted to energy being transferred.

Heat flows from one body to another as the result of a difference in temperature. If two bodies at different temperatures are brought together, heat flows from the hotter body to the colder body. As a result, in the absence of phase change such as melting or solidification, the temperature of the colder body increases and that of the hotter body decreases.

Since heat flow takes place in a system whenever there is a temperature gradient, it is essential to know the temperature distribution in order to compute the heat flow. The temperature distribution and the heat flow are of interest in many scientific and engineering applications, such as the design of heat exchangers, nuclear-reactor cores, heating and air-conditioning systems, and solar energy systems.

In studying heat flow it is customary to consider three distinct mechanisms of heat flow: *conduction, convection,* and *radiation.* Heat is transferred by conduction in a solid or fluid at rest. Conduction needs a medium in which to take place, whereas radiation can take place in a vacuum with no material carrier. Heat is transferred by convection in fluids in motion. In fact, conduction and radiation are the two basic modes of heat flow; convection can be regarded as conduction with fluid in motion.

In general, the temperature distribution in a medium is controlled by the combined effects of these different mechanisms of heat flow. It is not possible to completely isolate one from the others. However, when one mechanism is dominant, the others can be neglected, and with such a restriction we present below a brief qualitative description of each of these three different mechanisms of heat transfer.

1-1   CONDUCTION

Conduction is the mechanism of heat flow in which energy is transported from the region of high temperature to the region of low temperature by the drift of electrons, as in solids. Therefore, metals that are good conductors of electricity are also good conductors of heat.

The conduction law is based on experimental observations made by Biot and named after Fourier. The *Fourier law* states that the rate of heat flow by conduction in a given direction (say, the x direction) is proportional to:

The gradient of temperature in that direction, dT/dx

The area normal to the direction of heat flow, A

Then for heat flow in the x direction, we have

$$Q_x = -kA \frac{dT}{dx} \qquad \text{W} \qquad (1\text{-}1)$$

where Q_x is the rate of heat flow in the positive x direction, through area A normal to the x direction, and

$$\frac{dT}{dx} = \lim_{\Delta x \to 0} \frac{\Delta T}{\Delta x} \qquad (1\text{-}2)$$

is the gradient of temperature in that direction. The proportionality constant k, called the *thermal conductivity,* is a property of the material.

The reason for including the minus sign on Eq. (1-1), as illustrated in Fig. 1-1(a), is as follows: If temperature decreases in the positive x direction, dT/dx is negative; then Q_x becomes a positive quantity because of the presence of the negative sign, and hence the heat flow is in the positive x direction. Likewise, if temperature increases in the positive x direction, dT/dx is positive, and hence Q_x becomes negative and the heat flow is in the negative x direction, as

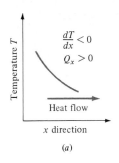

(a)

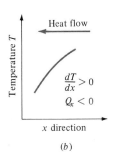

(b)

FIGURE 1-1
Sign convention for
the direction of heat
flow in the Fourier law
of heat conduction.

illustrated in Fig. 1-1(*b*). Thus the notation used in Eq. (1-1) implies that when Q_x is a *positive* quantity, the *heat flow is in the positive x direction*, and vice versa.

To illustrate the application of this concept, we consider a slab (i.e., a plate) with a linear temperature distribution within the body, as shown in Fig. 1-2. Equation (1-1) becomes

$$Q_x = -kA \frac{dT}{dx}$$

$$= -kA \frac{T_2 - T_1}{x_2 - x_1} = kA \frac{T_1 - T_2}{x_2 - x_1}$$

$$= kA \frac{\Delta T}{L} \tag{1-3}$$

where $x_2 - x_1 = L$, the thickness of the slab, is a positive quantity.

For the specific situation shown in Fig. 1-2, we have $T_1 > T_2$, and hence $\Delta T = T_1 - T_2$ is also a positive quantity. Thus the heat flow Q_x is in the positive *x* direction.

Conversely, if we had $T_2 > T_1$, then $T_1 - T_2$ would be negative and heat flow would be in the negative *x* direction.

The heat flow rate per unit area is called the *heat flux*. Therefore, Q_x divided by the area A,

$$\boxed{q_x = \frac{Q_x}{A}} \qquad \text{W/m}^2 \tag{1-4}$$

is the heat flux in the *x* direction. Thus q_x represents the *amount of heat flow per unit area, per unit time in the x direction*. When heat flow Q_x is in watts and heat flux q_x in watts per square meter, then the thermal conductivity k has the dimension W/(m · °C) or J/(m · s · °C).

FIGURE 1-2
Temperature distribution $T(x)$
and heat flow by conduction
through a slab.

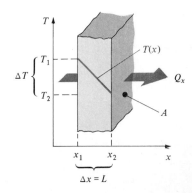

Example 1-1. A brick wall of thickness 25 cm and thermal conductivity 0.69 W/ (m · °C) is maintained at 20°C at one surface and 10°C at the other surface. Determine the heat flow rate across a 5-m^2 surface area of the wall.

Solution. As illustrated in the accompanying figure, the origin of the x coordinate is placed on the hot surface at $T_1 = 20°C$. Then the cold surface $T_2 = 10°C$ is located at $x_2 = L = 0.25$ m. Knowing the thermal conductivity of the wall $k = 0.69$ W/(m · °C), Eq. (1-3) is applied to determine the heat flow rate across the surface area of the wall $A = 5$ m^2.

$$Q_x = kA \frac{T_1 - T_2}{x_2 - x_1} = kA \frac{\Delta T}{L}$$

$$= 0.69 \text{ W/(m · °C)} \times 5 \text{ m}^2 \times \frac{(20 - 10)°C}{0.25 \text{ m}}$$

$$= 138 \text{ W} = 0.138 \text{ kW}$$

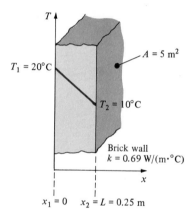

T

$A = 5$ m^2

$T_1 = 20°C$

$T_2 = 10°C$

Brick wall
$k = 0.69$ W/(m·°C)

x

$x_1 = 0$ $x_2 = L = 0.25$ m

FIGURE EXAMPLE 1-1
Heat conduction through the slab.

Example 1-2. The inside and outside surfaces of a window glass are at 20°C and −5°C, respectively. If the glass is 100 cm by 50 cm in size and 1.5 cm thick, with a thermal conductivity of 0.78 W/(m · °C), determine the heat loss through the glass over a period of 2 h.

Solution. The origin of the x coordinate is placed on the hot surface $x_1 = 0$; then $x_2 = L = 0.015$ m. Given that $T_1 = 20°C$, $T_2 = -5°C$, $k = 0.78$ W/(m · °C), and $A = 1$ m $\times 0.5$ m $= 0.5$ m^2, Eq. (1-3) is used:

$$Q_x = kA \frac{T_1 - T_2}{x_2 - x_1} = kA \frac{\Delta T}{L}$$

$$= 0.78 \text{ W/(m · °C)} \times (1 \times 0.5 \text{ m}^2) \times \frac{[20 - (-5)]°C}{0.015 \text{ m}}$$

$$= 650 \text{ W} = 0.650 \text{ kW}$$

Then the heat loss through the window glass over a period of $\Delta t = 2$ h becomes

$$\text{Heat loss} = Q_x \, \Delta t$$

$$= 0.650 \text{ kW} \times 2 \text{ h} = 1.30 \text{ kW · h}$$

Example 1-3. A fiberglass insulating board with thermal conductivity 0.05 W/ (m · °C) is to be used to limit the heat losses to 100 W/m^2 for a temperature difference of 150°C across the board. Determine the thickness of the insulating board.

Solution. Given $k = 0.05$ W/(m · °C), $\Delta T = 150$°C, and $q = Q/A = 100$ W/m^2, Eqs. (1-3) and (1-4) are applied to find L:

$$q = \frac{Q}{A} = k \frac{\Delta T}{L}$$

$$100 \text{ W/m}^2 = 0.05 \text{ W/(m} \cdot {}^\circ\text{C)} \frac{150{}^\circ\text{C}}{L}$$

$$L = 0.075 \text{ m} = 75 \text{ mm}$$

Thermal conductivity

There is a wide difference in the range of thermal conductivity among various engineering materials, as illustrated in Fig. 1-3. Between gases (e.g., air) and highly conducting metals (e.g., copper), k varies by a factor of about 10^4. Thermal conductivity also varies with temperature. For some materials the variation over certain temperature ranges is small enough to be neglected, but in many situations the variation of k with temperature is significant. At very

FIGURE 1-3
Typical range of thermal conductivity of various materials.

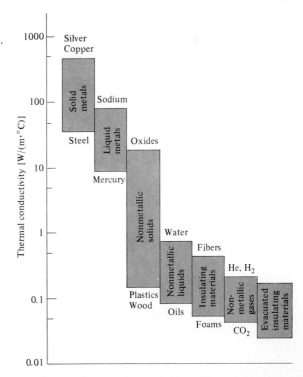

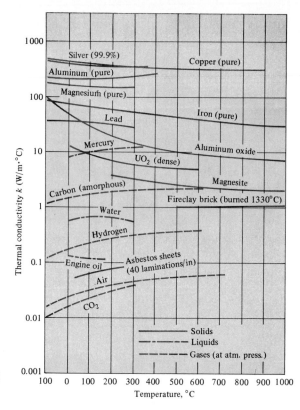

FIGURE 1-4
Effect of temperature on thermal conductivity of materials.

low temperatures, near absolute zero, k varies very rapidly with temperature. Actual values of thermal conductivity of various materials are given in Appendix B. Figure 1-4 illustrates the variation of thermal conductivity of some engineering materials with temperature.

1-2 CONVECTION

When a fluid flows over a solid, and the temperatures are different, heat transfer takes place between the fluid and the solid surface as a result of the motion of the fluid. This mechanism of heat flow is called *convection*, since the motion of the fluid plays a significant role in augmenting the heat transfer rate. Clearly, if there were no fluid motion, heat transfer would be by conduction. If the fluid motion is caused externally by a forcing mechanism, such as a fan, blower, pump, or wind, the mechanism of heat flow is said to be *forced convection*. If the fluid motion is set up by the buoyancy resulting from density differences caused by the temperature difference within the fluid, the mechanism of heat flow is said to be *free* (or *natural*) convection. Here the word *fluid* is used to describe both liquids and gases; for example, air and water are both fluids.

In engineering applications, to simplify the heat flow calculations, a quantity called the *heat transfer coefficient h* is defined. To illustrate the concept, consider the flow of a cold fluid at a temperature T_f over a hot body at temperature T_w, as illustrated in Fig. 1-5. Let q be the *heat flux* (in watts per square meter) *from the wall to the fluid*. Then the heat transfer coefficient h is defined as

$$q = h(T_w - T_f) \qquad (1\text{-}5a)$$

If the heat flux is in watts per square meter and the temperatures are in degrees Celsius (or Kelvins), then the heat transfer coefficient has the dimension $W/(m^2 \cdot {}^\circ C)$, and it is always a *positive quantity*.

If we rewrite Eq.(1-5a) in the form

$$q = h(T_f - T_w) \qquad (1\text{-}5b)$$

it implies that q is the *heat flux from the fluid to the wall*.

The expression given by Eq. (1-5a), first used as a law of cooling for the removal of heat from a hot body into a cold fluid flowing over it, is generally referred to as *Newton's law of cooling*.

In general, the determination of the heat transfer coefficient for convection problems is a very complicated matter because h is affected by:

The type of flow (i.e., laminar, turbulent, or transitional)

The geometry of the body

The physical properties of the fluid

The temperature difference

The position along the surface of the body

Whether the mechanism is forced or free convection

When the heat transfer coefficient varies with the position along the surface of the body, for convenience in engineering applications, its *mean (average)* value h_m over the surface is considered. Equations (1-5a and b) are also applicable in such cases if the local heat transfer coefficient h is replaced by its mean value h_m and the local heat flux q by its mean value q_m.

There is a wide range of values of the heat transfer coefficient for various

FIGURE 1-5
Heat flow by convection from a hot wall at T_w to a cold fluid at T_f.

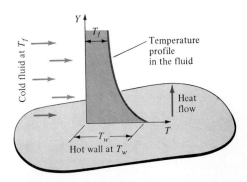

TABLE 1-1
Typical values of convective heat transfer coefficient h

Free convection, $\Delta T = 25°C$	h, W/(m$^2 \cdot$ °C)
0.25-m-*vertical plate in*:	
Atmospheric air	5.0
Engine oil	37.0
Water	440.0
0.02-m-*OD horizontal cylinder in*:	
Atmospheric air	8.0
Engine oil	62.0
Water	741.0
0.02-m-*diameter sphere in*:	
Atmospheric air	9.0
Engine oil	60.0
Water	606.0
Forced convection	
Atmospheric air at 25°C with velocity	
$v = 10$ m/s over a flat plate of	
length $L = 0.5$ m	17.0
Flow at $v = 5$ m/s across a	
1-cm-OD cylinder of:	
Atmospheric air	85.0
Engine oil	1,800.0
Flow of water at 1 kg/s inside	
2.5-cm-ID tube	10,500.0

applications. Table 1-1 lists typical values of h encountered in various applications. The heat transfer coefficient can be determined analytically for simple configurations, but an experimental approach becomes necessary for complex geometries. More recently, elaborate computer codes have been prepared that use purely numerical approaches to determine heat transfer coefficients, and some success has been recorded.

Example 1-4. Cold air at 10°C is forced to flow over a flat plate maintained at 40°C. The mean heat transfer coefficient is $h_m = 30$ W/(m$^2 \cdot$ °C). Find the heat flow rate from the plate to the air through a plate area of $A = 2$ m^2.

Solution. Given: $T_f = 10$°C, $T_w = 40$°C, and $h_m = 30$ W/(m$^2 \cdot$ °C). Equation (1-5a) is used to determine the average heat flux:

$$q_m = h_m(T_w - T_f)$$

$$= 30 \text{ W/(m}^2 \cdot \text{°C)} \times (40 - 10)\text{°C} = 900 \text{ W/m}^2$$

Then the heat flow through the area $A = 2$ m^2 becomes

$$Q = q_m A$$

$$= 900 \text{ W/m}^2 \times 2 \text{ m}^2 = 1800 \text{ W} = 1.8 \text{ kW}$$

Example 1-5. Atmospheric air at a temperature of 10°C flows with velocity 5 m/s across a 1-cm-outside-diameter (OD) and 5-m-long tube whose surface is maintained at 110°C, as illustrated in the accompanying figure. Determine the rate of heat flow from the tube surface to the atmospheric air.

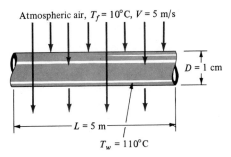

Atmospheric air, $T_f = 10°C$, $V = 5$ m/s

$D = 1$ cm

$L = 5$ m

$T_w = 110°C$

FIGURE EXAMPLE 1-5
Forced convection across the tube.

Solution. The heat transfer area A is the outside surface area of the tube.

$$P = \pi DL$$

$$= \pi \times 0.01 \text{ m} \times 5 \text{ m} = 0.05\pi = 0.157 \text{ m}^2$$

For forced convection across a tube of diameter $D = 0.01$ m with $V_\infty = 5$ m/s, the average heat transfer coefficient between the outside surface of the tube and the atmospheric air is obtainable from Table 1-1 as $h = 85$ W/(m² · °C). Equation (1-5a) is used to calculate the heat flux; taking $T_f = 10°C$, $T_w = 110°C$, and $h_m = 85$ W/(m² · °C), q_m is determined to be

$$q_m = h_m(T_w - T_f)$$

$$= 85 \text{ W/(m}^2 \cdot °C) \times (110 - 10)°C = 8500 \text{ W/m}^2$$

The heat flow rate through $A = 0.05\pi$ m² becomes

$$Q = q_m A$$

$$= 8500 \text{ W/m}^2 \times 0.05\pi \text{ m}^2 = 1335.18 \text{ W} = 1.335 \text{ kW}$$

Example 1-6. A 10-cm-diameter sphere is heated internally with a 100-W electric heater. The sphere dissipates heat by convection from its outer surface into the ambient air. Calculate the convection heat transfer coefficient between the air and the sphere if the temperature difference between the sphere surface and the ambient air is 50°C.

Solution. Equation (1-5a) can be used to calculate h, because $Q = 100$ W, $\Delta T = 50°C$, and $A = \pi D^2 = \pi(0.1)^2 = 0.001\pi$ m² are known.
We find

$$\frac{Q}{A} = q = h(T_w - T_f) \equiv h \, \Delta T$$

or

$$h = \frac{Q}{A \, \Delta T}$$

$$= \frac{100 \ W}{0.01\pi \text{ m}^2 \times 50°C} = 63.7 \text{ W/(m}^2 \cdot °C)$$

Example 1-7. A 2-cm-diameter cylinder having its surface maintained at 50°C is suspended in atmospheric air at 25°C. Determine the rate of heat loss from the cylinder by free convection per unit length.

Solution. Given: $T_w = 50°C$, $T_f = 25°C$, and $D = 0.02$ m. Equation (1-5a) can be used to calculate the average heat flux q_m over the surface of the cylinder if the free convection heat transfer coefficient h_m is known. We refer to Table 1-1 to find $h_m = 8.0$ W/(m² · °C). Then

$$q_m = h_m(T_w - T_f)$$

$$= 8.0 \text{ W/(m}^2 \cdot °C) \times (50 - 25)°C = 200 \text{ W/m}^2$$

and the heat flow rate Q from the cylinder to the air per unit length becomes

$$\frac{Q}{L} = q_m A = q_m(\pi D)$$

$$= 200 \text{ W/m}^2 \times \pi(0.02) \text{ m} = 12.57 \text{ W}$$

Example 1-8. The inside surface of an insulation layer is maintained at $T_1 = 200°C$ and the outside surface is dissipating heat by convection into air at $T_f = 20°C$. The insulating layer has a thickness of 5 cm and thermal conductivity of 1.5 W/(m · °C). What is the minimum value of the heat transfer coefficient at the outside surface, if the temperature T_2 at the outside surface should not exceed 100°C?

Solution. Under steady-state conditions, for an energy balance at the surface, the conduction heat flux through the insulation layer must be equal to the convection heat flux from the outer surface into the air, as illustrated in the accompanying figure. Then we have

$$q_{\text{conduction}} = q_{\text{convection}}$$

or

$$k \frac{T_1 - T_2}{L} = h(T_2 - T_f)$$

Given $T_1 = 200°C$, $T_2 = 100°C$, $T_f = 20°C$, $L = 0.05$ m, and $k = 1.5$ W/(m · °C), we obtain

$$1.5 \text{ W/(m} \cdot °C) \times \frac{(200 - 100)°C}{0.05 \text{ m}} = h(100 - 20)°C$$

or

$$h = 37.5 \text{ W/(m}^2 \cdot °C)$$

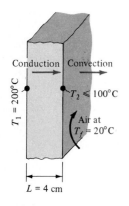

Conduction | Convection

$T_1 = 200°C$

$T_2 \leqslant 100°C$

Air at $T_f = 20°C$

$L = 4$ cm

FIGURE EXAMPLE 1-8
Conduction and convection through the insulation layer.

1-3 RADIATION

All bodies emit energy due to their temperature. The energy emitted by a body due to its temperature is called *thermal radiation*. The radiation energy leaving a body through its bounding surfaces actually originates from the interior of the region. Similarly, the radiation energy incident on the surface of a body

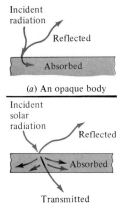

(a) An opaque body

(b) A semitransparent body (i.e., glass)

FIGURE 1-6
Concept of opaque and semitransparent bodies

penetrates into the depths of the body, where it is absorbed. If the emission and absorption of radiation take place within a very short distance from the surface, e.g., a few anstroms, then the radiation process is called *surface radiation*, and the body is said to be *opaque* to thermal radiation. For example, materials such as metals, wood, stone, paper, and numerous others are considered opaque to thermal radiation. On the other hand, a sheet of glass is said to the *semitransparent* to the solar radiation incident upon it, because part of the solar radiation is absorbed, part reflected, and the remainder transmitted by the glass. Figure 1-6 schematically illustrates the concepts of opaque and semitransparent bodies.

Radiation propagating in a medium is weakened as a result of absorption. It is only in a *vacuum* that radiation propagates with no weakening (attenuation) at all. Therefore, a vacuum is considered completely *transparent* to radiation. Also, the atmospheric air contained in a room is considered transparent to thermal radiation for all practical purposes, because the weakening of radiation by air is insignificant unless the air layer is several kilometers thick.

Emission of Radiation

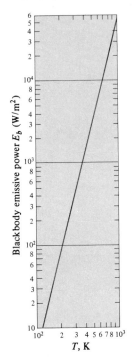

FIGURE 1-7
Blackbody emissive power, $E_b = \sigma T^4$.

The maximum radiation flux emitted by a body at temperature T is given by the *Stefan-Boltzmann law*,

$$E_b = \sigma T^4 \qquad \text{W/m}^2 \qquad (1\text{-}6)$$

where σ is the *Stefan-Boltzmann* constant $[\sigma = 5.6697 \times 10^{-8} \text{ W/(m}^2 \cdot \text{K}^4)]$, E_b is the *blackbody emissive power*, and T is the *absolute temperature* in kelvins,

$$\text{K} = {}^\circ\text{C} + 273.15 \qquad (1\text{-}7)$$

Only an *ideal radiator* or the so-called *blackbody* can emit the maximum radiation flux according to Eq. (1-6). The radiation flux q emitted by a *real body* at an absolute temperature T is always less than the blackbody emissive power E_b, and it is given by

$$q = \epsilon E_b = \epsilon \sigma T^4 \qquad (1\text{-}8)$$

where ϵ is the *emissivity* of the body, which is less than unity for all real bodies and equal to 1 only for a blackbody.

Figure 1-7 shows a plot of the blackbody emissive power E_b defined by Eq. (1-6) against the absolute temperature. The radiation flux rapidly increases with rising temperature, since E_b is proportional to the fourth power of the absolute temperature.

Absorption of Radiation

A radiation flux q_{inc} incident on a *blackbody* is completely absorbed by the blackbody; therefore

$$q_{abs} = q_{inc} \qquad (1\text{-}9)$$

where q_{abs} is the absorbed energy. On the other hand, a radiation flux q_{inc} incident on a *real body* is partly absorbed by the body, and the energy absorbed q_{abs} is given by

$$q_{abs} = \alpha q_{inc} \qquad (1\text{-}10)$$

where α is the *absorptivity*, which lies between zero and unity; only for a blackbody is it equal to unity.

The absorptivity α of a body is generally different from its emissivity ϵ.

Example 1-9. A blackbody at 20°C is heated to 100°C. Calculate the increase in its emissive power.

Solution. The Stefan-Boltzmann law given by Eq. (1-6) can be used to calculate the emissive power at a given temperature. At the temperature $T_1 = 20$°C, we have

$$E_{b_1} = \sigma T_1^4$$

$$= 5.6697 \times 10^{-8}\ \text{W/(m}^2 \cdot \text{K}^4) \times (20 + 273.15)^4\ \text{K}^4$$

$$= 5.6697 \times 10^{-8}\ \text{W/(m}^2 \cdot \text{K}^4) \times (2.9315)^4 \times 10^8\ \text{K}^4$$

$$= 418.72\ \text{W/m}^2$$

and at $T_2 = 100$°C, we have

$$E_{b_2} = \sigma T_2^4$$

$$= 5.6697 \times 10^{-8}\ \text{W/(m}^2 \cdot \text{K}^4) \times (100 + 273.15)^4\ \text{K}^4$$

$$= 5.6697 \times 10^{-8}\ \text{W/(m}^2 \cdot \text{K}^4) \times (3.7315)^4 \times 10^8\ \text{K}^4$$

$$= 1099.24\ \text{W/m}^2$$

Therefore, the increase in the emissive power becomes

$$E_{b_2} - E_{b_1} = 1099.24 - 418.72 = 680.5\ \text{W/m}^2$$

1-4 UNITS AND DIMENSIONS

When the dimensions of a physical quantity are to be expressed numerically, a consistent system of units is preferred. In engineering the two most commonly used system of units include (1) the *SI system* (*Système International d'Unités*) and (2) the *English engineering system*. The basic units for length, mass, time, and temperature in each of these systems are listed in Table 1-2. In this book the SI system is used throughout.

TABLE 1-2
Systems of units

Quantity	SI	English engineering system
Length	m	ft
Mass	kg	lb
Time	s	s
Temperature	K	R
Force	N	lb_f
Energy	J or N · m	Btu or ft · lb_f

In the SI system, one newton (that is, 1 N) is a force that accelerates a mass of one kilogram to one meter per second per second; or 1 N force is equal to $1 \, kg \cdot m/s^2$. This is apparent from Newton's second law of motion,

$$\text{Force} = \text{mass} \times \text{acceleration}$$

$$1 \, N = 1 \, kg \times 1 \, m/s^2 = 1 \, kg \cdot m/s^2 \tag{1-11a}$$

Energy is measured in joules (J) or newton-meters (N · m). Thus

$$1 \, J = 1 \, N \cdot m = 1 \, kg \cdot m^2/s^2 \tag{1-11b}$$

Power is measured in watts (W) or kilowatts (kW), or joules per second. Then

$$1 \, W = 1 \, J/s = 1 \, N \cdot m/s = 1 \, kg \cdot m^2/s^3 \tag{1-11c}$$

and

$$1 \, kW = 1000 \, W \tag{1-11d}$$

Pressure is measured in bars,

$$1 \, bar = 10^5 \, N/m^2 = 10^5 \, kg/(m \cdot s^2) \tag{1-11e}$$

and

$$1 \, atm = 1.01325 \, bar \tag{1-11f}$$

When the size of units becomes too large or too small, multiples in powers of 10 are formed using certain prefixes. Some of them are listed in Table 1-3. For example,

$$1000 \, W = 1 \, kW \, (\text{kilowatt})$$

$$1{,}000{,}000 \, W = 1 \, MW \, (\text{megawatt})$$

$$1000 \, m = 1 \, km \, (\text{kilometer})$$

$$10^{-3} \, m = 1 \, mm \, (\text{millimeter})$$

TABLE 1-3
Prefixes for multiplying factors

10^{-12} = pico (p)	10 = deca (da)
10^{-9} = nano (n)	10^2 = hecto (h)
10^{-6} = micro (μ)	10^3 = kilo (k)
10^{-3} = milli (m)	10^6 = mega (M)
10^{-2} = centi (c)	10^9 = giga (G)
10^{-1} = deci (d)	10^{12} = tera (T)

PROBLEMS

Conduction

1-1. Determine the heat flux across a 0.05-m-thick iron plate [with thermal conductivity of 70 W/(m · °C)] if one of its surfaces is maintained at 60°C and the other at 10°C. What is the heat flow rate across the plate over a surface area of $2 \, m^2$?
Answer: 140 kW

1-2. A temperature difference of 100°C is applied across a cork board 5 cm thick with thermal conductivity of 0.04 W/(m · °C). Determine the heat flow rate across a 3-m^2 area per hour.
Answer: 0.24 kW

1-3. Two large plates, one at 80°C and the other at 200°C, are 12 cm apart. If the space between them is filled with loosely packed rock wool of thermal conductivity 0.08 W/(m · °C), calculate the heat flux across the plates.
Answer: 80 W/m^2

1-4. The hot surface of a 5-cm-thick insulating material of thermal conductivity 0.1 W/(m · °C) is maintained at 200°C. If the heat flux across the material is 120 W/m^2, what is the temperature of the cold surface?
Answer: 140°C

1-5. The heat flow through a 4-cm-thick insulation layer for a temperature difference of 200°C across the surfaces is 500 W/m^2. What is the thermal conductivity of the insulating material?
Answer: 0.1 W/(m · °C)

1-6. A brick wall 30 cm thick with thermal conductivity 0.5 W/(m · °C) is maintained at 50°C at one surface and 20°C at the other surface. Determine the heat flow rate across a 1-m^2 surface area of the wall.
Answer: 0.05 kW

1-7. A window glass 0.5 cm thick with thermal conductivity 0.8 W/(m · °C) is maintained at 30°C at one surface and 20°C at the other surface. Determine the heat flow rate across a 1-m^2 surface area of the glass.
Answer: 1.6 kW

1-8. The heat flow through a 4-cm-thick cork board for a temperature difference of 100°C across the surfaces is 0.1 kW/m^2. What is the thermal conductivity of the cork board?

1-9. The heat flux through a 10-cm-thick layer of loosely packed rock wool for a temperature difference of 100°C is 75 W/m^2. What is the thermal conductivity of the packed rock wool?

Convection

1-10. Water at 20°C flows over a flat plate at 80°C. If the heat transfer coefficient is 200 W/(m^2 · °C), determine the heat flow per square meter of the plate (heat flux).
Answer: 12 kW/m^2

1-11. Air at 200°C flows over a flat plate maintained at 50°C. If the heat transfer coefficient for forced convection is 300 W/(m^2 · °C), determine the heat transferred to the plate through 5 m^2 over a period of 2 h.
Answer: 450 kW · h

1-12. Water at 50°C flows through a 5-cm-diameter, 2-m-long tube with an inside surface temperature maintained at 150°C. If the heat transfer coefficient between the water and inside surface of the tube is 1000 W/(m^2 · °C), determine the heat transfer rate between the tube and the water.
Answer: 31.4 kW

1-13. Cold air at 10°C flows over a 2-cm-OD tube, as illustrated in Fig. P1-13. The outside surface of the tube is maintained at 110°C. If the heat transfer coefficient between the outside surface of the tube and the air is 100 W/(m^2 · °C), determine the rate of heat flow to the air over the 5-m length of the tube.
Answer: 3.142 kW

Cold air, $T_f = 10°$C, $h = 100$ W/(m^2·°C)

$D = 2$ cm

$T_w = 100°$C

FIGURE PROBLEM 1-13

1-14. A 20-cm-diameter sphere with a surface temperature of 70°C is suspended in a stagnant gas at 20°C. If the free convection heat transfer coefficient between the sphere and the gas is 10 W/(m^2 · °C), determine the rate of heat loss from the sphere.
Answer: 62.8 W

1-15. A 2-cm-diameter sphere with a surface temperature maintained at 50°C is suspended in water at 25°C. Determine the heat flow rate by free convection from the sphere to the water.
Answer: 19.04 W

1-16. Cold air at 0°C is forced to flow over a flat plate maintained at 30°C. The mean heat transfer coefficient is 50 W/(m^2 · °C). Find the heat flow rate from the plate to the air per unit plate area.
Answer: 1.5 kW/m^2

1-17. Hot air at 75°C is forced to flow over a flat plate maintained at 20°C. The mean heat transfer coefficient is 60 W/(m^2 · °C). Find the heat flow rate from the air to the plate.
Answer: 3.3 kW/m^2

1-18. A 25-cm vertical plate with a surface temperature maintained at 50°C is suspended in atmospheric air at 25°C. Determine the heat flow rate by free convection from the plate to the air.
Answer: 125 W/m^2

1-19. A 5-cm-diameter sphere dissipates heat by convection from its outer surface into ambient air. Calculate the convection heat transfer coefficient between the air and the sphere if the sphere is heated internally with a 50-W electric heater and the temperature difference between the sphere surface and the air is 25°C.

1-20. A 2-cm-diameter sphere with a surface temperature maintained at 40°C is suspended in atmospheric air at 15°C. Determine the rate of heat loss from the sphere by free convection.

Radiation

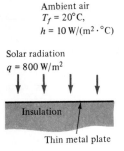

Ambient air
$T_f = 20°C$,
$h = 10 \text{ W/(m}^2 \cdot °C)$

Solar radiation
$q = 800 \text{ W/m}^2$

Insulation

Thin metal plate

FIGURE PROBLEM 1-21

1-21. A thin metal plate is insulated at the back surface and exposed to the sun at the front surface, as shown in Fig. P1-21. The front surface absorbs the solar radiation of 800 W/m² and dissipates it by convection to the ambient air at 20°C. If the convection heat transfer coefficient between the plate and the air is 10 W/(m² · °C), what is the temperature of the plate?
  Answer: 100°C

1-22. A hot surface at a temperature of 400 K has an emissivity of 0.8. Calculate the radiation flux emitted by the surface.
  Answer: 1.161 kW/m²

1-23. A radiation flux of 1000 W/m² is incident upon a surface that absorbs 80 percent of the incident radiation. Calculate the amount of radiation energy absorbed by a 4-m² area of the surface over a period of 2 h.
  Answer: 6.4 kW · h

1-24. The temperature T of a plate changes with time t as $T = T_o t^{1/4}$, where the constant T_o is in absolute temperature. How does the blackbody emissive power of the plate change with time?

1-25. The inside surface of an insulation layer is maintained at 150°C and the outside surface dissipates heat by convection into air at 15°C. The insulation layer has a thickness of 10 cm and a thermal conductivity of 1 W/(m · °C). What is the minimum value of the heat transfer coefficient at the outside surface, if the temperature at the outside surface should not exceed 75°C?
  Answer: 12.5 W/(m² · °C)

1-26. A blackbody at 30°C is heated to 80°C. Calculate the increase in its emissive power.
  Answer: 403.01 W/m²

Units, Dimensions, and Conversion Factors

1-27. Calculate the force of a mass of 10 kg attracted to the earth at a point where the gravitational acceleration is 9.6 m/s².
  Answer: 96 N

1-28. A steady force of 10 kN acts on a mass of 5 kg. What is the acceleration of this mass?
  Answer: 2000 m/s²

1-29. Consider a manometer containing a fluid with a density of 1000 kg/m³. The difference in height of the two columns is 500 mm. Calculate the pressure difference in kilopascals.
  Answer: 4.905 kPa

1-30. A plastic plate with an area of 0.1 m² and a thickness of 2 mm is found to conduct heat at a rate of 1 cal/s at steady-state with surface temperatures 29°C and 31°C. What is the thermal conductivity of plastic in W/(m · °C) at 30°C?

CHAPTER
2

UNSTEADY HEAT FLOW CONCEPT

The objective of this chapter is to introduce the concept of time-dependent behavior of temperature in a system. If the surface temperature of a solid body is suddenly altered, the temperature within the body begins to change with both position and time. However, there are many practical applications in which the variation of temperature within the body with position is negligible and the temperature is considered to vary only with time. Analysis of unsteady heat flow under such an assumption, generally referred to as *lumped system analysis*, greatly simplifies the handling of time-dependent problems. Obviously, the range of applicability of such a simple analysis is very limited, but it is helpful in understanding the concept of time-dependent behavior of temperature in a system, as well as being applicable in many practical situations. Therefore, in this chapter we explain the use of lumped system analysis to predict the variation of temperature of a solid body with time during transients and discuss the criterion for the validity of the concept.

2-1 LUMPED SYSTEM CONCEPT

Consider a cold solid of arbitrary shape, with mass m, initially at a uniform temperature T_o, suddenly immersed into a higher-temperature environment. As heat flows from the hot environment into the cold body, the temperature of the solid increases. It is assumed that the lumped system approximation is applicable, namely, that the distribution of temperature within the solid at any instant can be regarded as almost uniform (i.e., the temperature gradients

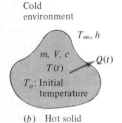

Hot environment

T_∞, h

m, V, c

$Q(t)$

$T(t)$

T_o: Initial temperature

(a) Cold solid

Cold environment

T_∞, h

m, V, c

$Q(t)$

$T(t)$

T_o: Initial temperature

(b) Hot solid

FIGURE 2-1
Energy balance for lumped system analysis.

within the solid are neglected). Figure 2-1a illustrates such a system. The energy balance on the system over a time interval dt can be stated as

$$\begin{pmatrix} \text{Increase of the internal} \\ \text{energy of the solid} \\ \text{over the time interval } dt \end{pmatrix} = \begin{pmatrix} \text{heat flow into the solid} \\ \text{through the outer surfaces} \\ \text{over the time interval } dt \end{pmatrix} \quad (2\text{-}1)$$

Let dT be the average temperature rise of the solid over the time interval dt resulting from the heat flow through the boundary surfaces. The left-hand side of Eq. (2-1) is determined as

$$\begin{pmatrix} \text{Increase of the} \\ \text{internal energy} \\ \text{of the solid over} \\ \text{the time interval } dt \end{pmatrix} = (\text{mass})c \, dT = \rho V c \, dT \quad (2\text{-}2)$$

where ρ is the density, c is the specific heat, V is the volume, and $T \equiv T(t)$ is the temperature of the body.

Let $Q(t)$ be the total rate of heat flowing into the body through its boundary surfaces at any instant t. Then the right-hand side of Eq. (2-1) becomes

$$\begin{pmatrix} \text{Heat flow into the solid} \\ \text{from the outer surfaces} \\ \text{over the time interval } dt \end{pmatrix} = Q(t) \, dt \quad (2\text{-}3)$$

Introducing Eqs. (2-2) and (2-3) into (2-1), we obtain

$$\rho V c \, dT = Q(t) \, dt$$

or

$$\frac{dT(t)}{dt} = \frac{Q(t)}{\rho V c} \quad (2\text{-}4)$$

We derived this equation with reference to a cold solid in a hot fluid, as illustrated in Fig. 2-1a. The same equation is also applicable to the cooling of a hot solid in a cold environment, as illustrated in Fig. 2-1b.

Equation (2-4) forms the basis for generating an ordinary differential equation for predicting the temperature $T(t)$ of the solid body as a function of time. That is, once the specific expression defining $Q(t)$ has been established, Eq. (2-4) provides an ordinary differential equation in the time variable for determining $T(t)$. The functional form of $Q(t)$ depends on the types of heat transfer process at the outer surface of the body. In the following sections we discuss the development of specific expressions for $Q(t)$ for various situations and determine the temperature $T(t)$ of the solid as a function of time.

2-2 LUMPED SYSTEM APPLICATION

In this section, we illustrate the use of lumped system analysis to predict the variation of the temperature of a solid with time. Our starting point in the analysis is Eq. (2-4). Once the functional form of $Q(t)$ is established for a specific problem, the resulting ordinary differential equation (2-4) can be solved and the variation of the temperature $T(t)$ of the solid as a function of time is determined. Here we examine two simple situations: (1) a solid subjected to convection over its entire boundary surface, and (2) a solid subjected to convection over part of its boundary surface, with a prescribed heat flux over the remaining portion of the boundary surface.

Convection Over All Boundary Surfaces

We consider a physical situation as illustrated in Fig. 2-1a. That is, a cold body initially at a uniform temperature T_o is suddenly immersed into a large volume of well-stirred fluid maintained at a uniform temperature T_∞. The mechanism of heat transfer between the fluid and the entire boundary surface of the solid is convection. We assume that the heat transfer coefficient h remains constant and uniform over the entire surface of the solid. Then, the total rate of heat flow $Q(t)$ from the fluid into the solid is given by the expression

$$Q(t) = Ah[T_\infty - T(t)] \tag{2-5}$$

When this expression is introduced into Eq. (2-4), the following ordinary differential equation for the temperature $T(t)$ results:

$$\frac{dT(t)}{dt} = \frac{Ah}{\rho Vc}[T_\infty - T(t)] \tag{2-6}$$

The solid is at a temperature T_o when it is immersed in the fluid. This information provides the *initial condition* for this differential equation.

Hence the mathematical formulation of the problem becomes

$$\frac{dT(t)}{dt} + \frac{Ah}{\rho Vc}[T(t) - T_\infty] = 0 \qquad \text{for } t > 0 \tag{2-7a}$$

$$T(t) = T_o \qquad \text{for } t = 0 \tag{2-7b}$$

For convenience in the analysis, we measure the temperature in excess of the ambient temperature T_∞; that is, we choose T_∞ as the reference temperature. Then the following new temperatures are defined:

$$\theta(t) = T(t) - T_\infty \tag{2-8a}$$

$$\theta_o = T_o - T_\infty \tag{2-8b}$$

and a quantity m is introduced:

$$m = \frac{Ah}{\rho cV} \tag{2-8c}$$

where m has the dimension of $(\text{time})^{-1}$. Then Eq. (2-7) becomes

$$\frac{d\theta(t)}{dt} + m\theta(t) = 0 \qquad \text{for } t > 0 \tag{2-9a}$$

$$\theta(t) = \theta_0 \qquad \text{for } t = 0 \tag{2-9b}$$

where

$$m = \frac{Ah}{\rho cV} \tag{2-9c}$$

Equations (2-9a to c) govern the variation of the temperature $\theta(t)$ of the solid as a function of time. They can be solved as now described.

The fundamental solution of Eq. (2-9a) is an exponential in the form

$$e^{-mt}$$

Then the solution for $\theta(t)$ is constructed by multiplying this solution by a constant:

$$\theta(t) = Ce^{-mt} \tag{2-10a}$$

The unknown constant C is determined by applying the initial condition (2-9b), namely, $\theta(0) = \theta_o$ for $t = 0$, to give

$$\theta_o = C \tag{2-10b}$$

Introducing the value of C into Eq. (2-10a), the solution for $\theta(t)$ becomes

$$\frac{\theta(t)}{\theta_o} = e^{-mt} \tag{2-11a}$$

Recalling the definitions of $\theta(t)$ and θ_o given by Eq. (2-8a and b), this solution is written as

$$\frac{\theta(t)}{\theta_o} \equiv \frac{T(t) - T_\infty}{T_o - T_\infty} = e^{-mt} \tag{2-11b}$$

where

$$m = Ah/\rho cV \tag{2-11c}$$

Solution (2-11) is developed for the heating of a cold body immersed in a hot environment, as illustrated in Fig. 2-1a. It is also applicable to the cooling of a hot body immersed in a cold environment, as shown in Fig. 2-1b.

Figure 2-2 shows a plot of the dimensionless temperature $\theta(t)/\theta_o$ given by Eq. (2-11) as a function of time. The temperature decays with time exponentially, and the rate of decay is dependent upon the magnitude of the exponent m. The larger the value of m, the faster the rate of decay. An examination of the definition of the parameter $m = hA/\rho cV$ reveals that increasing the surface area for a given volume or increasing the heat transfer coefficient increases m, which in turn increases the temperature response of a body for a change in environment temperature. On the other hand, increasing the density, specific heat, or volume decreases m, which in turn reduces the temperature response of the solid.

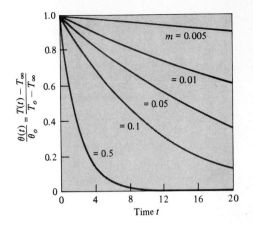

FIGURE 2-2
Dimensionless temperature $\theta(t)/\theta_0$ as a function of time (A plot of Eq. 2-11).

For certain common body shapes, the ratio A/V is readily determined; then we have the following specific formulae for the temperature $\theta(t)$ as obtained from Eq. (2-11).

For a solid sphere of radius R:

$$\frac{A}{V} = \frac{4\pi R^2}{\frac{4}{3}\pi R^3} = \frac{3}{R}$$

$$\boxed{\frac{\theta(t)}{\theta_o} = \exp\left(-\frac{3h}{\rho c R}\, t\right)} \tag{2-12}$$

For a solid cylinder of radius R, length L:

$$\frac{A}{V} = \frac{2\pi R(L + R)}{\pi R^2 L} = \frac{2(L + R)}{RL}$$

$$\boxed{\frac{\theta(t)}{\theta_o} = \exp\left(-\frac{2h(L + R)}{\rho c RL}\, t\right)} \tag{2-13a}$$

For a long cylinder, $L \gg R$, this expression reduces to

$$\boxed{\frac{\theta(t)}{\theta_o} = \exp\left(-\frac{2h}{\rho c R}\, t\right)} \tag{2-13b}$$

For a cube of side L:

$$\frac{A}{V} = \frac{6L^2}{L^3} = \frac{6}{L}$$

$$\boxed{\frac{\theta}{\theta_o} = \exp\left(-\frac{6h}{\rho c L}\, t\right)} \tag{2-14}$$

For a large plate of thickness L:
Let A^* denote the surface area of one side of the plate. Then

$$\frac{A}{V} = \frac{2A^*}{LA^*} = \frac{2}{L}$$

$$\boxed{\frac{\theta(t)}{\theta_o} = \exp\left(-\frac{2h}{\rho c L} t\right)} \qquad (2\text{-}15)$$

The conditions under which lumped system analysis is applicable are discussed in the next section. Here we state that for solids having such shapes as a slab, a cylinder, or a sphere, lumped system analysis is applicable if the *Biot number* Bi is less than about 0.1; that is, if

$$\boxed{\text{Bi} \equiv \frac{hL_s}{k_s} < 0.1} \qquad (2\text{-}16a)$$

where h is the heat transfer coefficient, k_s is the thermal conductivity of the solid, and L_s is the *characteristic length* of the solid body, defined as

$$L_s = \frac{\text{volume}}{\text{surface area}} \qquad (2\text{-}16b)$$

Example 2-1. Using lumped system analysis, determine the time required for a solid steel ball of radius $R = 2.5$ cm, $k = 54$ W/(m $\cdot$ °C), $\rho = 7833$ kg/m^3, and $c = 0.465$ kJ/(kg $\cdot$ °C) to cool from 850°C to 250°C if it is exposed to an air stream at 50°C having a heat transfer coefficient $h = 100$ W/(m$^2 \cdot$ °C).

Solution. This problem is that of a hot solid exposed to a cold environment system, as illustrated earlier in Fig. 2-1b. From Eq. (2-11) we have

$$\frac{T(t) - T_\infty}{T_o - T_\infty} = e^{-mt}$$

where $T(t) = 250$°C, $T_\infty = 50$°C, and $T_o = 850$°C.
    The volume-to-area ratio is

$$\frac{V}{A} = \frac{\frac{4}{3}\pi R^3}{4\pi R^2} = \frac{R}{3}$$

and

$$m = \frac{hA}{\rho c V} = \frac{h}{\rho c R/3} = \frac{100 \text{ W/(m}^2 \cdot \text{°C)}}{(7833 \text{ kg/m}^3)[465 \text{ J/(kg} \cdot \text{°C)}][0.025/3 \text{ m}]}$$

$$= 1/303.529 \text{ s}^{-1}$$

Then

$$\frac{(250 - 50)\text{°C}}{(850 - 50)\text{°C}} = \exp\left(-t/303.529\right)$$

or

$$4 = \exp{(t/303.529)}$$

$$t = \ln{(4)} \times 303.529$$

$$t \cong 420.78 \text{ s} \cong 7 \text{ min}$$

We compute the Biot number in order to check the validity of the lumped analysis.

$$\text{Bi} = \frac{hL_s}{k} = \frac{h}{k} \frac{R}{3} = \frac{100 \times (2.5 \times 10^{-2})}{54 \times 3} = 0.016 < 0.1$$

Hence the analysis is valid.

Example 2-2. A long steel bar of radius $R = 2.5$ cm, $k = 50$ W/(m·°C), $\rho = 7800$ kg/m^3, and $c = 0.5$ kJ/(kg·°C) is to be annealed by cooling slowly from 800°C to 100°C in an environment at temperature 50°C. If the heat transfer coefficient between the environment and the surface of the bars is $h = 45$ W/(m^2·°C), determine the time required for the annealing process using lumped system analysis.

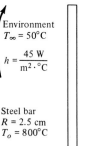

Environment
$T_\infty = 50°C$

$h = \dfrac{45 \text{ W}}{\text{m}^2 \cdot °C}$

L

Steel bar
$R = 2.5$ cm
$T_o = 800°C$

FIGURE EXAMPLE 2-2
Long steel bar exposed to a cold environment.

Solution. The problem is illustrated in the accompanying figure. From Eq. (2-11) we have

$$\frac{T(t) - T_\infty}{T_o - T_\infty} = e^{-mt}$$

where $T(t) = 100°C$, $T_\infty = 50°C$, and $T_o = 800°C$.
As discussed previously, for a long cylinder the volume-to-area ratio is

$$\frac{V}{A} \cong \frac{\pi R^2 L}{2\pi RL} = \frac{R}{2}$$

$$m = \frac{hA}{\rho cV} = \frac{h}{\rho c(R/2)} = \frac{45 \text{ W/(m} \cdot °C)}{(7800 \text{ kg/m}^3)[500 \text{ J/(kg} \cdot °C)](0.025/2 \text{ m})}$$

$$= 1/1083.33 \text{ s}^{-1}$$

Then

$$\frac{(100 - 50)°C}{(800 - 50)°C} = \exp{(-t/1083.33)}$$

$$15 = \exp{(t/1083.33)}$$

$$t = \ln{(15)} \times 1083.33$$

$$t = 2933.7 \text{ s} \cong 49 \text{ min}$$

A check of the Biot number indicates that the criterion Bi < 0.1 is also satisfied.

Example 2-3. A large aluminum plate [$k = 204$ W/(m·°C), $\rho = 2707$ kg/m^3, and $c = 0.896$ kJ/(kg·°C)] of thickness $L = 0.1$ m, that is initially at a uniform temperature of 250°C is cooled by exposing it to an air stream at temperature 50°C. Using lumped system analysis, determine the time required to cool the aluminum plate to 100°C if the heat transfer coefficient between the air stream and the plate surface is $h = 80$ W/(m^2·°C).

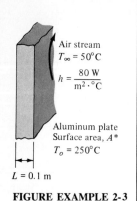

FIGURE EXAMPLE 2-3

Large aluminum plate exposed to a cold air stream.

Solution. The problem is illustrated in the accompanying figure. $T(t) = 100°C$, $T_\infty = 50°C$, and $T_o = 250°C$ are given.

The volume-to-area ratio is

$$\frac{V}{A} \cong \frac{A^*L}{2A^*} = \frac{L}{2}$$

where A^* is the surface area of one side of the plate. Then

$$m = \frac{hA}{\rho c V} = \frac{h}{\rho c L/2} = \frac{80 \text{ W}/(\text{m}^2 \cdot °C)}{(2707 \text{ kg/m}^3)[896 \text{ J}/(\text{kg} \cdot °C)](0.1/2 \text{ m})}$$

$$= 1/1515.92 \text{ s}^{-1}$$

and the time required for cooling is determined as

$$\frac{100 - 50}{250 - 50} = \exp\left(-t/1515.92\right)$$

Then
$$t = \ln(4) \times 1515.92$$

$$t = 2101.5 \text{ s} = 35 \text{ min}$$

Convection Over Part of the Boundary Surface, Prescribed Heat Flux Over the Remaining Surfaces

The lumped system analysis described previously is also applicable to situations involving a body having a part of its surface subjected to convection and the remainder to prescribed heat flux. To illustrate such applications, we consider below two specific problems, one involving a solid shaped as a slab and the other as a cube.

SLAB. Consider a slab of thickness L, initially at a uniform temperature T_o. Suddenly one of its surfaces is subjected to a uniform heat flux q_o, while the other surface is exposed to a cool environment at temperature T_∞ with a heat transfer coefficient h. Figure 2-3 illustrates the physical problem. Assuming that lumped system analysis is applicable, an equation for determining the temperature $T(t)$ of the plate as a function of time is now developed.

An energy balance on the slab results in an expression exactly the same as that given by Eq. (2-4), namely,

$$\frac{dT(t)}{dt} = \frac{Q(t)}{\rho V c} \tag{2-17a}$$

Here $Q(t)$ is the total rate of heat transfer into the body; for the specific problem considered here, it is determined as

$$Q(t) = \begin{pmatrix} \text{heat flow rate into} \\ \text{the body due to the} \\ \text{prescribed heat flux } q_0 \end{pmatrix} + \begin{pmatrix} \text{heat flow rate into the} \\ \text{body due to convection} \\ \text{from the environment} \end{pmatrix}$$

or

$$Q(t) = A^* q_o + A^* h[T_\infty - T(t)] \tag{2-17b}$$

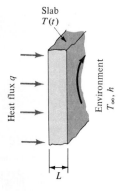

FIGURE 2-3

A slab with convection on one surface and prescribed heat flux on the other.

where A^* is the surface area on one side of the plate, q is the prescribed heat flux, and T_∞ is the environment temperature. Introducing Eq. (2-17b) into Eq. (2-17a), we obtain

$$\frac{dT(t)}{dt} + \frac{A^*h}{\rho cV}[T(t) - T_\infty] = \frac{A^*q_0}{\rho cV} \qquad \text{for } t > 0 \qquad (2\text{-}17c)$$

subject to the initial condition

$$T(t) = T_o \qquad \text{at } t = 0 \qquad (2\text{-}17d)$$

Equations (2-17c and d) represent the mathematical formulation of lumped system analysis for the determination of the temperature $T(t)$ of a slab under the conditions specified above.

We now measure the temperature in excess of the environment temperature T_∞ and define new temperatures $\theta(t)$ and θ_o as

$$\theta(t) = T(t) - T_\infty \qquad (2\text{-}18a)$$

$$\theta_o = T_o - T_\infty \qquad (2\text{-}18b)$$

and introduce a quantity m^*, defined as

$$m^* = \frac{hA^*}{\rho c_p V} = \frac{h}{\rho c_p L} \qquad (2\text{-}18c)$$

For the slab geometry considered here, we have

$$\frac{A^*}{V} = \frac{A^*}{A^*L} = \frac{1}{L}$$

where L is the plate thickness. Then Eqs. (2-17c and d) becomes

$$\frac{d\theta(t)}{dt} + m^*\theta(t) = m^*\frac{q_o}{h} \qquad \text{for } t > 0 \qquad (2\text{-}19a)$$

$$\theta(t) = \theta_o \qquad \text{at } t = 0 \qquad (2\text{-}19b)$$

where

$$m^* \equiv \frac{h}{\rho cL} \qquad (2\text{-}19c)$$

The difference between Eqs. (2-19a) and (2-9a) is the constant term on the right-hand side of Eq. (2-19a). Therefore, the solution for Eq. (2-19a) is written as a sum of the *homogeneous solution*, given by Eq. (2-10a) and a *particular solution* as

$$\theta(t) = C e^{-m^*t} + \frac{q_o}{h} \qquad (2\text{-}20)$$

where q_o/h is a particular solution.

The unknown constant C is determined by the application of the initial condition (2-19b) as

$$\theta_o = C + q_o/h$$

or

$$C = \theta_o - \frac{q_o}{h} \tag{2-21}$$

Substitution of Eq. (2-21) into Eq. (2-20) gives the solution to temperature $\theta(t)$ of the plate as a function of time as

$$\boxed{\theta(t) = \theta_o \, e^{-m^*t} + (1 - e^{-m^*t}) \frac{q_o}{h} \qquad \text{for } t > 0} \tag{2-22}$$

The steady-state temperature of the plate is obtained by setting $t \rightarrow \infty$.

$$\boxed{\theta(\infty) = \frac{q_o}{h}} \tag{2-23}$$

since the exponential terms in Eq. (2-22) vanish for $t \rightarrow \infty$.

Example 2-4. A large aluminum plate $[k = 204 \, \text{W/(m} \cdot {}^\circ\text{C)}, \, \rho = 2707 \, \text{kg/m}^3$, and $c = 0.896 \, \text{kJ/(kg} \cdot {}^\circ\text{C)}]$ of thickness 3 cm is initially at a uniform temperature of 50°C. Suddenly one of its surfaces is subjected to a uniform heat flux of $8000 \, \text{W/m}^2$, while the other surface is exposed to a cool air at a temperature of 20°C. The heat transfer coefficient between the air stream and the surface is $50 \, \text{W/(m}^2 \cdot {}^\circ\text{C)}$. Assuming that lumped system analysis is applicable, determine the temperature of the plate as a function of time and plot the temperature against time. Also calculate the steady-state temperature of the plate.

Solution. From Eq. (2-22) we have

$$\theta(t) = \theta_o \, e^{-m^*t} + (1 - e^{-m^*t}) \frac{q_o}{h} \qquad \text{for } t > 0$$

where

$$\theta_o \equiv T_o - T_\infty = 50 - 20 = 30°C$$

$$q_o = 8000 \, \text{W/m}^2$$

$$h = 50 \, \text{W/(m}^2 \cdot {}^\circ\text{C)}$$

$$m^* = \frac{hA^*}{\rho c V} = \frac{h}{\rho c L} = \frac{50}{2707 \times 896 \times 0.03}$$

$$= 1/1455.3 \, \text{s}^{-1}$$

Then

$$\theta(t) = 30 \exp(-t/1455.3) + (1 - \exp(-t/1455.3)) \frac{8000}{50}$$

$$= -130 \exp(-t/1455.3) + 160°C$$

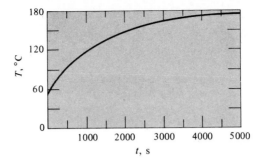

The temperature of
the plate as a function
of time.

and $T(t)$ is determined as

$$T(t) = \theta(t) + T_{\infty}$$

$$= 180 - 130 \exp(-t/1455.3)°C \qquad \text{for } t \geq 0$$

The accompanying figure shows a plot of the temperature $T(t)$ as a function
of time.

The steady-state temperature is obtained by setting $t \to \infty$, namely $T(\infty) = 180°C$.

CUBE. Consider a solid cube of side L that is initially at a uniform temperature
T_o. For times $T > 0$, two of the boundary surfaces are kept insulated, two are
subjected to uniform heating at a rate of q_o, and the remaining two dissipate
heat by convection into a cool environment at temperature T_{∞} with a heat
transfer coefficient h. The problem is illustrated in Fig. 2-4. Assuming that
lumped system analysis is applicable, an energy balance on the cube results in
the expression given by Eq. (2-4).

$$\frac{dT(t)}{dt} = \frac{Q(t)}{\rho V c} \qquad (2\text{-}24a)$$

where $Q(t)$ is the total heat transfer rate into the body through the boundary
surfaces. For the specific boundary conditions of the cube shown in Fig. 2-4,
$Q(t)$ is determined as:

$$Q(t) = \begin{pmatrix} \text{heat flow rate into the} \\ \text{body due to applied} \\ \text{heated flux } q_o \end{pmatrix} + \begin{pmatrix} \text{heat flow rate into the} \\ \text{body due to convection} \end{pmatrix}$$

$$= 2A_1 q_o + 2A_1 h[T_{\infty} - T(t)] \qquad (2\text{-}24b)$$

FIGURE 2-4
A solid cube subject to
convection, prescribed heat flux
and insulated boundary
conditions.

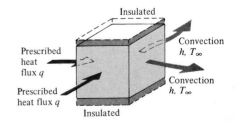

where A_1 is the surface area of one side of the cube and q_o is the prescribed surface heat flux. Introducing Eq. (2-24b) into Eq. (2-24a), we obtain

$$\frac{dT(t)}{dt} + \frac{2A_1 h}{\rho cV}\left[T(t) - T_\infty\right] = \frac{2A_1 q_o}{\rho cV} \qquad \text{for } t>0 \qquad (2\text{-}25a)$$

subject to the initial condition

$$T(t) = T_o \qquad \text{at } t=0 \qquad (2\text{-}25b)$$

We measure the temperature in excess of the environment temperature T_∞ and define new temperatures

$$\theta(t) = T(t) - T_\infty \qquad (2\text{-}26a)$$

$$\theta_o = T_o - T_\infty \qquad (2\text{-}26b)$$

and also introduce a quantity m_1:

$$m_1 = \frac{hA_1}{\rho c_p V} = \frac{h}{\rho c_p L} \qquad (2\text{-}26c)$$

since for the cube geometry we have

$$\frac{A_1}{V} = \frac{L^2}{L^3} = \frac{1}{L}$$

where L is the length of a side of the cube. Then Eq. (2-25) becomes

$$\frac{d\theta(t)}{dt} + 2m_1\theta(t) = 2m_1\frac{q_o}{h} \qquad \text{for } t>0 \qquad (2\text{-}27a)$$

$$\theta(t) = \theta_o \qquad \text{at } t=0 \qquad (2\text{-}27b)$$

where

$$m_1 = \frac{h}{\rho c_p L} \qquad (2\text{-}27c)$$

This problem is of exactly the same form as Eq. (2-19) except that m^* is replaced by $2m_1$. Therefore, the solution can be immediately obtained from Eq. (2-22) by replacing m^* by $2m_1$ in that equation. We find

$$\theta(t) = \theta_o\, e^{-2m_1 t} + (1 - e^{-2m_1 t})\frac{q_o}{h} \qquad \text{for } t>0 \qquad (2\text{-}28)$$

Example 2-5. Consider an aluminum cube [$k = 204$ W/(m · °C), $\rho = 2700$ kg/m³, and $c = 0.896$ kJ/(kg · °C)] of side 5 cm that is initially at a uniform temperature of 20°C. For times $t>0$, two of the boundary surfaces are kept insulated, two are subjected to uniform heating at a rate of 10,000 W/m², and the remaining two dissipate heat by convection into an environment at a temperature of 20°C with a heat transfer coefficient of 50 W/(m² · °C). Using lumped system analysis, determine the time required for the cube to reach 100°C. What is the steady-state temperature of the cube?

Solution. Given: $T_o = 20°C$, $T_\infty = 20°C$, $q_o = 10,000 \text{ W/m}^2$, $V/A = L = 0.05 \text{ m}$, and $h = 50 \text{ W/(m}^2 \cdot °C)$. We have

$$\theta_o = T_o - T_\infty = 0$$

$$m = \frac{h}{\rho c L} = \frac{50}{2700 \times 896 \times 0.05} = \frac{1}{2419} \text{ s}^{-1}$$

Then, using Eq. (2-28), the temperature of the cube as a function of time becomes

$$T(t) = T_\infty + (1 - e^{-2m_1 t}) \frac{q_o}{h} \qquad \text{for } t > 0$$

Substituting the numerical values, we have

$$T(t) = 20 + (1 - \exp(-2t/2419)) \frac{10,000}{50}$$

or

$$T(t) = 220 - 200 \exp(-2t/2419) \qquad °C$$

Then the time required for the cube temperature to reach $T(t) = 100°C$ becomes

$$t = \frac{2419}{2} \ln(10/6) \cong 618 \text{ s} = 10 \text{ min } 18 \text{ s}$$

The steady-state temperature of the cube is obtained by letting $t \to \infty$:

$$T_\infty = 220°C$$

2-3 CRITERIA FOR LUMPED SYSTEM

The lumped system analysis described previously is applicable only under very restricted conditions. As is the case with all approximate methods of analysis, the range of validity and the accuracy of the results obtained from lumped system analysis cannot be assessed without making comparisons with the exact solutions. Since the principal concept for the development of this method is the uniformity of temperature within the solid during transients, it is necessary to determine the conditions under which the temperature distribution within the solid during transients can be regarded as uniform. Such conditions can be established by examining the behavior of the temperature distribution in solids as a function of time and position as obtained from the exact solutions of the problem. For example, consider temperature transients in a solid that has a simple geometry, such as a slab, a long solid cylinder, or a sphere, initially at a uniform temperature T_o and for times $t > 0$ is subjected to convection with a heat transfer coefficient h into an ambient at a temperature T_∞. At any instant the maximum temperature difference within the solid occurs between the center and surface temperatures. It is found that this temperature difference remains less than about 5 percent of the center temperature if the *Biot number* remains less than about 0.1, that is, if

$$\boxed{\text{Bi} = \frac{hL_s}{k_s} < 0.1} \qquad (2\text{-}29)$$

where k_s is the thermal conductivity of the solid, h is the heat transfer coefficient at the outer surface, and L_s is the *characteristic length* of the solid body, defined as

$$L_s = \frac{\text{volume}}{\text{surface area}} \tag{2-30}$$

For example, for a solid sphere of radius R, the characteristic length becomes

$$L_s = \frac{\frac{4}{3}\pi R^3}{4\pi R^2} = \frac{R}{3} \tag{2-31}$$

Therefore, for solids having geometries that are not significantly different from a slab, sphere, or cylinder, the assumption of uniform temperature within the body is considered valid and lumped system analysis becomes applicable if the criterion given by Eq. (2-29) is satisfied. The physical significance of the Biot number is better envisioned if it is arranged in the form

$$\text{Bi} = \frac{h}{k_s/L_s} = \frac{\left(\begin{array}{c}\text{heat-transfer coefficient} \\ \text{at surface of solid}\end{array}\right)}{\left(\begin{array}{c}\text{internal unit conductance of} \\ \text{solid across length } L_s\end{array}\right)} \tag{2-32}$$

That is, the Biot number is the ratio of the heat transfer coefficient to the internal unit conductance of the solid in terms of the characteristic length.

In order to illustrate the role of the Biot number as the criterion for the validity of lumped system analysis, we rearrange Eq. (2-32) in the form

$$\text{Bi} = \frac{L_s/k_s}{1/h} = \frac{\left(\begin{array}{c}\text{specific internal resistance} \\ \text{of the solid to heat flow} \\ \text{across length } L_s\end{array}\right)}{\left(\begin{array}{c}\text{specific skin resistance} \\ \text{to heat flow at the outer} \\ \text{surface of the solid}\end{array}\right)} \tag{2-33}$$

Clearly, the criterion $\text{Bi} < 0.1$ implies that the specific internal resistance of the solid to heat flow should be small in comparison to the skin resistance to heat flow at the outer surface of the body. Such a condition is realized with a small value of L_s, a large value of k_s, or a small value of h.

Example 2-6. A 6-cm-diameter steel ball $[k = 61 \text{ W}/(\text{m} \cdot {}^\circ\text{C}), \rho = 7865 \text{ kg}/\text{m}^3$, and $c = 0.46 \text{ kJ}/(\text{kg} \cdot {}^\circ\text{C})]$ is at a uniform temperature of $800{}^\circ\text{C}$. It is to be hardened by suddenly dropping it into an oil bath at a temperature of $50{}^\circ\text{C}$. If the quenching occurs when the ball reaches a temperature of $100{}^\circ\text{C}$ and the heat transfer coefficient between the oil and the sphere is $500 \text{ W}/\text{m}^2 \cdot {}^\circ\text{C}$, determine how long the ball should be kept in the oil bath.

If 100 balls are to be quenched per minute, determine the rate at which heat must be removed from the oil bath in order to maintain the bath temperature at $50{}^\circ\text{C}$.

Solution. The problem can be solved with lumped system analysis with sufficient accuracy if Bi < 0.1. Therefore, we first check the magnitude of the Biot number. The characteristic length L_s is determined using Eq. (2-30) as

$$L_s = \frac{\text{volume}}{\text{area}} = \frac{V}{A} = \frac{R}{3} = 1 \text{ cm}$$

Then the Biot number becomes

$$\text{Bi} = \frac{hL_s}{k}$$

$$= \frac{500 \text{ W/(m}^2 \cdot {}^\circ\text{C)}(0.01 \text{ m})}{61 \text{ W/(m} \cdot {}^\circ\text{C)}} = 0.082$$

which is less than 0.1; hence lumped system analysis is applicable. From Eq. (2-11) we have

$$\frac{T(t) - T_\infty}{T_o - T_\infty} = e^{-mt}$$

where $T(t) = 100°C$, $T_\infty = 50°C$, $T_o = 800°C$, and

$$m = \frac{hA}{\rho cV} = \frac{h}{\rho cL_s}$$

$$= \frac{500}{7865 \times 460 \times 0.01} = \frac{1}{72.358} \text{ s}^{-1}$$

Then

$$\frac{100 - 50}{800 - 50} = \exp\left(-t/72.358\right)$$

or

$$t = 195.95 \text{ s} \cong 3.3 \text{ min}$$

The rate of heat removed from one ball is

$$Q_1 = mc \, \Delta T = \rho cV \, \Delta T$$

$$= 7865 \times 0.46 \times \tfrac{4}{3}\pi(0.03)^3 \times (800 - 100)$$

$$= 286.42 \text{ kJ/ball}$$

For 100 balls, we have

$$Q = 286.42 \times 100 = 28{,}642 \text{ kJ/min}$$

Example 2-7. A thermocouple is to be used to measure the temperature in a gas stream. The junction is approximated as a sphere with thermal conductivity of 25 W/(m · °C), density of 9000 kg/m³, and specific heat of 0.35 kJ/(kg · °C). The heat transfer coefficient between the junction and the gas is 250 W/(m² · °C). Calculate the diameter of the junction if the thermocouple should measure 95 percent of the applied temperature difference in 3 s.

Solution. The characteristic length in terms of diameter D is

$$L_s = \frac{V}{A} = \frac{R}{3} = \frac{D}{6}$$

From Eq. (2-11) we have

$$\frac{\theta(t)}{\theta_o} = \frac{T(t) - T_\infty}{T_o - T_\infty} = e^{-mt}$$

when T_∞ is regarded as the actual gas stream temperature; then $|T_o - T_\infty|$ becomes the applied temperature difference. If the thermocouple should measure 95 percent of this difference, we should have

$$\frac{\theta(t)}{\theta_o} = \frac{5}{100}$$

and the coefficient m becomes

$$m_1 = \frac{h}{\rho c L_s} = \frac{250}{9000 \times 350(D/6)} = \frac{0.000476}{D} \text{ s}$$

Then

$$\frac{5}{100} = \exp{(-0.000476/D \times 3)}$$

or

$$\frac{0.000476 \times 3}{D} = \ln 20 = 2.9957$$

$$D = \frac{0.000476 \times 3}{2.9957} = 0.000476 \text{ m} = 0.476 \text{ mm}$$

Here we applied lumped system analysis before checking the magnitude of the Biot number. For the diameter we calculated above, the Biot number becomes

$$\text{Bi} = \frac{hL_s}{k} = \frac{250 \times 0.000476/6}{25} = 0.00079$$

which is much less than 0.1; hence lumped system analysis is valid.

Example 2-8. Check the validity of the lumped system analysis used in Example 2-4.

Solution. The characteristic length is

$$L_s = \frac{\text{volume}}{\text{surface area}} = \frac{A^* L}{2 A^*} = \frac{L}{2}$$

$$= \frac{0.03 \text{ m}}{2} = 0.015 \text{ m}$$

where A^* is the surface area of one side of the plate. Then the Biot number becomes

$$\text{Bi} = \frac{hL_s}{k} = \frac{50 \times 0.015}{204} = 0.003675 < 0.1$$

therefore the lumped system analysis is applicable.

PROBLEMS

Lumped system concept

2-1. A solid steel ball of radius 3 cm $[k = 54 \text{ W}/(\text{m} \cdot °\text{C})]$ is exposed to an air stream having a heat transfer coefficient of 75 $\text{W}/(\text{m}^2 \cdot °\text{C})$. Calculate the Biot number.
Answer: 0.0139

2-2. A long solid steel bar of radius 3 cm $[k = 54 \text{ W}/(\text{m} \cdot °\text{C})]$ is exposed to an air stream having a heat transfer coefficient of 75 $\text{W}/(\text{m}^2 \cdot °\text{C})$. Calculate the Biot number.
Answer: 0.0208

2-3. A large steel plate of thickness 3 cm $[k = 54 \text{ W}/(\text{m} \cdot °\text{C})]$ is exposed to an air stream having a heat transfer coefficient of 75 $\text{W}/(\text{m}^2 \cdot °\text{C})$. Calculate the Biot number.

2-4. Check whether lumped system analysis is suitable for Examples 2-2 and 2-3.

Lumped system applications and criteria

2-5. A solid copper sphere of diameter 10 cm, initially at a uniform temperature of 250°C, is suddenly immersed in a well-stirred fluid that is maintained at a uniform temperature of 50°C. The heat transfer coefficient between the sphere and the fluid is 200 $\text{W}/(\text{m}^2 \cdot °\text{C})$. (*a*) Check whether lumped system analysis is applicable. (*b*) If it is applicable, determine the temperature of the copper block at times $t = 1$ min, $t = 2$ min, and $t = 5$ min after immersion in the cold fluid. [For copper, $k = 386 \text{ W}/(\text{m} \cdot °\text{C})$, $\rho = 8954 \text{ kg}/\text{m}^3$, and $c = 383 \text{ J}/(\text{kg} \cdot °\text{C})$.]

2-6. A solid copper cylinder of diameter 10 cm, initially at a uniform temperature of 200°C, is suddenly immersed in a well-stirred fluid that is maintained at a uniform temperature of 20°C. The heat transfer coefficient between the cylinder and the fluid is 150 $\text{W}/(\text{m}^2 \cdot °\text{C})$. (*a*) Check whether lumped system analysis is applicable. (*b*) If applicable, determine the temperature of the copper cylinder at time $t = 2$ min after immersion in the cold fluid. [For copper, $k = 386 \text{ W}/(\text{m} \cdot °\text{C})$, $\rho = 8954 \text{ kg}/\text{m}^3$, and $c = 383 \text{ J}/(\text{kg} \cdot °\text{C})$.]
Answer: (*b*) 165.9°C

2-7. A solid iron sphere of diameter 5 cm, initially at a uniform temperature of 700°C, is exposed to a cool air stream at 100°C. The heat transfer coefficient between the air stream and the surface of the iron sphere is 80 $\text{W}/(\text{m}^2 \cdot °\text{C})$. (*a*) Check whether lumped system analysis is applicable. (*b*) If applicable, determine the time required for the temperature of the sphere to reach 300°C. [For iron, $k = 60 \text{ W}/(\text{m} \cdot °\text{C})$, $\rho = 7800 \text{ kg}/\text{m}^3$, and $c = 460 \text{ J}/(\text{kg} \cdot °\text{C})$.]
Answer: (*b*) 6.84 min

2-8. A long cylindrical iron bar of diameter 5 cm, initially at a uniform temperature of 700°C, is exposed to a cool air stream at 100°C. The heat transfer coefficient between the air stream and the surface of the iron bar is 80 $\text{W}/(\text{m}^2 \cdot °\text{C})$. (*a*) Check whether lumped system analysis is applicable. (*b*) If applicable, determine the time required for the temperature of the rod to reach 300°C. [For iron, $k = 60 \text{ W}/(\text{m} \cdot °\text{C})$, $\rho = 7800 \text{ kg}/\text{m}^3$, and $c = 460 \text{ J}/(\text{kg} \cdot °\text{C})$.]

2-9. A large aluminum plate of thickness 3 cm is initially at a uniform temperature of 50°C. Suddenly it is subjected (both surfaces) to a cool air stream at 20°C. The heat transfer coefficient between the air stream and the surface is 50 $\text{W}/(\text{m}^2 \cdot °\text{C})$. (*a*) Check whether lumped system analysis is applicable. (*b*) If applicable, determine the time required for the temperature of the plate to reach 40°C. [For aluminum, $k = 204 \text{ W}/(\text{m} \cdot °\text{C})$, $\rho = 2707 \text{ kg}/\text{m}^3$, and $c = 896 \text{ J}/(\text{kg} \cdot °\text{C})$.]
Answer: (*b*) 4.8 min

2-10. A household electric iron has an aluminum base that weighs 1.5 kg. The base has an ironing surface of 0.03 m^2 that is heated from the inner surface with a 300-W heating element and dissipates heat from the outer surface by convection into the air. Initially the iron is at the same temperature as the ambient air at 20°C. How long will it take for the iron to reach a temperature of 120°C after it is turned on, if the heat transfer coefficient between the iron and the ambient air is $20 \text{ W/(m}^2 \cdot \text{°C)}$? [For aluminum, $k = 204 \text{ W/(m} \cdot \text{°C)}$, $\rho = 2707 \text{ kg/m}^3$, and $c = 896 \text{ J/(kg} \cdot \text{°C)}$.] What is the steady-state temperature of the iron if the control does not cut off the power supply?

Answer: 520°C

2-11. A household electric iron has a steel base that weighs 1 kg. The base has an ironing surface of 0.02 m^2 that is heated from the inner surface with a 300-W heating element and dissipates heat from the outer surface by convection into the ambient air at 20°C. Initially the iron is at a uniform temperature of 30°C. Suddenly the heating starts. The heat transfer coefficient between the iron and the ambient air is $40 \text{ W/(m}^2 \cdot \text{°C)}$. Calculate the temperature of the iron 3 min after the start of heating. [For steel, $k = 70 \text{ W/(m} \cdot \text{°C)}$, $\rho = 7840 \text{ kg/m}^3$, and $c = 450 \text{ J/(kg} \cdot \text{°C)}$.]

Answer: 129.6°C

2-12. A large steel frying pan of thickness 0.5 cm, initially at 10°C, is placed on the stove. The bottom of the pan is subjected to a uniform heat flux of 300 W/m^2 and the top exposed to cool ambient air at 10°C. The heat transfer coefficient between the pan and the ambient air is $30 \text{ W/(m}^2 \cdot \text{°C)}$. Calculate the temperature of the pan at 3 and 8 min after the start of heating. [For steel, $k = 70 \text{ W/(m} \cdot \text{°C)}$, $\rho = 7840 \text{ kg/m}^3$, and $c = 450 \text{ J/(kg} \cdot \text{°C)}$.]

2-13. A large aluminum frying pan of thickness 1 cm, initially at room temperature (20°C), is placed on the stove. The bottom of the pan is subjected to a uniform heat flux of 500 W/m^2 and the top exposed to cool air at 20°C. The heat transfer coefficient between the pan and the air is $50 \text{ kW/(m}^2 \cdot \text{°C)}$. Calculate the time required for the temperature of the pan to reach 200°C. [For aluminum, $k = 204 \text{ W/(m} \cdot \text{°C)}$, $\rho = 2707 \text{ kg/m}^3$, and $c = 896 \text{ J/(kg} \cdot \text{°C)}$.]

Answer: 94.5 s

2-14. A 3-cm-diameter aluminum sphere is initially at a temperature of 175°C. It is suddenly immersed in a well-stirred fluid at a temperature of 25°C. The temperature of the sphere is lowered to 100°C in 42 s. Calculate the heat transfer coefficient. Check whether lumped system analysis is applicable. [For aluminum, $k = 204 \text{ W/(m} \cdot \text{°C)}$, $\rho = 2707 \text{ kg/m}^3$, and $c = 896 \text{ J/(kg} \cdot \text{°C)}$.]

2-15. A grape of 1-cm diameter, initially at a uniform temperature of 20°C, is placed in a refrigerator in which the air temperature is 5°C. If the heat transfer coefficient between the air and the grape is $20 \text{ W/(m}^2 \cdot \text{°C)}$, determine the time required for the grape to reach 10°C. [For the grape, $k \approx 0.6 \text{ W/(m} \cdot \text{°C)}$, $\rho \approx 1100 \text{ kg/m}^3$, and $c = 4200 \text{ J/(kg} \cdot \text{°C)}$.]

Answer: 7.05 min

2-16. A 6-cm-diameter potato at a uniform temperature of 80°C is taken out of the oven and suddenly exposed to ambient air at 20°C. If the heat transfer coefficient between the air and the potato is $25 \text{ W/(m}^2 \cdot \text{°C)}$, determine the time required for the potato to reach 50°C. [For the potato, $k = 7 \text{ W/(m} \cdot \text{°C)}$, $\rho = 1300 \text{ kg/m}^3$, and $c = 4300 \text{ J/(kg} \cdot \text{°C)}$.]

Answer: 26 min

2-17. A metal water immersion heater $[k = 300 \text{ W}/(\text{m} \cdot \text{°C})$, $\rho = 8900 \text{ kg}/\text{m}^3$, and $c = 400 \text{ J}/(\text{kg} \cdot \text{°C})]$ dissipates electric energy at a rate of 50 W. The total volume of the heater is $1.5 \times 10^{-5} \text{ m}^3$, and the outside area is 0.003 m^2. The heater, initially at a uniform temperature of 20°C, is plugged in while in air. The heat transfer coefficient between the air and the heater is around $10 \text{ W}/(\text{m}^2 \cdot \text{°C})$. Calculate the time required for the heater to reach its melting temperature, 500°C.

Answer: 10.1 min

2-18. A large aluminum plate of thickness 3 cm is initially at a uniform temperature of 50°C. Suddenly one of its surfaces is subjected to a uniform heat flux of $7500 \text{ W}/\text{m}^2$ and the other surface is exposed to a cool air stream at a temperature of 30°C. The heat transfer coefficient between the air stream and the surface is $60 \text{ W}/(\text{m}^2 \cdot \text{°C})$. Using lumped system analysis, determine the temperature of the plate as a function of time and plot it against time. Calculate the steady-state temperature of the plate as time approaches infinity. [For aluminum, $k = 204 \text{ W}/(\text{m} \cdot \text{°C})$, $\rho = 2707 \text{ kg}/\text{m}^3$, and $c = 896 \text{ J}/(\text{kg} \cdot \text{°C})$.]

2-19. Consider an aluminum cube of side 3 cm that is initially at a uniform temperature of 50°C. Suddenly all its surfaces are exposed to cool air at 20°C. The heat transfer coefficient between the air stream and the surfaces is $50 \text{ W}/(\text{m}^2 \cdot \text{°C})$. Assuming that lumped system analysis is applicable, develop an expression for the temperature $T(t)$ of the cube as a function of time and plot the temperature of the solid against time. [For aluminum, $k = 204 \text{ W}/(\text{m} \cdot \text{°C})$, $\rho = 2707 \text{ kg}/\text{m}^3$, and $c = 896 \text{ J}/(\text{kg} \cdot \text{°C})$.]

2-20. Consider the aluminum cube of Problem 2-19, which is initially at a uniform temperature of 50°C. For times $t > 0$, one of the boundary surfaces is kept insulated, one is subjected to uniform heating at a rate $8000 \text{ W}/\text{m}^2$, and the remaining four dissipate heat by convection into an environment whose temperature is 20°C with a heat transfer coefficient of $50 \text{ W}/(\text{m}^2 \cdot \text{°C})$. Using lumped system analysis, develop an expression for the temperature of the cube as a function of time.

2-21. Consider a copper block of sides $2 \text{ cm} \times 2 \text{ cm} \times 3 \text{ cm}$, initially at a uniform temperature of 300°C, that is immersed in a fluid at 25°C. The heat transfer coefficient between the fluid and the surfaces is $80 \text{ W}/(\text{m}^2 \cdot \text{°C})$. Calculate the time required for the cube to cool to 50°C. Check the validity of the lumped system analysis. [For copper, $k = 386 \text{ W}/(\text{m} \cdot \text{°C})$, $\rho = 8954 \text{ kg}/\text{m}^3$, and $c = 383 \text{ J}/(\text{kg} \cdot \text{°C})$.]

Answer: 6.42 min

2-22. A solid copper cylindrical rod of 1 cm diameter and 2 cm height is initially at a uniform temperature of 300°C. Suddenly the surfaces are subjected to convection with a heat transfer coefficient of $20 \text{ W}/(\text{m}^2 \cdot \text{°C})$ into an ambient fluid at 25°C. Determine the temperature of the rod 2 min after the start of the cooling. [For copper, $k = 386 \text{ W}/(\text{m} \cdot \text{°C})$, $\rho = 8954 \text{ kg}/\text{m}^3$, and $c = 383 \text{ J}/(\text{kg} \cdot \text{°C})$.]

Answer: 218.8°C

2-23. A soldering iron has an outside area of 0.01 m^2, a mass of 0.6 kg, and a 100-W heating element. The iron, initially at 25°C, is covered with insulation and plugged in. Assuming that lumped system analysis is adequate, estimate the time required for the iron to reach 300°C. [For soldering iron, $k = 75 \text{ W}/(\text{m} \cdot \text{°C})$, $\rho = 8000 \text{ kg}/\text{m}^3$, and $c = 418 \text{ J}/(\text{kg} \cdot \text{°C})$.]

Answer: 11.5 min

2-24. A 0.1-cm-diameter long wooden stick at 15°C is suddenly exposed to 500°C gases with a surface heat transfer coefficient of 15 W/(m$^2 \cdot$°C) between the stick and the gases. If the ignition temperature of the wood is 315°C, find the exposure time before possible ignition. [For wood, $k = 0.14$ W/(m $\cdot$ °C), $\rho = 600$ kg/m^3, and $c = 250$ J/(kg $\cdot$ °C).]

       Answer: 2.41 s

2-25. A short, cylindrical aluminum bar of 1 cm diameter and 2 cm height is initially at a uniform temperature of 150°C. Suddenly the surfaces are subjected to convective cooling with a heat transfer coefficient of 15 W/(m$^2 \cdot$°C) into an ambient fluid at 30°C. Calculate the temperature of the cylinder 1 min after the start of the cooling. [For aluminum, $k = 204$ W/(m $\cdot$ °C), $\rho = 2707$ kg/m^3, and $c = 896$ J/(kg $\cdot$ °C).]

       Answer: 129.7°C

2-26. A column with cross section 2 cm by 2 cm is initially at a uniform temperature of 200°C. Suddenly, the surfaces are subjected to convective cooling with a heat transfer coefficient of 10 W/(m$^2 \cdot$°C) into ambient air at 25°C. Calculate the temperature of the column 10 min after the start of the cooling. Compare this temperature with that of a plane wall 2 cm thick of the same material and under the same conditions. [For the column, $k = 1.28$ W/(m $\cdot$ °C), $\rho = 1458$ kg/m^3, and $c = 880$ J/(kg $\cdot$ °C).]

2-27. A thermocouple junction, approximated as a sphere of constantan, is to be used to measure the temperature of a gas. The heat transfer coefficient between the gas and the thermocouple is 400 W/(m$^2 \cdot$°C). Calculate the maximum allowable diameter of the junction if the thermocouple should measure 95 percent of the applied temperature difference in 5 s. [For constantan, $k = 22.7$ W/(m $\cdot$ °C), $\rho = 8920$ kg/m^3, and $c = 410$ J/(kg $\cdot$ °C).]

       Answer: 1.09 mm

CHAPTER
3

ONE-DIMENSIONAL HEAT CONDUCTION CONCEPT

In this chapter, we develop the one-dimensional heat conduction equation in the rectangular, cylindrical, and spherical coordinate systems and discuss the appropriate boundary conditions. Our objective is to provide a good basis for understanding the physical significance of the steady and unsteady heat conduction equation and its boundary conditions, and hence prepare the necessary background for the mathematical formulation of heat conduction problems. Numerous illustrative examples are presented to demonstrate the mathematical formulation of typical physical problems. The basic equations of heat conduction and the concepts of boundary conditions developed in this chapter form the basis on which the mathematical formulation of heat conduction problems is constructed in the subsequent chapters.

3-1 HEAT CONDUCTION EQUATION

In the previous chapter we considered a simplified analysis of heat conduction in solids that assumed that temperature distribution within the solid, at every instant, can be regarded as uniform. However, for most problems of practical interest the spatial variation of temperature within the body cannot be ignored. In such situations, the energy equation considered previously is no longer applicable. We need to develop a mathematical formulation of heat conduction in solids that allows for both spatial and time variation of temperature within

37

the body. For simplicity in the analysis, we assume that temperature varies with *time* and only in *one direction*; and we consider the formulation in the rectangular, cylindrical, and spherical coordinate systems. We need to develop the heat conduction equation in different coordinate systems because of geometrical considerations arising from the body shapes. For example, for bodies in the form of a slab (i.e., plate), the equation is needed in the rectangular coordinate system. Similarly, for bodies in the form of a cylinder and sphere, the equations are needed in the cylindrical and spherical coordinate systems, respectively. The reason for using different coordinate systems is to ensure that the boundary surfaces of the region coincide with the coordinate surfaces. For example, in the cylindrical coordinate system, one of the coordinate surfaces is a cylinder; hence it coincides with the cylindrical surface of a body in the form of a cylinder.

To develop the heat conduction equation, we consider a solid whose temperature $T(X, t)$ depends on time t and varies only in one direction, say along the X *coordinate*. For generality in the analysis, it is assumed that the coordinate X represents the x axis in the rectangular coordinate system or the r axis in the cylindrical and spherical coordinate systems.

Then the Fourier law, given by Eq. 1-1, can be written in the X coordinate as

$$q = -k \frac{\partial T(X, t)}{\partial X} \qquad (3\text{-}1)$$

where q is the heat flux in the X direction in W/m^2 and k is the thermal conductivity of the solid in W/(m·°C). For generality, we assume an energy source within the solid of strength $g(X, t)$, W/m^3 that varies with the position X and time t. Here $g(X, t)$ is a specified quantity. For example, for a fuel element of a nuclear reactor, it represents the rate of energy generation within the fuel element due to nuclear fission. In the case of an electric current passing through a wire, it represents the ohm heating. When radioactive elements are present within a solid, it represents the rate of energy generation within the body due to the disintegration of the radioactive elements.

To develop the one-dimensional heat conduction equation, we consider a volume element of thickness ΔX and area A normal to the coordinate axis X, as illustrated in Fig. 3-1. The energy balance equation for this volume element

FIGURE 3-1

Nomenclature for the derivation of one-dimensional heat conduction equation.

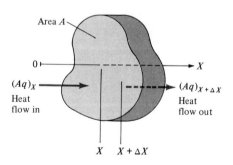

is stated as

$$
\begin{pmatrix} \text{Rate of} \\ \text{increase of} \\ \text{internal energy} \end{pmatrix} = \begin{pmatrix} \text{net rate of} \\ \text{heat gain} \\ \text{by conduction} \end{pmatrix} + \begin{pmatrix} \text{rate of} \\ \text{energy} \\ \text{generation} \end{pmatrix} \tag{3-2}
$$

Each of the terms in this statement is now evaluated.

The rate of increase of internal energy of the volume element resulting from the change of temperature with time is

$$
\begin{pmatrix} \text{Rate of} \\ \text{increase of} \\ \text{internal energy} \end{pmatrix} \equiv A\,\Delta X \rho c\,\frac{\partial T}{\partial t} \tag{3-3a}
$$

where A = surface area normal to the X axis, m^2
ΔX = thickness, m
ρ = density of the material, kg/m^3
c = specific heat of the material, J/(kg·°C)
T = temperature of the volume element, °C
t = time, s

The rate of heat flow into the element by conduction through area A at location X in the X direction is

$$(Aq)_X$$

where q is the conduction heat flux defined previously. Similarly, the rate of the heat flow out of the element by conduction at location $X + \Delta X$ in the X direction is written as

$$(Aq)_{X+\Delta X}$$

Then, the net rate of heat gain by conduction is the difference between the energy entering and leaving the element, that is,

$$
\begin{pmatrix} \text{Net rate of} \\ \text{heat gain} \\ \text{by conduction} \end{pmatrix} \equiv \underset{\text{IN}}{(Aq)_X} - \underset{\text{OUT}}{(Aq)_{X+\Delta X}} \tag{3-3b}
$$

The rate of energy generation in the volume element $A\,\Delta X$, for a specified energy generation rate per unit volume $g \equiv g(X, t)$ W/m^3, is given by

$$
\begin{pmatrix} \text{Rate of} \\ \text{energy} \\ \text{generation} \end{pmatrix} \equiv A\,\Delta X\,g \tag{3-3c}
$$

Introducing Eqs. (3-3a to c) into Eq. (3-2), we have

$$
A\,\Delta X \rho c\,\frac{\partial T}{\partial t} = (Aq)_X - (Aq)_{X+\Delta X} + A\,\Delta X\,g
$$

which is rearranged in the form

$$
\rho c\,\frac{\partial T}{\partial t} = -\frac{1}{A}\,\frac{(Aq)_{X+\Delta X} - (Aq)_X}{\Delta X} + g \tag{3-4a}
$$

As $\Delta X \rightarrow 0$, the first term on the right-hand side becomes the derivative of Aq with respect to X, that is,

$$\frac{(Aq)_{X+\Delta X} - (Aq)_X}{\Delta X} = \frac{\partial(Aq)}{\partial X} \qquad (3\text{-}4b)$$

hence Eq. (3-4a) takes the form

$$\rho c \frac{\partial T}{\partial t} = \frac{1}{A}\frac{\partial}{\partial X}\left(Ak\frac{\partial T}{\partial X}\right) + g \qquad (3\text{-}5)$$

where $T \equiv T(X, t)$
$\quad A \equiv A(X)$
$\quad k \equiv k(X)$
$\quad g \equiv g(X, t)$

In the above derivation of Eq. (3-5), we have not specified any particular coordinate system. Equation (3-5) is applicable in the rectangular, cylindrical, and spherical coordinate systems if proper cognizance is given to the variation of area A with the coordinate axis, as described below.

Rectangular Coordinates

We replace the coordinate X by the x *variable* and note that the area in the rectangular coordinate system does not vary with x. Then A cancels out, and Eq. (3-5) takes the form

$$\rho c \frac{\partial T(x, t)}{\partial t} = \frac{\partial}{\partial x}\left[k(x)\frac{\partial T}{\partial x}\right] + g(x, t) \qquad (3\text{-}6)$$

which is the one-dimensional, time-dependent heat conduction equation in the rectangular coordinate system.

For constant thermal conductivity $k(x) \equiv k$, Eq. (3-6) reduces to

$$\rho c \frac{\partial T(x, t)}{\partial t} = k \frac{\partial^2 T}{\partial x^2} + g(x, t) \qquad (3\text{-}7)$$

Cylindrical Coordinates

We replace the coordinate X by the r *variable* and note that area A varies linearly with the r variable, (i.e., $A = 2\pi r H$, where H is the cylinder length). Then Eq. (3-5) becomes

$$\rho c \frac{\partial T(r, t)}{\partial t} = \frac{1}{r}\frac{\partial}{\partial r}\left[rk(r)\frac{\partial T}{\partial r}\right] + g(r, t) \qquad (3\text{-}8)$$

which is the one-dimensional, time-dependent heat conduction equation in the cylindrical coordinate system.

For constant thermal conductivity $k(r) \equiv k$, Eq. (3-8) reduces to

$$\rho c \, \frac{\partial T(r, t)}{\partial t} = \frac{k}{r} \frac{\partial}{\partial r} \left(r \frac{\partial T}{\partial r} \right) + g(r, t) \qquad (3\text{-}9a)$$

or

$$\rho c \, \frac{\partial T(r, t)}{\partial t} = k \left(\frac{\partial^2 T}{\partial r^2} + \frac{1}{r} \frac{\partial T}{\partial r} \right) + g(r, t) \qquad (3\text{-}9b)$$

Spherical Coordinates

In the spherical coordinate system, area A is proportional to the square of the r variable in the form $A = 4\pi r^2$. Introducing this into Eq. (3-5) and replacing X by r, we obtain

$$\rho c \, \frac{\partial T(r, t)}{\partial t} = \frac{1}{r^2} \frac{\partial}{\partial r} \left[r^2 k(r) \frac{\partial T}{\partial r} \right] + g(r, t) \qquad (3\text{-}10)$$

which is the one-dimensional, time-dependent heat conduction equation in the spherical coordinate system.

For constant thermal conductivity $k(r) \equiv k$, Eq. (3-10) reduces to

$$\rho c \, \frac{\partial T(r, t)}{\partial t} = \frac{k}{r^2} \frac{\partial}{\partial r} \left(r^2 \frac{\partial T}{\partial r} \right) + g(r, t) \qquad (3\text{-}11)$$

or

$$\rho c \, \frac{\partial T(r, t)}{\partial t} = k \left[\frac{\partial^2 T}{\partial r^2} + \frac{2}{r} \frac{\partial T}{\partial r} \right] + g(r, t) \qquad (3\text{-}12)$$

3-2   EQUATIONS REVISITED

The heat conduction Eqs. (3-7), (3-9), and (3-11) can be written more compactly in the form of a single equation as

$$\rho c \, \frac{\partial T(X, t)}{\partial t} = \frac{1}{X^n} \frac{\partial}{\partial X} \left[X^n k(X) \frac{\partial T}{\partial X} \right] + g(X, t) \qquad (3\text{-}13)$$

and for the case of constant thermal conductivity $k(X) = k = \text{constant}$, this equation reduces to

$$\frac{1}{\alpha} \frac{\partial T(X, t)}{\partial t} = \frac{1}{X^n} \frac{\partial}{\partial X}\left(X^n \frac{\partial T}{\partial X}\right) + \frac{1}{k} g(X, t)$$

(3-14)

where $X \equiv x$ and $n = 0$ for rectangular coordinates
$\quad\quad X \equiv r$ and $n = 1$ for cylindrical coordinates
$\quad\quad X \equiv r$ and $n = 2$ for spherical coordinates

and $\alpha \equiv k/\rho c$ is the *thermal diffusivity* of the material in meters squared per second. The physical significance of thermal diffusivity is associated with the propagation of heat into the medium during changes of temperature with time. The higher the thermal diffusivity, the faster the propagation of heat into the medium. Table 3-1 lists the thermal diffusivities of typical materials. There are order of magnitude differences in the values of thermal diffusivity for different materials. For example, α varies from about $170 \times 10^{-6}\ \mathrm{m}^2/\mathrm{s}$ for silver to $0.077 \times 10^{-6}\ \mathrm{m}^2/\mathrm{s}$ for soft rubber. The larger the thermal diffusivity, the shorter the time required for heat to penetrate into the solid. Therefore, under a given applied temperature condition, the penetration of heat into silver is much faster than the penetration into soft rubber.

TABLE 3-1
Thermal diffusivity of typical materials

Material	Average temperature, °C	Thermal diffusivity, $\alpha \times 10^6\ \mathrm{m}^2/\mathrm{s}$
Metals		
Aluminum	0	85.9
Copper	0	114.1
Gold	20	120.8
Iron, pure	0	18.1
Lead	21.1	25.5
Nickel	0	15.5
Silver	0	170.4
Steel, mild	0	12.4
Zinc	0	41.3
Nonmetals		
Asbestos	0	0.258
Brick, fireclay	204.4	0.516
Cork, ground	37.8	0.155
Glass, Pyrex	—	0.594
Granite	0	1.291
Ice	0	1.187
Oak, across grain	29.4	0.160
Pine, across grain	29.4	0.152
Quartz sand, dry	—	0.206
Rubber, soft	—	0.077
Water	0	0.129

Steady-State

The steady-state condition implies that the temperature within the solid does not vary with time, but may vary with the position. For such situations, the time derivative of temperature vanishes and the heat conduction Eq. (3-13) reduces to

$$\frac{1}{X^n}\frac{d}{dX}\left[X^n k(X)\frac{dT(X)}{dX}\right] + g(X) = 0 \qquad (3\text{-}15)$$

As discussed previously, this equation is of interest in different coordinate systems.

For rectangular coordinates, we set $X = x$ and $n = 0$:

$$\frac{d}{dx}\left[k(x)\frac{dT(x)}{dx}\right] + g(x) = 0 \qquad (3\text{-}16a)$$

For cylindrical coordinates, we set $X = r$ and $n = 1$:

$$\frac{1}{r}\frac{d}{dr}\left[rk(r)\frac{dT(r)}{dr}\right] + g(r) = 0 \qquad (3\text{-}16b)$$

For spherical coordinates, we set $X = r$ and $n = 2$:

$$\frac{1}{r^2}\frac{d}{dr}\left[r^2 k(r)\frac{dT(r)}{dr}\right] + g(r) = 0 \qquad (3\text{-}16c)$$

For the case of constant thermal conductivity, $k(X) = k$, Eqs. (3-16) become:

For rectangular coordinates:

$$\frac{d^2 T(x)}{dx^2} + \frac{g(x)}{k} = 0 \qquad (3\text{-}17a)$$

For cylindrical coordinates:

$$\frac{1}{r}\frac{d}{dr}\left[r\frac{dT(r)}{dr}\right] + \frac{g(r)}{k} = 0 \qquad (3\text{-}17b)$$

For spherical coordinates:

$$\frac{1}{r^2}\frac{d}{dr}\left[r^2\frac{dT(r)}{dr}\right] + \frac{g(r)}{k} = 0 \qquad (3\text{-}17c)$$

For steady-state heat conduction with no energy sources within the body and constant thermal conductivity, Eqs. (3-17a) to (3-17c) simplify to:
For rectangular coordinates:

$$\frac{d^2 T(x)}{dx^2} = 0 \qquad (3\text{-}18a)$$

For cylindrical coordinates:

$$\frac{d}{dr}\left[r \frac{dT(r)}{dr} \right] = 0 \qquad (3\text{-}18b)$$

For spherical coordinates:

$$\frac{d}{dr}\left[r^2 \frac{dT(r)}{dr} \right] = 0 \qquad (3\text{-}18c)$$

3-3   BOUNDARY-CONDITION CONCEPT

When the temperature varies with the space variable, the energy equation involves a second derivative in the space variable, as is apparent from the equations given above. In such situations, two boundary conditions are needed in addition to the initial condition to solve the heat conduction problem.

The boundary conditions specify the thermal conditions imposed at the boundary surfaces of the solid. For example, at a given surface one may specify either the temperature or the heat flux or convection into an ambient at a specified temperature.

We present below the mathematical formulation of three different commonly used types of boundary conditions.

Prescribed Temperature Boundary Condition

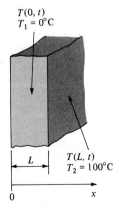

$T(0, t)$
$T_1 = 0°C$

L
$T(L, t)$
$T_2 = 100°C$

0 x

FIGURE 3-2
Prescribed temperature at the boundaries.

There are numerous applications in which the temperature of the boundary surface is considered to be known. For example, a boundary surface in contact with melting ice is at a uniform temperature 0°C, or a boundary surface in contact with boiling water is at the saturation temperature of the water at a given pressure. When the value of the temperature at the boundary surface is specified, the boundary condition is said to be of the *first kind*.

Consider a plate of thickness L, as illustrated in Fig. 3-2. Suppose that the boundary surface at $x = 0$ is maintained at $T_1 = 0°C$ and the boundary surface at $x = L$ is maintained at $T_2 = 100°C$ (e.g., boiling water temperature at atmospheric pressure). These boundary conditions are expressed as

$$T(0, t) = T_1 = 0°C \qquad (3\text{-}19a)$$

$$T(L, t) = T_2 = 100°C \qquad (3\text{-}19b)$$

Prescribed Heat Flux Boundary Condition

There are situations in which the rate at which heat is supplied to or removed from a boundary surface is known. For example, at an electrically heated surface, the rate of heat flow entering the solid is known; at a thermally insulated surface, the heat flux is zero; at a surface subjected to solar irradiation, the rate of energy absorbed can be estimated, and hence the heat flux is regarded as known. Also, there are situations in which the surface heat flux may vary with time. When the magnitude of the heat flux at a boundary surface is specified, the boundary condition is said to be of the *second kind*.

Consider a plate of thickness L, as illustrated in Fig. 3-3a. Suppose that there is a heat supply into the body at a rate of $f_1 = 100 \ W/m^2$ through the boundary surface at $x = 0$. The mathematical formulation of this boundary condition is obtained by considering an energy balance at the surface $x = 0$, as

$$\begin{pmatrix} \text{External heat} \\ \text{supply, } f_1 \ \text{W/m}^2 \\ \text{at } x = 0 \end{pmatrix} = \begin{pmatrix} \text{heat flow by} \\ \text{conduction} \\ \text{into the body} \\ \text{at } x = 0 \end{pmatrix}$$

that is,

$$100 \ \text{W/m}^2 = -k \left. \frac{\partial T(x, t)}{\partial x} \right|_{\substack{\text{evaluated} \\ \text{at } x = 0}}$$

or, more compactly,

$$-k \left. \frac{\partial T(x, t)}{\partial x} \right|_{x=0} = 100 \ \text{W/m}^2 \qquad (3\text{-}20a)$$

Next, we consider a heat supply into the body at the rate of $f_2 = 200 \ \text{W/m}^2$ through the boundary surface at $x = L$, as illustrated in Fig. 3-3b. The energy balance at the surface at $x = L$ yields

$$\begin{pmatrix} \text{Heat flow by} \\ \text{conduction} \\ \text{into the body} \\ \text{at } x = L \end{pmatrix} = \begin{pmatrix} \text{external heat} \\ \text{supply, } f_2 \ \text{W/m}^2 \\ \text{into the body} \\ \text{at } x = L \end{pmatrix}$$

FIGURE 3-3
Prescribed heat flux at the boundaries.

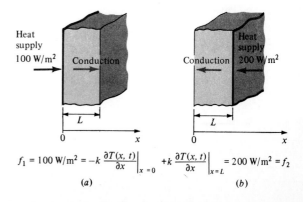

$$f_1 = 100 \ \text{W/m}^2 = -k \left. \frac{\partial T(x, t)}{\partial x} \right|_{x=0} \qquad +k \left. \frac{\partial T(x, t)}{\partial x} \right|_{x=L} = 200 \ \text{W/m}^2 = f_2$$

(a)   (b)

that is,

$$+k \left. \frac{\partial T(x, t)}{\partial x} \right|_{x=L} = 200 \text{ W/m}^2 \tag{3-20b}$$

Equations (3-20a) and (3-20b) are the mathematical representation of the prescribed heat flux boundary condition for heat supply into the body. If heat is to be removed from the body, f_1 and f_2 assume negative values, e.g., $f_1 = -100 \text{ W/m}^2$ and $f_2 = -200 \text{ W/m}^2$. Then Eqs. (3-20a) and (3-20b) become

$$-k \left. \frac{\partial T(x, t)}{\partial x} \right|_{x=0} = -100 \text{ W/m}^2 \tag{3-21a}$$

$$+k \left. \frac{\partial T(x, t)}{\partial x} \right|_{x=L} = -200 \text{ W/m}^2 \tag{3-21b}$$

Therefore, without specifying whether we have heat supply to or heat removal from the boundary surface, we can express the prescribed heat flux boundary conditions at $x = 0$ and $x = L$ as

$$\boxed{\begin{aligned} -k \frac{\partial T(x, t)}{\partial x} &= f_1 \qquad \text{at } x = 0 \\[2mm] +k \frac{\partial T(x, t)}{\partial x} &= f_2 \qquad \text{at } x = L \end{aligned}}$$

$$\tag{3-22a}$$
$$\tag{3-22b}$$

where f_1 and f_2 are *positive for heat supply* into the body and *negative for heat removal* from the body.

Convection Boundary Condition

In most practical problems, heat flow at the boundary surface takes place by convection with a known heat transfer coefficient h into an environment at a specified temperature.

Consider a plate of thickness L, as illustrated in Fig. 3-4a. A fluid with a

FIGURE 3-4
Convection at the boundaries.

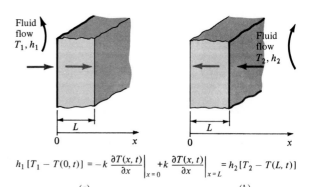

$$h_1 [T_1 - T(0, t)] = -k \left. \frac{\partial T(x, t)}{\partial x} \right|_{x=0} \qquad +k \left. \frac{\partial T(x, t)}{\partial x} \right|_{x=L} = h_2 [T_2 - T(L, t)]$$

(a)   (b)

heat transfer coefficient h_1 at a temperature T_1 flows over the surface of the plate at $x = 0$. The mathematical representation of this convection boundary condition is obtained by considering an energy balance at the surface $x = 0$.

$$\begin{pmatrix} \text{Convection heat flux} \\ \text{from the fluid at } T_1 \\ \text{to the surface at } x = 0 \end{pmatrix} = \begin{pmatrix} \text{conduction heat} \\ \text{flow into the body} \\ \text{at } x = 0 \end{pmatrix}$$

that is,

$$h_1[T_1 - T(0, t)] = -k \left. \frac{\partial T(x, t)}{\partial x} \right|_{x=0}$$

or, rearranging,

$$-k \left. \frac{\partial T(x, t)}{\partial x} \right|_{x=0} = h_1[T_1 - T(0, t)] \qquad (3\text{-}23a)$$

At the other surface, we consider a fluid with a heat transfer coefficient h_2 at a temperature T_2 flowing over the surface $x = L$, as illustrated in Fig. 3-4b. The energy balance for this surface becomes

$$\begin{pmatrix} \text{Conduction heat} \\ \text{flow into the body} \\ \text{at } x = L \end{pmatrix} = \begin{pmatrix} \text{convection heat flux} \\ \text{from the fluid at } T_2 \\ \text{to the surface at } x = L \end{pmatrix}$$

that is,

$$+k \left. \frac{\partial T(x, t)}{\partial x} \right|_{x=L} = h_2[T_2 - T(L, t)] \qquad (3\text{-}23b)$$

The convection boundary conditions given by Eqs. (3-23a) and (3-23b) are rewritten in the form

$$-k \left. \frac{\partial T(x, t)}{\partial x} \right|_{x=0} + h_1 T(0, t) = h_1 T_1$$

$$+k \left. \frac{\partial T(x, t)}{\partial x} \right|_{x=L} + h_2 T(L, t) = h_2 T_2$$

or

$$-k \frac{\partial T(x, t)}{\partial x} + h_1 T(x, t) = h_1 T_1 \qquad \text{at } x = 0 \qquad (3\text{-}24a)$$

$$+k \frac{\partial T(x, t)}{\partial x} + h_2 T(x, t) = h_2 T_2 \qquad \text{at } x = L \qquad (3\text{-}24b)$$

where the left-hand sides are in terms of the unknown surface temperature of the body and the right-hand sides are in terms of the known fluid temperatures.

Example 3-1. A thick-walled tube has inside radius r_1 and outside radius r_2. A hot gas at temperature T_1 flows inside the tube, and a cold gas at temperature T_2 flows outside. The heat transfer coefficients for flow inside and outside the tube are specified as h_1 and h_2, respectively. Write the boundary conditions.

Solution. The convection boundary conditions for this problem, as illustrated in the accompanying figure, are similar to those given previously by Eqs. (3-23a) and (3-23b) for a plate. Therefore, by replacing the coordinate x in Eq. (3-23a) and (3-23b) by the radial variable r, we obtain

$$-k\frac{\partial T}{\partial r} = h_1(T_1 - T) \qquad \text{at } r = r_1$$

$$+k\frac{\partial T}{\partial r} = h_2(T_2 - T) \qquad \text{at } r = r_2$$

where T is the temperature and k is the thermal conductivity of the tube.

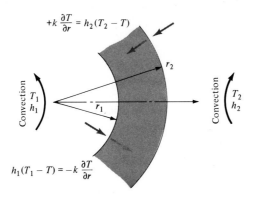

FIGURE EXAMPLE 3-1
Convection boundary conditions.

Example 3-2. A plane wall, confined to the region $0 \le x \le L$, is subjected to a heat supply at a rate of q_0 W/m^2 at the boundary surface $x = 0$ and dissipates heat by convection with a heat transfer coefficient h_∞ W/(m^2·°C) into the ambient air at a temperature T_∞ from the boundary conditions at $x = 0$ and $x = L$.

Solution. The boundary conditions at $x = 0$ and $x = L$ for this problem are similar to those given by Eqs. (3-22a) and (3-23b), respectively. Therefore, by setting $f_1 \equiv q_0$ W/m^2, $h_2 \equiv h_\infty$, and $T_2 \equiv T_\infty$ in these equations, we obtain

$$-k\frac{\partial T}{\partial x} = q_0 \qquad \text{W/m}^2 \qquad \text{at } x = 0$$

$$-k\frac{\partial T}{\partial x} = h_\infty(T - T_\infty) \qquad \text{W/m}^2 \qquad \text{at } x = L$$

or these boundary conditions can be rearranged in the form

$$-K\frac{\partial T}{\partial x} = q_0 \qquad \text{W/m}^2 \qquad \text{at } x = 0$$

$$k\frac{\partial T}{\partial x} + h_\infty T = h_\infty T_\infty \qquad \text{W/m}^2 \qquad \text{at } x = L$$

Example 3-3. Consider a cylindrical wall with inside radius r_1 and outside radius r_2. The inside surface is heated uniformly at a rate of q_1 W/m^2, and the outside surface dissipates heat by convection with a heat transfer coefficient h_2 W/(m^2·°C) into the environment at zero temperature. Write the boundary conditions for the surfaces at $r = r_1$ and $r = r_2$.

Solution. Given: $f_1 = q_1$ W/m^2, h_2 and $T_\infty = 0$°C. These boundary conditions are also obtainable from the boundary conditions given by Eqs. (3-22a) and (3-23b), respectively, by setting $f_1 \equiv q_1$, $T_2 = 0$, $x = r$, and $L = r_2$.

$$-k \frac{\partial T}{\partial r} = q_1 \qquad \text{at } r = r_1$$

$$k \frac{\partial T}{\partial r} + h_2 T = 0 \qquad \text{at } r = r_2$$

Thermal and Geometric Symmetry

If a thermal and geometric symmetry coexist about a common symmetry axis (or plane), the temperature distribution is symmetrical about the same axis (or plane). Then the temperature gradient at the symmetry axis in the direction perpendicular to the axis vanishes, and hence we have

$$\frac{\partial T}{\partial X} = 0 \qquad \text{at the symmetry axis} \qquad (3\text{-}25)$$

When such a symmetry condition exists within the medium, the mathematical formulation of the heat conduction problem is simplified, as illustrated below.

Consider a slab of thickness L which is initially (i.e., at time $t = 0$) at a specified temperature and for times $t > 0$ is subjected to convection at the boundaries $x = 0$ and $x = L$ as illustrated in Fig. 3-5a. The heat transfer coefficient h_∞ and the environment temperature T_∞ are the same for both boundary surfaces. The boundary conditions become

$$-k \frac{\partial T}{\partial x} = h_\infty (T_\infty - T) \qquad \text{at } x = -L/2 \qquad (3\text{-}26a)$$

$$+k \frac{\partial T}{\partial x} = h_\infty (T_\infty - T) \qquad \text{at } x = +L/2 \qquad (3\text{-}26b)$$

To determine the temperature distribution within the solid, the heat conduc-

FIGURE 3-5
Thermal and geometric symmetry.

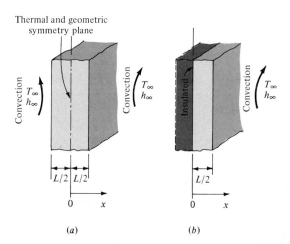

tion equation for this problem should be solved within the region $-L/2 < x < L/2$ subject to the boundary conditions (3-26) and a specified initial condition.

Suppose the initial distribution within the medium is symmetrical about $x = 0$. Then the problem has both geometrical and thermal symmetry at $x = 0$, and the gradient of temperature at $x = 0$ vanishes, that is,

$$\frac{\partial T}{\partial x} = 0 \qquad \text{at } x = 0 \tag{3-27}$$

When the problem has both geometrical and thermal symmetry about $x = 0$, it is more convenient to solve the problem over the half of the region $0 < x < L/2$ subject to the boundary conditions given by

$$\frac{\partial T}{\partial x} = 0 \qquad \qquad \text{at } x = 0 \tag{3-28a}$$

$$k \frac{\partial T}{\partial x} = h_\infty(T_\infty - T) \qquad \text{at } x = +\frac{L}{2} \tag{3-28b}$$

rather than solving it over the full domain $-L/2 < x < L/2$ subject to the boundary conditions given by Eqs. (3-26a) and (3-26b). Figure 3-5b illustrates the physical situation associated with such a symmetry consideration. Clearly, solving the heat conduction problem subject to the boundary conditions given by Eqs. (3-28a) and (3-28b) is much easier than solving the problem over the entire region subject to the boundary conditions given by Eqs. (3-26a) and (3-26b).

Example 3-4. Consider a solid bar of radius $r = R$ in which energy is generated at a constant rate g_0 W/m³. The bar is cooled by convection from its lateral surfaces into ambient air at a temperature T_∞ with a heat transfer coefficient h_∞. Write the boundary conditions needed for the solution of this problem.

Solution. The problem possesses both geometric and thermal symmetry about the axis of the bar. Therefore, we have the symmetry boundary condition at $r = 0$ and the convection boundary condition at $r = R$. Then the heat conduction equation for this problem is to be solved in the region $0 \le r \le R$, subject to the boundary conditions

$$\frac{\partial T}{\partial r} = 0 \qquad \qquad \text{at } r = 0$$

$$+k \frac{\partial T}{\partial r} = h_\infty(T_\infty - T) \qquad \text{at } r = R$$

3-4   FORMULATION OF HEAT CONDUCTION PROBLEMS

The analysis of heat conduction problems begins with the development of an appropriate mathematical model to represent the actual physical situation under consideration. The solution of the heat conduction equation subject to appropriate boundary and initial conditions gives the distribution of temperature within the solid as a function of time and position. Once the temperature distribution $T(X, t)$ in the solid is known, the heat flow rate anywhere in the medium can be determined from the Fourier law, given previously.

Therefore, the first step in the mathematical formulation of the heat conduction problem is choosing the appropriate heat conduction equation according to whether the problem is a steady-state or a transient one. The geometry of the body establishes the coordinate system to be used. The appropriate heat conduction equation can readily be obtained from the equations developed previously.

One-dimensional problems require only two boundary conditions. The physical situation at any one of the surfaces may be a prescribed temperature, a prescribed heat flux, or convection into an ambient at a specified temperature. Such boundary conditions can readily be formulated as discussed previously.

In the case of a time-dependent problem, an initial condition is also needed.

In this section we illustrate with examples the mathematical formulation of typical steady-state and transient heat conduction problems, but the solutions are not considered. The development of solutions is the subject of the following chapters.

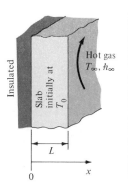

FIGURE EXAMPLE 3-5
Schematic representation of the plane wall.

Example 3-5. A plane wall of thickness L with constant thermal properties is initially at a uniform temperature T_0. At time $t = 0$, the surface at $x = L$ is subjected to heating by the flow of a hot gas at temperature T_∞, while the surface at $x = 0$ is kept insulated. The heat transfer coefficient between the hot gas and the surface is h_∞. There is no energy generation within the place. Develop the mathematical formulation of this unsteady heat conduction problem.

Solution. The physical problem considered is schematically illustrated in the accompanying figure. This is a one-dimensional, time-dependent problem with constant thermal properties and no energy generation. Therefore, the governing differential equation is immediately obtained from Eq. (3-7) by setting $g(x, t) = 0$ in that equation. The boundary condition at $x = 0$ can be obtained from Eq. (3-22a) by setting $f_1 = 0$, and the boundary condition at $x = L$ can be obtained from Eq. (3-24b) by setting $h_2 = h_\infty$ and $T_2 = T_\infty$. Then the mathematical formulation of the stated problem with the specified initial condition is given by

$$\rho c \frac{\partial T(x, t)}{\partial t} = k \frac{\partial^2 T(x, t)}{\partial x^2} \qquad \text{in } 0 < x < L, t > 0$$

$$\frac{\partial T(x, t)}{\partial x} = 0 \qquad \text{at } x = 0, t > 0$$

$$k \frac{\partial T(x, t)}{\partial x} = h_\infty [T_\infty - T(x, t)] \qquad \text{at } x = L, t > 0$$

$$T(x, t) = T_0 \qquad \text{for } t = 0$$

or the formulation can be rewritten in the form

$$\frac{1}{\alpha} \frac{\partial T}{\partial t} = \frac{\partial^2 T}{\partial x^2} \qquad \text{in } 0 < x < L, t > 0$$

$$\frac{\partial T}{\partial x} = 0 \qquad \text{at } x = 0, t > 0$$

$$k \frac{\partial T}{\partial x} + h_\infty T = h_\infty T_\infty \qquad \text{at } x = L, t > 0$$

$$T(x, t) = T_0 \qquad \text{for } t = 0$$

where $T \equiv T(x, t)$.

Example 3-6. Energy is generated at a constant rate of g_o W/m³ in a copper rod of radius $r = R$ by the passage of an electric current. The heat dissipation is by convection from the boundary surface at $r = R$ into the ambient air at temperature T_∞ with the heat transfer coefficient h_∞. The heat transfer process has continued for a sufficiently long time that the transients have passed and the steady-state condition is established in the rod. Develop the mathematical formulation of this heat conduction problem for the determination of the temperature distribution within the rod.

Solution. The problem is one-dimensional, steady-state heat conduction in the cylindrical coordinate system with constant thermal properties. The heat conduction equation is immediately available from Eq. (3-17b) by setting $g(r) = g_o =$ constant.

$$\frac{1}{r}\frac{d}{dr}\left(r\frac{dT}{dr}\right) + \frac{g_o}{k} = 0 \quad \text{in } 0 < r < R$$

The boundary condition at $r = 0$ involves the symmetry condition and that at $r = R$ involves convection. Hence we have

$$\frac{dT}{dr} = 0 \qquad \text{at } r = 0$$

$$-k\frac{dT}{dr} = h_\infty(T - T_\infty) \qquad \text{at } r = R$$

where $T \equiv T(r)$.

Example 3-7. A solid sphere of radius R with thermal conductivity k that is initially at a uniform temperature T_0 is suddenly dropped into an environment at a constant temperature T_∞. The heat transfer coefficient between the solid and the fluid is h_∞. Lumped system analysis is not applicable, and hence the problem has to be formulated as a one-dimensional transient heat conduction problem. Develop the mathematical formulation of this heat conduction problem.

Solution. This is a one-dimensional, time-dependent heat conduction problem in the spherical coordinate system with constant thermal properties and energy generation. Therefore, the heat conduction equation is readily available from Eq. (3-11) by setting $g(r, t) = 0$ or from Eq. (3-14) by setting $n = 2$, $x = r$, and $g(r, t) = 0$, to give

$$\frac{1}{\alpha}\frac{\partial T}{\partial t} = \frac{1}{r^2}\frac{\partial}{\partial r}\left(r^2\frac{\partial T}{\partial r}\right) \quad \text{in } 0 < r < R, t > 0$$

The boundary conditions involve symmetry at $r = 0$ and convection at $r = R$, and the solid is initially at a uniform temperature T_0. Hence the boundary and initial conditions for the problem become

$$\frac{\partial T}{\partial r} = 0 \qquad \text{at } r = 0$$

$$k\frac{\partial T}{\partial r} + h_\infty T = h_\infty T_\infty \qquad \text{at } r = R$$

$$T = T_0 \qquad \text{for } t = 0$$

where $T \equiv T(r, t)$.

Example 3-8. Develop the mathematical formulation of one-dimensional, steady-state heat conduction for a hollow cylinder with constant thermal conductivity in the region $r_1 \leq r \leq r_2$ for the following boundary conditions: Heat is supplied into the cylinder at a rate q_o W/m^2 from the boundary surface at $r = r_1$ and dissipated by convection from the boundary surface at $r = r_2$ into a medium at zero temperature with a heat transfer coefficient h_∞.

Solution. This is a one-dimensional, steady-state heat conduction problem in the cylindrical coordinate system with no energy generation in the medium and constant thermophysical properties. The accompanying figure schematically illustrates the geometry and the coordinates for this problem. The governing differential equation is obtained from Eq. (3-16b) by setting $g(r) = 0$.

The prescribed heat flux boundary condition at $r = r_1$ is obtained from Eq. (3-22a) by replacing x by r and f_1 by q_o. The convection boundary condition at $r = r_2$ is obtained from Eq. (3-24b) by replacing x by r and setting the right-hand side equal to zero. Hence the complete mathematical formulation of the problem becomes

$$\frac{d}{dr}\left(r\frac{dT}{dr}\right) = 0 \qquad \text{in } r_1 < r < r_2$$

subject to the boundary conditions

$$-k\frac{dT}{dr} = q_0 \qquad \text{at } r = r_1$$

$$k\frac{dT}{dr} + h_\infty T = 0 \qquad \text{at } r = r_2$$

where $T \equiv T(r)$.

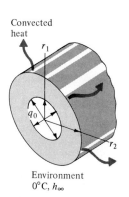

Convected heat

r_1

q_0

r_2

Environment
0°C, h_∞

FIGURE EXAMPLE 3-8
Schematic representation of the hollow cylinder.

PROBLEMS

3-1 The thermal conductivity k, the density ρ, and the specific heat c of steel are 61 W/(m·°C), 7865 kg/m^3, and 0.46 kJ(kg·°C), respectively. Calculate the thermal diffusivity α m^2/s. Compare the calculated value with the thermal diffusivity value given in Table 3-1 for mild steel at 0°C.
Answer: 12.4×10^{-6} m^2/s

3-2 The thermal conductivity k, the density ρ, and the specific heat c of an aluminum plate are 160 W/(m·°C), 2790 kg/m^3, and 0.88 kJ(kg · °C), respectively. Calculate the thermal diffusivity α m^2/s. Compare the calculated value with the thermal diffusivity value given in Table 3-1 at 0°C.
Answer: 65.2×10^{-6} m^2/s

3-3 Consider three plates of the same thickness L made of copper, lead, and asbestos. Initially the three plates are at the same uniform temperature T_0. Suddenly, each plate is subjected to a uniform heat flux q_o W/m^2 at one of its boundary surfaces while the other surface is kept insulated. Which plate will give rise to a faster increase in temperature at the insulated surface? Explain the reason.

3-4 Consider two very thick solids, which can be regarded as semi-infinite mediums extending from $x = 0$ to $x \rightarrow \infty$. Initially they are at a uniform temperature 100°C. Suddenly the temperature of the surfaces at $x = 0$ is lowered to 0°C and maintained at that temperature while the temperatures at locations 30 cm from the boundary surfaces are monitored. It takes 10 min for one of the solids and 120 min for the other to reach a temperature of 50°C at this location. Which solid has larger thermal diffusivity?

Heat Conduction Equation

3-5 Consider a plate fuel element of thickness $2L$ for a gas-cooled nuclear reactor. The energy generation in the fuel element can be approximated by a cosine distribution $q = q_0 \cos x$, where x is the coordinate measured from the plate center. Thermal conductivity of the plate material is assumed constant. Write the steady-state heat conduction equation governing the temperature distribution in the fuel element.

3-6 A plate fuel element of thickness $2L$ has thermal conductivity that varies with temperature, $k(T)$. The following two different types of energy generation are considered: (*a*) $g = g_o$; (*b*) $g = g_o \cos x/L$, where x is the coordinate measured from the plate center. Write the steady-state heat conduction governing the temperature distribution in the fuel element for each of these types of energy generation.

3-7 Consider a plate fuel element of thickness $2L$, initially at a uniform temperature T_∞, in which one of the following three different types of energy generation in the element is suddenly possible: (*a*) uniform, $g = g_0$; (*b*) increasing exponentially, $g = g_o(1 - e^{-bt})$; (*c*) oscillating as $g = g_0(1 + \alpha \sin \omega t)$. Write the unsteady heat conduction equation for the temperature distribution for each of these three cases.

3-8 Consider a long tube of inside radius a and outside radius b. Suddenly, energy is generated in the tube at a constant rate g_0 W/m^3 by the passage of electric current. Write the heat conduction equation for the determination of the unsteady temperature distribution in the tube.

3-9 Consider a hollow sphere of inside radius a and outside radius b. Suddenly, energy is generated in the sphere at a constant rate g_o W/m^3. Write the heat conduction equation for the determination of the unsteady temperature distribution in the sphere.

3-10 A copper bar of radius b is suddenly heated by the passage of electric current, which generates heat in the rod at a rate $g_o e^{-\gamma t}$. The thermal conductivity of the rod varies with the radius, $k = k(r)$. Write the unsteady heat conduction equation governing the temperature distribution in the rod.

Boundary Conditions

3-11 Consider a tube of inside radius r_1 and outside radius r_2. Boiling water at a saturation temperature of 100°C flows inside the tube while heat is dissipated from the outside surface by convection with a heat transfer coefficient of 15 W/(m^2·°C) into ambient air at a temperature of 20°C. Write the boundary conditions for the boundary surfaces at r_1 and r_2.

3-12 A spherical shell has an inside radius r_1 and an outside radius r_2. The inside surface is electrically heated at a rate of q_1 W/m^2, while the outside surface dissipates heat by convection with a heat transfer coefficient h_2 into ambient air at a temperature $T_{\infty 2}$. Write the boundary conditions for the inner and outer surfaces.

3-13 Consider a slab of thickness L. The boundary at $x = 0$ is subjected to forced convection with a heat transfer coefficient h into an ambient at temperature T_∞. The boundary at $x = L$ is absorbing solar radiation at a rate of q_o W/m^2. Write the boundary conditions for both surfaces.

3-14 One of the surfaces of a marble slab [with $k = 2$ W/(m·°C)] is maintained at 200°C, while the other boundary is subjected to a constant heat flux of 5000 W/m^2. Write the boundary conditions.

Formulation of Heat Conduction Problems

3-15 Consider a slab of thickness L in which energy is generated at a constant rate of g_o W/m^3. The surfaces of the slab are kept at a fixed temperature T_w. Does the problem have both geometric and thermal symmetry? Develop the mathematical formulation of this heat conduction problem by taking into account the symmetry condition at the center.

3-16 A wood board of thickness 1 cm, initially at a uniform temperature of 20°C, is suddenly dropped into boiling water. Develop the mathematical formulation of the problem of determining the temperature distribution within the plate for times $t > 0$.

3-17 A long cylindrical iron bar of diameter 30 cm, initially at a temperature of 20°C, is exposed to a hot fluid at 90°C. The heat transfer coefficient between the hot fluid and surface of the iron bar is 200 W/(m$^2 \cdot$°C). (a) Check whether lumped system analysis is suitable. (b) If it is not, develop the mathematical formulation of the problem of determining the temperature distribution within the bar for times $t > 0$. [For iron, $k = 60$ W/(m·°C), $\rho = 7800$ kg/m^3, and $c = 460$ J/°C.]

3-18 One surface of a plate of thickness L is suddenly subjected to radiant heat flux q_o, as illustrated in the accompanying figure. The initial temperature of the plate is equal to the ambient temperature T_∞. The heat transfer coefficient h is the same for both surfaces. Develop the mathematical formulation of the problem of determining the unsteady temperature distribution in the plate.

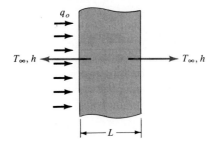

FIGURE PROBLEM 3-18

3-19 An orange of diameter D and thermal conductivity k is initially at a uniform temperature T_i. It is placed in a refrigerator in which the air temperature is T_∞. The heat transfer coefficient between the air and the surface of the orange is h_∞. Develop the mathematical formulation of the problem of determining the temperature distribution within the orange for times $t > 0$.

3-20 A potato with diameter D and thermal conductivity k, initially at a uniform temperature T_i, is suddenly dropped into boiling water at T_o. The heat transfer coefficient between the water and the surface of the potato is very large. Develop the mathematical formulation of the problem of determining the temperature distribution within the potato for times $t > 0$.

3-21 Consider a slab of thickness L and thermal conductivity k that is initially at a uniform temperature T_i. Suddenly, the boundary at $x = 0$ is subjected to forced convection with a heat transfer coefficient h into an ambient at temperature T_∞, and the boundary at $x = L$ is absorbing solar radiation at a rate of q_o W/m^2. Develop the mathematical formulation of the problem of determining the temperature distribution within the slab for times $t > 0$.

3-22 Consider a tube of inside radius r_1, outside radius r_2, and thermal conductivity k. The inside surface is kept at a temperature of 100°C by boiling water, while the outside surface dissipates heat by convection with a heat transfer coefficient of 15 W/(m²·°C) into ambient air at a temperature of 20°C. Develop the mathematical formulation of the problem of determining the steady-state distribution within the tube.

3-23 A spherical shell has an inside radius r_1, an outside radius r_2, and thermal conductivity k. The inside surface is electrically heated at a rate q_1 W/m², while the outside surface dissipates heat by convection with a heat transfer coefficient h_2 into the ambient air at temperature $T_{\infty 2}$. Develop the mathematical formulation of the problem of determining the steady-state temperature distribution within the shell.

3-24 Consider a plate fuel element of thickness $2L$ and thermal conductivity k. The energy generation in the fuel element can be approximated by a cosine distribution $g = g_0 \cos x/L$, with the coordinate x measured from the plate center. The boundaries at $x = \mp L$ are subjected to cooling by forced convection with a heat transfer coefficient h into the cooling liquid at T_∞. Develop the mathematical formulation of the problem of determining the steady-state temperature distribution within the fuel element.

3-25 Consider a long cylindrical fuel element of diameter D and thermal conductivity k in which energy is generated at a constant rate g_0. The boundary surface is assumed to be maintained at a constant temperature T_w. Develop the mathematical formulation of the problem of determining the steady-state temperature distribution within the fuel element.

3-26 Consider the steady-state heat conduction in a plate of thickness L in which energy is generated at a constant rate g_0. The boundary surface at $x = 0$ is maintained at a constant temperature T_0, while the boundary surface at $x = L$ dissipates heat by convection with a heat transfer coefficient h into an ambient at temperature T_∞. Develop the mathematical formulation of the problem.

3-27 Consider the steady-state heat conduction in a slab of thickness L in which energy is generated at a constant rate g_0 and the thermal conductivity varies with temperature, $k(T)$. The boundary surface at $x = 0$ is insulated, while the boundary surface at $x = L$ is maintained at T_0. Develop the mathematical formulation of the problem.

3-28 Consider a slab of thickness L, initially with a temperature distribution $F(x)$. For $t > 0$ the boundary surfaces at $x = 0$ and $x = L$ dissipate heat by convection with a heat transfer coefficient h into an ambient at temperature T_∞. Develop the mathematical formulation of the problem of determining the steady-state temperature distribution within the plate for times $t > 0$.

CHAPTER

4

STEADY CONDUCTION WITHOUT GENERATION: THERMAL RESISTANCE CONCEPT

This chapter is devoted to the determination of steady-state temperature distribution and heat flow in solids with such shapes as a plane wall (i.e., a slab), a long hollow cylinder, and a hollow sphere that have constant thermal conductivity and *no energy generation* within the medium. In engineering applications, the *thermal resistance* concept is frequently used for calculating heat transfer through solids under such conditions. Therefore, the physical significance of this concept and its application in the analysis of heat flow through composite parallel layers are discussed. The use of conduction shape factors in the determination of heat flow through bodies having more complicated shapes is presented.

4-1 SINGLE LAYER

The determination of steady-state temperature and heat flow rate through bodies with such shapes as a plane wall, a long hollow cylinder, and a hollow sphere is of interest in numerous engineering applications. Typical examples include heat flow through a window glass, the walls of a building, the wall of a tube, and many others. In such applications there is no energy generation within the solid. Furthermore, the thermal properties of the solid can be regarded as constant, since the temperature variation across the solid is not large. In such situations the temperature distribution within the solid is governed by the heat conduction equations, Eqs. (3-18), which can be written more compactly in the form

$$\frac{d}{dX}\left[X^n \frac{dT(X)}{dX}\right] = 0 \tag{4-1}$$

where

$$X \equiv x \text{ and } n = 0 \text{ for rectangular coordinates}$$

$$X \equiv r \text{ and } n = 1 \text{ for cylindrical coordinates}$$

$$X \equiv r \text{ and } n = 2 \text{ for spherical coordinates}$$

Figure 4-1 shows the geometry and the coordinates for a plane wall, a hollow cylinder, and a hollow sphere. The solution of Eq. (4-1) over the thickness of the solid subject to appropriate boundary conditions at both surfaces gives the distribution of temperature in the body.

Once the temperature distribution $T(X)$ in the solid has been established, the heat flux $q(X)$ anywhere in the solid can be determined from the Fourier law

$$q(X) = -k \frac{dT(X)}{dX} \qquad \text{W/m}^2 \tag{4-2}$$

When we know the heat flux, the total heat flow rate Q through a surface area A can be readily determined.

In engineering applications the concept of *thermal resistance* is generally used to determine one-dimensional heat flow through solids having such shapes as a plane wall, a long hollow cylinder, or a hollow sphere. In this section we first develop the thermal resistances associated with heat flow through a plane wall and cylindrical and spherical walls, then illustrate the use of this concept in determining the heat flow through a solid.

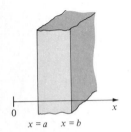

(a) Plane wall

(b) Hollow cylinder

(c) Hollow sphere

FIGURE 4-1
Coordinates for one-dimensional steady heat conduction problems.

Plane Wall

Consider a slab of thickness L, as illustrated in Fig. 4-2. The boundary surfaces at $x = 0$ and $x = L$ are maintained at constant but different temperatures T_1 and T_2, respectively. There is no energy generation in the solid, and the thermal conductivity is assumed to be constant. To develop the thermal resistance concept for this problem, we first solve the heat conduction problem and determine the expression for the heat flow, then recast this result in a form analogous to Ohm's law in electricity in order to establish the equivalent thermal resistance. The procedure is as follows.

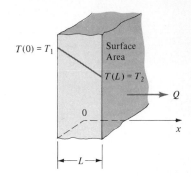

FIGURE 4-2
Temperature distribution for a
slab with prescribed temperature
at both surfaces.

The mathematical formulation of this heat conduction problem is given
by

$$\frac{d^2 T(x)}{dx^2} = 0 \quad \text{in } 0 < x < L \tag{4-3a}$$

$$T(x) = T_1 \quad \text{at } x = 0 \tag{4-3b}$$

$$T(x) = T_2 \quad \text{at } x = L \tag{4-3c}$$

The integration of Eq. (4-3a) twice yields

$$T(x) = C_1 x + C_2 \tag{4-4}$$

We have two unknown integration constants C_1 and C_2 and two boundary
conditions, Eqs. (4-3b) and (4-3c), for their determination.

The application of the first boundary condition, Eq. (4-3b), gives

$$C_2 = T_1 \tag{4-5a}$$

and the application of the second boundary condition results in

$$C_1 = \frac{T_2 - C_2}{L} = \frac{T_2 - T_1}{L} \tag{4-5b}$$

Then the solution becomes

$$T(x) = (T_2 - T_1)\frac{x}{L} + T_1 \tag{4-6}$$

Equation (4-6) demonstrates that for one-dimensional steady heat conduction
through a slab having a constant thermal conductivity and no energy generation, the temperature distribution $T(x)$ is a linear function of x.

Knowing the temperature $T(x)$, the heat flux through the slab is determined from the Fourier law. The differentiation of Eq. (4-6) with respect to x
gives

$$\frac{dT(x)}{dx} = \frac{T_2 - T_1}{L}$$

and from the Fourier law we have

$$q = -k \frac{dT(x)}{dx}$$

Then the heat flux becomes

$$q = k \frac{T_1 - T_2}{L} \qquad \text{W/m}^2 \qquad (4\text{-}7)$$

When $T_1 > T_2$, the right-hand side is positive, and hence the heat flow is in the positive x direction.

The total heat flow rate Q through an area A of the slab normal to the direction of the heat flow becomes

$$Q = Aq = \frac{T_1 - T_2}{L/Ak} \equiv \frac{\Delta T}{R_{\text{slab}}} \qquad (4\text{-}8a)$$

where

$$\boxed{R_{\text{slab}} = \frac{L}{Ak}} \qquad (4\text{-}8b)$$

and

$$\Delta T = T_1 - T_2$$

The quantity R_{slab} is called the *thermal reisitance* for heat flow through the slab of thickness L, area A, thermal conductivity k and subjected to prescribed temperature boundary conditions at both surfaces.

The thermal resistance concept developed above can be generalized to include situations involving convection boundary conditions as illustrated below.

PLANE WALL, PRESCRIBED TEMPERATURE AT $x = 0$, CONVECTION AT $x = L$. We consider a plane wall of thickness L, with the boundary surface at $x = 0$ kept at constant temperature T_1 and that at $x = L$ subjected to convection with a heat transfer coefficient $h_{\infty 2}$ into an ambient at temperature $T_{\infty 2}$, as illustrated in Fig. 4-3.

FIGURE 4-3
Temperature distribution and heat flow for slab with a prescribed temperature at one surface and convection at the other.

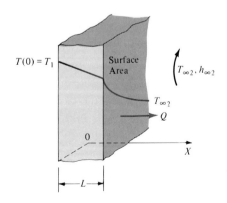

   An examination of the heat flow path shown in this figure reveals that the heat conducted through the slab Q must be equal to the heat convected from the surface at $x = L$ into the ambient at temperature $T_{\infty 2}$. Therefore, an energy balance equation can be stated as

$$Q = \begin{pmatrix} \text{conduction through} \\ \text{the solid from the} \\ \text{surface at } x = 0 \\ \text{to } x = L \end{pmatrix} = \begin{pmatrix} \text{convection from the} \\ \text{surface at } x = L \text{ into} \\ \text{the fluid at } T_{\infty 2} \end{pmatrix} \qquad (4\text{-}9)$$

We utilize the results given by Eqs. (4-8) and (1-5), respectively, to determine the heat flows by conduction and convection. Then Eq. (4-9) becomes

$$Q = Ak \frac{T_1 - T(L)}{L} = Ah_{\infty 2}[T(L) - T_{\infty 2}] \qquad (4\text{-}10)$$

where $T(L)$ represents the temperature at the boundary surface $x = L$ of the slab. This expression can be rearranged in the form of thermal resistances as

$$Q = \frac{T_1 - T(L)}{L/Ak} = \frac{T(L) - T_{\infty 2}}{1/Ah_{\infty 2}} \qquad (4\text{-}11)$$

By adding the numerators and denominators, this result can be rewritten as

$$Q = \frac{T_1 - T_{\infty 2}}{L/Ak + 1/Ah_{\infty 2}} \qquad (4\text{-}12)$$

or

$$Q = \frac{T_1 - T_{\infty 2}}{R_{\text{slab}} + R_{\infty 2}} \qquad (4\text{-}13)$$

where we defined

$$R_{\text{slab}} = \frac{L}{Ak} \qquad R_{\infty 2} = \frac{1}{Ah_{\infty 2}} \qquad (4\text{-}14)$$

   The quantity R_{slab}, as discussed previously, is the thermal resistance for the slab itself, and $R_{\infty 2}$ is the *thermal resistance for convective heat flow* from the boundary surface at $x = L$ to the ambient at $T_{\infty 2}$.

PLANE WALL, CONVECTION AT BOTH BOUNDARIES. We consider a slab of thickness L subjected to convection at both boundaries, as illustrated in Fig. 4-4. That is, the boundary surfaces at $x = 0$ and $x = L$ exchange heat by convection with heat transfer coefficients $h_{\infty 1}$ and $h_{\infty 2}$ with fluids at temperatures $T_{\infty 1}$ and $T_{\infty 2}$, respectively. Since there are no energy sources or sinks anywhere in the medium, the heat flow rate Q remains constant throughout the

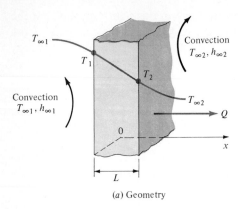

$T_{\infty 1}$

T_1

T_2

Convection
$T_{\infty 2}, h_{\infty 2}$

Convection
$T_{\infty 1}, h_{\infty 1}$

$T_{\infty 2}$

Q

0

x

L

(a) Geometry

$Q \longrightarrow$ $T_{\infty 1}$ T_1 T_2 T_3 $\longrightarrow Q$

$$R_{\infty 1} \equiv \frac{1}{Ah_{\infty 1}}, R_{\text{slab}} \equiv \frac{L}{Ak}, R_{\infty 2} \equiv \frac{1}{Ah_{\infty 2}}$$

(b) Thermal resistance

FIGURE 4-4

Temperature
distribution and
equivalent thermal
resistance concept for
heat flow through a
slab with convection at
both boundaries.

path of heat flow; hence we write

$$Q = \begin{pmatrix} \text{convection} \\ \text{from the fluid} \\ \text{at } T_{\infty 1} \text{ to the} \\ x = 0 \end{pmatrix} = \begin{pmatrix} \text{conduction} \\ \text{through the} \\ \text{solid from} \\ x = 0 \text{ to } x = L \end{pmatrix} = \begin{pmatrix} \text{convection from} \\ \text{the surface at} \\ x = L \text{ into the} \\ \text{fluid at } T_{\infty 2} \end{pmatrix} \quad (4\text{-}15)$$

Equivalent mathematical expressions are written for each term:

$$Q = Ah_{\infty 1}(T_{\infty 1} - T_1) = Ak\,\frac{T_1 - T_2}{L} = Ah_{\infty 2}(T_2 - T_{\infty 2}) \quad (4\text{-}16)$$

This expression is rearaanged as

$$Q = \frac{T_{\infty 1} - T_1}{1/Ah_{\infty 1}} = \frac{T_1 - T_2}{L/Ak} = \frac{T_2 - T_{\infty 2}}{1/Ah_{\infty 2}} \quad (4\text{-}17)$$

By adding the numerators and the denominators, we obtain

$$Q = \frac{T_{\infty 1} - T_{\infty 2}}{(1/Ah_{\infty 1}) + (L/Ak) + (1/Ah_{\infty 2})} \quad (4\text{-}18)$$

These results can be rearranged in the form of thermal resistances as

$$Q = \frac{T_{\infty 1} - T_{\infty 2}}{R_{\infty 1} + R_{\text{slab}} + R_{\infty 2}} \quad (4\text{-}19)$$

where the various thermal resistances are

$$R_{\infty 1} = \frac{1}{Ah_{\infty 1}} \qquad R_{\text{slab}} = \frac{L}{Ak} \qquad R_{\infty 2} = \frac{1}{Ah_{\infty 2}} \quad (4\text{-}20)$$

Clearly, the quantity R_{slab} is the thermal resistance for the slab itself, and $R_{\infty 1}$ and $R_{\infty 2}$ are the *thermal resistances for convection* at the boundary surfaces $x = 0$ and $x = L$, respectively.

Example 4-1. Determine the steady heat flux through a 0.2-m-thick brick wall $[k = 0.69 \ W/(m \cdot {}^\circ C)]$ with one surface at a temperature of 30°C and the other at −20°C.

Solution. Given: $L = 0.2 \ m$, $T_1 = 30°C$, $T_2 = -20°C$, and $k = 0.69 \ W/(m \cdot {}^\circ C)$. The heat flux is calculated using Eq. (4-7) as

$$q = k \frac{T_1 - T_2}{L}$$

$$= 0.69 \ W/(m \cdot {}^\circ C) \frac{[30 - (-20)]°C}{0.2 \ m}$$

$$= 172.5 \ W/m^2$$

Example 4-2. A large window glass 0.5 cm thick $[k = 0.78 \ W/(m \cdot {}^\circ C)]$ is exposed to warm air at 25°C over its inner surface, and the heat transfer coefficient for the inside air is $15 \ W/(m^2 \cdot {}^\circ C)$. The outside air is at −15°C, and the heat transfer coefficient associated with the outside surface is $50 \ W/(m^2 \cdot {}^\circ C)$. Determine the temperatures of the inner and outer surfaces of the glass.

Solution. Given: $L = 0.005 \ m$, $k = 0.78 \ W/(m \cdot {}^\circ C)$, $T_{\infty 1} = 25°C$, $h_{\infty 1} = 15 \ W/(m^2 \cdot {}^\circ C)$, $T_{\infty 2} = -15°C$, and $h_{\infty 2} = 50 \ W/(m^2 \cdot {}^\circ C)$.
Various thermal resistances are determined for $A = 1 \ m^2$. From Eq. (4-20),

$$R_{\infty 1} = \frac{1}{Ah_{\infty 1}} = \frac{1}{15} = 0.06666 \qquad (m^2 \cdot {}^\circ C)/W$$

$$R_{\infty 2} = \frac{1}{Ah_{\infty 2}} = \frac{1}{50} = 0.02000 \qquad (m^2 \cdot {}^\circ C)/W$$

$$R_{slab} = \frac{L}{Ak} = \frac{0.005}{0.78} = 0.00641 \qquad (m^2 \cdot {}^\circ C)/W$$

Then from Eq. (4-19) we have

$$Q = \frac{T_{\infty 1} - T_{\infty 2}}{R_{\infty 1} + R_{slab} + R_{\infty 2}} = \frac{25 + 15}{0.09307} = 429.78 \ W/m^2$$

Knowing Q, Eq. (4-17) can be used to calculate the surface temperatures T_1 and T_2.

$$T_1 = T_{\infty 1} - QR_{\infty 1} = 25 - (429.78 \times 0.0666) = -3.65°C$$

$$T_2 = T_1 - QR_{slab} = -3.65 - (429.78 \times 0.00641) = -6.40°C$$

Long Hollow Cylinder

We consider a long hollow cylinder of inner radius $r = a$ and outer radius $r = b$. The inner and outer surfaces are kept at uniform temperatures T_1 and T_2, respectively, as illustrated in Fig. 4-5. There is no energy generation within the solid, and thermal conductivity k is constant.

To develop an expression for the thermal resistance R_{cyl} of the cylinder wall to heat flow, we solve the heat transfer problem, determine the heat flow rate Q, then recast the result for the heat flow in a form similar to Ohm's law, as described below.

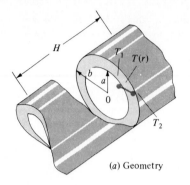

(a) Geometry

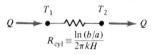

(b) Thermal resistance

FIGURE 4-5
Thermal resistance concept for heat flow through a hollow cylinder.

The mathematical formulation of this heat conduction problem is given by

$$\frac{d}{dr}\left[r\,\frac{dT(r)}{dr}\right] = 0 \qquad \text{in } a < r < b \tag{4-21a}$$

$$T(r) = T_1 \qquad \text{at } r = a \tag{4-21b}$$

$$T(r) = T_2 \qquad \text{at } r = b \tag{4-21c}$$

Integration of Eq. (4-21a) twice gives

$$T(r) = C_1 \ln r + C_2 \tag{4-22}$$

and the application of the boundary conditions (4-21b) and (4-21c), respectively, results

$$T_1 = C_1 \ln a + C_2 \tag{4-23a}$$

$$T_2 = C_1 \ln b + C_2 \tag{4-23b}$$

The constants C_1 and C_2 are obtained from the simultaneous solution of those two equations:

$$C_1 = \frac{T_2 - T_1}{\ln(b/a)} \tag{4-24a}$$

$$C_2 = T_1 - (T_2 - T_1)\frac{\ln a}{\ln(b/a)} \tag{4-24b}$$

Introducing these coefficients into the equation for $T(r)$, the temperature distribution in the wall is expressed as

$$\frac{T(r) - T_1}{T_2 - T_1} = \frac{\ln(r/a)}{\ln(b/a)} \tag{4-25}$$

The heat flux $q(r)$ and the heat flow rate Q over a legnth H of the cylinder are determined from their definition as

$$q(r) = -k \frac{dT(r)}{dr}$$

$$= -\frac{k(T_2 - T_1)}{r \ln (b/a)} \qquad \text{W/m}^2 \qquad (4\text{-}26)$$

and

$$Q = A(r)q(r)$$

$$= (2\pi r H)\left[-\frac{k(T_2 - T_1)}{r \ln (b/a)} \right]$$

$$= \frac{2\pi k H}{\ln (b/a)} (T_1 - T_2) \qquad \text{W} \qquad (4\text{-}27)$$

This expression for Q is now rearranged in the form

$$\boxed{Q = \frac{T_1 - T_2}{R_{\text{cyl}}}} \qquad (4\text{-}28)$$

where

$$\boxed{R_{\text{cyl}} = \frac{\ln (b/a)}{2\pi k H}} \qquad (4\text{-}29)$$

Here R_{cyl} is called the *thermal resistance to heat flow through the hollow cylinder wall* across a temperature potential $T_1 - T_2$, as illustrated in Fig. 4-5b.

The thermal resistance concept developed above for a hollow cylinder subjected to constant surface temperatures can be generalized to situations involving convection at the boundary surfaces.

HOLLOW CYLINDER, CONVECTION AT BOTH BOUNDARIES. We consider a hollow cylinder subjected to convection at both boundary surfaces, as illustrated in Fig. 4-6. That is, a fluid at temperature $T_{\infty 1}$ with a heat transfer coefficient $h_{\infty 1}$ flows over the inner surface at $r = a$, and another fluid at temperature $T_{\infty 2}$ with a heat transfer coefficient $h_{\infty 2}$ flows over the outer surface at $r = b$ of the cylinder. By noting that the total radial heat transfer rate Q through the cylinder wall remains constant, we write the following energy conservation equation:

$$Q = \begin{pmatrix} \text{convection from} \\ \text{the fluid at } T_{\infty 1} \\ \text{to the inner} \\ \text{surface at } r = a \end{pmatrix} = \begin{pmatrix} \text{conduction} \\ \text{through the} \\ \text{cylinder wall} \\ \text{from } r = a \\ \text{to } r = b \end{pmatrix} = \begin{pmatrix} \text{convection from} \\ \text{the outer wall} \\ \text{at } r = b \text{ to the} \\ \text{fluid at } T_{\infty 2} \end{pmatrix} \qquad (4\text{-}30a)$$

We utilize the results given by Eqs. (4-28) and (1-5), respectively, to determine

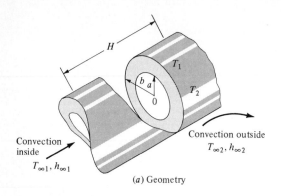

FIGURE 4-6
Thermal resistance concept for radial heat flow through a hollow cylinder with convection at the boundaries.

(a) Geometry

(b) Thermal resistance

the heat flows by conduction and convection. Then Eq. (4-30a) becomes

$$Q = A_a h_{\infty 1}(T_{\infty 1} - T_1) = \frac{2\pi kH}{\ln(b/a)}(T_1 - T_2) = A_b h_{\infty 2}(T_2 - T_{\infty 2}) \qquad (4\text{-}30b)$$

This expression is rearranged as

$$Q = \frac{T_{\infty 1} - T_1}{1/A_a h_{\infty 1}} = \frac{T_1 - T_2}{\ln(b/a)/2\pi kH} = \frac{T_2 - T_{\infty 2}}{1/A_b h_{\infty 2}} \qquad (4\text{-}30c)$$

By adding the numerators and the denominators, we obtain

$$Q = \frac{T_{\infty 1} - T_{\infty 2}}{1/A_a h_{\infty 1} + \ln(b/A)/2\pi kH + 1/A_b h_{\infty 2}} \qquad (4\text{-}31)$$

This result can be rewritten as

$$Q = \frac{T_{\infty 1} - T_{\infty 2}}{R_{\infty 1} + R_{cyl} + R_{\infty 2}} \qquad (4\text{-}32)$$

where various thermal resistances are defined as

$$R_{\infty 1} = \frac{1}{A_a h_{\infty 1}} = \frac{1}{2\pi a H h_{\infty 1}} \qquad (4\text{-}33a)$$

$$R_{cyl} = \frac{\ln(b/a)}{2\pi kH} \qquad (4\text{-}33b)$$

$$R_{\infty 2} = \frac{1}{A_b h_{\infty 2}} = \frac{1}{2\pi b H h_{\infty 2}} \qquad (4\text{-}33c)$$

Example 4-3. A cylindrical insulation for a steam pipe has an inside radius of 5 cm, an outside radius of 10 cm, and a thermal conductivity of 0.5 W/(m · °C). The inside surface of the insulation is at a temperature of 200°C, and the outside surface is at 20°C. Determine the heat loss per meter length of the insulation.

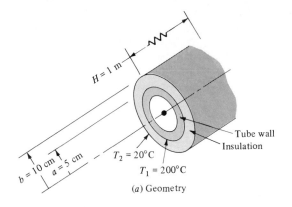

$T_1 = 200°C \quad T_2 = 20°C$

$Q \longrightarrow \bullet\!\!-\!\!\text{w}\!\!-\!\!\bullet \longrightarrow Q$

$$R_{cyl} \equiv \frac{\ln (b/a)}{2\pi kH}$$

(b) Thermal resistance

(a) Geometry

FIGURE EXAMPLE 4-3
Thermal insulation for
the steam pipe and the
equivalent thermal
resistance network.

Solution. The problem is illustrated in the accompanying figure. Given: $a = 5$ cm, $b = 10$ cm, $k = 0.5$ W/(m·°C), $T_1 = 200°C$, and $T_2 = 20°C$. The heat loss per meter length, $H = 1$ m, is determined using Eq. (4-28):

$$Q = \frac{T_1 - T_2}{R_{cyl}} = \frac{T_1 - T_2}{\ln (b/a)/2\pi kH}$$

$$= \frac{200 - 20}{\ln (10/5)/(2\pi \times 0.5 \times 1)}$$

$$= 815.8 \text{ W/m length}$$

Example 4-4. A long hollow cylinder has an inner radius of 10 cm, an outer radius of 20 cm, and a thermal conductivity $k = 50$ W/(m·°C). The inner surface is heated uniformly at a constant rate $q_a = 1.16 \times 10^5$ W/m², and the outer surface is maintained at zero temperature. Calculate the temperature of the inner surface.

Solution. Given that $q_a = 1.16 \times 10^5$ W/m² at the inner surface, the heat flow rate Q over a length H is determined by

$$Q = A(r)q(r) = A_a q_a$$

$$= 2\pi aHq_a$$

and from Eq. (4-27) we have

$$Q = \frac{2\pi kH}{\ln (b/a)} (T_1 - T_2)$$

By combining these two results, we obtain

$$2\pi aHq_a = \frac{2\pi kH}{\ln (b/a)} (T_1 - T_2)$$

which is now solved for the inner surface temperature T_1:

$$T_1 = \frac{aq_a}{k} \ln (b/a) + T_2$$

$$= \frac{(0.1)(1.16 \times 10^5)}{50} \ln \left(\frac{20}{10}\right) + 0$$

$$= 160.8°C$$

Example 4-5. A steam pipe of outside diameter 10 cm, maintained at 130°C, is covered with a 3-cm-thick asbestos insulation $[k = 0.1 \, \text{W}/(\text{m} \cdot {}^\circ\text{C})]$. The ambient air temperature is 30°C, and the heat transfer coefficient for convection at the outer surface of the asbestos insulation is $25 \, \text{W}/(\text{m}^2 \cdot {}^\circ\text{C})$. Using the thermal resistance concept, calculate the rate of heat loss from the pipe per meter length of the pipe.

Solution. Given: $a = 5 \, \text{cm}$, $b = 8 \, \text{cm}$, $T_1 = 130°C$, $k = 0.1 \, \text{W}/(\text{m} \cdot {}^\circ\text{C})$, $T_{\infty 2} = 30°C$, and $h_{\infty 2} = 25 \, \text{W}/(\text{m}^2 \cdot {}^\circ\text{C})$. The heat flow rate Q through the pipe is calculated by using Eq. (4-32) and noting that the thermal resistance to heat flow at the inside surface vanishes (i.e., $R_{\infty 1} = 0$). Then

$$Q = \frac{T_1 - T_{\infty 2}}{R_{\text{cyl}} + R_{\infty 2}} = \frac{T_1 - T_{\infty 2}}{\ln (b/a)/2\pi kH + 1/2\pi b H h_{\infty 2}}$$

$$= \frac{130 - 30}{\ln (8/5)/[(2\pi)(0.1)] + [(2\pi)(0.08)(1)(25)]}$$

$$= 120.8 \, \text{W/m length}$$

Hollow Sphere

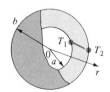

(a) Geometry

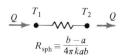

$$R_{\text{sph}} \equiv \frac{b - a}{4\pi kab}$$

(b) Thermal resistance

FIGURE 4-7
Thermal resistance concept for radial heat flow through a hollow sphere.

We now consider radial heat flow through a hollow sphere of inner radius $r = a$ and outer radius $r = b$, as illustrated in Fig. 4-7. The inner and outer surfaces are maintained at uniform temperatures T_1 and T_2, respectively. There is no energy generation in the solid, and the thermal conductivity k is constant. To develop the thermal resistance R_{sph} for the hollow sphere, we first solve this heat conduction problem and determine the expression for radial heat flow rate Q through the hollow sphere. Then this result is recast in a form similar to Ohm's law in order to determine the equivalent thermal resistance for the hollow sphere, as described below.

The mathematical formulation of this heat conduction problem is given by

$$\frac{d}{dr}\left[r^2 \frac{dT(r)}{dr}\right] = 0 \quad \text{in } a < r < b \tag{4-34a}$$

$$T(r) = T_1 \quad \text{at } r = a \tag{4-34b}$$

$$T(r) = T_2 \quad \text{at } r = b \tag{4-34c}$$

Equation (4-34a) is integrated twice:

$$T(r) = -\frac{C_1}{r} + C_2 \tag{4-35}$$

and the boundary conditions (4-34b) and (4-34c) are applied:

$$T_1 = -\frac{C_1}{a} + C_2 \tag{4-36a}$$

$$T_2 = -\frac{C_1}{b} + C_2 \tag{4-36b}$$

The constants C_1 and C_2 are determined from the simultaneous solution of these two equations:

$$C_1 = -\frac{ab}{b-a}(T_1 - T_2) \tag{4-37a}$$

$$C_2 = \frac{bT_2 - aT_1}{b-a} \tag{4-37b}$$

Then the temperature distribution $T(r)$ becomes

$$T(r) = \frac{a}{r}\left(\frac{b-r}{b-a}\right)T_1 + \frac{b}{r}\left(\frac{r-a}{b-a}\right)T_2 \tag{4-38}$$

The heat flux $q(r)$ is determined from

$$q(r) = -k\frac{dT(r)}{dr}$$

$$= k\frac{ab}{b-a}(T_1 - T_2) \qquad \text{W/m}^2 \tag{4-39}$$

and the total heat flow rate Q through the sphere becomes

$$Q = A(r)q(r)$$

$$= (4\pi r^2)\left[k\frac{ab}{b-a}(T_1 - T_2)\right]$$

$$= 4\pi k\frac{ab}{b-a}(T_1 - T_2) \qquad \text{W} \tag{4-40}$$

This expression for Q is now rearranged in the form

$$Q = \frac{T_1 - T_2}{R_{\text{sph}}} \tag{4-41}$$

where we defined

$$R_{\text{sph}} = \frac{b-a}{4\pi kab} \tag{4-42}$$

and R_{sph} is called *the thermal resistance to heat flow for a hollow sphere* across a temperature potential $T_1 - T_2$, as illustrated in Fig. 4-7b.

The thermal resistance concept developed above can be generalized to situations involving convection at the boundary surface.

HOLLOW SPHERE, CONVECTION AT THE OUTER BOUNDARY SURFACE. We consider a hollow sphere maintained at a uniform temperature T_1 at the inner surface $r = a$ and subjected to convection with a heat transfer coefficient $h_{\infty 2}$ into ambient air of temperature $T_{\infty 2}$. The physical situation is illustrated in Fig. 4-8.

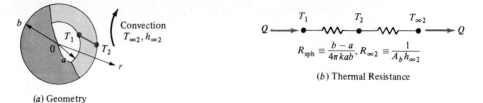

(a) Geometry

(b) Thermal Resistance

FIGURE 4-8
Thermal resistance concept for a hollow sphere with a prescribed temperature at the inner surface and convection at the outer surface.

Noting that the total radial heat flow rate through the sphere remains constant, we write

$$Q = \begin{pmatrix} \text{conduction} \\ \text{through the} \\ \text{wall from} \\ r = a \text{ to } r = b \end{pmatrix} = \begin{pmatrix} \text{convection from} \\ \text{the outer wall at} \\ r = b \text{ to the fluid} \\ \text{at } T_{\infty 2} \end{pmatrix} \tag{4-43}$$

We utilize the results given by Eqs. (4-40) and (1-5), respectively, to determine the heat flows by conduction and convection. Then Eq. (4-43) becomes

$$Q = 4\pi k \frac{ab}{b-a}(T_1 - T_2) = 4\pi b^2 h_{\infty 2}(T_2 - T_{\infty 2}) \tag{4-44}$$

This expression is rearranged as

$$Q = \frac{T_1 - T_2}{(b-a)/4\pi kab} = \frac{T_2 - T_{\infty 2}}{1/4\pi b^2 h_{\infty 2}} \tag{4-45}$$

By adding the numerators and denominators, we obtain

$$Q = \frac{T_1 - T_{\infty 2}}{(b-a)/4\pi kab + 1/4\pi b^2 h_{\infty 2}} \tag{4-46}$$

This result can be rewritten as

$$\boxed{Q = \frac{T_1 - T_{\infty 2}}{R_{\text{sph}} + R_{\infty 2}}} \tag{4-47}$$

where various thermal resistances are defined as

$$\boxed{R_{\text{sph}} = \frac{b-a}{4\pi kab}, \qquad R_{\infty 2} = \frac{1}{4\pi b^2 h_{\infty 2}}} \tag{4-48}$$

Example 4-6. Determine the heat flow rate Q through a copper spherical shell $[k = 386 \text{ W/(m} \cdot {}^\circ\text{C)}]$ of inner radius 2 cm and outer radius 6 cm, if the inner surface is kept at 200°C and the outer surface at 100°C.

Solution. Given: $a = 2$ cm, $b = 6$ cm, $k = 386$ W/(m $\cdot$ °C), $T_1 = 200$°C, and $T_2 = 100$°C. The heat flow rate is calculated by Eq. (4-41):

$$Q = \frac{T_1 - T_2}{R_{sph}} = \frac{T_1 - T_2}{(b - a)/4\pi kab}$$

$$= \frac{200 - 100}{(0.06 - 0.02)/[(4\pi)(386)(0.06)(0.02)]}$$

$$= 14.55 \text{ kW}$$

Example 4-7. Consider an aluminum hollow sphere $[k = 200$ W/(m $\cdot$ °C)] of inside radius 2 cm and outside radius 6 cm. The inside surface is kept at a uniform temperature of 100°C, and the outside surface dissipates heat by convection with a heat transfer coefficient of 80 W/m² into ambient air at 20°C. Determine the outside surface temperature of the sphere and the rate of heat flow from the sphere by using thermal resistance concept.

Solution. Given: $a = 2$ cm, $b = 6$ cm, $k = 200$ W/(m $\cdot$ °C), $T_1 = 100$°C, $T_{\infty 2} = 80$°C, and $h_{\infty 2} = 80$ W/m². The heat flow rate Q is calculated using Eq. (4-46):

$$Q = \frac{T_1 - T_{\infty 2}}{(b - a)/4\pi kab + 1/4\pi b^2 h_{\infty 2}}$$

$$= \frac{100 - 20}{(0.06 - 0.02)/[(4\pi)(200)(0.02)(0.06)] + 1/[(4\pi)(0.06)^2(80)]}$$

$$= 276.3 \text{ W}$$

Knowing the heat flow rate Q, the outside surface temperature T_2 is calculated from Eqs. (4-41) and (4-42) as

$$Q = \frac{T_1 - T_2}{R_{sph}} = \frac{T_1 - T_2}{(b - a)/4\pi kab}$$

$$276.3 = \frac{100 - T_2}{(0.06 - 0.02)/[(4\pi)(200)(0.02)(0.06)]}$$

Solving for T_2, we obtain

$$T_2 = 96.3°C$$

4-2 MULTILAYERS

In the previous section we considered one-dimensional heat flow through a single layer. However, the problems encountered in most engineering applications involve heat flow through a medium consisting of several parallel layers—for example, the wall of a building consisting of several layers, or a pipe with several layers of insulation. In such problems, we assume no energy generation within the medium and the thermal conductivity of the layers constant but different for each layer.

Then the thermal resistance concept developed in the previous section can readily be applied to determine the heat flow rate through such a multilayer medium. We study this problem for the cases of multilayer plane walls, hollow cylinders, and hollow spheres.

Multilayer Plane Wall

Consider a multilayer plane wall consisting of three parallel layers in perfect thermal contact, as illustrated in Fig. 4-9. Let Q be the heat flow rate through a surface area A of this composite medium. Applying the thermal resistance concept to each of the individual layers, including resistance to convection at the outer surfaces, we write

$$Q = \frac{T_{\infty 1} - T_0}{R_{\infty 1}} = \frac{T_0 - T_1}{R_1} = \frac{T_1 - T_2}{R_2} = \frac{T_2 - T_3}{R_3} = \frac{T_3 - T_{\infty 2}}{R_{\infty 2}} \quad (4\text{-}49a)$$

where various thermal resistances are defined as

$$R_{\infty 1} = \frac{1}{Ah_{\infty 1}} \quad R_1 = \frac{L_1}{Ak_1} \quad R_2 = \frac{L_2}{Ak_2} \quad R_3 = \frac{L_3}{Ak_3} \quad R_{\infty 2} = \frac{1}{Ah_{\infty 2}} \quad (4\text{-}49b)$$

Summing the numerators and the denominators of the individual ratios in Eq. (4-49a), the result is written as

$$Q = \frac{T_{\infty 1} - T_{\infty 2}}{R_T} \quad \text{W} \quad (4\text{-}50a)$$

where the *total thermal resistance* R_T is defined as

$$R_T = R_{\infty 1} + R_1 + R_2 + R_3 + R_{\infty 2} \quad (4\text{-}50b)$$

This result is analogous to the result for a single-layer slab, except that the quantity R_T represents the sum of all the individual thermal resistances in the path of the heat flow through an area A from the temperature $T_{\infty 1}$ to the temperature $T_{\infty 2}$.

Example 4-8. An industrial furnace is made of fireclay brick of thickness 0.2 m and thermal conductivity 1.0 W/(m·°C). The outside surface is to be insulated with an insulation material of thermal conductivity 0.05 W/(m·°C). Determine the thickness of the insulation layer needed to limit the heat loss from the furnace wall to 900 W/m² when the inside surface of the wall is at 930°C and the outside surface is at 30°C.

FIGURE 4-9
Thermal resistance concept for a multilayer plane wall.

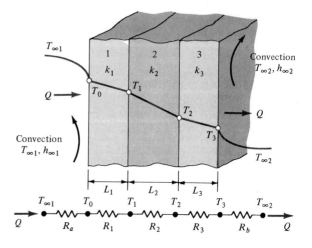

Solution. The problem is illustrated in the accompanying figure. The equivalent resistance for the two layers is determined from

$$R_T = R_1 + R_2 = \frac{L_1}{Ak_1} + \frac{L_2}{Ak_2}$$

Then the heat transfer rate Q through an area A is given by

$$Q = \frac{T_1 - T_3}{R_T} = \frac{T_1 - T_3}{L_1/Ak_1 + L_2/Ak_2} = \frac{A(T_1 - T_3)}{L_1/k_1 + L_2/k_2}$$

and the heat flux q becomes

$$q = \frac{Q}{A} = \frac{T_1 - T_3}{L_1/k_1 + L_2/k_2}$$

Introducing the numerical values, we find

$$900 = \frac{930 - 30}{0.2/1 + L_2/0.05}$$

$$L_2 = 0.04 \text{ m}$$

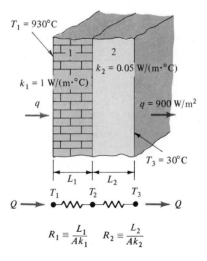

$$R_1 \equiv \frac{L_1}{Ak_1} \qquad R_2 \equiv \frac{L_2}{Ak_2}$$

FIGURE EXAMPLE 4-8
Equivalent thermal resistance network.

Example 4-9. A thermopane window consists of two 5-mm-thick sheets of glass separated by a stagnant air space of thickness 10 mm. The thermal conductivity of the glass is $0.78 \text{ W/(m} \cdot \text{°C)}$, and that of air is $0.025 \text{ W/(m} \cdot \text{°C)}$. The convection heat transfer coefficients for the inside and outside air are $10 \text{ W/(m}^2 \cdot \text{°C)}$ and $50 \text{ W/(m}^2 \cdot \text{°C)}$, respectively.

(*a*) Determine the rate of heat loss per square meter of the glass surface for a temperature difference of 60°C between the inside and outside air.
(*b*) Compare the result with the heat loss if the window had only a single sheet of glass of thickness 5 mm instead of the thermopane.
(*c*) Compare the result with the heat loss if the window had no stagnant air (i.e., a double sheet of glass of thickness 10 mm).

Solution. The physical situation is similar to that illustrated in Fig. 4-9. Various quantities are specified as $L_1 = L_3 = 0.005$ m, $L_2 = 0.01$ m, $k_1 = k_3 = 0.78$ W/(m $\cdot$ °C), $k_2 = 0.025$ W/(m $\cdot$ °C), $h_{\infty 1} = 10$ W/(m^2 $\cdot$ °C), $h_{\infty 2} = 50$ W/(m^2 $\cdot$ °C), and $\Delta T = T_{\infty 1} - T_{\infty 2} = 0$°C.

(*a*) The total thermal resistance, according to Eq. (4-50*b*), is

$$R_T = R_{\infty 1} + R_1 + R_2 + R_3 + R_{\infty 2}$$

$$= \frac{1}{Ah_{\infty 1}} + \frac{L_1}{Ak_1} + \frac{L_2}{Ak_2} + \frac{L_3}{Ak_3} + \frac{1}{Ah_{\infty 2}}$$

Then the rate of heat loss per square meter of the glass surface (i.e., the heat flux) becomes

$$q = \frac{Q}{A} = \frac{T_{\infty 1} - T_{\infty 2}}{AR_T} = \frac{\Delta T}{1/h_{\infty 1} + L_1/k_1 + L_2/k_2 + L_3/k_3 + 1/h_{\infty 2}}$$

$$= \frac{60}{1/10 + 0.005/0.78 + 0.01/0.025 + 0.005/0.78 + 1/50}$$

$$= 112.6 \text{ W/m}^2$$

(*b*) If the window has a single sheet of glass with no stagnant air space, the total thermal resistance is

$$R_T = R_{\infty 1} + R_1 + R_{\infty 2}$$

$$= \frac{1}{Ah_{\infty 1}} + \frac{L_1}{Ak_1} + \frac{1}{Ah_{\infty 2}}$$

Then the heat flux becomes

$$q = \frac{Q}{A} = \frac{T_{\infty 1} - T_{\infty 2}}{AR_T} = \frac{\Delta T}{1/h_{\infty 1} + L_1/k_1 + 1/h_{\infty 2}}$$

$$= \frac{60}{1/10 + 0.005/0.78 + 1/50}$$

$$= 474.65 \text{ W/m}^2$$

The heat loss is about four times larger than that of the previous case.

(*c*) If the sheets of glass are touching each other (i.e., double glass) with no stagnant air layer between them, the total resistance is given by

$$R_T = R_{\infty 1} + R_1 + R_3 + R_{\infty 2}$$

$$= \frac{1}{Ah_{\infty 1}} + \frac{L_1}{Ak_1} + \frac{L_3}{Ak_3} + \frac{1}{Ah_{\infty 2}}$$

Then the heat flux becomes

$$q = \frac{\Delta T}{1/h_{\infty 1} + L_1/k_1 + L_3/k_3 + 1/h_{\infty 2}} = \frac{\Delta T}{1/h_{\infty 1} + 2L_1/k_1 + 1/h_{\infty 2}}$$

$$= 451.74 \text{ W/m}^2$$

We note that doubling the glass thickness reduced the heat loss very little; it is the stagnant air space between the two sheets of glass that causes a significant reduction in the heat loss or gain.

Multilayer Hollow Cylinder

To illustrate the basic concept in the analysis of heat flow through a multilayer hollow cylinder, we consider a two-layer cylinder, as shown in Fig. 4-10. A hot fluid at temperature $T_{\infty 1}$ with a heat transfer coefficient $h_{\infty 1}$ flows inside the tube, while a cold fluid at temperature $T_{\infty 2}$ with a heat transfer coefficient $h_{\infty 2}$ flows outside the tube. Let Q be the total heat flow through the cylinder over a length H. Applying the thermal resistance concept to each individual layer, we write

$$Q = \frac{T_{\infty 1} - T_0}{R_{\infty 1}} = \frac{T_0 - T_1}{R_1} = \frac{T_1 - T_2}{R_2} = \frac{T_2 - T_{\infty 2}}{R_{\infty 2}} \qquad (4\text{-}51)$$

where $R_{\infty 1}$ and $R_{\infty 2}$ are the thermal resistances to convection at the inner and outer surfaces, respectively. R_1 and R_2 are the thermal resistances for the inner and outer cylinders, respectively; they can be determined using relation (4-29). Summing the numerators and the denominators of the individual equalities in Eq. (4-51), we obtain

$$Q = \frac{T_{\infty 1} - T_{\infty 2}}{R_{\infty 1} + R_1 + R_2 + R_{\infty 2}} = \frac{T_{\infty 1} - T_{\infty 2}}{R_T} \qquad \text{W} \qquad (4\text{-}52a)$$

where the various thermal resistances are given by

$$R_{\infty 1} = \frac{1}{2\pi r_0 H h_{\infty 1}} \qquad R_1 = \frac{1}{2\pi H k_1} \ln\left(\frac{r_1}{r_0}\right)$$

$$R_2 = \frac{1}{2\pi H k_2} \ln\left(\frac{r_2}{r_1}\right) \qquad R_{\infty 2} = \frac{1}{2\pi r_2 H h_{\infty 2}} \qquad (4\text{-}52b)$$

FIGURE 4-10
Thermal resistance concept for a two-layer hollow cylinder.

and $R_T = R_{\infty 1} + R_1 + R_2 + R_{\infty 2}$ is the *total thermal resistance* to heat flow between the temperatures $T_{\infty 1}$ and $T_{\infty 2}$.

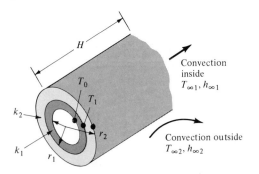

(*a*) Geometry

$$T_{\infty 1} \qquad T_0 \qquad T_1 \qquad T_2 \qquad T_{\infty 2}$$

$$Q \longrightarrow \bullet \!-\!\!\text{\Large w}\!\!-\! \bullet \!-\!\!\text{\Large w}\!\!-\! \bullet \!-\!\!\text{\Large w}\!\!-\! \bullet \!-\!\!\text{\Large w}\!\!-\! \bullet \longrightarrow Q$$

$$R_{\infty 1} \equiv \frac{1}{2\pi r_0 H h_{\infty 1}}, R_1 \equiv \frac{\ln(r_1/r_0)}{2\pi H k_1}, R_2 \equiv \frac{\ln(r_2/r_1)}{2\pi H k_2}, R_{\infty 2} \equiv \frac{1}{2\pi r_2 H h_{\infty 2}}$$

(*b*) Thermal Resistance

Example 4-10. A steel tube $[k = 15 \text{ W}/(\text{m} \cdot {}^\circ\text{C})]$ of outside diameter 7.6 cm and thickness 1.3 cm is covered with an insulation material $[k = 0.2 \text{ W}/(\text{m} \cdot {}^\circ\text{C})]$ of thickness 2 cm. A hot gas at 320°C with a heat transfer coefficient of 200 W/$(\text{m}^2 \cdot {}^\circ\text{C})$ flows inside the tube. The outer surface of the insulation is exposed to cooler air at 20°C with a heat transfer coefficient of 50 W/$(\text{m}^2 \cdot {}^\circ\text{C})$. Calculate

(a) the heat loss from the tube to the air for a 5-m length of the tube;
(b) the temperature drops due to the thermal resistances of the hot gas flow, the steel tube, the insulation layer, and the outside air.

Solution. (a) The radial heat flow through the tube is given by Eq. (4-52) as

$$Q = \frac{T_{\infty 1} - T_{\infty 2}}{R_{\infty 1} + R_1 + R_2 + R_{\infty 2}} \quad \text{W}$$

where R_1 and R_2 are the thermal resistances across the steel tube and the insulation, respectively, and $R_{\infty 1}$ and $R_{\infty 2}$ are the thermal resistances to flow inside and outside the tube.

Given: $r_0 = 0.025 \text{ m}$, $r_1 = 0.038 \text{ m}$, $r_2 = 0.058 \text{ m}$, $H = 5 \text{ m}$, $k_1 = 15 \text{ W}/(\text{m} \cdot {}^\circ\text{C})$, $k_2 = 0.2 \text{ W}/(\text{m} \cdot {}^\circ\text{C})$, $h_{\infty 1} = 200 \text{ W}/(\text{m}^2 \cdot {}^\circ\text{C})$, and $h_{\infty 2} = 50 \text{ W}/(\text{m}^2 \cdot {}^\circ\text{C})$. The various thermal resistances are calculated as

$$R_{\infty 1} = \frac{1}{2\pi r_0 H h_{\infty 1}} = \frac{1}{(2\pi)(0.025)(5)(200)} = 6.37 \times 10^{-3} \, {}^\circ\text{C}/\text{W}$$

$$R_1 = \frac{1}{2\pi H k_1} \ln\left(\frac{r_1}{r_0}\right) = \frac{1}{(2\pi)(5)(15)} \ln\left(\frac{3.8}{2.5}\right) = 0.89 \times 10^{-3} \, {}^\circ\text{C}/\text{W}$$

$$R_2 = \frac{1}{2\pi H k_2} \ln\left(\frac{r_2}{r_1}\right) = \frac{1}{(2\pi)(5)(0.2)} \ln\left(\frac{5.8}{3.8}\right) = 67.3 \times 10^{-3} \, {}^\circ\text{C}/\text{W}$$

$$R_{\infty 2} = \frac{1}{2\pi r_2 H h_{\infty 2}} = \frac{1}{(2\pi)(0.058)(50)} = 10.98 \times 10^{-3} \, {}^\circ\text{C}/\text{W}$$

Then the total thermal resistance becomes

$$R_T = R_{\infty 1} + R_1 + R_2 + R_{\infty 2} = 85.54 \times 10^{-3} \, {}^\circ\text{C}/\text{W}$$

and the total heat loss from the tube is determined as

$$Q = \frac{320 - 20}{85.54 \times 10^{-3}} = 3.507 \text{ kW}$$

(b) Knowing Q, various temperature drops are calculated according to Eqs. (4-52a) and (4-52b).

$$\Delta T_{\text{hot gas}} = Q \times R_{\infty 1} = 22.74 {}^\circ\text{C}$$

$$\Delta T_{\text{tube wall}} = Q \times R_1 = 3.18 {}^\circ\text{C}$$

$$\Delta T_{\text{insulation}} = Q \times R_2 = 240.27 {}^\circ\text{C}$$

$$\Delta T_{\text{outside air}} = Q \times R_{\infty 2} = 39.20 {}^\circ\text{C}$$

Clearly the smallest temperature drop occurs across the steel tube and the largest across the insulation material.

Example 4-11. A steam pipe with an outside radius of 4 cm is covered with a layer of asbestos insulation $[k = 0.15 \text{ W/(m} \cdot \text{°C)}]$ 1 cm thick, which is in turn covered with fiberglass insulation $[k = 0.05 \text{ W/(m} \cdot \text{°C)}]$ 3 cm thick. The outside surface of the steam pipe is at a temperature of 330°C, and the outside surface of the fiberglass insulation is at 30°C. Determine the heat transfer rate per meter length of the pipe and the interface temperature between the asbestos and the fiberglass insulation.

Solution. The accompanying figure illustrates the physical situation. The heat flow rate per meter length of the pipe, Q/H, can be determined using Eqs. (4-52a) and (4-52b) by noting that $R_{\infty 1} = 0$ and $R_{\infty 2} = 0$. Then we have

$$\frac{Q}{H} = \frac{T_0 - T_2}{H(R_1 + R_2)} = \frac{T_0 - T_2}{\ln(r_1/r_0)/2\pi k_1 + \ln(r_2/r_1)/2\pi k_2}$$

$$= \frac{330 - 30}{\ln(5/4)/[(2\pi)(0.15)] + \ln(8/5)/[(2\pi)(0.05)]}$$

$$= 173.1 \text{ W/m length}$$

The interface temperature T_1 between the asbestos and the fiberglass insulation is determined from the relation

$$Q = \frac{T_0 - T_2}{R_1 + R_2} = \frac{T_0 - T_1}{R_1}$$

or

$$\frac{Q}{H} = \frac{T_0 - T_1}{HR_1} = \frac{T_0 - T_1}{\ln(r_1/r_0)/2\pi k_1}$$

$$173.1 = \frac{330 - T_2}{\ln(5/4)/[(2\pi)(0.15)]}$$

Hence

$$T_2 = 289°C$$

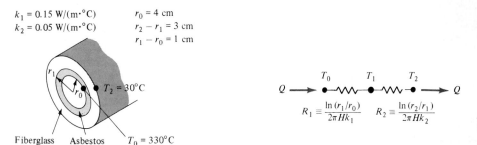

$k_1 = 0.15 \text{ W/(m} \cdot \text{°C)}$
$k_2 = 0.05 \text{ W/(m} \cdot \text{°C)}$

$r_0 = 4 \text{ cm}$
$r_2 - r_1 = 3 \text{ cm}$
$r_1 - r_0 = 1 \text{ cm}$

$T_2 = 30°C$

Fiberglass   Asbestos $T_0 = 330°C$

$$R_1 \equiv \frac{\ln(r_1/r_0)}{2\pi H k_1} \qquad R_2 \equiv \frac{\ln(r_2/r_1)}{2\pi H k_2}$$

FIGURE EXAMPLE 4-11
Illustration of the physical system.

Multilayer Hollow Sphere

To illustrate the basic thermal resistance concept in the analysis of heat flow through multilayered hollow spheres, we consider a two-layer hollow sphere, as shown in Fig. 4-11. The radial heat flow Q through the sphere is determined

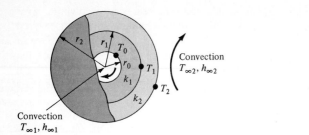

Convection
$T_{\infty 1}, h_{\infty 1}$

FIGURE 4-11
Thermal resistance concept for a multilayer hollow sphere.

by applying the thermal resistance concept to each individual layer:

$$Q = \frac{T_{\infty 1} - T_0}{R_{\infty 1}} = \frac{T_0 - T_1}{R_1} = \frac{T_1 - T_2}{R_2} = \frac{T_2 - T_{\infty 2}}{R_{\infty 2}} \qquad (4\text{-}53)$$

Summing the numerators and denominators of these individual equalities, we obtain

$$Q = \frac{T_{\infty 1} - T_{\infty 2}}{R_{\infty 1} + R_1 + R_2 + R_{\infty 2}} \equiv \frac{T_{\infty 1} - T_{\infty 2}}{R_T} \qquad (4\text{-}54)$$

where the various thermal resistances are given by

$$R_{\infty 1} = \frac{1}{4\pi r_0^2 h_{\infty 1}} \qquad R_1 = \frac{1}{4\pi k_1}\left(\frac{r_1 - r_0}{r_1 r_0}\right)$$

$$ \qquad (4\text{-}55)$$

$$R_2 = \frac{1}{4\pi k_2}\left(\frac{r_2 - r_1}{r_2 r_1}\right) \qquad R_{\infty 2} = \frac{1}{4\pi r_2^2 h_{\infty 2}}$$

and $R_T = R_{\infty 1} + R_1 + R_2 + R_{\infty 2}$ is called the *total thermal resistance* to heat flow between the temperatures $T_{\infty 1}$ and $T_{\infty 2}$.

Overall Heat Transfer Coefficient Concept

In engineering applications the heat flow rate Q through multilayer media is also expressed in the form

$$\boxed{Q = (UA)\,\Delta T} \qquad (4\text{-}56a)$$

where U is called the *overall heat transfer coefficient*. Recalling the definition of the total thermal resistance R_T (i.e., $Q = \Delta T/R_T$) in the path of the heat flow, U is related to R_T by the relation

$$R_T = \frac{1}{UA} \qquad (4\text{-}56b)$$

For multilayer hollow cylinder and multilayer hollow spherical systems, the area A changes in the radial direction. Therefore, when relating the total thermal resistance to the overall heat transfer coefficient, the surface area on which U is based should be specified. For example, Eq. (4-56b) can be written as

$$R_T = \frac{1}{U_a A_a} = \frac{1}{U_b A_b} \tag{4-57}$$

where U_a is the overall heat transfer coefficient based on the inner surface A_a and U_b is the overall heat transfer coefficient based on the outer surface A_b.

Then the total heat flow rate Q, in terms of the overall heat transfer coefficient, is written as

$$\boxed{Q = U_a A_a \, \Delta T = U_b A_b \, \Delta T} \tag{4-58}$$

It does not matter which area the overall heat transfer coefficient U is based on as long as it is specified in the definition.

In the case of heat flow through a composite layer of plane wall, the area does not change in the direction of heat flow, and hence U is the same whether it is based on the outer or inner surface.

For example, the thermal resistance to heat flow through a tube subjected to convection at the inner and outer boundary surfaces, as illustrated in Fig. 4-6, can be obtained from Eq. (4-31) as

$$R = \frac{1}{A_a h_{\infty 1}} + \frac{b-a}{k(A_b - A_a)/\ln(A_b/A_a)} + \frac{1}{A_b h_{\infty 2}} \tag{4-59}$$

since the thermal resistance for the tube given by Eq. (4-33b) is rearranged as

$$R_{cyl} = \frac{\ln(b/a)}{2\pi kH} = \frac{(b-a)\ln[(2\pi bH)/(2\pi aH)]}{(b-a)(2\pi kH)} = \frac{(b-a)\ln(A_b/A_a)}{k(A_b - A_a)}$$

$$= \frac{b-a}{k(A_b - A_a)/\ln(A_b/A_a)}$$

where A_a and A_b are the inner and outer surface areas of the tube, respectively, and $(b-a)$ represents the tube thickness. Then the *overall heat transfer coefficient U_b* based on the outside surface area, becomes

$$U_b = \frac{1}{A_b R_t} = \frac{1}{(A_b/A_a)(1/h_{\infty 1}) + (A_b/A_{\ln,m})[(b-a)/k] + (1/h_{\infty 2})} \tag{4-60a}$$

where

$$A_{\ln,m} = \frac{A_b - A_a}{\ln(A_b/A_a)} \equiv \text{logarithmic mean area} \tag{4-60b}$$

Similarly, the overall heat transfer coefficient U_a based on the *inside surface area* of the tube is defined as

$$U_a = \frac{1}{A_a R} = \frac{1}{1/h_{\infty 1} + (A_a/A_{\ln,m})[(b-a)/k] + (A_a/A_b)(1/h_{\infty 2})} \tag{4-61}$$

Example 4-12. Calculate the overall heat transfer coefficient for Example 4-5, based on the outside surface area of the asbestos insulation.

Solution. Thermal resistance to the heat flow is

$$R = R_{cyl} + R_{\infty 2} = \frac{\ln (b/a)}{2\pi kH} + \frac{1}{2\pi bHh_{\infty 2}}$$

$$= \frac{\ln (A_b/A_a)(b-a)}{k(A_b - A_a)} + \frac{1}{A_b h_{\infty 2}} = 0.83$$

Then the overall heat transfer coefficient based on the outside surface of the asbestos becomes

$$U_b = \frac{1}{A_b R} = \frac{1}{2\pi bHR}$$

$$= \frac{1}{(2\pi)(0.08)(1)(0.83)} = 2.4 \text{ W/(m}^2 \cdot {}^\circ\text{C)}$$

4-3   CONTACT RESISTANCE CONCEPT

Consider two solids, each having a plane flat surface, brought together and one surface pressed against the other. The actual direct contact between the surfaces takes place at only a limited number of spots because the surfaces are not perfectly smooth but possess some micro roughness; as a result, the surfaces are not in perfect thermal contact. The voids between the surfaces are filled with the surrounding fluid, which usually is the air. The heat flow across such an interface takes place by conduction both through the thin fluid layer filling the voids and through the spots in direct metal-to-metal contact. Since the thermal conductivity of the air is less than that of the metal, the thin layer of air filling the voids acts as a thermal resistance to heat flow. Since this resistance is confined to a very thin layer between the surfaces, it is called *thermal contact resistance*. Figure 4-12 illustrates the thermal contact resistance concept and the temperature profile through the solids. There is a sudden drop in temperature across the interface. Unless the two metals are welded, there is always a thermal resistance across such an interface. The magnitude of this thermal resistance depends on the surface roughness, the type of material, the interface pressure, the interface temperature, and the type of fluid filling the voids. A considerable amount of experimental work in predicting thermal contact resistance has been done. The inverse of the interface thermal contact resistance is called the *interface thermal contact conductance*, which has the dimensions of W/(m$^2 \cdot {}^\circ$C). Figure 4-13 illustrates the effects of surface roughness, the interface pressure, and interface temperature on the thermal contact conductance for a stainless steel–to–stainless steel joint with air as the interfacial fluid. The contact conductance increases with increasing interface pressure, increasing interface temperature, and decreasing surface roughness. If an aluminum–to–aluminum joint were used instead of the stainless steel joint, the contact conductance would be higher, because aluminum is softer than steel and allows more direct contact between the surfaces. Experimental data

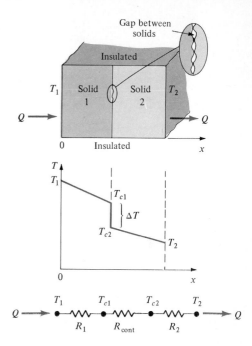

FIGURE 4-12
The concept of thermal contact
resistance.

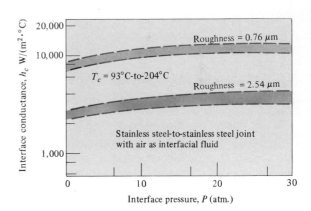

FIGURE 4-13
Effect of interface pressure,
contact temperature and
roughness on interface
conductance, h_c. (Data from
Barzelay, Tong and Holloway).

for contact conductance, such as those shown in Fig. 4-13, for different surface
roughnesses and different combinations of metals are available in the
literature.

Example 4-13. Consider two stainless steel blocks [$k = 20$ W/(m · °C)], each having a thickness of 1 cm, length of 8 cm, and width of 6 cm, that are pressed
together with a pressure of 20 atm. The surfaces have a roughness of about
0.76 μm. The outside surfaces of the blocks are at 120°C and 70°C. Calculate

(*a*) the heat flow rate across the blocks;
(*b*) the temperature drop at the interface.

Solution. (*a*) The total heat flow rate Q across the blocks is determined by the application of the thermal resistance concept:

$$Q = \frac{\Delta T}{R_1 + R_c + R_2} \quad \text{W}$$

where $\Delta T = T_1 - T_2 = 120 - 70 = 50°C$. The thermal resistances of the blocks are equal, and they are determined by

$$R_1 = R_2 = \frac{L}{kA} = \frac{0.01}{(20)(0.08 \times 0.06)} = 0.1042°C/W$$

The contact conductance of the interface is determined from Fig. 4-13. For a mean interface temperature of $T_m \cong (120 + 70)/2 = 90°C$, a surface roughness of $0.76 \, \mu m$, and interface pressure of $20 \, atm$, the contact conductance becomes $h_c = 10{,}000 \, W/(m^2 \cdot °C)$. Then the thermal contact resistance of the interface is given by

$$R_c = \frac{1}{hA} = \frac{1}{(10{,}000)(0.08 \times 0.06)} = 0.0208°C/W$$

Then the heat flow rate across the blocks is

$$Q = \frac{50}{0.1042 + 0.02083 + 0.1042} = 218 \, W$$

(*b*) The temperature drop across the interface becomes

$$\Delta T_c = \frac{R_c}{R_1 + R_c + R_1} \Delta T$$

$$= \frac{0.0208}{0.1042 + 0.0208 + 0.1042} \times 50 = 4.54°C$$

4-4 CONDUCTION SHAPE FACTORS

So far we have discussed the determination of one-dimensional, steady-state heat flow through bodies having such shapes as a plane wall, cylinder, or sphere. The analysis of heat flow through bodies with more complicated geometries is generally rather involved. However, heat transfer calculations have been performed for a number of specific geometries, and the total heat transfer rate Q through the body is related to the temperature difference ΔT and a quantity called the *conduction shape factor* S by

$$\boxed{\begin{aligned} Q &= Sk \, \Delta T \\ &= Sk(T_1 - T_2) \quad \text{W} \end{aligned}} \tag{4-62}$$

where k is the thermal conductivity of the body and T_1 and T_2 are the boundary surface temperatures across which heat flow takes place. It is assumed that there is no energy generation within the body.

Clearly, once the conduction shape factor S is known for a specific geometry, the total heat flow rate through the body can be calculated from the simple formula (4-62) for a given thermal conductivity of the solid and temperature difference between the boundary surfaces.

The physical significance of the shape factor S can be better envisioned if Eq. (4-62) is rearranged as

$$Q = \frac{\Delta T}{R} \tag{4-63a}$$

where

$$R = \frac{1}{Sk} \tag{4-63b}$$

Clearly, the shape factor S is related to the thermal resistance of the body. In the case of one-dimensional heat flow, the shape factors for such geometries as a plane wall, a hollow cylinder, and a hollow sphere are immediately obtained by utilizing the results developed previously.

For a plane wall, from Eqs. (4-8b) and (4-63b) we have

$$S = \frac{A}{L} \tag{4-64a}$$

For a hollow cylinder, from Eqs. (4-29) and (4-63b) we write

$$S = \frac{2\pi H}{\ln (b/a)} \tag{4-64b}$$

For a hollow sphere, Eqs. (4-42) and (4-63b) give

$$S = \frac{4\pi ab}{b - a} \tag{4-64c}$$

Heat transfer calculations have been performed and the corresponding conduction shape factors determined for a number of more complicated geometries. Table 4-1 lists a number of these shape factors. The results in this table can be used in Eq. (4-62) to determine the heat flow rate between surfaces maintained at temperatures T_1 and T_2.

Consider, for example, case 4 of Table 4-1. It illustrates an isothermal sphere of radius R, maintained at a uniform temperature T_1, placed with its center at a distance z from the surface of a semi-infinite domain. The medium has a thermal conductivity k, and its surface is maintained at a uniform temperature T_2. Using the shape factor concept, the total heat transfer rate Q from the sphere is given by

$$Q = Sk(T_1 - T_2) \quad \text{W} \tag{4-65a}$$

where the shape factor S for this particular case is

$$S = \frac{4\pi R}{1 - R/(2z)} \tag{4-65b}$$

Applications involving other configurations shown in this table are treated in a similar manner.

TABLE 4-1
Conduction shape factors; S defined by $Q = Sk(T_1 - T_2)$

1. One-dimensional heat conduction through a slab

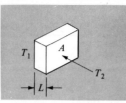

$$S = \frac{A}{L}$$

2. One-dimensional heat conduction through a long hollow cylinder

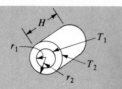

$$S = \frac{2\pi H}{\ln(r_2/r_1)}$$

3. One-dimensional heat conduction through a hollow sphere

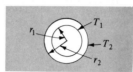

$$S = \frac{4\pi r_1 r_2}{r_2 - r_1}$$

4. An isothermal sphere at T_1 placed in a semi-infinite medium having a surface temperature T_2

$$S = \frac{4\pi R}{1 - R/(2z)}$$

5. An isothermal sphere at T_1 placed near the insulated boundary of a semi-infinite medium at T_2

$$S = \frac{4\pi R}{1 + R/(2z)}$$

6. Isothermal cylinder of length L at T_1 placed horizontally in a semi-infinite medium having a surface at T_2

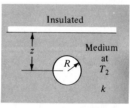

$$S = \frac{2\pi L}{\cosh^{-1}(z/R)}$$
for $L \gg R$

7. Thin, circular disk at T_1 placed horizontally in a semi-infinite medium having a surface at T_2

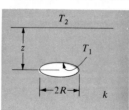

$S = 8R$, for $L \gg R$
More accurate
$$S = \frac{8.88R}{1 - R/(2.83z)}$$

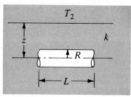
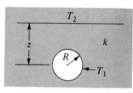

TABLE 4-1 (*continued*)
Conduction shape factors; S defined by $Q = Sk(T_1 - T_2)$

8. Isothermal cylinder of length L, at T_1 placed vertically in a semi-infinite medium having a surface at T_2

$$S = \frac{2\pi L}{\ln(2L/R)}$$

$$L \gg 2R$$

9. Two parallel, isothermal cylinders of length L at T_1 and T_2 placed in an infinite medium

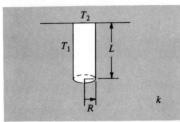

$$S = \frac{2\pi L}{\cosh^{-1}[(z^2 - R_1^2 - R_2^2)/(2R_1 R_2)]}$$

for $L \gg R_1, R_2$

$$L \gg z$$

10. Circular hole centered in a square solid of length L

$$S = \frac{2\pi L}{\ln(0.54W/R)}$$

for $L \gg W$

11. Eccentric circular hole in a cylindrical solid of length L

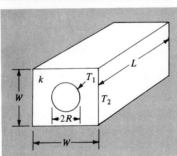

$$S = \frac{2\pi L}{\cosh^{-1}[(R_1^2 + R_2^2 - z^2)/(2R_1 R_2)]}$$

for $L \gg R_2$

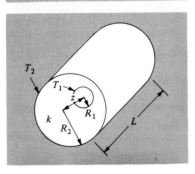

Example 4-14. A spherical tank of radius 1 m containing radioactive material is buried in the earth. The distance between the earth's surface and the tank center is 4 m. Heat release due to radioactive decay in the tank is 500 W. Calculate the steady temperature of the tank's surface if the earth's surface is at 20°C and the thermal conductivity of the earth at this location is $k = 1.0 \, \text{W/(m} \cdot {}^{\circ}\text{C)}$.

Solution. The physical situation and the configuration for this problem are similar to case 4 in Table 4-1. Given $R = 1$ m and $z = 4$ m, the shape factor S is determined as

$$S = \frac{4 \pi R}{1 - R/(2z)}$$

$$= \frac{4 \pi \times 1}{1 - 1/[(2)(4)]} = 14.36 \, \text{m}$$

For the earth's surface at $T_2 = 20°C$, the steady temperature of the tank's surface T_1 is calculated from Eq. (4-62) as

$$Q = Sk(T_1 - T_2)$$

$$500 = (14.36)(1)(T_1 - 20)$$

or

$$T_1 = 54.8°C$$

Example 4-15. A hot pipe with an outside radius of 2 cm passes through the hole at the center of a long concrete block 50 cm × 50 cm in cross section, as illustrated in the accompanying figure. The pipe surface is at 80°C, and the outer surface of the concrete block is at 20°C. The thermal conductivity of the concrete may be taken as $k = 0.76 \, \text{W/(m} \cdot {}^{\circ}\text{C)}$. Determine the rate of heat loss from the pipe per meter of length.

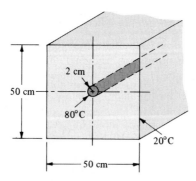

FIGURE EXAMPLE 4-15
The hot water pipe inside a long square block.

Solution. The physical situation and the configuration for this problem are similar to case 10 in Table 4-1. Given $W = 0.4$ m and $R = 0.02$ m, the shape factor S for a 1-m length of the pipe, $L = 1$ m,

$$S = \frac{2 \pi L}{\ln{(0.54W/R)}}$$

$$= \frac{2 \pi \times 1}{\ln{[0.54(0.4/0.02)]}}$$

$$= 2.64 \, \text{m}$$

For the pipe surface at $T_1 = 80°C$ and the outside surface of the concrete at $T_2 = 20°C$, the heat loss from the pipe is calculated using Eq. (4-62) as

$$Q = Sk(T_1 - T_2)$$
$$= (2.64)(0.76)(80 - 20)$$
$$= 120.4 \text{ W/m length}$$

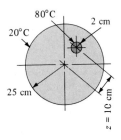

FIGURE EXAMPLE 4-1
The hot water pipe inside a long cylinder block.

Example 4-16. A hot water pipe with an outside radius of 2 cm is embedded eccentrically inside a long cylindrical concrete block of radius 25 cm, as illustrated in the accompanying figure. The distance between the center of the pipe and the center of the cylinder is 10 cm. The surface of the hot water pipe is at 80°C, and the outside surface of the concrete cylinder is at 20°C. The thermal conductivity of concrete may be taken as $k = 0.76 \text{ W/(m} \cdot °C)$. Determine the rate of heat loss from the pipe per meter of length.

Solution. The physical situation and the configuration for this problem are similar to case 11 in Table 4-1. Given $R_1 = 0.02$ m, $R_2 = 0.25$ m, and $z = 0.1$ m, the shape factor S for a 1-m length of the pipe, $L = 1$ m, is

$$S = \frac{2\pi L}{\cosh^{-1}[(R_1^2 + R_2^2 - z^2)/(2R_1R_2)]}$$
$$= \frac{(2\pi)(1)}{\cosh^{-1}\{[(0.02)^2 + (0.25)^2 - (0.10)^2]/(2)(0.02)(0.25)\}}$$
$$= \frac{2\pi}{\cosh^{-1}(5.29)}$$
$$= 2.67$$

and the heat loss from the pipe is calculated using Eq. (4-62) as

$$Q = Sk(T_1 - T_2)$$
$$= (2.67)(0.76)(80 - 20)$$
$$= 122 \text{ W}$$

4-5 FINS

The rate of heat removal by convection from a surface is increased by increasing the surface area for heat transfer by using extended surfaces called *fins*. A familiar example of a finned surface, as illustrated in Fig. 4-14, is a metal spoon in hot water. Heat conducted through the spoon causes the handle to become warmer than the surrounding air. Heat is then transferred from the spoon handle to the air by convection. In industry, fins are used in numerous applications, such as car radiators, double-pipe heat exchangers, electronic equipment cooling, and compressors. The determination of heat flow through a finned surface requires knowledge of the temperature distribution in the fin.

FIGURE 4-14
A metal spoon in hot water
acting as a fin.

Fins of Uniform Cross Section

Figure 4-15 illustrates the geometry, the coordinates, and the nomenclature for the development of the one-dimensional, steady-state energy equation for fins of uniform cross section. Consider a small volume element Δx and write the steady-state energy balance for the volume element as

$$\begin{pmatrix} \text{Net rate of heat gain} \\ \text{by conduction in} \\ x \text{ direction into} \\ \text{volume element } \Delta x \end{pmatrix} + \begin{pmatrix} \text{net rate of heat gain} \\ \text{by convection through} \\ \text{lateral surfaces into} \\ \text{volume element } \Delta x \end{pmatrix} = 0$$

$$-\frac{d(qA)}{dx}\,\Delta x + h[T_\infty - T(x)]P\,\Delta x = 0 \qquad (4\text{-}66a)$$

$$\frac{d^2T(x)}{dx^2} - \frac{hP}{Ak}\,[T(x) - T_\infty] = 0 \qquad (4\text{-}66b)$$

FIGURE 4-15
Nomenclature for the
derivation of one-
dimensional fin
equation.

where the cross-sectional area A, the perimeter P, the heat transfer coefficient h, and the thermal conductivity of the fin material k are constant. This

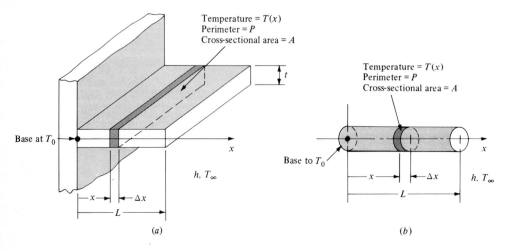

Temperature $= T(x)$
Perimeter $= P$
Cross-sectional area $= A$

Temperature $= T(x)$
Perimeter $= P$
Cross-sectional area $= A$

Base at T_0

x

t

Base to T_0

x

h, T_∞

x

Δx

x

Δx

h, T_∞

L

L

(a)

(b)

equation is written more compactly in the form

$$\frac{d^2\theta(x)}{dx^2} - m^2\theta(x) = 0 \qquad (4\text{-}67a)$$

where

$$m^2 \equiv \frac{hP}{Ak} \qquad \theta(x) = T(x) - T_\infty \qquad (4\text{-}67b)$$

Equation (4-67) is a linear, homogeneous, second-order ordinary differential equation with constant coefficients. Its general solution may be taken in the form

$$\theta(x) = C_1 \cosh m(L - x) + C_2 \sinh m(L - x) \qquad (4\text{-}68)$$

where the two constants C_1 and C_2 are to be determined from the boundary conditions for the problem. The solution given by Eq. (4-68) is more convenient to use in the analysis of fins of finite length, as will soon be apparent in the solution that follows. Customarily, the temperature distribution at the fin base $x = 0$ is considered known; that is,

$$\theta(0) = T_0 - T_\infty \equiv \theta_0 \qquad (4\text{-}69)$$

where T_0 is the fin base temperature. Equation (4-69) provides one of the boundary conditions needed for the determination of the unknown constants in Eq. (4-68). The second boundary condition at the fin tip and the determination of the constants are now described.

FINS WITH NEGLIGIBLE HEAT LOSS AT THE TIP. The heat transfer area at the fin tip is generally small compared with the lateral area of the fin. For such a case, the heat loss from the tip of the fin is negligible, and the boundary condition for the fin tip (i.e., $x = L$) can be taken as no heat loss (i.e., insulated).

$$\frac{d\theta(x)}{dx} = 0 \qquad \text{at } x = L \qquad (4\text{-}70)$$

The boundary condition (4-70) requires that $C_2 = 0$, and the application of the boundary condition (4-69) gives $C_1 = \theta_0 / \cosh mL$. Then the solution given by Eq. (4-68) becomes

$$\frac{\theta(x)}{\theta_0} = \frac{T(x) - T_\infty}{T_0 - T_\infty} = \frac{\cosh m(L - x)}{\cosh mL} \qquad (4\text{-}71)$$

The heat flow Q through the fin is determined by evaluating the conductive heat flow at the fin base according to the relation

$$Q = -Ak \left. \frac{d\theta(x)}{dx} \right|_{x=0}$$

which gives

$$Q = Ak\theta_0 m \tanh mL = \theta_0 \sqrt{PhkA} \tanh mL \qquad (4\text{-}72)$$

FIN EFFICIENCY. In industrial applications, fins having geometries that are more complicated than those shown in Fig. 4-15 are frequently used. The analysis of heat transfer through such fins is a complicated matter, and the resulting expressions for heat flow are very involved for use in practice. To alleviate this difficulty and simplify the calculation of heat transfer through fins, the concept of *fin efficiency* has been introduced. The fin efficiency η is defined by

$$\eta = \frac{\text{actual heat transfer through fin}}{\substack{\text{ideal heat transfer through fin} \\ \text{if entire fin surface were at} \\ \text{fin base temperature } T_0}} = \frac{Q_{\text{fin}}}{Q_{\text{ideal}}} \qquad (4\text{-}73a)$$

Here Q_{ideal} is given by

$$Q_{\text{ideal}} = a_f h \theta_0 \qquad (4\text{-}73b)$$

where a_f = surface area of fin
h = heat transfer coefficient
$\theta_0 = T_0 - T_\infty$

Thus, if the fin efficiency η is known, the heat transfer Q_{fin} through the fin is determined from

$$Q_{\text{fin}} = \eta Q_{\text{ideal}} = \eta a_f h \theta_0 \qquad (4\text{-}73c)$$

To illustrate how η can be determined, we consider heat flow through a fin having a uniform cross section and negligible heat loss from the fin tip. The heat transfer through the fin is given by Eq. (4-72) as

$$Q_{\text{fin}} = \theta_0 \sqrt{PhkA} \tanh mL$$

For a fin of length L and perimeter P, the heat transfer area a_f for the fin is taken as $a_f = PL$. Then Q_{ideal} is determined by Eq. (4-73b) as

$$Q_{\text{ideal}} = PLh\theta_0$$

Introducing these results into the definition of the fin efficiency η, we obtain an explicit expression for the efficiency of a rectangular fin:

$$\eta = \frac{\theta_0 \sqrt{PhkA} \tanh mL}{\theta_0 PLh} = \frac{\tanh mL}{mL} \qquad (4\text{-}74a)$$

where

$$mL = L \sqrt{\frac{Ph}{Ak}} = L \sqrt{\frac{2h}{kt}} \qquad (4\text{-}74b)$$

since $P/A \cong 2/t$, t being the fin thickness. The fin efficiency η given by Eq. (4-74a) is plotted in Fig. 4-16 against the parameter $mL = L\sqrt{2h/kt}$.

Equations for fin efficiency have been developed for fins having other geometries. Figure 4-17 shows the fin efficiency for a circular disc fin of constant cross section plotted against the parameter $mL = L\sqrt{2h/kt}$ for several different values of the radius ratio r_o/r_i. Groups such as this are available for fins with other geometries.

So far we have discussed heat transfer through a single fin. In practice, however, heat transfer through a finned surface consists of heat transfer through the fins and through the bare area between the fins. Let

$a_f =$ surface area of the fins

$a_b =$ bare (unfinned) area between the fins

$h =$ heat transfer coefficient, which is assumed to be the same for both the fin surface and the bare surface

Then the total heat transfer rate Q_t through the fin assembly is determined from

$$Q_t = \left(\begin{array}{c}\text{heat transfer through}\\ \text{the fins}\end{array}\right) + \left(\begin{array}{c}\text{heat transfer through}\\ \text{the bare surface}\\ \text{between the fins}\end{array}\right) \qquad (4\text{-}75a)$$

$$Q_t = \eta a_f h\theta_0 + a_b h\theta_0 \qquad (4\text{-}75b)$$

$$Q_t = h\theta_0(\eta a_f + a_b) \qquad (4\text{-}75c)$$

Knowing the areas a_f and a_b, the fin efficiency η, the heat transfer coefficient h, and the temperature difference θ_0 between the fin base and the ambient, the total heat transfer rate Q_t can be determined from Eq. (4-75c).

FIGURE 4-16
A plot of the efficiency given by Eq. (4-74).

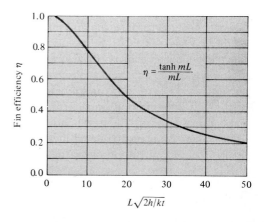

$\eta = \dfrac{\tanh mL}{mL}$

Fin efficiency η

$L\sqrt{2h/kt}$

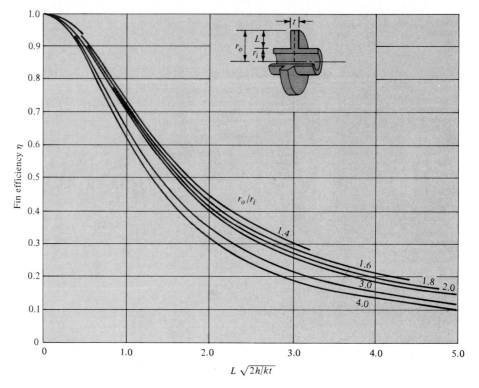

FIGURE 4-17
Efficiency of circular disc fins of constant thickness (From Gardner).

Example 4-17. Copper-plate fins of rectangular cross section having a thickness $t = 1$ mm, height $L = 10$ mm, and thermal conductivity $k = 380$ W/(m · °C) are attached to a plane wall maintained at a temperature $T_0 = 230$°C. The fins dissipate heat by convection into ambient air at $T_\infty = 30$°C with a heat transfer coefficient $h = 40$ W/(m² · °C). Assuming negligible heat loss from the fin tip, determine the fin efficiency.

Solution. To determine the fin efficiency η, we first calculate the parameter mL as follows:

$$mL = (14.51)(0.01) = 0.1451$$

where

$$\frac{P}{A} = \frac{2(b + t)}{bt} \cong \frac{2b}{bt} = \frac{2}{t}$$

$$m = \sqrt{\frac{Ph}{kA}} = \sqrt{\frac{2h}{kt}}$$

$$= \sqrt{\frac{(2)(40)}{380 \times 10^{-3}}} = 14.51 \text{ m}^{-1}$$

Then, the efficiency η from Eq. (4-74) becomes

$$\eta = \frac{\tanh mL}{mL}$$

$$= \frac{\tanh 0.1451}{0.1451} = \frac{0.144}{0.1451} = 0.993$$

Example 4-18. Circular aluminum disk fins of constant rectangular profile (shown in Fig. 4-17) are attached to a tube having $D = 2.5$ cm outside diameter. The fins have a thickness $t = 1$ mm, height $L = 15$ mm, and thermal conductivity $k = 200$ W/(m · °C). The tube wall is maintained at a temperature $T_0 = 190°C$, and the fins dissipate heat by convection into ambient air at $T_\infty = 40°C$ with a heat transfer coefficient $h_\infty = 80$ W/(m² · °C). Determine the fin efficiency.

Solution. The parameters mL and the radius ratio r_o/r_i are

$$mL = L\sqrt{\frac{2h}{kt}} = 0.015\sqrt{\frac{(2)(80)}{200 \times 10^{-3}}} = 0.424$$

$$\frac{r_o}{r_i} = \frac{12.5 + 15}{12.5} = 2.2$$

and from Fig. 4-16 we read the fin efficiency as

$$\eta = 0.99$$

Example 4-19. A steel rod of diameter $D = 2$ cm, length $L = 10$ cm, and thermal conductivity $k = 50$ W/(m · °C) is exposed to ambient air at $T_\infty = 20°C$ with a heat transfer coefficient $h_\infty = 30$ W/(m² · °C). If one end of the rod is maintained at a temperature of 70°C, calculate the heat loss from the rod.

Solution. This problem is equivalent to the problem of heat transfer through a fin of constant cross section. The parameter mL is calculated as follows:

$$m^2 = \frac{hP}{Ak} = \frac{h\pi D}{(\pi/4)D^2 k} = \frac{4h}{kD} = \frac{(4)(30)}{(50)(0.02)} = 120$$

Therefore $m = 10.95$ and $mL = (10.95)(0.10) = 1.095$.
Then the heat flow rate is calculated using Eq. (4-72):

$$Q = \theta_0 \sqrt{PhkA} \tanh mL$$

$$= \theta_0 \sqrt{(\pi D)hk\left(\frac{\pi}{4}D^2\right)} \tanh mL$$

$$= (70 - 20)\frac{\pi}{2}\sqrt{(0.02)^3(50)(30)} = 8.6 \text{ W}$$

PROBLEMS

Single Layers

4-1. Consider a slab of thickness 10 cm. One surface is kept at 20°C and the other at 100°C. Determine the heat flow rate across the slab if the slab is made of pure copper [$k = 387$ W/(m · °C)], pure aluminum [$k = 202$ W/(m · °C)], and pure iron [$k = 62$ W/(m · °C)].

4-2. A brick wall [$k = 0.69$ W/(m · °C)] 5 cm thick is exposed to cool air at 10°C with a heat transfer coefficient of 10 W/(m² · °C) at one of its surfaces, while the other surface is kept at 70°C. What is the temperature of the surface that is exposed to cool air?

Answer: 44.9°C

4-3. Consider a furnace wall [$k = 1$ W/(m · °C)] with the inside surface at 1000°C and the outside surface at 400°C. If the heat flow through the wall should not exceed 2000 W/m², what is the minimum wall thickness L?

Answer: 30 cm

4-4. Consider a plane wall 25 cm thick. The inner surface is kept at 400°C, and the outer surface is exposed to an environment at 800°C with a heat transfer coefficient of 10 W/(m² · °C). If the temperature of the outer surface is 685°C, calculate the thermal conductivity of the wall. What might the material be?

4-5. Consider a pipe with an inner radius of 5 cm and an outer radius of 7 cm. The inner surface is kept at 100°C, and the outer surface at 80°C. Determine the heat loss per meter length of the pipe if the pipe is made of pure copper [$k = 387$ W/(m · °C)], pure aluminum [$k = 200$ W/(m · °C)], and pure iron [$k = 62$ W/(m · °C)].

4-6. A metal pipe with an outside diameter (OD) of 12 cm is covered with an insulation material [$k = 0.07$ W/(m · °C)] of 2.5 cm thick. If the outer pipe wall is at 100°C and the outer surface of the insulation is at 20°C, find the heat loss from the pipe per meter length.

Answer: 101 W/m length

4-7. A metal pipe of 10 cm OD is covered with a 2-cm-thick insulation [$k = 0.07$ W/(m · °C)]. The heat loss from the pipe is 100 W per meter of length when the pipe surface is at 100°C. What is the temperature of the outer surface of the insulation?

Answer: 23.5°C

4-8. Consider a hollow cylinder with inner radius a and outer radius b. The inner surface is subjected to a heat supply at a constant rate of q_1 W/m², while the outer surface is maintained at a uniform temperature T_2. Develop an expression for the steady temperature distribution $T(r)$ in the cylinder.

4-9. A 5-cm-OD and 0.5-cm-thick copper pipe [$k = 386$ W/(m · °C)] has hot gas flowing inside at a temperature of 200°C with a heat transfer coefficient of 30 W/(m² · °C). The outer surface dissipates heat by convection into the ambient air at 20°C with a heat transfer coefficient of 15 W/(m² · °C). Determine the heat loss from the pipe per meter of length.

Answer: 261 W/m length

4-10. A brass condenser tube [$k = 115$ W/(m · °C)] with an outside diameter of 2 cm and a thickness of 0.2 cm is used to condense steam on its outer surface at 50°C with a heat transfer coefficient of 2000 W/(m² · °C). Cooling water at 20°C with a heat transfer coefficient of 5000 W/(m² · °C) flows inside.

(*a*) Determine the heat flow rate from the steam to the cooling water per meter of length of the tube.

(*b*) What would be the heat transfer rate per meter of length of the tube if the outer and the inner surfaces of the tube were at 50°C and 20°C, respectively? Compare this result with (*a*), and explain the reason for the difference between the two results.

4-11. Consider a hollow sphere with an inner radius of 5 cm and an outer radius of 6 cm. The inner surface is kept at 100°C, and the outer surface at 50°C.

Determine the heat loss from the sphere if it is made of pure copper $[k = 387 \text{ W/(m} \cdot {}^\circ\text{C)}]$, pure aluminum $[k = 200 \text{ W/(m} \cdot {}^\circ\text{C)}]$, and pure iron $[k = 62 \text{ W/(m} \cdot {}^\circ\text{C)}]$.

Answers: 72.95 kW, 37.7 kW, 11.69 kW

4-12. Consider a hollow sphere of inner radius $r = a$ and outer radius $r = b$. The inner surface is heated uniformly at a constant rate of q_a W/m^2, while the outer surface is exposed to an environment at T_∞ with a heat transfer coefficient h_∞. Develop an expression for the steady temperature distribution $T(r)$ in the sphere.

4-13. A 6-cm-OD, 2-cm-thick copper hollow sphere $[k = 386 \text{ W/(m} \cdot {}^\circ\text{C)}]$ is uniformly heated at the inner surface at a rate of 150 W/m^2. The outer surface is cooled with air at 20°C with a heat transfer coefficient of 10 W/(m$^2 \cdot {}^\circ$C). Calculate the temperature of the outer surface.

Answer: 21.7°C

4-14. Steam at 150°C flows inside a pipe $[k = 40 \text{ W/(m} \cdot {}^\circ\text{C)}]$ with an inside radius of 4 cm and an outside radius of 5 cm. The heat transfer coefficient between the steam and the inside surface is 2000 W/(m$^2 \cdot {}^\circ$C). The outside surface is exposed to atmospheric air at 30°C with a heat transfer coefficient of 15 W/(m$^2 \cdot {}^\circ$C). Find the heat flow rate across the pipe per meter length of the pipe.

Answer: 557.9 W/m length

Multilayers

4-15. A container made of 2-cm-thick iron plate $[k = 62 \text{ W/(m} \cdot {}^\circ\text{C)}]$ is insulated with a 1-cm-thick asbestos layer $[k = 0.1 \text{ W(m} \cdot {}^\circ\text{C)}]$. If the inner surface of the iron plate is exposed to hot gas at 530°C with a heat transfer coefficient of 100 W/(m$^2 \cdot {}^\circ$C) and the outer surface of the asbestos is in contact with cool air at 30°C with a heat transfer coefficient 20 W/(m$^2 \cdot {}^\circ$C), calculate: (*a*) the heat flow rate across the layers per square meter of surface area, and (*b*) the interface temperature between the layers.

Answers: (*a*) 3119 W/m^2, (*b*) 497.8°C

4-16. Steam at 180°C flows inside a pipe $[k = 40 \text{ W/(m} \cdot {}^\circ\text{C)}]$ that has an inside radius of 3 cm and an outside radius of 3.5 cm. The heat transfer coefficient between the steam and the inside surface of the pipe is 3000 W/(m$^2 \cdot {}^\circ$C). To reduce the heat loss from the steam pipe to the atmospheric air at 30°C, a 2-cm-thick layer of magnesia insulation $[k = 0.06 \text{ W/(m} \cdot {}^\circ\text{C)}]$ is added on the outer surface. If the heat transfer coefficient between the atmospheric air and the outside surface of the insulation material is 10 W/(m$^2 \cdot {}^\circ$C), calculate the heat loss to the air.

Answer: 100.6 W/m length

4-17. An industrial furnace is made of fireclay brick 0.3 m thick with a thermal conductivity of 1.5 W/(m $\cdot {}^\circ$C). The outside surface is to be insulated with a material that has a thermal conductivity of 0.01 W/(m $\cdot {}^\circ$C) and a thickness of 0.2 m. The inner surface of the furnace is kept at 600°C, while the outer surface of the insulation material is exposed to cool air at 30°C with a heat transfer coefficient of 15 W/(m$^2 \cdot {}^\circ$C). Calculate the heat flow rate across the layers per square meter of surface area and the outer surface temperature of the furnace.

Answer: −28.12 W/m^2, 31.9°C

4-18. A steel tube $[k = 15 \text{ W/(m} \cdot {}^\circ\text{C)}]$ with an outside diameter of 4 cm and a thickness of 1 cm is covered with insulation material 1.5 cm thick. The inside temperature of the tube is kept at 200°C, and the outside surface temperature of the tube is measured at 199°C. Calculate the outside surface temperature of the insulation material, if the thermal conductivity of the insulation is 0.2 W/(m $\cdot {}^\circ$C).

4-19. A hollow sphere $[k = 15\ \mathrm{W/(m \cdot °C)}]$ with an outside diameter of 8 cm and a thickness of 2 cm is covered with an insulation material $[k = 0.2\ \mathrm{W/(m \cdot °C)}]$ 2 cm thick. Inside the sphere energy is generated at a rate of $3 \times 10^5\ \mathrm{W/m^3}$. The temperature of the interface between the outer surface of the sphere and the insulation is measured to be 300°C. Calculate the outside surface temperature of the insulation material.

  Answer: 266.7°C

4-20. The wall of a building consists of 10 cm of brick $[k = 0.69\ \mathrm{W/(m \cdot °C)}]$, 1.25 cm of Celotex $[k = 0.048\ \mathrm{W/(m \cdot °C)}]$, 8 cm of glass wool $[k = 0.038\ \mathrm{W/(m \cdot °C)}]$, and 1.25 cm of asbestos cement board $[k = 0.74\ \mathrm{W/(m \cdot °C)}]$. If the outside surface of the brick is at 5°C and the inside surface of the cement board is at 20°C, calculate the heat flow rate per square meter of wall surface.

  Answer: $-5.94\ \mathrm{W/m^2}$

4-21. A plane wall of thickness L_1 and thermal conductivity k_1 is covered with an insulation layer of thermal conductivity k_2. The inside surface of the wall is maintained at a uniform temperature T_1, and the outside surface of the insulation layer is exposed to air at temperature $T_{\infty 2}$ with a heat transfer coefficient $h_{\infty 2}$. Develop an expression for determining the thickness L of the insulation layer needed to reduce the heat loss from the uninsulated wall by 30 percent.

4-22. A wall of a building is made of 8 cm of building brick $[k = 0.69\ \mathrm{W/(m \cdot °C)}]$, 2 cm of Celotex $[k = 0.048\ \mathrm{W/(m \cdot °C)}]$, and 2 cm of an asbestos cement board $[k = 0.74\ \mathrm{W/(m \cdot °C)}]$ Glass wool $[k = 0.038\ \mathrm{W/(m \cdot °C)}]$ is to be added between the Celotex and Asbestos to reduce the heat flow rate through the wall by 50 percent. Determine the thickness of the cement board.

  Answer: 2.1 cm

4-23. Determine the interface temperature T_1 and the surface temperature T_3 of the composite wall shown in Fig. P4-23.

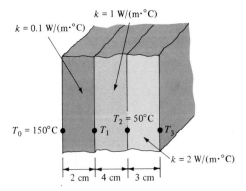

FIGURE P4-23

4-24. Consider a multilayer plane wall consisting of three parallel layers in perfect thermal contact, as illustrated in Fig. 4-9. Develop expressions for the determination of the interface temperatures T_1 and T_2.

4-25. A steel tube $[k = 15\ \mathrm{W/(m \cdot °C)}]$ with an outside diameter of 7.6 cm and a thickness of 1.3 cm is covered with an insulation material $[k = 0.2\ \mathrm{W/(m \cdot °C)}]$ 2 cm thick. A hot gas at 330°C with a heat transfer coefficient of $400\ \mathrm{W/(m^2 \cdot °C)}$ flows inside the tube. The outer surface of the insulation is exposed to cooler air at 30°C with a heat transfer coefficient of $60\ \mathrm{W/(m^2 \cdot °C)}$. Calculate the heat loss from the tube to the air for a 10-m length of the tube.

  Answer: 7453 W

4-26. Consider a pipe of inside radius $r_1 = 2$ cm, outside radius $r_2 = 4$ cm, and thermal conductivity $k_1 = 10$ W/(m · °C). The inside surface is maintained at a uniform temperature $T_1 = 300$°C, and the outside surface is to be insulated with an insulation material of thermal conductivity $k_2 = 0.1$ W/(m · °C). The outside surface of the insulation material is exposed to an environment at $T_{\infty 2} = 20$°C with a heat transfer coefficient $h_{\infty 2} = 10$ W/(m² · °C). Develop an expression for determining the thickness L of the insulation material needed to reduce the heat loss by 30 percent of that of the uninsulated pipe exposed to the same environmental conditions.

Answer: 0.0062 m

4-27. Consider a steel pipe [$k = 10$ W/(m · °C)] with an inside radius of 5 cm and an outside radius of 10 cm. The outside surface is to be insulated with a fiberglass insulation [$k = 0.05$ W/(m · °C)] to reduce the heat flow rate through the pipe wall by 50 percent. Determine the thickness of the fiberglass.

Answer: 0.0347 cm

4-28. Consider a hollow sphere with an inside radius $r_1 = 2$ cm, an outside radius $r_2 = 4$ cm, and thermal conductivity $k_1 = 10$ W/(m · °C). The inside surface is maintained at a uniform temperature of $T_1 = 300$°C, and the outside surface is to be insulated with an insulating material of thermal conductivity $k_2 = 0.1$ W/(m · °C). The outside surface of the insulating material is exposed to an environment at $T_{\infty 2} = 20$°C with a heat transfer coefficient $h_{\infty 2} = 10$ W/(m² · °C). Develop an expression for determining the thickness of the insulation needed to reduce the heat loss by 30 percent of that of the uninsulated hollow sphere exposed to the same environmental conditions.

4-29. Consider a hollow steel sphere [$k = 10$ W/(m · °C)] with an inside radius of 5 cm and an outside radius of 10 cm. The outside surface is to be insulated with a fiberglass insulation [$k = 0.05$ W/(m · °C)] to reduce the heat flow rate through the sphere wall by 50 percent. Determine the thickness of the fiberglass.

Answer: 0.05 cm

4-30. A steel tube [$k = 15$ W/(m · °C)] with an outside diameter of 7.6 cm and a thickness of 1.3 cm is covered with an insulation [$k = 0.2$ W/(m · °C)] 2 cm thick. A hot gas with a heat transfer coefficient of 400 W/(m² · °C) flows inside the tube, and the outer surface of the insulation is exposed to cooler air with a heat transfer coefficient of 60 W/(m² · °C). Calculate the overall heat transfer coefficient U based on the outside surface area of the insulation.

4-31. The thickness of the insulation in Prob. 4-30 is increased to 4 cm. Calculate the overall heat transfer coefficient U based on the outside surface area of the insulation material.

Answer: 3.26 W/(m² · °C)

4-32. For Problem 4-30, calculate the overall heat transfer coefficient based on the inside surface area of the insulation material.

4-33. Consider a brass tube [$k = 115$ W/(m · °C)] with an outside radius of 4 cm and a thickness of 0.5 cm. The inside surface of the tube is kept at uniform temperature, and the outside surface is covered with two layers of insulation, each 1 cm thick, with thermal conductivities of 0.1 W/(m · °C) and 0.05 W/(m · °C), respectively. Calculate the overall heat transfer coefficient based on the outside surface area of the outer insulation.

Answer: 2.83 W/(m² · °C)

4-34. An iron plate 2.5 cm thick [$k = 62$ W/(m · °C)] is in contact with asbestos insulation 1 cm thick [$k = 0.2$ W/(m · °C)] on one side and exposed to hot gas with a heat transfer coefficient of 200 W/(m$^2 \cdot$°C) on the other surface. If the outer surface of the asbestos is exposed to cool air with a heat transfer coefficient of 40 W/(m$^2 \cdot$°C), calculate the overall heat transfer coefficient U and the heat flow rate across the composite wall per square meter of the surface for a ΔT of 200°C between the hot gas and cool air.

Answer: 12.43 W/(m$^2 \cdot$°C), 2.48 kW/m^2

4-35. A brass condenser tube [$k = 115$ W/(m $\cdot$ C)] with an outside diameter of 3 cm and a thickness of 0.25 cm is used to condense steam on its outer surface with a heat transfer coefficient of 4000 W/(m$^2 \cdot$°C). Cooling water with a heat transfer coefficient of 5000 W/(m$^2 \cdot$°C) flows inside. Find the overall heat transfer coefficient based on the outside surface area of the condenser tube, and calculate the heat transfer rate per 10-m length of the tube for a $\Delta T = 10$°C between the condensing steam and the cooling water.

Answer: 18.34 kW

4-36. A steel tube [$k = 15$ W/(m $\cdot$ °C)] with an outside diameter of 8 cm and a thickness of 2 cm is covered with an insulation [$k = 0.07$ W/(m $\cdot$ °C)] 2 cm thick. A hot gas with a heat transfer coefficient of 500 W/(m$^2 \cdot$°C) flows inside the tube. The outer surface of the insulation is exposed to air with a heat transfer coefficient of 50 W/(m$^2 \cdot$°C). Find the overall heat transfer coefficient based on the inside surface area of the steel tube.

4-37. Consider a hollow brass sphere [$k = 115$ W/(m $\cdot$ °C)] with an outside radius of 4 cm and a thickness of 0.5 cm. The outside surface is covered with an insulation [$k = 0.05$ W/(m $\cdot$ °C)] 1 cm thick. The inside surface of the sphere is kept at a uniform temperature. The outer surface of the insulation is exposed to air with a heat transfer coefficient of 60 W/(m$^2 \cdot$°C). Calculate the overall heat transfer coefficient based on the outside surface area of the insulation and the total heat transfer rate for a $\Delta T = 200$°C between the inner surface of the sphere and the outside air temperatures.

Contact Resistance Concept

4-38. Two stainless steel pipes are pressed together with a pressure of 10 atm. The surfaces have a roughness of about 2.54 μm and are kept at a mean temperature of 100°C. Determine the interface contact conductance. What would be the contact conductance for a roughness of about 0.76 μm?

4-39. Consider two stainless steel plates of thicknesses 0.5 cm and 1 cm, pressed together with a pressure of 10 atm. The surfaces have roughnesses of about 2.54 μm. The outside surfaces of the walls are at 200°C and 100°C. Calculate the heat flow rate per square meter of wall area. What would be the heat flow rate for an interface roughness of 0.76 μm?

4-40. Consider two stainless steel slabs [$k = 20$ W/(m $\cdot$ °C)] with thicknesses of 1 cm and 1.5 cm that are pressed together with a pressure of 20 atm. The surfaces have roughnesses of about 0.76 μm. The outside surfaces of the blocks are at 100°C and 150°C. Calculate the heat flow rate across the slabs and the temperature drop at the interface.

Answer: 3.7°C

Conduction Shape Factor

4-41. Consider a spherical heat source with an outside diameter of 0.5 m buried 3 m below the surface of a semi-infinite soil [$k = 0.5$ W/(m $\cdot$ °C)]. If the surface of the source is at 50°C and the surface of the soil is at 20°C, determine the steady heat flow rate from the sphere to the soil.

4-42. A spherical heat source with an outside diameter of 30 cm is buried 150 cm below the surface of a semi-infinite soil [$k = 0.5$ W/(m · °C)] at 20°C. If the surface of the source is kept at 50°C, and the surface of the soil is insulated, calculate the steady heat flow from the sphere to the soil.

4-43. A pipe with an outer radius of 5 cm, carrying a hot fluid at 200°C, is buried in the earth [$k = 1.3$ W/(m · °C)]. The distance between the pipe center and the earth's surface is 2 m. The pipe is covered with a 2-cm-thick fiberglass layer [$k = 0.04$ W/(m · °C)]. The surface of the earth is at 15°C. Determine the heat loss per meter length of the pipe.

4-44. A cylindrical heat source and a cylindrical heat sink, each having a radius of 10 cm and a length of 8 m, are buried in an infinite medium of fireclay brick [$k = 1$ W/(m · °C)], 15 cm apart and parallel to each other. The surfaces of the cylinders are measured to be 500°C and 400°C, respectively. Determine the steady heat flow from one cylinder to the other.

Answer: 2611 W

4-45. A cylindrical storage tank with a radius of 1 m and a length of 3 m is buried in each [$k = 1.3$ W/(m · °C)] with its axis parallel to the surface of the earth. The distance between the tank axis and the earth's surface is 2 m. The tank is kept at 70°C, and the earth's surface is at 15°C. Determine the heat loss from the tank.

Answer: 10230 W

4-46. A spherical tank with a radius of 2 m containing some radioactive material is buried in earth [$k = 1.6$ W/(m · °C)]. The center of the sphere is 5 m from the earth's surface. The tank is at 200°C, and the earth's surface is at 25°C. Determine the heat loss from the sphere.

Answer: 8798 W

4-47. A hot pipe with an outside radius of 10 cm passes through the hole at the center of a long concrete block [$k = 0.76$ W/(m · °C)] 1 m × 1 m in cross section. The pipe surface is at 100°C, and the outer surface of the block is at 25°C. Determine the heat loss from the pipe per meter length of the pipe.

Answer: 212.6 W/m length

4-48. A hot water pipe with an outside radius of 1 cm is embedded eccentrically inside a long cylindrical concrete block radius of 10 cm. The distance between the center of the pipe and the cylinder is 7 cm. The outside surface of the concrete is maintained at 25°C. The heat loss from the hot water to the concrete is 100 W/m length. Determine the wall temperature of the pipe.

Answer: 58.7°C

Fins

4-49. Aluminum fins of rectangular profile are attached on a plane wall. The fins have thickness $t = 1$ mm, length $L = 10$ mm, and thermal conductivity $k = 200$ W/(m · °C). The wall is maintained at a temperature $T_0 = 200$°C, and the fins dissipate heat by convection into the ambient air at $T_\infty = 40$°C with a heat transfer coefficient $h_\infty = 50$ W/(m^2 · °C). Determine the fin efficiency.

Answer: 0.98

4-50. Circular aluminum fins of constant rectangular profile are attached to a tube of outside diameter $D = 5$ cm. The fins have thickness $t = 2$ mm, height $L = 15$ mm, and thermal conductivity $k = 200$ W/(m · °C). The tube surface is maintained at a uniform temperature $T_0 = 200°C$, and the fins dissipate heat by convection into the ambient air at $T_\infty = 30°C$ with a heat transfer coefficient $h_\infty = 50 \, W/(m^2 \cdot °C)$. Determine the fin efficiency.

Answer: 0.96

4-51. An iron rod of length $L = 30 \, cm$, diameter $D = 1 \, cm$, and thermal conductivity $k = 65 \, W/(m \cdot °C)$ is attached horizontally to a large tank at temperature $T_0 = 200°C$, as illustrated in the accompanying figure. The rod is dissipating heat by convection into ambient air at $T_\infty = 20°C$ with a heat transfer coefficient $h_\infty = 15 \, W/(m^2 \cdot °C)$. What is the temperature of the rod at distances of 10 and 20 cm from the tank surface?

Answers: 90.1°C, 50.1°C

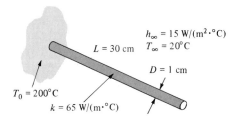

$h_\infty = 15 \, W/(m^2 \cdot °C)$
$T_\infty = 20°C$
$L = 30 \, cm$
$D = 1 \, cm$
$T_0 = 200°C$
$k = 65 \, W/(m \cdot °C)$

FIGURE P4-51

4-52. An aluminum fin of rectangular profile has a thickness $t = 2 \, mm$, length $L = 20 \, mm$, and thermal conductivity $k = 200 \, W/(m \cdot °C)$. Heat is dissipated from the fin by convection into ambient air at $T_\infty = 20°C$ with a heat transfer coefficient $h_\infty = 40 \, W/(m^2 \cdot °C)$. If the fin base is at $T_0 = 150°C$, calculate the heat loss from the fin into the ambient air.

4-53. An iron rod of length $L = 20 \, cm$, diameter $D = 2 \, cm$, and thermal conductivity $k = 65 \, W/(m \cdot °C)$ is attached to a large surface at $T_0 = 150°C$. The rod has dissipated heat into the ambient air at $T_\infty = 20°C$ with a heat transfer coefficient $h_\infty = 15 \, W/(m^2 \cdot °C)$. Calculate the heat transfer rate from the rod to the ambient.

4-54. A low carbon steel rod of length $L = 40 \, cm$, diameter $D = 10 \, mm$, and thermal conductivity $k = 40 \, W/(m \cdot °C)$ is placed in a medium where one of its end temperatures reaches to $T_0 = 400°C$ and is kept at that temperature. The ambient is at $T_\infty = 30°C$, and the heat transfer coefficient is $10 \, W/(m^2 \cdot °C)$. Determine (*a*) the temperature profile, (*b*) the fin efficiency, and (*c*) the heat transfer rate from the fin to the ambient.

4-55. An aluminum circular disc fin $[k = 200 \, W/m \cdot °C)]$ of constant thickness $t = 0.4 \, cm$ and diameter of 12 cm is placed on a copper tube of 6 cm outside diameter. The tube surface is maintained at a constant temperature $T_0 = 230°C$. Heat is transferred by convection from the fin surface into the ambient air at $T_\infty = 30°C$ with a heat transfer coefficient $h = 80 \, W/(m^2 \cdot °C)$. Calculate the rate of heat transfer from the fin surface to the surrounding air.

REFERENCES

1. Barzelay, M. E., K. N. Tong, and G. F. Holloway: "Effect of Pressure on Thermal Conductance of Contact Joints," NACA Tech. Note 3295, May 1955.
2. Gardner, K. A.: "Efficiency of Extended Surfaces," *Trans. ASME* **67**: 621–631 (1945).

CHAPTER
5

STEADY CONDUCTION WITH GENERATION: ANALYTIC AND NUMERICAL SOLUTIONS

Heat conduction problems involving energy generation in the solid come up in numerous engineering applications, including heat removal from the fuel elements of nuclear reactors, heat dissipation from an electrically heated source, and many others. In this chapter we present analytic and numerical solution techniques for determining temperature distribution and heat flow in one-dimensional steady-state heat conduction with energy generation in solids having simple shapes, such as a plane wall, a long cylinder, or a sphere. Problems of this type cannot be solved with the thermal resistance concept presented previously.

The methodology for analytic solution of such problems is systematically presented and illustrated with examples. The finite-difference formulation of heat conduction problems is introduced, numerical solution techniques are discussed, and computer applications are illustrated.

5-1 ANALYTIC SOLUTIONS

In this section we are concerned with the analytic solution of the one-dimensional steady-state heat conduction equation with energy generation and the determination of temperature distribution and/or heat flow rate in the medium. The problem of heat flow in simple geometries, such as a plane wall, a long cylinder, and a sphere, is studied, and illustrative examples are given.

Plane Wall with Energy Generation and Constant Surface Temperatures

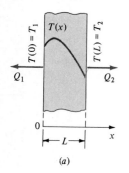

(a)

Consider a plane wall of thickness L, as illustrated in Fig. 5-1a. The boundary surfaces at $x = 0$ and $x = L$ are maintained at constant but different temperatures T_1 and T_2, respectively. The thermal conductivity of the slab k is constant, and within the slab energy is generated at a constant rate of g W/m³. The governing differential equation is Eq. (3-17a). Then the mathematical formulation of this heat conduction problem becomes

$$\frac{d^2T(x)}{dx^2} + \frac{g}{k} = 0 \qquad \text{in } 0 < x < L \tag{5-1a}$$

$$T(x) = T_1 \qquad \text{at } x = 0 \tag{5-1b}$$

$$T(x) = T_2 \qquad \text{at } x = L \tag{5-1c}$$

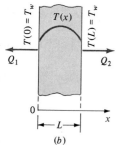

(b)

The temperature distribution $T(x)$ in the slab can readily be determined by direct integration of the differential equation and then application of the boundary conditions, as now described. Integrating Eq. (5-1a) twice yields

$$T(x) = -\frac{g}{2k}x^2 + C_1 x + C_2 \tag{5-2}$$

The application of the first boundary condition gives

$$C_2 = T_1 \tag{5-3}$$

and the application of the second boundary condition results in

$$T_2 = -\frac{g}{2k}L^2 + C_1 L + C_2 \tag{5-4a}$$

or

$$C_1 = \frac{gL}{2k} + \frac{T_2 - T_1}{L} \tag{5-4b}$$

Knowing C_1 and C_2, the temperature distribution in Eq. (5-2) becomes

$$T(x) = \frac{gL^2}{2k}\left[\frac{x}{L} - \left(\frac{x}{L}\right)^2\right] + (T_2 - T_1)\frac{x}{L} + T_1 \tag{5-5}$$

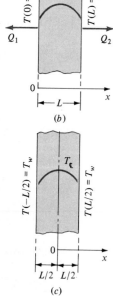

(c)

FIGURE 5-1
A plate with energy generation. (a) Asymmetric temperature distribution, (b) symmetric temperature distribution, and (c) shift of the origin for the symmetric case.

which is a *parabolic temperature distribution*. Here, the first term on the right-hand side is the contribution of the energy generation in the medium to the temperature distribution, and the second and third terms are the contributions of the temperatures of the boundary surfaces.

We consider some special cases of the result given by Eq. (5-5).

(a) No energy generation. For the case of no energy generation, we set $g = 0$, and Eq. (5-5) reduces to

$$T(x) = (T_2 - T_1)\frac{x}{L} + T_1 \tag{5-6}$$

which is a *linear temperature distribution*.

(b) Boundary surface temperatures are equal. If the boundary surface temperatures T_1 and T_2 are equal, $T_1 = T_2 \equiv T_w$, Eq. (5-5) reduces to

$$T(x) = \frac{gL^2}{2k}\left[\frac{x}{L} - \left(\frac{x}{L}\right)^2\right] + T_w \tag{5-7}$$

which is a *parabolic and symmetrical temperature* distribution having a maximum temperature at the symmetry axis $x = L/2$, as illustrated in Fig. 5-1b. For this particular case the problem has both geometric and thermal symmetry about the centerline of the plane. In such situations it is convenient to formulate the problem by shifting the origin of the x coordinate to the centerline of the plate, as illustrated in Fig. 5-1c. Then the mathematical formulation of the problem is considered only for one-half of the plate, $0 < x < L/2$, as given below:

$$\frac{d^2T(x)}{dx^2} + \frac{1}{k}g = 0 \qquad \text{in } 0 < x < \frac{L}{2} \tag{5-8a}$$

$$\frac{dT(x)}{dx} = 0 \qquad \text{at } x = 0 \tag{5-8b}$$

$$T(x) = T_w \qquad \text{at } x = \frac{L}{2} \tag{5-8c}$$

This problem is solved by integrating Eq. (5-8a) twice and applying the boundary conditions (5-8b) and (5-8c). The resulting expression for the temperature becomes

$$T(x) = \frac{gL^2}{2k}\left[\frac{1}{4} - \left(\frac{x}{L}\right)^2\right] + T_w \tag{5-9}$$

where the origin of the x-coordinate is at the centerline of the slab, as shown in Fig. 5-1c.

The temperature at the centerline, $x = 0$, is obtained from Eq. (5-9) by setting $x = 0$.

$$T_{\updownarrow} = \frac{gL^2}{8k} + T_w \tag{5-10}$$

ENERGY LEAVING THE PLATE. The energy leaving the plate from its boundary surfaces is a quantity of practical interest that can be determined from the definition of the heat flux. We consider the symmetrical problem illustrated in Fig. 5-1c and the corresponding temperature profile given by Eq. (5-9).

The energy leaving the slab from the right boundary surface at $x = L/2$ is determined from

$$q\left(\frac{L}{2}\right) = -k\left.\frac{dT(x)}{dx}\right|_{x=\frac{L}{2}} \tag{5-11a}$$

Introducing the temperature given by Eq. (5-9) into Eq. (5-11a) and setting $x = L/2$, the heat flux at the boundary surface $x = L/2$ becomes

$$q\left(\frac{L}{2}\right) = \frac{gL}{2} \qquad \text{W/m}^2 \tag{5-11b}$$

Since the quantity $gL/2$ is positive, the heat flux $q(L/2)$ is positive. This implies that the heat flow at the boundary surface $x = L/2$ is in the positive x direction, or outward.

Similarly, the heat flux at the left boundary surface $x = -L/2$ is determined from

$$q\left(-\frac{L}{2}\right) = -k \left.\frac{dT(x)}{dx}\right|_{x=-\frac{L}{2}} \tag{5-12a}$$

Introducing the temperature profile Eq. (5-9) into Eq. (5-12a) and setting $x = -L/2$, the heat flux at $x = -L/2$ becomes

$$q\left(-\frac{L}{2}\right) = -\frac{gL}{2} \qquad \text{W/m}^2 \tag{5-12b}$$

Here, the quantity $-gL/2$ is negative, and so $q(-L/2)$ is also negative. This implies that the heat flow at the boundary surface $x = -L/2$ is in the negative x direction, or outward.

PLANE WALL WITH ENERGY GENERATION AND CONVECTION. To illustrate the solution of a heat conduction problem for a plane wall with energy generation at a constant rate g W/m^3 and with a convection boundary condition, we consider the physical problem illustrated in Fig. 5-2. Here, the boundary surface at $x = 0$ is insulated (i.e., adiabatic), and the boundary surface at $x = L$ dissipates heat by convection with a heat transfer coefficient h_∞ into an ambient at temperature T_∞. The mathematical formulation of the problem is given by

$$\frac{d^2T(x)}{dx^2} + \frac{g}{k} = 0 \qquad \text{in } 0 < x < L \tag{5-13a}$$

$$\frac{dT(x)}{dx} = 0 \qquad \text{at } x = 0 \tag{5-13b}$$

$$-k\frac{dT(x)}{dx} = h_\infty[T(x) - T_\infty] \qquad \text{at } x = L \tag{5-13c}$$

FIGURE 5-2
A plate with energy generation
subject to convective and
insulated boundary condition.

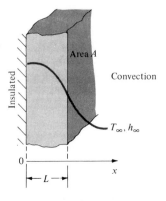

To solve this heat conduction problem, Eq. (5-13a) is integrated twice. The first integration gives

$$\frac{dT(x)}{dx} = -\frac{g}{k}x + C_1 \qquad (5\text{-}14)$$

and the application of the boundary condition Eq. (5-13b) establishes C_1 as

$$0 = 0 + C_1 \qquad \text{or} \qquad C_1 = 0 \qquad (5\text{-}15)$$

Then

$$\frac{dT(x)}{dx} = -\frac{g}{k}x \qquad (5\text{-}16)$$

The integration of Eq. (5-14) with $C_1 = 0$ results in

$$T(x) = -\frac{g}{2k}x^2 + C_2 \qquad (5\text{-}17)$$

and the application of the second boundary condition gives

$$-k\left(-\frac{g}{k}L\right) = h_\infty\left(-\frac{g}{2k}L^2 + C_2 - T_\infty\right)$$

or

$$C_2 = \frac{gL^2}{2k} + \frac{gL}{h_\infty} + T_\infty \qquad (5\text{-}18)$$

Then the temperature distribution becomes

$$T(x) = \frac{gL^2}{2k}\left[1 - \left(\frac{x}{L}\right)^2\right] + \frac{gL}{h_\infty} + T_\infty \qquad (5\text{-}19)$$

We now examine the physical significance of the two limiting cases, $h_\infty \to \infty$ and $h_\infty \to 0$, of the solution (5-19).

(a) $h_\infty \to \infty$. For very large values of the heat transfer coefficient, $h_\infty \to \infty$, the second term on the right vanishes and Eq. (5-19) reduces to

$$T(x) = \frac{gL^2}{2k}\left[1 - \left(\frac{x}{L}\right)^2\right] + T_\infty \qquad (5\text{-}20)$$

This result implies that as $h_\infty \to \infty$, the thermal resistance between the wall surface and the fluid is zero, and hence the surface temperature at $x = L$ equals the ambient temperature T_∞.

(b) $h_\infty \to 0$. For very small values of the heat transfer coefficient, $h_\infty \to 0$, the solution (5-19) shows that the temperature $T(x)$ becomes infinite. The physical significance of this particular situation is better envisioned by recalling the boundary condition (5-13c). As $h_\infty \to 0$, the boundary conditions (5-13c) reduces to

$$\frac{dT(x)}{dx} \to 0 \qquad \text{as } h_\infty \to 0$$

This result implies that both boundaries of the plate, $x = 0$ and $x = L$, are insulated while energy is continuously generated within the medium. As the generated energy has no way to escape through the insulated boundaries, the temperature of the slab continuously increases, or *the problem does not have a steady-state solution.*

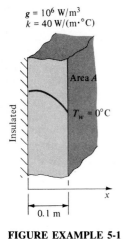

$g = 10^6$ W/m^3
$k = 40$ W/(m·°C)

Area A

Insulated

$T_w = 0$°C

x

0.1 m

FIGURE EXAMPLE 5-1

Example 5-1. Consider a slab of thickness 0.1 m. One of the boundary surfaces, that at $x = 0$, is kept insulated, and the other, at $x = 0.1$ m, is kept at 0°C. There is uniform energy generation at a rate of 10^6 W/m^3 in the solid, and the thermal conductivity is constant [$k = 40$ W/(m · °C)]. Determine the temperature of the insulated surface.

Solution. The problem is illustrated in the accompanying figure. Given $L = 0.1$ m, $k = 40$ W/(m · °C), $g = 10^6$ W/m^3, and $T_w = 0$°C, Eq. (5-20) can be used to determine the temperature distribution in the slab. We find

$$T(x) = \frac{gL^2}{2k}\left[1 - \left(\frac{x}{L}\right)^2\right] + T_w$$

$$= \frac{10^6 \times (0.1)^2}{2 \times 40}\left[1 - \left(\frac{x}{0.1}\right)^2\right] + 0$$

$$= 125(1 - 100x^2)$$

and the temperature of the insulated surface is calculated by setting $x = 0$:

$$T(0) = 125(1 - 100 \times 0)$$

$$= 125°C$$

Solid Cylinder with Energy Generation and Constant Surface Temperature

Consider a long solid cylinder of radius $r = b$, as illustrated in Fig. 5-3. The boundary surface is maintained at a uniform temperature T_w. There is energy generation in the solid at a constant rate of g W/m^3, and the thermal conductivity is constant. We wish to determine the temperature distribution and the heat flow rate in the solid.

The heat conduction equation is given by Eq. (3-17b), and the boundary condition at the outer surface is as specified above. The boundary condition at the center, $r = 0$, is established from the fact that the problem possesses geometric and thermal symmetry about the centerline, and hence dT/dr must be zero at $r = 0$. With this consideration, the mathematical formulation of the

FIGURE 5-3
A long solid cylinder with energy generation.

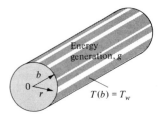

Energy generation, g

b

0

r

$T(b) = T_w$

problem is given by

$$\frac{1}{r}\frac{d}{dr}\left[r\frac{dT(r)}{dr}\right] + \frac{g}{k} = 0 \qquad \text{in } 0 < r < b \qquad (5\text{-}21a)$$

$$\frac{dT(r)}{dr} = 0 \qquad \text{at } r = 0 \qquad (5\text{-}21b)$$

$$T(r) = T_w \qquad \text{at } r = b \qquad (5\text{-}21c)$$

This problem can be solved by direct integration, as we did for the case of the plate problem. The first integration of Eq. (5-21a) gives

$$r\frac{dT(r)}{dr} = -\frac{g}{2k}r^2 + C_1$$

or

$$\frac{dT(r)}{dr} = -\frac{g}{2k}r + \frac{C_1}{r} \qquad (5\text{-}22)$$

The application of the boundary condition (5-21b) requires

$$C_1 = 0$$

so that

$$\frac{dT(r)}{dr} = -\frac{g}{2k}r \qquad (5\text{-}23)$$

Integration of Eq. (5-23) gives

$$T(r) = -\frac{g}{4k}r^2 + C_2 \qquad (5\text{-}24)$$

and the application of the boundary condition (5-21c) establishes C_2:

$$T_w = -\frac{g}{4k}b^2 + C_2$$

or

$$C_2 = \frac{gb^2}{4k} + T_w \qquad (5\text{-}25)$$

Introducing Eq. (5-25) into (5-24) yields the temperature distribution in the cylinder as

$$T(r) = \frac{gb^2}{4k}\left[1 - \left(\frac{r}{b}\right)^2\right] + T_w \qquad (5\text{-}26)$$

The heat flux anywhere in the medium is determined from its definition,

$$q(r) = -k\frac{dT(r)}{dr} \qquad (5\text{-}27)$$

Introducing the temperature profile equation, Eq. (5-26) into Eq. (5-27), we obtain

$$q(r) = \frac{gr}{2} \qquad (5\text{-}28)$$

Generally, the heat flux at the boundary surface is a quantity of practical interest. It is immediately determined by setting $r = b$ in Eq. (5-28),

$$q(b) = \frac{gb}{2} \qquad \text{W/m}^2 \tag{5-29}$$

Here, since $gb/2$ is a positive quantity, the heat flux $q(b)$ is positive, and hence the heat flow is in the positive r direciton or outward. This result is expected from physical considerations.

In the problem considered here, the highest temperature occurs at the center of the cylinder, and the centerline temperature $T_{\cent}$ is obtained from Eq. (5-26) by setting $r = 0$.

$$T_{\cent} = \frac{gb^2}{4k} + T_w \tag{5-30}$$

Clearly, the larger the radius, the larger the generation rate, or the smaller the thermal conductivity, the larger the center temperature is.

SOLID CYLINDER WITH ENERGY GENERATION AND CONVECTION. In the above discussion we considered a problem involving a solid cylinder subjected to a constant temperature boundary condition at the outer surface. However, in most practical applications energy is dissipated by convection from the outer surface into an ambient at a specified constant temperature T_∞, as illustrated in Fig. 5-4. Therefore, we consider heat conduction in a solid cylinder of radius $r = b$ with energy generation at a constant rate of g W/m^3, but subjected to convection at the outer surface with a heat transfer coefficient h_∞ into an ambient at a temperature T_∞. The mathematical formulation of this problem is given by

$$\frac{1}{r}\frac{d}{dr}\left[r\frac{dT(r)}{dr}\right] + \frac{g}{k} = 0 \qquad \text{in } 0 < r < b \tag{5-31a}$$

$$\frac{dT(r)}{dr} = 0 \qquad \text{at } r = 0 \tag{5-31b}$$

$$-k\frac{dT(r)}{dr} = h_\infty[T(r) - T_\infty] \qquad \text{at } r = b \tag{5-31c}$$

The solution for the temperature distribution $T(r)$ is developed by following the general procedure described previously. That is, the integration of Eq.

FIGURE 5-4
A long solid cylinder with
energy generation subject to
convection boundary condition.

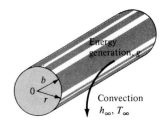

(5-31a) and the application of the boundary condition (5-31b) gives $C_1 = 0$; then

$$\frac{dT(r)}{dr} = -\frac{g}{2k}r \qquad (5\text{-}32)$$

Integration of this equation gives

$$T(r) = -\frac{g}{4k}r^2 + C_2 \qquad (5\text{-}33a)$$

and then the application of the boundary condition (5-31c) gives the integration constant C_2 as

$$C_2 = \frac{g}{4k}b^2 + \frac{g}{2h_\infty}b + T_\infty \qquad (5\text{-}33b)$$

Knowing C_2 the temperature distribution $T(r)$ in the cylinder becomes

$$T(r) = \frac{gb^2}{4k}\left[1 - \left(\frac{r}{b}\right)^2\right] + \frac{gb}{2h_\infty} + T_\infty \qquad (5\text{-}34)$$

We now examine the physical significance of two limiting cases, $h_\infty \to \infty$ and $h_\infty \to 0$, of the solution (5-34).

(a) $h_\infty \to \infty$. For very large values of the heat transfer coefficient, $h_\infty \to \infty$, the second term on the right-hand side vanishes, and Eq. (5-34) reduces to

$$T(r) = \frac{gb^2}{4k}\left[1 - \left(\frac{r}{b}\right)^2\right] + T_\infty \qquad (5\text{-}35)$$

which is identical to Eq. (5-26) for the case of constant surface temperature. This is expected, because thermal resistance between the surface and the ambient vanishes as $h_\infty \to \infty$, and the surface temperature is equal to the ambient temperature T_∞.

(b) $h_\infty \to 0$. For very small values of the heat transfer coefficient, $h_\infty \to 0$, the solution (5-34) shows that the temperature $T(r)$ becomes infinite. The physical significance of this particular situation becomes apparent from the boundary condition (5-31c), which becomes

$$\frac{dT(r)}{dr} \to 0 \qquad \text{as } h_\infty \to 0$$

Then we have a physical problem involving a solid cylinder with energy generation in the solid and an insulated outer boundary. As the energy generated has no way to escape from the insulated boundary, the temperature continuously increases; hence *the problem has no steady-state solution.*

The heat flux $q(r)$ anywhere in the cylinder is determined from the definition

$$q(r) = -k\frac{dT(r)}{dr} \qquad (5\text{-}36)$$

Introducing the solution (5-34) into Eq. (5-36), we obtain

$$q(r) = \frac{g}{2} r \qquad (5\text{-}37)$$

which is identical with the result given by Eq. (5-28) for a cylinder with constant surface temperature.

Example 5-2. A 3-mm-diameter chrome-nickel wire of thermal conductivity 20 W/(m · °C) is heated electrically by the passage of electric current, which generates energy within the wire at a uniform rate of 10^9 W/m³. If the surface of the wire is maintained at 100°C, determine the center temperature of the wire.

Solution. Given: $b = 0.0015$ m, $k = 20$ W/(m · °C), $g = 10^9$ W/m³, and $T_w = 100$°C. Equation (5-26) or (5-34) with $h_\infty \to \infty$ can be used to determine the temperature distribution $T(r)$ as

$$T(r) = \frac{gb^2}{4k} \left[1 - \left(\frac{r}{b} \right)^2 \right] + T_w$$

$$= \frac{10^9 \times (0.0015)^2}{4 \times 20} \left[1 - \left(\frac{r}{0.0015} \right)^2 \right] + 100$$

$$= 28.12(1 - 4.44 \times 10^5 \, r^2) + 100$$

The temperature of the center, $r = 0$, becomes

$$T(0) = 128.12°C$$

Example 5-3. In a cylindrical fuel element for a gas-cooled nuclear reactor, the energy generation from within the fuel element due to fission can be approximated by the relation

$$g(r) = g_0 \left[1 - \left(\frac{r}{b} \right)^2 \right] \qquad \text{W/m}^3$$

where b is the radius of the fuel element and g_0 is constant. The boundary surface at $r = b$ is maintained at a uniform temperature T_w.

(a) Assuming one-dimensional steady heat flow, develop a relation for the temperature drop from the centerline to the surface of the fuel element.

(b) For radius $b = 1$ cm, thermal conductivity $k = 10$ W/(m · °C), and $g = 1.6 \times 10^8$ W/m³, calculate the temperature drop from the centerline to the surface.

Solution.
(a) The mathematical formulation of the problem is given as

$$\frac{1}{r} \frac{d}{dr} \left[r \frac{dT(r)}{dr} \right] + \frac{g(r)}{k} = 0 \qquad \text{in } 0 < r < b$$

$$\frac{dT(r)}{dr} = 0 \qquad \text{at } r = 0$$

$$T(r) = T_w \qquad \text{at } r = b$$

where

$$g(r) = g_0 \left[1 - \left(\frac{r}{b} \right)^2 \right] \qquad \text{W/m}^3$$

Substituting the energy generation rate $g(r)$ into the differential equation and rearranging results:

$$\frac{d}{dr}\left(r\frac{dT}{dr}\right) = -\frac{g}{k}\left(r - \frac{r^3}{b^2}\right)$$

The first integration of this equation gives

$$r\frac{dT}{dr} = -\frac{g}{k}\left(\frac{r^2}{2} - \frac{r^4}{4b^2}\right) + C_1$$

and the application of the boundary condition at $r = 0$ results in $C_1 = 0$. Then with $C_1 = 0$, we have

$$\frac{dT}{dr} = -\frac{g}{k}\left(\frac{r}{2} - \frac{r^3}{4b^2}\right)$$

Integration of this equation gives

$$T(r) = -\frac{g}{k}\left(\frac{r^2}{4} - \frac{r^4}{16b^2}\right) + C_2$$

and then the application of the boundary condition at $r = b$ gives C_2 as

$$C_2 = \frac{3gb^2}{16k} + T_w$$

Therefore, the temperature distribution in the fuel element becomes

$$T(r) = \frac{gr^2}{4k}\left[1 - \left(\frac{r}{2b}\right)^2\right] + \frac{3gb^2}{16k} + T_w$$

The center temperature $T_\mathcal{L}$ is obtained by setting $r = 0$ in this result. Then the temperature drop from the center to the surface, $T_\mathcal{L} - T_w$, becomes

$$T_\mathcal{L} - T_w = \frac{3gb^2}{16k}$$

(b) For $b = 1$ cm, $k = 10$ W/(m · °C), and $g = 1.6 \times 10^8$ W/m^3, the temperature drop is calculated as

$$T_\mathcal{L} - T_w = \frac{3 \times (1.6 \times 10^8) \times (1 \times 10^{-2})^2}{16 \times 10}$$

$$= 300°C$$

Example 5-4. A hollow cylindrical fuel element of inner radius a and outer radius b is heated uniformly within the entire volume at a constant rate of g W/m^3 as a result of disintegration of radioactive elements. The inner and outer surfaces of the fuel element are at zero temperature, and the thermal conductivity of the material is constant. Develop an expression for the one-dimensional steady temperature distribution $T(r)$ within the cylinder.

Solution. The mathematical formulation of the problem is given as

$$\frac{1}{r}\frac{d}{dr}\left[r\frac{dT(r)}{dr}\right] + \frac{g}{k} = 0 \qquad \text{in } a < r < b$$

$$T(r) = 0 \qquad \text{at } r = a$$

$$T(r) = 0 \qquad \text{at } r = b$$

Integrating the differential equation twice, we have

$$T(r) = -\frac{g}{4k} r^2 + C_1 \ln r + C_2$$

and the application of the boundary conditions gives the integration constants C_1 and C_2 as

$$C_1 = \frac{g}{4k} \frac{b^2 - a^2}{\ln (b/a)}$$

$$C_2 = \frac{g}{4k} \left[a^2 - \frac{b^2 - a^2}{\ln (b/a)} \ln a \right]$$

After rearranging, the temperature distribution $T(r)$ becomes

$$T(r) = \frac{g}{4k} \left[a^2 - r^2 + \frac{(b^2 - a^2)}{\ln (b/a)} \ln \frac{r}{a} \right]$$

Solid Sphere with Energy Generation and Convection

We are now concerned with the problem of steady-state heat conduction in a solid sphere of radius b, having constant thermal conductivity k and uniform energy generation throughout the volume at a constant rate of g W/m^3. Heat is dissipated by convection from the outer surface into an ambient at temperature T_∞ with a heat transfer coefficient h_∞. We wish to determine the temperature distribution $T(r)$ in the sphere.

The mathematical formulation of the problem is given by

$$\frac{1}{r^2} \frac{d}{dr} \left[r^2 \frac{dT(r)}{dr} \right] + \frac{g}{k} = 0 \qquad \text{in } 0 < r < b \qquad (5\text{-}38a)$$

$$\frac{dT(r)}{dr} = 0 \qquad \text{at } r = 0 \qquad (5\text{-}38b)$$

$$-k \frac{dT(r)}{dr} = h_\infty [T(r) - T_\infty] \qquad \text{at } r = b \qquad (5\text{-}38c)$$

This problem is readily solved by following the methodology developed previously.

The first integration of the differential equation (5-38a) gives

$$r^2 \frac{dT}{dr} = -\frac{g}{3k} r^3 + C_1 \qquad (5\text{-}39a)$$

and the application of the boundary condition at $r = 0$ yields

$$C_1 = 0 \qquad (5\text{-}39b)$$

The integration of Eq. (5-39a) with $C_1 = 0$ results in

$$T(r) = -\frac{g}{6k} r^2 + C_2 \qquad (5\text{-}40a)$$

and the application of the boundary condition at $r = b$ gives the constant C_2 as

$$-\frac{gb}{3} + h_\infty\left(-\frac{gb^2}{6k} + C_2\right) = hT_\infty$$

or

$$C_2 = \frac{gb}{3h_\infty} + \frac{g_0 b^2}{6k} + T_\infty \qquad (5\text{-}40b)$$

Then the temperature distribution in the sphere becomes

$$T(r) = \frac{gb^2}{6k}\left[1 - \left(\frac{r}{b}\right)^2\right] + \frac{gb}{3h_\infty} + T_\infty \qquad (5\text{-}41)$$

The two special cases of this solution for $h_\infty \to \infty$ and $h_\infty \to 0$ are similar to those discussed for the cases of the plate and solid cylinder. That is, as $h_\infty \to \infty$, the solution (5-41) reduces to

$$T(r) = \frac{gb^2}{6k}\left[1 - \left(\frac{r}{b}\right)^2\right] + T_\infty \qquad (5\text{-}42)$$

which corresponds to the sphere problem subject to a constant surface temperature $T_{\text{wall}} = T_\infty$ at $r = b$.

For the case of $h_\infty \to 0$, the surface at $r = b$ becomes insulated, and the problem has no steady-state solution because the energy generated has no way to escape from the medium.

Example 5-5. A solid sphere of radius 5 cm and thermal conductivity 20 W/ (m · °C) is heated uniformly throughout its volume at a rate of 2×10^6 W/m³, and heat is dissipated by convection to ambient air at 25°C with a heat transfer coefficient of 100 W/(m² · °C). Determine the steady temperatures at the center and the outer surface of the sphere.

Solution. Given: $b = 5\,\text{cm} = 0.05\,\text{m}$, $k = 20\,\text{W/(m · °C)}$, $g = 2 \times 10^6\,\text{W/m}^3$, $T_\infty = 25°\text{C}$, and $h_\infty = 100\,\text{W/(m}^2 · °\text{C)}$. Equation (5-41) is used to determine the center temperature $T_{\mathcal{C}}$ by setting $r = 0$:

$$T_{\mathcal{C}} = \frac{gb^2}{6k} + \frac{gb}{3h_\infty} + T_\infty$$

$$= \frac{(2 \times 10^6) \times (0.05)^2}{6 \times 20} + \frac{2 \times 10^6 \times 0.05}{3 \times 100} + 25$$

$$= 400°\text{C}$$

and the outer surface temperature T_w is detemined by setting $r = b$:

$$T_w = \frac{gb}{3h_\infty} + T_\infty$$

$$= \frac{2 \times 10^6 \times 0.05}{3 \times 100} + 25$$

$$= 358.33°\text{C}$$

5-2 NUMERICAL ANALYSIS

The analytic methods of solution become very difficult and even impossible to use in many practical problems that arise in engineering applications, such as heat conduction in complex geometries, nonlinear problems, thermal systems involving coupling between the elements, and many others. The numerical techniques are most powerful for solving such complicated problems. In this section we illustrate the basic concepts of numerical analysis for solving simple heat conduction problems using the finite-difference scheme.

When a heat conduction problem is solved exactly by an analytic method, the resulting solution satisfies the governing differential equation at every point in the region as well as at the boundaries. When a numerical scheme such as finite-differences is used, the differential equation of heat conduction is transformed into a set of algebraic equations that are satisfied only at a selected number of discrete nodes over the region.

Therefore, our starting point in finite-difference analysis is the development of the finite-difference equations for the nodes selected over the region. Such equations can be developed either by finite-differencing the derivatives in the differential equation of heat conduction by Taylor series expansion or by writing an energy balance for a differential volume element about a nodal point. Here we prefer the latter approach because it gives better insight into the physical nature of finite-difference formulation of heat conduction problems.

Once the finite-difference form of such equations has been developed, the solution of the heat conduction problem is transformed to the solution of a system of algebraic equations, the number of which is equal to the number of nodal points chosen over the region. Such a system of algebraic equations can readily be solved with a digital computer by using the standard subroutines for solving algebraic equations.

In this section we describe the finite-difference formulation of one-dimensional steady-state heat conduction problems for a plane wall, cylinder, and sphere, and illustrate the application with typical examples.

Plane Wall—Finite-Difference Formulation

Consider steady-state heat conduction in a plane wall (slab) of thickness L and constant thermal conductivity k, confined to a region $0 \leq x \leq L$ in which energy is generated at a rate of $g(x)$ W/m^3, that varies across the thickness of the slab. Figure 5-5a shows the geometry and the coordinates.

The temperature distribution $T(x)$ in the slab is governed by the steady-state heat conduction equation (3-17a); that is

$$\frac{d^2 T(x)}{dx^2} + \frac{1}{k}\, g(x) = 0 \qquad \text{in } 0 < x < L \qquad (5\text{-}43)$$

which should be solved subject to appropriate boundary conditions at $x = 0$ and $x = L$.

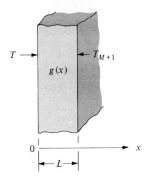

(a) Geometry and coordinate

$$T_1 \quad T_2 \quad T_3 \qquad T_{m-1} \quad T_m \quad T_{m+1} \qquad\qquad T_M \quad T_{M+1}$$

$$1 \quad 2 \quad \underbrace{}_{3\ \Delta x} \qquad m-1 \quad m \quad m+1 \quad \underbrace{}_{\Delta x} \qquad M \quad M+1$$

$$x = 0 \qquad\qquad (m - \tfrac{1}{2})\ (m + \tfrac{1}{2}) \qquad\qquad\qquad x = L$$

(b) Selection of nodes

FIGURE 5-5
Nomenclature for one-dimensional finite difference formulation of plane wall.

To solve this problem numerically with finite-differences, we need the finite-difference form of this equation written for a selected number of nodal points within the medium. Therefore, the first step in the analysis is to divide the region $0 \leq x \leq L$ into M equal subregions, each of size Δx, given by

$$\Delta x = \frac{L}{M} \tag{5-44}$$

Hence there are $M + 1$ nodes, from $m = 1$ to $m = M + 1$, as illustrated in Fig. 5-5b. In this notation, the node m corresponds to a location whose coordinate x is given by

$$x = (m - 1)\,\Delta x \qquad m = 1, 2, 3, \ldots, M + 1 \tag{5-45}$$

or $\qquad\qquad x + \Delta x = m\,\Delta x$

where $m = 1$ corresponds to the boundary surface at $x = 0$ and $m = M + 1$ corresponds to the boundary surface at $x = L$. Then the temperature $T(x)$ at the node m is denoted by

$$T(x) = T[(m - 1)\,\Delta x] \equiv T_m \tag{5-46}$$

The region $0 \leq x \leq b$ contains $M + 1$ nodes. Those for $m = 2, 3, \ldots, M$ are called the *internal nodes*, and $m = 1$ and $M + 1$ are called the *boundary nodes*.

To develop the finite-difference form of the energy equation (5-43) for an internal node m, we consider a differential volume element Δx around the node m, as illustrated in Fig. 5-6. The steady-state energy balance equation is stated as

$$\begin{pmatrix} \text{Rate of heat} \\ \text{entering} \\ \text{by conduction} \end{pmatrix} + \begin{pmatrix} \text{rate of} \\ \text{energy} \\ \text{generation} \end{pmatrix} = 0 \tag{5-47}$$

The right-hand side of this equation is zero because the steady-state conditions are assumed; hence the rate of increase of internal energy is zero.

The rate at which heat enters the element by conduction through the

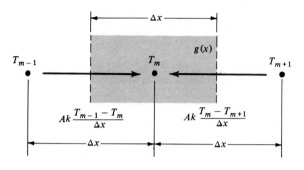

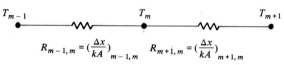

FIGURE 5-6
Energy balance at an internal
node m.

surfaces of the volume element is given by

$$\begin{pmatrix} \text{Rate of heat} \\ \text{entering} \\ \text{by conduction} \end{pmatrix} = Ak \, \frac{T_{m-1} - T_m}{\Delta x} + Ak \, \frac{T_{m+1} - T_m}{\Delta x}$$

$$= \frac{Ak}{\Delta x} (T_{m-1} - 2T_m + T_{m+1}) \qquad (5\text{-}48)$$

and the energy generation term becomes

$$\begin{pmatrix} \text{Rate of} \\ \text{energy} \\ \text{generation} \end{pmatrix} = A \, \Delta x g_m \qquad (5\text{-}49)$$

where g_m is the energy generation rate per unit volume at the node m
corresponding to the location $x = (m - 1) \, \Delta x$.

Introducing Eqs. (5-48) and (5-49) into Eq. (5-47), we obtain

$$\frac{Ak}{\Delta x} (T_{m-1} - 2T_m + T_{m+1}) + A \, \Delta x g_m = 0$$

or, after rearranging,

$$\boxed{(T_{m-1} - 2T_m + T_{m+1}) + \frac{(\Delta x)^2 g_m}{k} = 0} \qquad \text{for } m = 2, 3, \ldots, M \qquad (5\text{-}50)$$

Equation (5-50) is the finite-difference form of the steady-state heat
conduction equation (5-43) for an *internal node m*.

Clearly, the region contains $M + 1$ unknown node temperatures T_m
$(m = 1, 2, \ldots, M + 1)$, but Eq. (5-50) provides $M - 1$ algebraic equations. To
solve such a system, we need two more equations, which are obtained from the
consideration of the boundary conditions at the nodes $m = 1$ and $m = M + 1$.
We examine below the development of these two additional relations from the

boundary conditions for the cases of prescribed temperature, prescribed heat flux, and convection at the boundaries.

(a) Prescribed Temperature Boundary Condition. Suppose the temperatures at the boundaries $x = 0$ and $x = L$ are specified as f_1 and f_{M+1}, respectively. Then we have

$$\boxed{T_1 = f_1 \qquad T_{M+1} = f_{M+1}} \qquad (5\text{-}51a,b)$$

which provide the two additional relations that are needed to solve the system (5-50), since f_1 and f_{M+1} are known quantities.

(b) Prescribed Heat Flux Boundary Condition. Suppose the heat flux q_0 W/m^2, entering the plane wall through the boundary surface at $x = 0$, is prescribed. To develop an additional finite-difference equation for the node at $x = 0$, we write an energy balance equation for a volume element of thickness $\Delta x/2$ at the node $m = 1$, as illustrated in Fig. 5-7a.

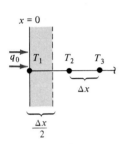

$x = 0$

Δx

$\dfrac{\Delta x}{2}$

(a) Node at $x = 0$

$$\begin{pmatrix} \text{Rate of heat} \\ \text{entering through} \\ \text{boundary surface} \end{pmatrix} + \begin{pmatrix} \text{rate of heat} \\ \text{entering by} \\ \text{conduction} \end{pmatrix} + \begin{pmatrix} \text{rate of} \\ \text{energy} \\ \text{generation} \end{pmatrix} = 0 \qquad (5\text{-}52)$$

If we consider an area A, the mathematical expressions for each of these three terms are written as

$$q_0 A + kA \frac{T_2 - T_1}{\Delta x} + \frac{\Delta x}{2} A g_1 = 0$$

This result is rearranged in the form

$$\boxed{2T_2 - 2T_1 + \frac{(\Delta x)^2 \, g_1}{k} + \frac{2\,\Delta x \, q_0}{k} = 0} \qquad \text{for } m = 1 \qquad (5\text{-}53a)$$

which is the finite-difference form of the prescribed heat flux boundary condition at the surface $x = 0$, for the node $m = 1$.

Suppose the heat flux q_L W/m^2 entering the wall through the boundary surface at $x = L$ is prescribed. We write an energy balance equation for a differential volume element $\Delta x/2$ at the node $m = M + 1$, as illustrated in Fig. 5-7b, and obtain

$$q_L A + kA \frac{T_M - T_{M+1}}{\Delta x} + \frac{\Delta x}{2} A g_{M+1} = 0$$

This result is rearranged in the form

$$\boxed{2T_M - 2T_{M+1} + \frac{(\Delta x)^2 \, g_{M+1}}{k} + \frac{2\,\Delta x \, q_L}{k} = 0} \qquad \text{for } m = M + 1 \qquad (5\text{-}53b)$$

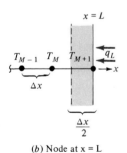

$x = L$

Δx

$\dfrac{\Delta x}{2}$

(b) Node at $x = L$

FIGURE 5-7
Finite differencing of prescribed heat flux boundary conditions.

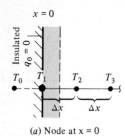

(a) Node at x = 0

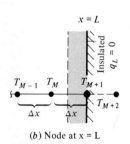

(b) Node at x = L

FIGURE 5-8
Finite differencing of insulated boundary conditions.

which is the finite-difference form of the prescribed heat flux boundary condition at the node $m = M + 1$.

For the *insulated boundary* (or geometric and thermal symmetry conditions), we have

$$q_0 = 0 \qquad \text{or/and} \qquad q_L = 0 \qquad (5\text{-}54)$$

Then Eqs. (5-53a) and (5-53b), respectively, reduce to

$$2T_2 - 2T_1 + \frac{(\Delta x)^2\, g_1}{k} = 0 \qquad (5\text{-}55a)$$

$$2T_M - 2T_{M+1} + \frac{(\Delta x)^2\, g_{M+1}}{k} = 0 \qquad (5\text{-}55b)$$

Note that Eq. (5-55a) can be obtained from Eq. (5-50) by setting $m = 1$ and taking $T_0 = T_2$, where T_0 represents the mirror image of T_2 because of symmetry, as illustrated in Fig. 5-8a. Similarly, Eq. (5-55b) can be obtained from Eq. (5-50) by setting $m = M$ and taking $T_M = T_{M+2}$, where T_M represents the mirror image of T_{M+2} because of the symmetry, as illustrated in Fig. 5-8b.

It is to be noted that the heat conduction problem in a slab with energy generation in the medium and prescribed heat fluxes at both boundaries *does not have a steady-state solution* unless the sum of the energy generation and the heat entering from one of the boundaries is equal to the heat leaving the slab from the other boundary.

(c) Convection Boundary Condition. Suppose the boundary surface at $x = 0$ is subjected to convection with a heat transfer coefficient $h_{\infty 1}$ into an ambient at temperature $T_{\infty 1}$, as illustrated in Fig. 5-9a. We consider an energy balance for a volume element of thickness $\Delta x/2$ at the node $m = 1$:

$$\begin{pmatrix} \text{Rate of heat} \\ \text{entering through} \\ \text{the surface} \\ \text{by convection} \end{pmatrix} + \begin{pmatrix} \text{rate of heat} \\ \text{entering by} \\ \text{conduction} \end{pmatrix} + \begin{pmatrix} \text{rate of} \\ \text{energy} \\ \text{generation} \end{pmatrix} = 0 \qquad (5\text{-}56)$$

Considering an area A, the mathematical expressions for each of the terms are written as

$$h_{\infty 1} A (T_{\infty 1} - T_1) + kA \frac{T_2 - T_1}{\Delta x} + \frac{\Delta x}{2} A g_1 = 0$$

After rearrangement, the finite-difference representation of the convection boundary condition at the node $m = 1$ becomes

$$2T_2 - \left(2 + \frac{2\,\Delta x\, h_{\infty 1}}{k}\right) T_1 + \frac{(\Delta x)^2\, g_1}{k} + \frac{2\,\Delta x\, h_{\infty 1}}{k} T_{\infty 1} = 0 \qquad \text{at } m = 1$$

$$(5\text{-}57a)$$

If the boundary surface at $x = L$ is subjected to convection with a heat transfer coefficient $h_{\infty 2}$ into an ambient at temperature $T_{\infty 2}$, the energy balance

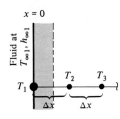

(a) Node at x = 0

equation (5-52) is applied to the node $m = M + 1$ to obtain

$$h_{\infty 2}A(T_{\infty 2} - T_{M+1}) + kA \frac{T_M - T_{M+1}}{\Delta x} + \frac{\Delta x}{2} A g_{M+1} = 0$$

After rearrangement, the finite-difference form of the convection boundary condition at the node $m = M + 1$ takes the form

$$2T_M - \left(2 + \frac{2\,\Delta x\,h_{\infty 2}}{k}\right)T_{M+1} + \frac{(\Delta x)^2\,g_{M+1}}{k} + \frac{2\,\Delta x\,h_{\infty 2}}{k}\,T_{\infty 2} = 0$$

at $m = M + 1$ (5-57b)

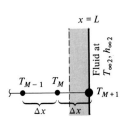

(b) Node at x = L

FIGURE 5-9
Finite differencing of convection boundary conditions.

We now summarize the basic steps discussed above for the finite-difference representation of a steady-state heat conduction problem for a plane wall with energy generation.

The region $0 \le x \le L$ is divided into M equal subregions, containing, in general, $M + 1$ unknown node temperatures T_m ($m = 1, 2, \ldots, M + 1$). Equation (5-50) provides $M - 1$ equations; the additional two equations are obtained from the boundary conditions for the problem. The finite-difference form of the boundary conditions are given by Eqs. (5-51), (5-53), and (5-57) for the prescribed temperature, prescribed heat flux, and convection boundary conditions, respectively. Then the problem involves $M + 1$ algebraic equations for the $M + 1$ unknown node temperatures T_m ($m = 1, 2, \ldots, M + 1$), which can readily be solved with a digital computer.

Example 5-6. Consider a slab of thickness 0.1 m with a thermal conductivity $k = 40$ W/(m · °C) in which energy is generated at a constant rate of 10^6 W/m³. The boundary surface at $x = 0$ is insulated, and the one at $x = 0.1$ m is subjected to convection with a heat transfer coefficient of 200 W/(m² · °C) into an ambient at a temperature of 150°C. The slab is subdivided into five equal subregions, as illustrated in the accompanying figure. Develop the finite-difference equations for the problem.

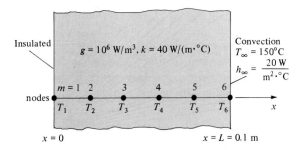

FIGURE EXAMPLE 5-6

Solution. Given: $L = 0.1$ m, $k = 40$ W/(m · °C), $g = 10^6$ W/m³, and $M = 5$. The size of the subdivision is

$$\Delta x = \frac{L}{M} = \frac{0.1}{5} = 0.02 \text{ m}$$

For the internal nodes, $m = 2$ to 5, the finite-difference equations are obtained from Eqs. (5-50) as

$$(T_{m-1} - 2T_m + T_{m+1}) + \frac{(0.02)^2 \times 10^6}{40} = 0$$

or

$$(T_{m-1} - 2T_m - T_{m+1}) + 10 = 0 \qquad \text{for } m = 2, 3, 4, 5$$

For the insulated boundary at $x = 0$, the finite-difference equation is obtained from Eq. (5-55a) as

$$2T_2 - 2T_1 + \frac{(0.02)^2 \times 10^6}{40} = 0$$

or

$$2T_2 - 2T_1 + 10 = 0 \qquad \text{for } m = 1$$

For the convection boundary at $x = L$, the finite-difference equation is obtained from Eq. (5-57b) as

$$2T_5 - \left(2 + \frac{2 \times 0.02 \times 200}{40}\right)T_6 + \frac{(0.02)^2 \times 10^6}{40} + \frac{2 \times 0.02 \times 200}{40} 150 = 0$$

or

$$2T_5 - 2.2T_6 + 40 = 0 \qquad \text{for } m = 6$$

Thus we have six algebraic equations for the six unknown node temperatures T_m $(m = 1, 2, \ldots, 6)$. We summarize these equations as

$$-2T_1 + 2T_2 = -10 \qquad \text{for } m = 1$$

$$T_1 - 2T_2 + T_3 = -10 \qquad \text{for } m = 2$$

$$T_2 - 2T_3 + T_4 = -10 \qquad \text{for } m = 3$$

$$T_3 - 2T_4 + T_5 = -10 \qquad \text{for } m = 4$$

$$T_4 - 2T_5 + T_6 = -10 \qquad \text{for } m = 5$$

$$2T_5 - 2.2T_6 = -40 \qquad \text{for } m = 6$$

These equations are written in the matrix form as

$$
\begin{bmatrix}
-2 & 2 & 0 & 0 & 0 & 0 \\
1 & -2 & 1 & 0 & 0 & 0 \\
0 & 1 & -2 & 1 & 0 & 0 \\
0 & 0 & 1 & -2 & 1 & 0 \\
0 & 0 & 0 & 1 & -2 & 1 \\
0 & 0 & 0 & 0 & 2 & -2.2
\end{bmatrix}
\begin{Bmatrix}
T_1 \\
T_2 \\
T_3 \\
T_4 \\
T_5 \\
T_6
\end{Bmatrix}
=
\begin{Bmatrix}
-10 \\
-10 \\
-10 \\
-10 \\
-10 \\
-40
\end{Bmatrix}
$$

Note that this is a tridiagonal matrix. Summarizing, with the finite-difference approach, the heat conduction equation and its boundary conditions are replaced by a set of algebraic equations. In this example the slab is divided into five subdivisions, resulting in six unknown node temperatures to be determined. If more subdivisions are considered, the result will be a larger number of algebraic equations, which can be solved by a digital computer using the standard sub-routines available for solving systems of algebraic equations.

Cylinder—Finite-Difference Formulation

The finite-difference formulation of the heat conduction problem for a plane wall, given previously, is not applicable to radial heat flow in cylindrical bodies, because the area A normal to the path of heat flow is not constant but varies linearly with radial position in the cylindrical coordinate system. The general procedure for developing the finite-difference equations for a cylindrical body is essentially similar to that for a plane wall, but the radial variation of the area needs to be included in the analysis.

Consider steady-state radial heat conduction in a long solid cylinder of radius $r = b$, in which energy is generated at a rate of $g(r)$ W/m^3. The temperature distribution $T(r)$ in the cylinder is governed by the heat conduction equation (3-17b), that is,

$$\frac{1}{r}\frac{d}{dr}\left[r\frac{dT(r)}{dr}\right] + \frac{1}{k}g(r) = 0 \qquad \text{in } 0 < r < b \qquad (5\text{-}58)$$

which should be solved subject to appropriate boundary conditions at $r = 0$ and $r = b$.

To develop the finite-difference form of this energy equation, the region $0 \le r \le b$ is divided into M cylindrical subregions, each of thickness Δr, given by

FIGURE 5-10
Nomenclature for one-dimensional finite-difference formulation for a long solid cylinder.

$$\Delta r = \frac{b}{M} \qquad (5\text{-}59)$$

as illustrated in Fig. 5-10a. Then the region $0 \le r \le b$ contains $M + 1$ nodes at

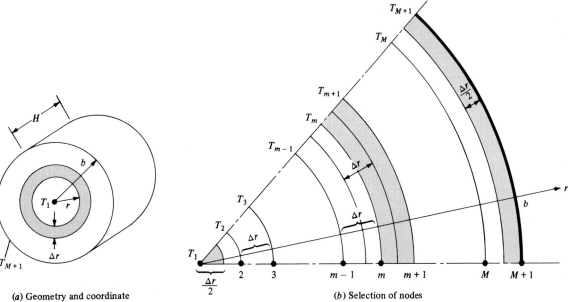

(a) Geometry and coordinate

(b) Selection of nodes

the locations

$$r = (m - 1)\,\Delta r \qquad m = 1, 2, \ldots, M + 1 \tag{5-60}$$

with the nodes $m = 1$ and $m = M + 1$ corresponding to the *center* and the *outer boundary surface* of the solid cylinder, respectively, and the nodes $m = 2, 3, \ldots, M$ being the *internal nodes* of the region. The problem involves $M + 1$ node temperatures, denoted by

$$T(r) = T[(m - 1)\,\Delta r] \equiv T_m \qquad m = 1, 2, \ldots, M + 1 \tag{5-61}$$

The finite-difference form of the heat conduction equation for the internal nodes is developed by considering the energy balance equation, Eq. (5-47). We apply this energy balance for a cylindrical volume element of thickness Δr about the node m illustrated in Fig. 5-10b. The conduction term becomes

$$\begin{pmatrix} \text{Rate of heat} \\ \text{entering} \\ \text{by conduction} \end{pmatrix} = A_{m-1,m}\,k\,\frac{T_{m-1} - T_m}{\Delta r} + A_{m+1,m}\,k\,\frac{T_{m+1} - T_m}{\Delta r}$$

$$= 2\pi Hk[(m - 1 - \tfrac{1}{2})T_{m-1} - 2mT_m + (m - 1 + \tfrac{1}{2})T_{m+1}] \tag{5-62a}$$

where

$$A_{m-1,m} = 2\pi\left[(m - 1)\,\Delta r - \frac{1}{2}\,\Delta r\right]H$$

$$A_{m+1,m} = 2\pi\left[(m - 1)\,\Delta r + \frac{1}{2}\,\Delta r\right]H$$

with H being the length of the cylinder. The rate of energy generation within the volume element is given by

$$\begin{pmatrix} \text{Rate of} \\ \text{energy} \\ \text{generation} \end{pmatrix} = A_m\,\Delta r g_m$$

$$= 2\pi H(m - 1)(\Delta r)^2\,g_m \tag{5-62b}$$

where $A_m = 2\pi H(m - 1)\,\Delta r$ and g_m are evaluated at the node m corresponding to the location $r = (m - 1)\,\Delta r$.

Introducing Eqs. (5-62) into the energy balance equation (5-47) and rearranging, we obtain the finite-difference form of the heat conduction equation (5-58) for the internal nodes of the cylindrical region

$$\boxed{\left[1 - \frac{1}{2(m - 1)}\right]T_{m-1} - 2T_m + \left[1 + \frac{1}{2(m - 1)}\right]T_{m+1} + \frac{(\Delta r)^2\,g_m}{k} = 0}$$

$$\text{for } m = 2, 3, \ldots, M \tag{5-63}$$

To determine the finite-difference equation for the central node $r = 0$ or $m = 1$, we apply the energy balance equation (5-47) for the volume element of radius $\Delta r/2$ about the node $m = 1$ shown in Fig. 5-10b. Various terms are evaluated as

$$\begin{pmatrix} \text{Rate of heat} \\ \text{entering} \\ \text{by conduction} \end{pmatrix} = A_{2,1} k \frac{T_2 - T_1}{\Delta r}$$

$$= \pi H k (T_2 - T_1) \qquad (5\text{-}64a)$$

The rate of energy generation in the volume element is

$$\begin{pmatrix} \text{Rate of} \\ \text{energy} \\ \text{generation} \end{pmatrix} = \pi \left(\frac{\Delta r}{2} \right)^2 H g_1 \qquad (5\text{-}64b)$$

Introducing Eqs. (5-64) into the energy balance equation (5-47), we obtain the finite-difference equation for the center node $m = 1$ as

$$4(T_2 - T_1) + \frac{(\Delta r)^2 g_1}{k} = 0 \qquad \text{for } m = 1 \qquad (5\text{-}65)$$

The system given by Eqs. (5-63) and (5-65) involves $M + 1$ unknown node temperatures T_m ($m = 1, 2, \ldots, M + 1$), but only M equations. An additional equation is obtained from the boundary condition at $r = b$, which can be a prescribed temperature, prescribed heat flux, or convection boundary condition. Each of these cases is now examined.

(a) Prescribed Temperature Boundary Condition. If the temperature of the boundary surface at the node M is specified as, say, f_{M+1}, we have

$$T_{M+1} = f_{M+1} \qquad \text{at } m = M + 1 \qquad (5\text{-}66a)$$

(b) Prescribed Heat Flux Boundary Condition. For one-dimensional heat conduction in a solid cylinder, no steady-state can be established under a prescribed heat flux condition at the boundary surface $r = b$. The reason for this is that heat entering the cylinder has no other way to escape; hence it will build up continuously. However, in a hollow cylinder, heat can escape from the inner boundary while entering from the boundary at $r = b$, or vice versa; hence the steady-state solution is possible. For this reason, the finite-difference form of the prescribed heat flux boundary condition at $r = b$ has practical application only for hollow cylinder problems, which will be discussed next. With such a restriction on its application, we now present the finite-difference form of the prescribed heat flux condition at $r = b$.

Let q_b W/m^2, the heat flux entering the cylinder from the outer boundary surface at $r = b$, be prescribed. The application of the energy balance equation

(5-52) to the boundary node $M + 1$ at $r = b$ yields

$$A_{M,M+1}k\frac{T_M - T_{M+1}}{\Delta r} + A_{M+1}q_{M+1} + A^*_{M+1}\frac{\Delta r}{2}g_{M+1} = 0$$

where

$$A_{M,M+1} = 2\pi\left(M\,\Delta r - \frac{1}{2}\,\Delta r\right)H$$

$$A_{M+1} = 2\pi(M\,\Delta r)H$$

$$A^*_{M+1} = 2\pi\left(M\,\Delta r - \frac{1}{4}\,\Delta r\right) \cong 2\pi(M\,\Delta r)H$$

and H is the length of the cylinder.

Introducing the area terms into the above expression and rearranging, we obtain

$$\boxed{\left(1 - \frac{1}{2M}\right)T_M - \left(1 - \frac{1}{2M}\right)T_{M+1} + \frac{\Delta r\,q_b}{k} + \frac{(\Delta r)^2\,g_{M+1}}{2k} = 0}$$

$$\text{for } m = M + 1 \qquad (5\text{-}66b)$$

(c) Convection Boundary Condition. If the outer boundary surface is subjected to convection with a heat transfer coefficient h_∞ into an ambient at temperature T_∞, we apply an energy balance equation similar to that given by Eq. (5-56) for the node M and obtain

$$A_{M,M+1}k\frac{T_M - T_{M+1}}{\Delta r} + A_{M+1}h_\infty(T_\infty - T_{M+1}) + A^*_{M+1}\frac{\Delta r}{2}g_{M+1} = 0$$

where

$$A_{M,M+1} = 2\pi\left(M\,\Delta r - \frac{1}{2}\,\Delta r\right)H$$

$$A_{M+1} = 2\pi(M\,\Delta r)H$$

$$A^*_{M+1} = 2\pi\left(M\,\Delta r - \frac{1}{4}\,\Delta r\right) \cong 2\pi(M\,\Delta r)H$$

where H is the length of the cylinder.

After introducing the area terms into the energy balance equation and some rearrangement, we obtain

$$\boxed{\left(1 - \frac{1}{2M}\right)T_M - \left[\left(1 - \frac{1}{2M}\right) + \frac{\Delta r\,h_\infty}{k}\right]T_{M+1} + \frac{\Delta r\,h_\infty}{k}T_\infty + \frac{(\Delta r)^2\,g_{M+1}}{2k} = 0}$$

$$\text{for } m = M + 1 \qquad (5\text{-}66c)$$

Summarizing Eqs. (5-63) and (5-65) together with the finite-difference form of the boundary condition at $r = b$ provides $M + 1$ algebraic equations for the determination of $M + 1$ node temperatures T_m ($m = 1, 2, \ldots, M + 1$) in a solid cylinder.

HOLLOW CYLINDER. In the case of a hollow cylinder with an inner radius of $r = a$ and outer radius of $r = b$, in which energy is generated at a rate of $g(r)$ W/m^3, the region $a \leq r \leq b$ is divided into M cylindrical subregions, each of thickness Δr, given by

$$\Delta r = \frac{b-a}{M}$$

as illustrated in Fig. 5-11. Then we have $M+1$ unknown node temperatures T_m ($m = 1, 2, \ldots, M+1$) and we need $M+1$ relations to determine them. The general procedure for obtaining such relations is similar to that described for the solid cylinder.

For the internal nodes $m = 2, 3, \ldots, M$, the energy balance equation (5-47) gives

$$A_{m-1,m} k \frac{T_{m-1} - T_m}{\Delta r} + A_{m+1,m} k \frac{T_{m+1} - T_m}{\Delta r} + A_m \Delta r \, g_m = 0$$

where

$$A_{m-1,m} = 2\pi \left[a + (m-1)\Delta r - \frac{1}{2}\Delta r \right] H$$

$$A_{m+1,m} = 2\pi \left[a + (m-1)\Delta r + \frac{1}{2}\Delta r \right] H$$

$$A_m = 2\pi [a + (m-1)\Delta r] H$$

FIGURE 5-11
Nomenclature for one-dimensional finite difference formulation for a long hollow cylinder.

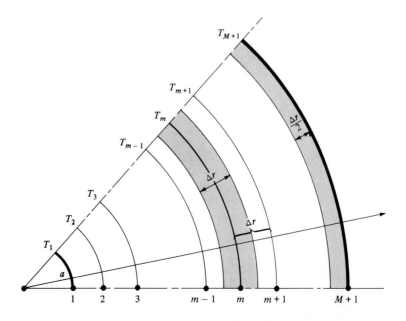

which is rearranged as

$$
\left[\frac{a}{(m-1)\,\Delta r}+1-\frac{1}{2(m-1)}\right]T_{m-1}-2\left[\frac{a}{(m-1)\,\Delta r}+1\right]T_{m}
$$
$$
+\left[\frac{a}{(m-1)\,\Delta r}+1+\frac{1}{2(m-1)}\right]T_{m+1}+\frac{[(a/(m-1)\,\Delta r)+1](\Delta r)^{2}g_{M}}{k}=0
$$

$$\text{for } 2, 3, \ldots, M \qquad (5\text{-}67)$$

If the outer boundary surfaces at $r = a$ and $r = b$ are subjected to convection, the finite-difference equations are obtained by the application of the energy balance given by Eq. (5-56).

For the node $m = 1$ at the inner boundary surface $r = a$, we have

$$
A_{1}h_{\infty 1}(T_{\infty 1}-T_{1})+A_{(2,1)}k\,\frac{T_{2}-T_{1}}{\Delta r}+A_{1}^{*}\,\frac{\Delta r}{2}\,g_{0}=0 \qquad \text{for } m = 0
$$

where

$$A_{1}=2\pi aH$$

$$A_{(2,1)}=2\pi\left(a+\frac{\Delta r}{2}\right)H$$

$$A_{1}^{*}\cong 2\pi aH$$

which is rearranged as

$$
\frac{ah_{\infty 1}}{k}\,T_{\infty 1}-\left[\frac{a}{\Delta r}\left(1+\frac{\Delta r\,h_{\infty 1}}{k}\right)+\frac{1}{2}\right]T_{1}+\left(\frac{a}{\Delta r}+\frac{1}{2}\right)T_{2}+\frac{a\,\Delta r}{2k}\,g_{1}=0
$$

$$\text{for } m = 1 \qquad (5\text{-}68)$$

For the node $m = M + 1$ at the outer boundary surface $r = b$, we have

$$
A_{M,M+1}k\,\frac{T_{M}-T_{M+1}}{\Delta r}+A_{M+1}h_{\infty 2}(T_{\infty 2}-T_{M+1})+A_{M+1}^{*}\,\frac{\Delta r}{2}\,g_{M+1}=0
$$

where

$$A_{M,M+1}=2\pi\left(b-\frac{1}{2}\,\Delta r\right)H$$

$$A_{M+1}=2\pi bH$$

$$A_{M+1}^{*}\cong 2\pi bH$$

which is rearranged as

$$
\left(\frac{b}{\Delta r}-\frac{1}{2}\right)T_{M}-\left[\frac{b}{\Delta r}\left(1+\frac{\Delta r h_{\infty 2}}{k}\right)-\frac{1}{2}\right]T_{M+1}+\frac{bh_{\infty 2}}{k}\,T_{\infty 2}+\frac{b\,\Delta r}{2k}\,g_{M+1}=0
$$

$$\text{for } m = M \qquad (5\text{-}69)$$

Example 5-7. A 10-cm-diameter solid chrome-nickel rod with thermal conductivity of 20 W/(m · °C) is heated electrically by the passage of an electric current which generates energy within the rod at a uniform rate of 10^7 W/m^3. The surface of the rod is subjected to convection with a heat transfer coefficient $h = 200$ W/(m · °C) into an ambient at 30°C. Determine the finite-difference equations if the radius of the rod is subdivided into five equal intervals.

Solution. Given: $b = 0.05$ m, $k = 20$ W/(m · °C), $g = 10^7$ W/m, $T_\infty = 30°C$, $h_\infty = 200$ W/(m^2 · °C), and $M = 5$. Then we have

$$\Delta r = \frac{b}{M} = \frac{0.05}{5} = 0.01 \text{ m}$$

$$\frac{(\Delta r)^2 g}{k} = \frac{(0.01)^2 (10^7)}{20} = 50$$

$$\frac{\Delta r h_\infty}{k} = \frac{0.01 \times 200}{20} = \frac{1}{10}$$

The finite-difference equation for the central node $m = 1$ is obtained from Eq. (5-65):

$$4(T_2 - T_1) + 50 = 0 \qquad \text{for } m = 1$$

and the finite-difference equations for the internal nodes $m = 2$ to 5 are obtained from Eq. (5-63):

$$\left[1 - \frac{1}{2(m-1)}\right] T_{m-1} - 2T_m + \left[1 + \frac{1}{2(m-1)}\right] T_{m+1} + 50 = 0 \quad \text{for } m = 2, 3, 4, 5$$

Finally, the finite-difference equation for the convection boundary condition at the node $m = M + 1 = 6$ is obtained from Eq. (5-66c):

$$\left(1 - \frac{1}{10}\right) T_5 - \left[\left(1 - \frac{1}{10}\right) + \frac{1}{10}\right] T_6 + \left(\frac{1}{10}\right)(30) + 25 = 0$$

or

$$0.9 T_5 - T_6 + 28 = 0 \qquad \text{for } m = 6$$

Summarizing, we have six algebraic equations for the six node temperatures T_m, $m = 1$ to 6:

$$-4T_1 + 4T_2 = -50 \qquad \text{for } m = 1$$

$$0.5 T_1 - 2T_2 + 1.5 T_3 = -50 \qquad \text{for } m = 2$$

$$0.75 T_2 - 2T_3 + 1.25 T_4 = -50 \qquad \text{for } m = 3$$

$$0.8333 T_3 - 2T_4 + 1.1666 T_5 = -50 \qquad \text{for } m = 4$$

$$0.875 T_4 - 2T_5 + 1.125 T_6 = -50 \qquad \text{for } m = 5$$

$$0.9 T_5 - T_6 = -28 \qquad \text{for } m = 6$$

which are expressed in the matrix form as

$$\begin{bmatrix} -4 & 4 & 0 & 0 & 0 & \\ 0.5 & -2 & 1.5 & 0 & 0 & 0 \\ 0 & 0.75 & -2 & 1.25 & 0 & 0 \\ 0 & 0 & 0.8333 & -2 & 1.1666 & 0 \\ 0 & 0 & 0 & 0.875 & -2 & 1.125 \\ 0 & 0 & 0 & 0 & 0.9 & -1 \end{bmatrix} \begin{Bmatrix} T_1 \\ T_2 \\ T_3 \\ T_4 \\ T_5 \\ T_6 \end{Bmatrix} = \begin{Bmatrix} -50 \\ -50 \\ -50 \\ -50 \\ -50 \\ -28 \end{Bmatrix}$$

Sphere—Finite-Difference Formulation

The basic principles of finite-differencing described for the cylinder are applicable for the sphere, except that the radial variation of the area is proportional to the square of the radius. We consider radial heat conduction in a solid sphere of radius $r = b$, in which energy is generated at a rate of $g(r)$ W/m^3. The region $0 \leq r \leq b$ is divided into M spherical shells, each of thickness

$$\Delta r = \frac{b}{M} \tag{5-70}$$

The nomenclature for the subdivisions is similar to that shown in Fig. 5-10b for a solid cylinder, except that in this case the geometry is a solid sphere.

For the internal nodes, $m = 2, 3, \ldots, M$, the energy balance equation (5-47) gives

$$A_{m-1,m} k \frac{T_{m-1} - T_m}{\Delta r} + A_{m+1,m} k \frac{T_{m+1} - T_m}{\Delta r} + A_m \, \Delta r \, g_m = 0$$

where

$$A_{m-1,m} = 4\pi \left[(m-1)\,\Delta r - \frac{\Delta r}{2} \right]^2 = 4\pi (m-1)^2 (\Delta r)^2 \left[1 - \frac{1}{2(m-1)} \right]^2$$

$$A_{m+1,m} = 4\pi \left[(m-1)\,\Delta r + \frac{\Delta r}{2} \right]^2 = 4\pi (m-1)^2 (\Delta r)^2 \left[1 + \frac{1}{2(m-1)} \right]^2$$

$$A_m = 4\pi \left[(m-1)\,\Delta r \right]^2$$

After rearranging, we obtain

$$\left[1 - \frac{1}{2(m-1)} \right]^2 (T_{m-1} - T_m) + \left[1 + \frac{1}{2(m-1)} \right]^2 (T_{m+1} - T_m) + \frac{(\Delta r)^2 \, g_m}{k} = 0$$

$$\text{for } m = 2, 3, \ldots, M \tag{5-71a}$$

For the values of $m \geq 2$, the following approximations can be made:

$$\left[1 - \frac{1}{2(m-1)} \right]^2 \cong 1 - \frac{1}{(m-1)}$$

$$\left[1 + \frac{1}{2(m-1)} \right]^2 \cong 1 + \frac{1}{(m-1)}$$

Then Eq. (5-71a) reduces to

$$\left(1 - \frac{1}{m-1} \right) T_{m-1} - 2T_m + \left(1 + \frac{1}{m-1} \right) T_{m+1} + \frac{(\Delta r)^2 \, g_m}{k} = 0$$

$$\text{for } m = 2, 3, \ldots, M \tag{5-71b}$$

For the center node $m = 1$, the energy balance equation (5-47) is also applicable, and yields

$$A_{2,1}k\frac{T_2 - T_1}{\Delta r} + \frac{4}{3}\pi\left(\frac{\Delta r}{2}\right)^3 g_1 = 0 \qquad \text{for } m = 1$$

where

$$A_{2,1} = 4\pi\left(\frac{\Delta r}{2}\right)^2$$

and after rearrangement we obtain

$$\boxed{6(T_2 - T_1) + \frac{(\Delta r)^2\, g_1}{k} = 0} \qquad \text{for } m = 1 \qquad (5\text{-}72)$$

For the node M at the boundary surface $r = b$, the finite-difference equation depends on the type of boundary condition. If the boundary surface at $r = b$ is subjected to convection with a heat transfer coefficient h_∞ into an ambient at temperature T_∞, application of the energy balance equation (5-56) gives

$$A_{M,M+1}k\frac{T_{M-1} - T_M}{\Delta r} + A_{M+1}h_\infty(T_\infty - T_{M+1}) + A^*_{M+1}\frac{\Delta r}{2}g_{M+1} = 0$$

where

$$A_{M,M+1} = 4\pi\left(M\,\Delta r - \frac{\Delta r}{2}\right)^2$$

$$A_{M+1} = 4\pi(M\,\Delta r)^2$$

$$A^*_{M+1} \cong 4\pi(M\,\Delta r)^2$$

After rearrangement we obtain

$$\boxed{\begin{aligned} &2\left(1 - \frac{1}{2M}\right)^2 T_M - \left[2\left(1 - \frac{1}{2M}\right)^2 + \frac{2h_\infty\,\Delta r}{k}\right]T_{M+1} \\ &\qquad + \frac{2h_\infty\,\Delta r}{k}T_\infty + \frac{(\Delta r)^2\, g_{M+1}}{k} = 0 \end{aligned}}$$

$$\text{for } m = M + 1 \qquad (5\text{-}73a)$$

For $M > 1$, the following approximations can be made:

$$\left(1 - \frac{1}{2M}\right)^2 \cong 1 - \frac{1}{M}$$

$$\left(1 + \frac{1}{2M}\right)^2 \cong 1 + \frac{1}{M}$$

Then Eq. (5-73a) reduces to

$$2\left(1 - \frac{1}{M}\right)T_M - \left[2\left(1 - \frac{1}{M}\right) + \frac{2h_\infty \Delta r}{k}\right]T_{M+1} + \frac{2h_\infty \Delta r}{k}T_\infty + \frac{(\Delta r)^2\, g_{M+1}}{k} = 0$$

for $m = M + 1$ (5-73b)

5-3   COMPUTER SOLUTIONS

As shown in the previous sections, the finite-differencing of one-dimensional steady-state heat conduction problems results in a triagonal system of algebraic equations for the node temperatures T_m $(m = 1, 2, \ldots, M + 1)$. Such a system can be expressed in matrix form as

$$[A]\{X\} = \{D\} \qquad (5\text{-}74a)$$

which can be written more explicitly as

$$\begin{bmatrix} b_1 & c_1 & 0 & & 0 & 0 \\ a_2 & b_2 & c_2 & & & 0 \\ & & & & & \\ 0 & & a_{N-1} & b_{N-1} & c_{N-1} \\ 0 & 0 & & 0 & a_N & b_N \end{bmatrix} \begin{Bmatrix} x_1 \\ x_2 \\ \\ x_{N-1} \\ x_N \end{Bmatrix} = \begin{Bmatrix} d_1' \\ d_2' \\ \\ d_{N-1}' \\ d_N' \end{Bmatrix} \qquad (5\text{-}74b)$$

In Table 5-1 we present a computer program written in Fortran IV for solving a tridiagonal system of N algebraic equations of the form shown in Eqs. (5-74a) and (5-74b). The main features of this program and the significance of various statements are as follows.

[A] is the $N \times N$ coefficient matrix, which is a banded tridiagonal matrix. In *lines 13 to 16* the program stores the coefficient matrix [A] and the vector $\{D\}$ in band storage mode in matrix $A(I, J)$ as

$$A(I, J) = \begin{bmatrix} 0 & b_1 & c_1 & d_1' \\ a_2 & b_2 & c_2 & d_2' \\ a_3 & b_3 & c_3 & d_3' \\ & & & \\ a_{N-1} & b_{N-1} & c_{N-1} & d_{N-1}' \\ a_N & b_N & 0 & d_N' \end{bmatrix}$$

where $I = 1$, N and $J = 1$, 4 and the vector $\{D\}$ is in $A(I, 4)$. *Lines 19 to 23* print the input data $A(I, J)$ for checking purposes. *Line 25* calls subroutine TRIDG, which solves tridiagonal equations. The subroutine itself is presented in *lines 44 to 59*. *Lines 32 to 40* print the node temperatures. The subroutine TRIDG will return the solution in the column vector $A(I, 4)$. In *line 13* the value of 6 was assigned to N because the specific example considered here (Example 5-6) involved only six equations. If the number of equations is different from six, change the assigned value. If the number of equations is more than 50, not only should the assigned value of N in line 8 be changed, but also the value of $N = 50$ appearing in the dimension statement of *line 7* should be changed to the proper value.

TABLE 5-1
Computer program for solving a system of N simultaneous tridiagonal equations

```
 1   C****  COMPUTER PROGRAM FOR SOLVING A SYSTEM OF N
 2   C       SIMULTANEOUS EQUATIONS IN ONE-DIMENSIONAL
 3   C       HEAT CONDUCTION PROBLEMS, WITH A TRIDIAGONAL
 4   C       COEFFICIENT MATRIX
 5   C****  EXAMPLE 5-6 ****
 6          IMPLICIT REAL*8(A-H,O-Z)
 7          DIMENSION A(50,4)
 8          NDIM = 50
 9
10   C****  READ THE MATRIX [A] AND THE VECTOR [D]
11   C       (IN BAND STORAGE MODE) AND STORE THEM IN [A].
12   C       N IS THE NUMBER OF ROWS.
13          N = 6
14          DO 10 I = 1,N
15             READ (5,*) (A(I,J), J=1,4)
16       10 CONTINUE
17
18   C****  PRINT THE INPUT DATA FOR CHECKING PURPOSES
19          WRITE (6,20) 'EXAMPLE 5-6'
20       20 FORMAT (1X,A)
21          DO 40 I = 1,N
22             WRITE (6,30) (A(I,J), J=1,4)
23       30    FORMAT (1X,/T7,3F9.3,G12.4)
24       40 CONTINUE
25          CALL TRIDG(A,N,NDIM)
26   C****  WHEN RETURNED, THE COLUMN VECTOR {A(I,4)}
27   C       CONTAINS THE SOLUTION.
28          WRITE (6,50)
29       50 FORMAT (1X///T10,'NODE #',5X,'TEMPERATURE'/
30                      17X,'FINITE DIFFERENCE'/)
31   C
32          DO 70 I = 1,N
33             IF (A(1,2)+A(1,3) .EQ. 0.0D0) THEN
34                JP = I
35             ELSE
36                JP = I + 1
37             ENDIF
38             WRITE (6,60) JP, A(I,4)
39       60    FORMAT (1X,T8,I5,7X,G13.5)
40       70 CONTINUE
41          STOP
42          END
43   C
44   C****  SUBROUTINE TO SOLVE A TRIDIAGONAL OF EQUATIONS
45          SUBROUTINE TRIDG(A,N,NDIM)
46          IMPLICIT REAL*8(A-H,O-Z)
47          DIMENSION A(NDIM,4)
48          DO 80 I = 2,N
49             A(I,2) = A(I,2) - A(I,1) / A(I-1,2) * A(I-1,3)
50             A(I,4) = A(I,4) - A(I,1) / A(I-1,2) * A(I-1,4)
51       80 CONTINUE
52          NM1 = N - 1
53          A(N,4) = A(N,4) / A(N,2)
54          DO 90 I = 1,NM1
55             M = N - I
56             A(M,4) = (A(M,4) - A(M,3) * A(M+1,4)) / A(M,2)
57       90 CONTINUE
58          RETURN
59          END
```

Example 5-8. Compare the numerical values of the finite-difference solution of Example 5-6, shown in Table 5-2, with the exact solution to the problem.

Solution. The finite-difference solution of Example 5-6 has $N = 6$ equations. The band storage mode of the matrix $A(6, 4)$, which represents the input data, is read by rows, and this matrix and the resulting numerical values of this matrix are shown in Table 5-2. The exact solution to this heat conduction problem is given by Eq. (5-19) as

$$T(x) = \frac{gL^2}{2k}\left[1 - \left(\frac{x}{L}\right)^2\right] + \frac{gL}{h_\infty} + T_\infty$$

The temperatures at $x = 0$, 0.2, 0.4, 0.6, 0.8, and 1.0, evaluated by using this

TABLE 5-2
Numerical solution to Example 5-6

0.000	-2.000	2.000	-10.00
1.000	-2.000	1.000	-10.00
1.000	-2.000	1.000	-10.00
1.000	-2.000	1.000	-10.00
1.000	-2.000	1.000	-10.00
2.000	-2.200	0.000	-40.00

NODE #	TEMPERATURE	
	FINITE DIFFERENCE	EXACT
1	775.00	775.00
2	770.00	770.00
3	755.00	755.00
4	730.00	730.00
5	695.00	695.00
6	650.00	650.00

expression, are also listed in this table. The results obtained using the numerical solution and those obtained using the exact solution are the same.

Example 5-9. Solve the problem considered in Example 5-7 using finite-differences by subdividing the region into five equal intervals.

Solution. The finite-difference solution of Example 5-7 involves $N = 6$ equations which are solved using the computer program shown in Table 5-1. The resulting finite-difference solution is given in Table 5-3, along with the exact solution for the problem. We note that for the cylinder problem the finite-difference solution is not in as good agreement with the exact results as it was in the previous example for plane geometry. The accuracy of the numerical solution can be improved by using a finer mesh size.

TABLE 5-3
Numerical solution to Example 5-7

0.000	-4.000	4.000	-50.00
0.500	-2.000	1.500	-50.00
0.750	-2.000	1.250	-50.00
0.833	-2.000	1.167	-50.00
0.875	-2.000	1.125	-50.00
0.900	-1.000	0.000	-28.00

NODE #	TEMPERATURE	
	FINITE DIFFERENCE	EXACT
1	1605.0	1592.5
2	1592.5	1580.0
3	1555.0	1542.5
4	1492.5	1480.0
5	1405.0	1392.5
6	1292.5	1280.0

PROBLEMS

Analytical Solutions

5-1. Develop an expression for one-dimensional, steady-state temperature distribution $T(x)$ in a plane wall (slab) of thickness L when the boundary surface at $x = 0$ is kept at a uniform temperature T_0 and the boundary surface at $x = L$ dissipates heat by convection with a heat transfer coefficient h into the ambient air at temperature T_∞. The thermal conductivity k is constant, and there is energy generation at a constant rate of g W/m^3.

5-2. Develop an expression for the one-dimensional, steady-state temperature distribution $T(x)$ in a slab of thickness L when the boundary surface at $x = 0$ is kept insulated and the boundary surface at $x = L$ is kept at zero temperature. The thermal conductivity k is constant, and there is energy generation at a rate of $g(x) = g_0 \cos \pi x/2L$ W/m^3, where g_0 is the energy generation rate per unit volume at $x = 0$. Give the relation for the temperature of the insulated boundary.

5-3. Consider a slab of thickness $L = 0.1$ m. One of the boundary surfaces, that at $x = 0$ is kept insulated, and the boundary surface at $x = L$ dissipates heat by convection with a heat transfer coefficient of 200 W/(m$^2 \cdot$ °C) into the ambient air at 150°C. The thermal conductivity of the wall is 10.0 W/(m $\cdot$ °C), and within the wall energy is generated at a constant rate of 10^6 W/m^3. Determine the boundary surface temperatures.

    Answers: 1150°C, 650°C

5-4. Consider a slab of thickness 0.1 m. One of the boundary surfaces, that at $x = 0$, is kept insulated, and the boundary surface at $x = L$ dissipates heat by convection with a heat transfer coefficient h into the ambient air at 150°C. The thermal conductivity of the wall is 40 W/(m $\cdot$ °C), and within the wall energy is generated at a constant rate of 10^6 W/m^3. Plot the temperature profiles in the slab for the heat transfer coefficients $h = 200$, 350, and 500 W/(m$^2 \cdot$ °C).

5-5. Develop an expression for the steady-state temperature distribution in a slab of thickness L when the boundary surface at $x = 0$ is kept insulated and the boundary surface at $x = L$ is kept at zero temperature. The thermal conductivity of the wall k is constant, and within the wall energy is generated at a rate of $g(x) = g_0 x^2$ W/m^3. Give the expressions for the temperature of the insulated surface at $x = 0$.

    Answer: $g_0 L^4/12k$

5-6. A nuclear reactor pressure vessel is approximated as a large plane wall of thickness L. The inside surface is insulated, the outside surface at $x = L$ is maintained at a uniform temperature T_2, and the gamma-ray heating of the plate can be represented as energy generation in the form

$$g(x) = g_0 e^{-\gamma x} \qquad \text{W/m}^3$$

where g_0 and γ are constants and x is measured from the insulated inside surface. Develop an expression for the temperature distribution in the plate.

5-7. Develop an expression for the steady temperature distribution $T(r)$ in a cylinder of radius b in which energy is generated at a rate of

$$g(r) = g_0\left(1 - \frac{r}{b}\right) \qquad \text{W/m}^2$$

where g_0 is a constant, and the boundary surface at $r = b$ is maintained at a uniform temperature T_w. Assuming one-dimensional steady heat flow, develop a relation for the temperature drop from the centerline to the surface of the fuel element. For radius $b = 2$ cm, thermal conductivity $k = 10$ W/(m · °C), and $g_0 = 4 \times 10^7$ W/m^3, calculate the temperature drop from the centerline to the surface.
Answer: $5g_0 b^2 / 36k$, 222.2°C

5-8. A hollow cylindrical fuel element of inner radius a and outer radius b is heated uniformly within the entire volume at a constant rate of g W/m^3 as a result of disintegration of radioactive material. The inner surface of the element is at zero temperature, and the outer surface is subjected to dissipation of heat by convection with a heat transfer coefficient h into the ambient air at temperature T_∞. The thermal conductivity of the cylinder is constant. Develop an expression for the one-dimensional steady-state temperature distribution $T(r)$ within the cylinder.

5-9. Determine the one-dimensional temperature distribution in the hollow cylindrical fuel element of Problem 5-8 for the case in which the heat generation rate $g_0(r)$ is a function of the radial position in the form

$$g(r) = g_0(1 + Ar) \qquad W/m^2$$

where g_0 and A are constants.

5-10. A long cylindrical rod of radius $b = 5$ cm and $k = 20$ W/(m · °C) contains radioactive material which generates energy uniformly within the cylinder at a constant rate of $g_0 = 2 \times 10^5$ W/m^3. The rod is cooled by convection from its cylindrical surface into the ambient air at $T_\infty = 20$°C with a heat transfer coefficient $h_\infty = 50$ W/(m^2 · °C). Determine the temperatures at the center and the outer surface of this cylindrical rod.
Answers: 126.3°C, 120°C

5-11. A tube of inner radius r_i, outer radius r_0, and thermal conductivity k is heated by passing an electric current through the tube. The passage of current generates energy at a constant rate of g_0 W/m^3. The outer surface of the pipe is insulated, and the inner surface is kept at zero temperature. Develop an expression for the steady-state temperature distribution $T(r)$ within the tube and for the temperature of the insulated outer surface.

5-12. An electric resistance wire of radius $a = 1 \times 10^{-3}$ m with thermal conductivity $k = 25$ W/(m · °C) is heated by the passage of electric current, which generates heat within the wire at a constant rate of $g_0 = 2 \times 10^9$ W/m^3. Determine the centerline temperature rise above the surface temperature of the wire if the surface is maintained at a constant temperature.
Answer: 20°C

5-13. A 3-mm-diameter chrome-nickel wire of thermal conductivity $k = 20$ W/(m · °C) is heated electrically by the passage of an electric current, which generates heat within the wire at a constant rate of $g_0 = 10^7$ W/m^3. If the outer surface of the wire is maintained at 50°C, determine the center temperature of the wire.
Answer: 50.28°C

5-14. Heat is generated at a constant rate of $g_0 = 10^6$ W/m^3 in a copper rod of radius $r = 0.5$ cm and thermal conductivity $k = 386$ W/(m · °C). The rod is cooled by convection from its cylindrical surface into an ambient at 20°C with a heat transfer coefficient $h = 1400$ W/(m^2 · °C). Determine the surface temperature of the rod.
Answer: 21.8°C

5-15. Develop an expression for the steady temperature distribution $T(r)$ in a solid sphere of radius b in which heat is generated at a rate of

$$g(r) = g_0\left(1 - \frac{r}{b}\right) \qquad W/m^3$$

where g_0 is a constant, and the boundary surface at $r = b$ is maintained at a uniform temperature T_w.

5-16. Determine the centerline temperature rise above the surface temperature of the sphere in by Problem 5-15.

5-17. Heat is generated at a constant rate of $g_0 = 2 \times 10^8$ W/m^3 in a copper sphere of radius 1 cm and thermal conductivity $k = 386$ W/(m · °C). The sphere is cooled by convection from its cylindrical surface into an ambient at 10°C with a heat transfer coefficient of $h = 2000$ W/(m^2 · °C). Determine the surface temperature and the center temperature of the sphere.
   Answers: 343.3°C, 352°C

5-18. Determine the temperature distributions in a hollow sphere of inner radius a and outer radius b in which energy is generated: (a) at a constant rate of g_0 W/m^3, and (b) as a function of radial position in the form $g(r) = g_0(1 + Ar)$ W/m^3, where g_0 and A are constants, if the outer surface at $r = b$ is kept at a constant temperature T_0 and the inner surface at $r = a$ is insulated.

5-19. A hollow sphere of inner radius a and outer radius b is heated uniformly within the entire volume at a constant rate of g_0 W/m^3. The inner surface is at zero temperature, and the outer surface is dissipating heat by convection with a heat transfer coefficient h into the ambient air at temperature T_∞. The thermal conductivity of the sphere is constant. Develop an expression for the steady temperature distribution within the sphere.

5-20. A plane wall of thickness L is maintained at a constant temperature T_0 at $x = 0$ and T_L at $x = L$ while energy is generated in the medium at a constant rate of g W/m^3.
   (a) Develop an expression for the steady-state temperature distribution in the wall.
   (b) Calculate the maximum temperature in the wall for the case in which $T_0 = 50$°C, $T_L = 0$°C, $k = 30$ W/(m · °C), $g = 5 \times 10^4$ W/m^3, and $L = 1$ m.

5-21. A plane wall of thickness L is insulated at the surface $x = 0$ and maintained at a constant temperature T_L at the surface $x = L$. Energy is generated in the wall at a constant rate of g W/m^3.
   (a) Develop an expression for the steady-state temperature distribution in the slab.
   (b) Determine the maximum amount of energy that can be generated in the wall if the temperature of the insulated surface should not exceed 200°C when plate thickness $L = 15$ cm, plate thermal conductivity $k = 30$ W/(m · °C), and the temperature of the surface at $x = L$ is $T_L = 50$°C.
   Answer: 6×10^4 W/m^3

5-22. A plane wall of thickness $L = 10$ cm is insulated at the surface $x = 0$ and maintained at a constant temperature $T_L = 200$°C at the surface $x = L$. The wall has a thermal conductivity $k = 20$ W/(m · °C), and energy is generated at a constant rate $g = 20000$ W/m^3. Calculate the temperature of the insulated surface.
   Answer: 205°C

5-23. A copper bar of radius $r = 5$ mm and thermal conductivity $k = 360$ W/(m · °C) is heated at a rate of $g = 10^4$ W/m^3 by the passage of an electric current. If the outer surface of the wire dissipates heat by convection into an ambient at $T_\infty = 25$°C with a heat transfer coefficient of $h = 200 \, \text{W}/(\text{m}^2 \cdot {}^\circ\text{C})$, calculate the center temperature of the wire.

  Answer: 25.1°C

5-24. An aluminum wire of radius $r = 2 \, \text{mm}$ and thermal conductivity $k = 200 \, \text{W}/(\text{m} \cdot {}^\circ\text{C})$ is heated at a rate of $g = 10^5 \, \text{W}/\text{m}^3$ by the passage of an electric current. If the outer surface of the wire is maintained at a constant temperature $T_\infty = 30{}^\circ\text{C}$, calculate the center temperature of the wire.

5-25. A plane wall of thickness L and constant thermal conductivity k has both its boundary surfaces kept at *zero* temperature. Energy is generated in the plate at a constant rate of $g = Ax^2 \, \text{W}/\text{m}^3$, where A is a constant.
(*a*) Develop an expression for the temperature distribution $T(x)$ in the slab.
(*b*) Calculate the maximum temperature in the slab for $k = 30 \, \text{W}/(\text{m} \cdot {}^\circ\text{C})$, $g = (1000)x^2 \, \text{W}/\text{m}^3$, $L = 50 \, \text{cm}$, and $g = 100 \, \text{W}/\text{m}^3$.

  Answer: 0.082°C

5-26. Electric current $I = 500 \, \text{A}$ flows through a stainless steel conductor of diameter $D = 5 \, \text{mm}$ that has an electric resistance $R = 5 \times 10^{-4} \, \Omega/\text{m}$. Energy generated as a result of the passage of the electric current is dissipated by convection into an ambient at temperature $T_\infty = 0{}^\circ\text{C}$ with a heat transfer coefficient $h = 50 \, \text{W}/(\text{m}^2 \cdot {}^\circ\text{C})$. The thermal conductivity of the conductor is $k = 60 \, \text{W}/(\text{m} \cdot {}^\circ\text{C})$. Calculate the center and surface temperatures of the cable.

Note:
$$g = \frac{RI^2}{(\pi D^2/4) \times (1)} \qquad \text{W}/\text{m}^3$$

5-27. An electric current $I = 200 \, \text{A}$ passes through a copper bar with a diameter $D = 2 \, \text{cm}$, thermal conductivity $k = 360 \, \text{W}/(\text{m} \cdot {}^\circ\text{C})$, and an electric resistance $R = 9 \times 10^{-3} \, \Omega/\text{m}$ that is exposed to air at temperature $T_\infty = 20{}^\circ\text{C}$. The heat transfer coefficient between the surface and the air is $h = 40 \, \text{W}/(\text{m} \cdot {}^\circ\text{C})$. Calculate the surface and center temperatures of the bar.

Note:
$$g = \frac{RI^2}{(\pi D^2/4) \times (1)} \qquad \text{W}/\text{m}^3$$

5-28. Energy is generated at a rate of $g \, \text{W}/\text{m}^3$ in a cylindrical fuel element for a nuclear reactor. The fuel element has a diameter D and thermal conductivity k, and is cooled by a fluid at a temperature T_∞ with a heat transfer coefficient $h = 1000 \, \text{W}/(\text{m}^2 \cdot {}^\circ\text{C})$.
(*a*) Develop an expression for the center temperature of the fuel element,
(*b*) Calculate the center temperature for $D = 1 \, \text{cm}$, $g = 6 \times 10^8 \, \text{W}/\text{m}^3$, $T_\infty = 200{}^\circ\text{C}$, and $k = 40 \, \text{W}/(\text{m} \cdot {}^\circ\text{C})$.

  Answer: 1793.8°C

Numerical Analysis

5-29. Consider the following one-dimensional, steady-state heat conduction problem:

$$\frac{d^2 T(x)}{dx^2} + \frac{g}{k} = 0 \qquad\qquad \text{in } 0 < x < L$$

$$\frac{dT(x)}{dx} = 0 \qquad\qquad \text{at } x = 0$$

$$-k \frac{dT(x)}{dx} = h_\infty[T(x) - T_\infty] \qquad \text{at } x = L$$

(*a*) Write the finite-difference formulation of this heat conduction problem by dividing the region $0 < x < L$ into five equal parts.

(*b*) Compute the node temperatures for $k = 10$ W/(m · °C), $h_\infty = 200$ W/(m² · °C), $T_1 = 0$°C, $T_\infty = 100$°C, $g = 10^6$ W/m³, and $L = 5$ cm.

(*c*) Compare the numerical solution at the nodes with the exact solution.

5-30. Consider the following one-dimensional, steady-state heat conduction problem:

$$\frac{d^2 T(x)}{dx^2} + \frac{g}{k} = 0 \qquad \text{in } 0 < x < L$$

$$T(x) = T_1 \qquad \text{at } x = 0$$

$$-k \frac{dT(x)}{dx} = h[T(x) - T_\infty] \qquad \text{at } x = L$$

Write the finite-difference formulation of this heat conduction problem by dividing the region $0 < x < L$ into (*a*) four equal parts and (*b*) eight equal parts.

(*c*) Compute the node temperatures for $k = 20$ W/(m · °C), $h_\infty = 400$ W/(m² · °C), $T_\infty = 150$°C, $g = 2 \times 10^6$ W/m³, and $L = 10$ cm.

(*d*) Compare the numerical solutions at the nodes with the exact solution.

5-31. Derive the finite-difference formulation for Problem 5-5 by dividing the slab into four equal parts.

(*a*) Evaluate the numerical values for $k = 20$ W/(m · °C), $g_0 = 10^8$ W/m³, and $L = 20$ cm.

(*b*) Compare the numerical solution at the nodes with the exact solution.

5-32. Derive the finite-difference formulation for Problem 5-6 by dividing the slab into four equal parts.

(*a*) Evaluate the numerical values for $k = 50$ W/(m · °C), $T_2 = 100$°C, $g_0 = 10^6$ W/m³, $L = 5$ cm, and $\gamma = 0.2$ m⁻¹.

(*b*) Compare the numerical solution at the nodes with the exact solution.

5-33. Consider the following one-dimensional steady-state heat conduction problem:

$$\frac{d^2 T(x)}{dx^2} + \frac{g_0}{k} \cos \frac{\pi x}{2L} = 0 \qquad 0 < x < L$$

$$\frac{dT(x)}{dx} = 0 \qquad \text{at } x = 0$$

$$T(x) = T_w \qquad \text{at } x = L$$

(*a*) Write the finite-difference formulation of the problem by dividing the region $0 < x < L$ into five equal parts.

(*b*) Compute the node temperatures for $k = 10$ W/(m · °C), $T_w = 500$°C, $g_0 = 10^6$ W/m³, and $L = 0.1$ m.

5-34. For Problem 5-7, write the finite-difference formulation of the problem by dividing the region $0 < r < b$ into five equal parts, and for $T_w = 0$°C compare the numerical values obtained at the nodes with the exact solutions.

5-35. Write the finite-difference formulation for Problem 5-10 by dividing the region $0 < r < b$ into five equal parts, and compare the resulting node temperatures with the exact solution.

5-36. Write the finite-difference formulation for Problem 5-12 by dividing the region $0 < r < b$ into five equal parts, and compare the results with the exact solution.

5-37. Write the finite-difference formulation for Problem 5-14 by dividing the region $0 < r < b$ into four and eight equal parts and compare these two results for the node temperatures with the exact solution.

CHAPTER
6

UNSTEADY
HEAT
CONDUCTION

In the previous two chapters we studied steady-state heat conduction in solids in which temperature within the body varied with position but not with time. We also described a simplified analysis of temperature variation within the solids as a function of time based on the lumped system approach, which assumes uniform temperature within the solid at any instant. However, in many situations the temperature variations within the solid are no longer negligible; hence lumped system analysis is no longer applicable. Then the analysis of heat conduction problems involves the determination of the temperature distribution within the solid as a function of both time and position, and it is a complicated matter. Various methods of analysis for solving such problems are discussed in advanced texts on heat conduction. Here we are concerned only with the application of the results obtained from such analysis.

Transient heat conduction in solids having simple shapes, such as a semi-infinite region, a plate, a long solid cylinder, and a solid sphere, can be solved analytically, and the transient temperature distribution and heat flow obtained from such solutions can be presented in the form of transient temperature and heat flow charts. In this chapter we present such charts and illustrate their use in predicting temperature variation within a solid as a function of time and position, as well as the heat transfer rate.

There are numerous situations in which either the geometry is complicated or the problem is nonlinear, so that the analytic solution to the problem is either not possible or too elaborate to use in practice. In such situations purely numerical schemes are useful for solving the problem. Therefore, we also describe here a simple explicit finite-difference scheme for solving one-dimensional transient heat conduction problems numerically.

6-1 TRANSIENT TEMPERATURE AND HEAT FLOW IN A SEMI-INFINITE SOLID

The concept of a one-dimensional semi-infinite solid refers, mathematically, to a region that has a single boundary surface and extends to infinity in one direction. However, in practice, it implies a plate that is sufficiently thick that any temperature disturbance applied to one of its surfaces has negligible effect, for all practical purposes, on the other surface during the period of observation of temperature transients.

We consider a semi-infinite solid confined to the region $x \geq 0$ and initially at a uniform temperature T_i. There is no energy generation within the medium. Suppose, at time $t = 0$, that the thermal condition at the boundary surface $x = 0$ is suddenly changed. The effects of this thermal disturbance will be gradually felt at the interior regions of the body; hence, for times $t > 0$, the temperature of the solid will vary with both position and time. Here we examine the temperature transients within a semi-infinite solid resulting from each of the following three different types of thermal disturbances applied at the boundary surface $x = 0$:

Sudden change in surface temperature

Suddenly imposed surface heat flux

Suddenly imposed convection

Sudden Change in Surface Temperature

Consider a semi-infinite solid that is initially at a uniform temperature T_i and confined to the domain $x \geq 0$. There is no energy generation within the medium. At time $t = 0$, the temperature of the boundary surface at $x = 0$ is suddenly changed to T_0. Figure 6-1 illustrates the geometry and coordinates. The mathematical formulation of this one-dimensional heat conduction problem is given by

$$\frac{\partial^2 T(x, t)}{\partial x^2} = \frac{1}{\alpha} \frac{\partial T(x, t)}{\partial t} \qquad \text{in } x \geq 0, t > 0 \qquad (6\text{-}1a)$$

FIGURE 6-1
Geometry and coordinate for transient temperature in a semi-infinite solid.

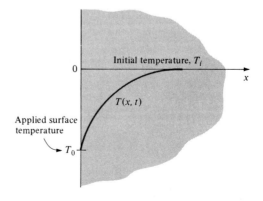

subject to the boundary conditions

$$T(x, t) = T_0 \qquad \text{at } x = 0, \ t > 0 \qquad (6\text{-}1b)$$

$$T(x, t) \rightarrow T_i \qquad \text{as } x \rightarrow \infty, \ t > 0 \qquad (6\text{-}1c)$$

and the initial condition

$$T(x, t) = T_i \qquad \text{for } t = 0, \text{ in } x \geq 0 \qquad (6\text{-}1d)$$

The transient heat conduction problem given by Eqs. (6-1a) to (6-1d) has been solved for $T(x, t)$ and the dimensionless temperature $\theta(x, t)$:

$$\theta(x, t) \equiv \frac{T(x, t) - T_0}{T_i - T_0} \qquad (6\text{-}2a)$$

is expressed in terms of the dimensionless parameter ξ

$$\xi \equiv \frac{x}{2\sqrt{\alpha t}} \qquad (6\text{-}2b)$$

with the following expression

$$\boxed{\theta(x, t) = \text{erf}\,(\xi)} \qquad (6\text{-}3)$$

Here "erf (ξ)" is called the *error function* of argument ξ, and its numerical values are tabulated as a function of ξ in Appendix D. Figure 6-2 shows a plot

FIGURE 6-2
Temperature distribution $T(x, t)$ in a semi-infinite solid which is initially at T_i and for $t > 0$ the boundary surface at $x = 0$ is maintained at T_0.

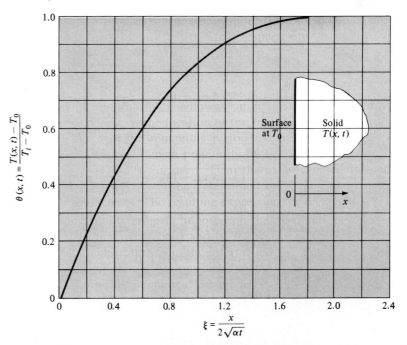

of the dimensionless temperature $\theta(x, t)$ as a function of the parameter $x/2\sqrt{\alpha t}$. The physical significance of this graph is as follows:

For a given value of x, the graph represents the variation in temperature with time at that particular location x. Conversely, for a given value of t, the graph represents the variation of temperature with position within the solid at that particular time t.

In engineering applications, the heat flux at the boundary surface $x = 0$ is also of interest. The heat flux at any position is obtained from its definition,

$$q(x, t) = -k \frac{\partial T}{\partial x} \tag{6-4}$$

Then the heat flux at the boundary surface $x = 0$ becomes

$$\boxed{q_s(t) = \frac{k(T_0 - T_i)}{\sqrt{\pi \alpha t}}} \quad \text{W/m}^2 \tag{6-5}$$

Suddenly Imposed Surface Heat Flux

Consider a semi-infinite solid, initially at a uniform temperature $T = T_i$ and confined to the domain $x \geq 0$. There is no energy generation within the medium. At time $t = 0$, the boundary surface at $x = 0$ is subjected to a constant heat flux q_0 W/m^2, and maintained so for times $t > 0$.

The mathematical formulation of this heat conduction problem is given by

$$\frac{\partial^2 T(x, t)}{\partial x^2} = \frac{1}{\alpha} \frac{\partial T(x, t)}{\partial t} \quad \text{in } x \geq 0, \, t > 0 \tag{6-6a}$$

subject to the boundary conditions

$$-k \frac{\partial T(x, t)}{\partial x} = q_0 \quad \text{at } x = 0, \, t > 0 \tag{6-6b}$$

$$T(x, t) \to T_i \quad \text{as } x \to \infty, \, t > 0 \tag{6-6c}$$

and the initial condition

$$T(x, t) = T_i \quad \text{for } t = 0, \text{ in } x \geq 0 \tag{6-6d}$$

The transient heat conduction problem given by Eqs. (6-6a) to (6-6d) has been solved, and the temperature distribution $T(x, t)$ within the solid is determined as a function of position and time as

$$\boxed{T(x, t) = T_i + \frac{2q_0}{k} (\alpha t)^{1/2} \left[\frac{1}{\sqrt{\pi}} e^{-\xi^2} + \xi \, \text{erf} \, (\xi) - \xi \right]} \tag{6-7a}$$

where the parameter ξ is defined as

$$\xi = \frac{x}{2\sqrt{\alpha t}} \qquad (6\text{-}7b)$$

We note that the temperature continues to change with time as long as heat flux q_0 is maintained at the boundary surface.

Suddenly Imposed Convection

We now consider a semi-infinite solid, initially at a uniform temperature T_i; for $t > 0$, the boundary surface at $x = 0$ is subjected to convection into a fluid at a constant temperature T_∞ with a known heat transfer coefficient h.

For this case we define a dimensionless temperature $\theta(x, t)$ as

$$\theta(x, t) = \frac{T(x, t) - T_i}{T_\infty - T_i} \qquad (6\text{-}8)$$

Figure 6-3 shows a plot of the quantity $1 - \theta(x, t)$ as a function of the dimensionless parameter $x/2\sqrt{\alpha t}$ for several different values of the dimensionless parameter $h\sqrt{\alpha t}/k$.

The special case $h \to \infty$ is equivalent to the problem with a constant temperature T_∞ at the boundary surface $x = 0$.

FIGURE 6-3

Transient temperature $T(x, t)$ in a semi-infinite solid subjected to convection at the boundary surface at $x = 0$, for $t > 0$, which is initially at T_i. (From Schnieder).

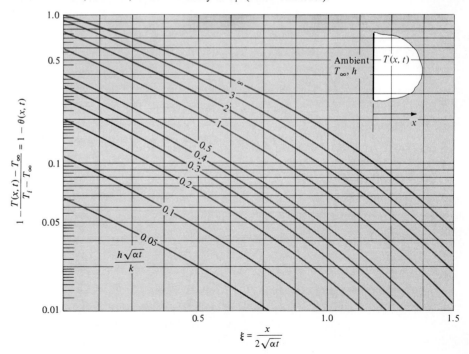

Example 6-1. A thick stainless-steel place $[\alpha = 1.6 \times 10^{-5}\, \text{m}^2/\text{s}$ and $k = 61\, \text{W}/(\text{m} \cdot {}^\circ\text{C})]$ is initially at a uniform temperature of 150°C. Suddenly one of its surfaces is lowered to 20°C and maintained at that temperature for times $t > 0$. By treating the plate as a semi-infinite solid, determine the temperature at a depth 2 cm from the surface and the surface heat flux at a time 1 min after lowering the surface temperature.

Solution. Given $\alpha = 1.6 \times 10^{-5}\, \text{m}^2/\text{s}$, $x = 0.02\, \text{m}$, and $t = 60\, \text{s}$, the parameter ξ becomes

$$\xi = \frac{x}{2\sqrt{\alpha t}} = \frac{0.02}{2\sqrt{1.6 \times 10^{-5} \times 60}} = 0.32$$

Knowing ξ, the dimensionless temperature is obtained either from the analytic expression, Eq. (6-3), or from the graph given in Fig. 6-2. If the analytic expression is used, we have for $\xi = 0.32$, from Appendix D, erf $(\xi) = 0.34913$. Then Eqs. (6-2a) and (6-3) give

$$\theta(x, t) = \frac{T(x, t) - T_0}{T_i - T_0} = \text{erf} (\xi) = 0.34913$$

If Fig. 6-2 is used, this result could also be obtained for $\xi = 0.32$, but not so accurately. Now, knowing $T_i = 150°C$ and $T_0 = 20°C$, the temperature at 2 cm depth and after 1 min is determined:

$$T(0.02\, \text{m}, 60\, \text{s}) = 0.34913(T_i - T_0) + T_0$$

$$= 0.34913(150 - 20) + 20 = 65.4°C$$

The heat flux at the surface at time $t = 1\, \text{min} = 60\, \text{s}$ is calculated from Eq. (6-5):

$$q_s(t) = \frac{k(T_0 - T_i)}{\sqrt{\pi \alpha t}}$$

$$= \frac{61(20 - 150)}{\sqrt{\pi \times 1.6 \times 10^{-5} \times 60}} = -144.4\, \text{kW}/\text{m}^2$$

where the minus sign indicates that the heat flow is out from the slab.

Example 6-2. A thick concrete slab $[\alpha = 7 \times 10^{-7}\, \text{m}^2/\text{s}$ and $k = 1.37\, \text{W}/(\text{m} \cdot {}^\circ\text{C})]$ is initially at a uniform temperature of 340°C. Suddenly one of its surfaces is subjected to convective cooling with a heat transfer coefficient of 100 W/(m² · °C) into an ambient at 40°C. Calculate the temperature at a depth 10 cm from the surface at the time 1 h after the start of cooling.

Solution. Given $\alpha = 7 \times 10^{-7}\, \text{m}^2/\text{s}$, $x = 0.1\, \text{m}$, and $t = 3600\, \text{s}$, then the parameter ξ becomes

$$\xi = \frac{x}{2\sqrt{\alpha t}} = \frac{0.1}{2\sqrt{7 \times 10^{-7} \times 3600}} = 1.0$$

Given $h = 100\, \text{W}/(\text{m}^2 \cdot {}^\circ\text{C})$ and $k = 1.37\, \text{W}/(\text{m} \cdot {}^\circ\text{C})$, the dimensionless parameter $h\sqrt{\alpha t}/k$ appearing in Fig. 6-3 becomes

$$\frac{h\sqrt{\alpha t}}{k} = \frac{100\sqrt{7 \times 10^{-7} \times 3600}}{1.37} = 3.66$$

Then from Fig. 6-3 we have

$$1 - \theta(x, t) = 0.11$$

and given $T_i = 340°C$ and $T_\infty = 40°C$, the temperature at 10 cm depth after 1 h is determined to be

$$1 - \theta(x, t) = 1 - \frac{T(x, t) - T_i}{T_\infty - T_i} = 0.11$$

$$T(0.1, 3600) = 0.11(40 - 340) + 340 = 307°C$$

6-2 TRANSIENT TEMPERATURE AND HEAT FLOW IN A SLAB, CYLINDER, AND SPHERE

Analytic solutions are available for temperature distribution and heat flow as a function of time and position in solids that have simple shapes, such as a slab, a long solid cylinder, or a sphere. Therefore transient-temperature and heat flow charts can be constructed from such solutions for ready reference in practical applications.

Here we consider the situation in which the solid is initially at a uniform temperature T_i and there is no energy generation in the medium. At time $t = 0$ the boundary surfaces are suddenly subjected to convection with a known heat transfer coefficient h into an ambient at a specified constant temperature T_∞. The resulting transient-temperature and heat flow charts are presented below, and their use is explained with examples.

Transient-Temperature Chart for Slab

Consider a slab (i.e., a plane wall) of thickness $2L$, confined to the region $-L \leq x \leq L$. Initially the slab is at a uniform temperature T_i. Suddenly, at time $t = 0$, both boundary surfaces of the slab are subjected to convection with a heat transfer coefficient h into ambients at a temperature T_∞, and they are maintained at that condition for $t > 0$. Figure 6-4a illustrates the geometry coordinates and the boundary conditions for this problem. It is apparent that the problem possesses both geometrical and thermal symmetry about the plane $x = 0$; therefore we need to consider this heat conduction problem for only half the region, say $0 \leq x \leq L$. Then the mathematical formulation of this heat conduction problem over the region $0 \leq x \leq L$ with a symmetry boundary condition at $x = 0$ and a convection boundary condition at $x = L$ becomes

$$\frac{\partial^2 T}{\partial x^2} = \frac{1}{\alpha} \frac{\partial T}{\partial t} \qquad \text{in } 0 < x < L, \text{ for } t > 0 \qquad (6\text{-}9a)$$

$$\frac{\partial T}{\partial x} = 0 \qquad \text{at } x = 0, \text{ for } t > 0 \qquad (6\text{-}9b)$$

$$k \frac{\partial T}{\partial x} + hT = hT_\infty \qquad \text{at } x = L, \text{ for } t > 0 \qquad (6\text{-}9c)$$

$$T = T_i \qquad \text{for } t = 0, \text{ in } 0 \leq x \leq L \qquad (6\text{-}9d)$$

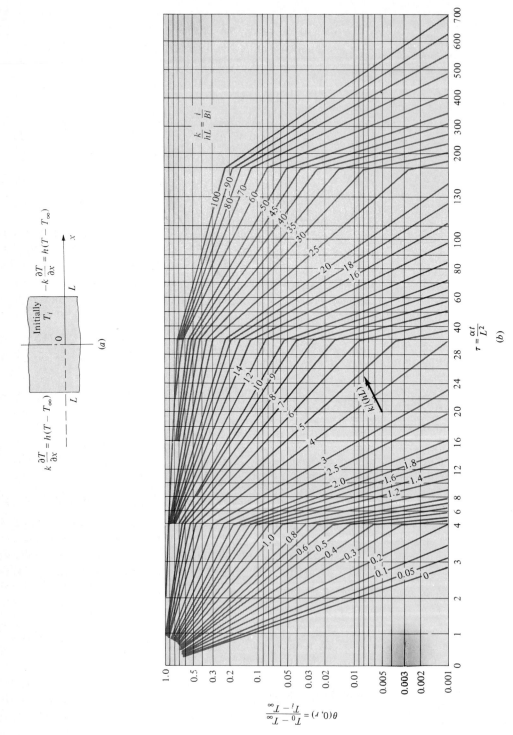

FIGURE 6-4

Transient temperature chart for a slab of thickness $2L$ subjected to convection at both boundary surfaces. (From Heisler) (a) Geometry, coordinates and boundary conditions for the physical problem; (b) Dimensionless temperature $\theta(0, \tau)$ at the center plane, $x = 0$;

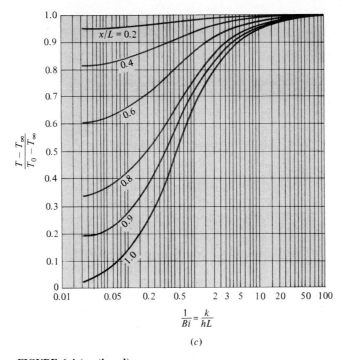

$$\frac{1}{Bi} = \frac{k}{hL}$$

(c)

FIGURE 6-4 (continued)
(c) Position correction for use with part (b).

A scrutiny of this formulation reveals that the temperature $T(x, t)$ depends on too many parameters, that is, t, x, L, T_i, T_∞, h, k, and α; therefore, it is not practical to present the solution for $T(x, t)$ in graphical form if the effects of all these parameters on temperature are to be included. To alleviate such a difficulty, the following dimensionless parameters are introduced:

$$\theta = \frac{T(x, t) - T_\infty}{T_i - T_\infty} = \text{dimensionless temperature} \qquad (6\text{-}10a)$$

$$X = \frac{x}{L} = \text{dimensionless coordinate} \qquad (6\text{-}10b)$$

$$\text{Bi} = \frac{hL}{k} = \text{Biot number} \qquad (6\text{-}10c)$$

$$\tau = \frac{\alpha t}{L^2} = \text{dimensionless time or Fourier number} \qquad (6\text{-}10d)$$

Then the above heat conduction problem becomes

$$\frac{\partial^2 \theta}{\partial X^2} = \frac{\partial \theta}{\partial \tau} \qquad \text{in } 0 < X < 1, \text{ for } \tau > 0 \qquad (6\text{-}11a)$$

$$\frac{\partial \theta}{\partial X} = 0 \qquad \text{at } X = 0, \text{ for } \tau > 0 \qquad (6\text{-}11b)$$

$$\frac{\partial \theta}{\partial X} + \text{Bi } \theta = 0 \qquad \text{at } X = 1, \text{ for } \tau > 0 \qquad (6\text{-}11c)$$

$$\theta = 1 \qquad \text{in } 0 \le X \le 1, \text{ for } \tau = 0 \qquad (6\text{-}11d)$$

In this dimensionless formulation the number of dimensionless parameters affecting $\theta(X, \tau)$ is reduced to three, namely, τ, Bi and X; hence it becomes practical to present the solution for temperature in graphical form. Before presenting some results, it is instructive to examine the physical significance of the dimensionless parameters τ and Bi that appear in the above equations. The dimensionless time τ (i.e., the Fourier number) given by Eq. (6-10d) is rearranged in the form

$$\tau = \frac{\alpha t}{L^2} = \frac{k(1/L)L^2}{\rho c_p L^3 / t} = \frac{\left(\begin{array}{l}\text{rate of heat conduction} \\ \text{across } L \text{ in volume } L^3, \text{W}/^\circ\text{C}\end{array}\right)}{\left(\begin{array}{l}\text{rate of heat storage} \\ \text{in volume } L^3, \text{W}/^\circ\text{C}\end{array}\right)} \qquad (6\text{-}12a)$$

Thus, the Fourier number is a measure of the rate of heat conduction compared with the rate of heat storage in a given volume element. Therefore, the larger the Fourier number, the deeper the penetration of heat into a solid over a given time.

The physical significance of the Biot number can be better understood if Eq. (6-10c) is rearranged in the form

$$\text{Bi} = \frac{hL}{k} = \frac{h}{k/L} = \frac{\left(\begin{array}{l}\text{heat transfer coefficient} \\ \text{at the surface of solid}\end{array}\right)}{\left(\begin{array}{l}\text{internal conductance of} \\ \text{solid across length } L\end{array}\right)} \qquad (6\text{-}12b)$$

That is, the Biot number is the ratio of the heat transfer coefficient to the unit conductance of the solid over the characteristic dimension.

The problem defined by Eqs. (6-11) has been solved, and the results for the dimensionless temperature are presented in Fig. 6-4b and c. Figure 6-4b gives the midplane temperature T_0 or $\theta(0, \tau)$ at $X = 0$ as a function of the dimensionless time τ for several different values of the parameter 1/Bi. The curve for $1/\text{Bi} = 0$ corresponds to the case in which $h \to \infty$, or the surfaces of the plate are maintained at the ambient temperature T_∞. For large values of 1/Bi, the Biot number is small, or the internal conductance of the solid is large in comparison with the heat transfer coefficient at the surface. This, in turn, implies that the temperature distribution within the solid is sufficiently uniform, and hence lumped system analysis becomes applicable. To illustrate this, we refer to Fig. 6-4c, which relates the temperatures at different locations within the slab to the midplane temperature T_0. Given T_0, temperatures at different locations within the slab can be determined. An examination of Fig. 6-4c reveals that for values of 1/Bi larger than 10, or Bi < 0.1, the temperature distribution within the slab may be considered uniform with an error of less than about 5 percent. We recall that Bi < 0.1 was used previously as the criterion for lumped system analysis to be applicable.

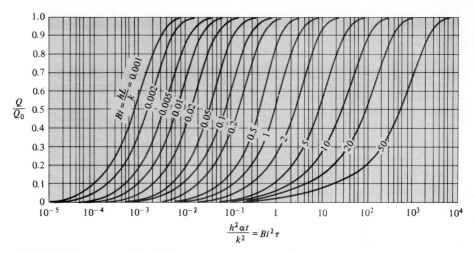

FIGURE 6-5
Dimensionless heat transferred Q/Q_0 for a slab of thickness $2L$. (From Gröber, Erk and Grigull).

The amount of heat transfer to or from the plate over a given period of time is also a quantity of practical interest.

Figure 6-5 shows the dimensionless heat transferred Q/Q_0 as a function of dimensionless time for several different values of the Biot number for a slab of thickness $2L$. Here Q represents the total amount of energy which is lost by the plate up to any time t during the transient heat transfer. The quantity Q_0 for the slab volume $V = (2L)(\text{depth})$, defined as

$$Q_0 = \rho c_p V(T_i - T_\infty) \qquad \text{W} \cdot \text{s} \qquad (6\text{-}13)$$

represents the initial energy of the slab relative to the ambient temperature. Knowing Q_0 and determining the ratio Q/Q_0 from Fig. 6-5, the amount of heat transfer Q is calculated.

Example 6-3. A steel plate $[\alpha = 1.2 \times 10^{-5}\,\text{m}^2/\text{s}, \quad k = 43\,\text{W}/(\text{m} \cdot {}^\circ\text{C}), \quad c_p = 465\,\text{J}/(\text{kg} \cdot {}^\circ\text{C})$, and $\rho = 7833\,\text{kg}/\text{m}^3]$ of thickness 10 cm, initially at a uniform temperature of 240°C, is suddenly immersed in an oil bath at 40°C. The convection heat transfer coefficient between the fluid and the surface is 600 W/$(\text{m}^2 \cdot {}^\circ\text{C})$.

(a) How long will it take for the center-plane to cool to 100°C?
(b) What is the temperature at a depth 3 cm from the outer surface?
(c) Calculate the energy removed from the plate during this time.

Solution. (a) Given $2L = 10\,\text{cm} = 0.1\,\text{m}$ or $L = 0.05\,\text{m}$ $k = 43\,\text{W}/(\text{m} \cdot {}^\circ\text{C})$, and $h = 600\,\text{W}/(\text{m}^2 \cdot {}^\circ\text{C})$, we have

$$\frac{1}{\text{Bi}} = \frac{k}{hL} = \frac{43}{600 \times 0.05} = 1.43 \qquad \text{or} \qquad \text{Bi} = 0.7$$

and given $T_0 = 100°C$, $T_\infty = 40°C$, and $T_i = 240°C$, we calculate

$$\theta(0, \tau) = \frac{T_0 - T_\infty}{T_i - T_\infty} = \frac{100 - 40}{240 - 40} = 0.3$$

Hence, Fig. 6-4b, for $\theta(0, \tau) = 0.3$ and $1/\text{Bi} = 0.7$ we read the dimensionless time $\tau \approx 1.4$. Knowing τ, we calculate the time as

$$\tau = \frac{\alpha t}{L^2} \qquad \text{or} \qquad t = \frac{1.4(0.05)^2}{1.2 \times 10^{-5}} = 292 \text{ s}$$

$$t = 4 \text{ min } 52 \text{ s}$$

(*b*) To determine the temperature at a location 3 cm from the surface (i.e., 2 cm from the centerline), we compute the dimensionless location x/L as

$$\frac{x}{L} = \frac{0.02}{0.05} = 0.4$$

For $1/\text{Bi} = 1.43$ and $x/L = 0.4$, from Fig. 6-4c we have

$$\frac{T - T_\infty}{T_0 - T_\infty} \cong 0.95$$

Hence T is determined as

$$T = T_\infty + (T_0 - T_\infty)(0.95)$$

$$= 40 + (100 - 40)(0.95) = 97°C$$

(*c*) The energy removed from the plate per square meter (including both sides) during $t = 4 \text{ min } 52 \text{ s}$ is determined as follows:
From Fig. 6-5, for $\text{Bi} = 0.7$ and for $\text{Bi}^2 \cdot \tau = (0.7)^2(1.4) = 0.686$, we have

$$\frac{Q}{Q_0} \approx 0.6$$

For the plate of thickness $2L = 0.1 \text{ m}$, Q_0 is determined from Eq. (6-13) as

$$Q_0 = \rho c_p V(T_i - T_\infty)$$

$$= (7833)(465)(0.1)(240 - 40)$$

$$= 72.847 \text{ MJ}$$

Knowing Q_0, the energy removed from the plate per square meter in 4 min 52 s becomes

$$Q = 0.6Q_0 = 43.7 \text{ MJ}$$

Transient-Temperature Chart for Long Cylinder

Consider one-dimensional transient heat conduction in a long cylinder of radius b which is initially at a uniform temperature T_i. Suddenly, at time $t = 0$, the boundary surface at $r = b$ is subjected to convection with a heat transfer coefficient h into an ambient at temperature T_∞ and maintained so for times $t > 0$. Figure 6-6a illustrates the geometry, the coordinates, and the boundary condition. The mathematical formulation of this heat conduction problem is similar to that given by Eqs. (6-9a), to (6-9d), except that the x variable is

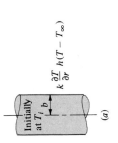

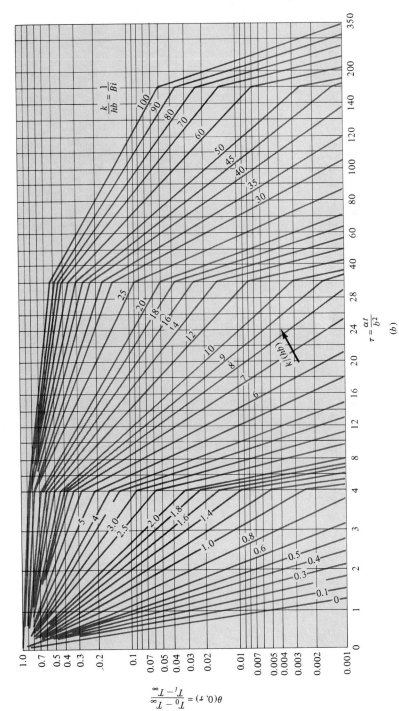

FIGURE 6-6

Transient temperature chart for a long solid cylinder of radius $r = b$ subjected to convection at the boundary surface $r = b$. (From Heisler). (a) Geometry, coordinates and boundary condition for the physical problem; (b) Dimensionless temperature $\theta(0, \tau)$ at the axis of the cylinder.

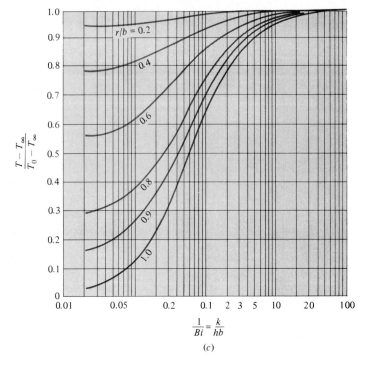

FIGURE 6-6
(c) Position correction for use with part (b).

replaced by the r variable, and the Laplacian term $\partial^2 T/\partial x^2$ is replaced by $(1/r)(\partial/\partial r)(r\,\partial T/\partial r)$. For the reasons stated previously, it is desirable to represent the formulation in the dimensionless form. Therefore, the following dimensionless quantities are defined:

$$\text{Bi} = \frac{hb}{k} = \text{Biot number} \tag{6-14a}$$

$$\tau = \frac{\alpha t}{D^2} = \text{dimensionless time, or Fourier number} \tag{6-14b}$$

$$\theta = \frac{T(r, t) - T_\infty}{T_i - T_\infty} = \text{dimensionless temperature} \tag{6-14c}$$

$$R = \frac{r}{b} = \text{dimensionless radial coordinate} \tag{6-14d}$$

Then the mathematical formulation of this heat conduction problem, in the dimensionless form, is given by

$$\frac{1}{R}\frac{\partial}{\partial R}\left(R\frac{\partial\theta}{\partial R}\right) = \frac{\partial\theta}{\partial\tau} \qquad \text{in } 0 < R < 1, \text{ for } \tau > 0 \tag{6-15a}$$

$$\frac{\partial\theta}{\partial R} = 0 \qquad \text{at } R = 0, \text{ for } \tau > 0 \tag{6-15b}$$

$$\frac{\partial \theta}{\partial R} + \text{Bi } \theta = 0 \qquad \text{at } R = 1, \text{ for } \tau > 0 \qquad (6\text{-}15c)$$

$$\theta = 1 \qquad \text{in } 0 \le R \le 1, \text{ for } \tau = 0 \qquad (6\text{-}15d)$$

This problem has been solved, and the results for the dimensionless center temperature $\theta(0, \tau)$ (or the center temperature T_0) at $R = 0$ are plotted in Fig. 6-6b as a function of the dimensionless time τ for several different values of the parameter $1/\text{Bi}$. Figure 6-6c relates the temperatures at different locations within the cylinder to the center temperature T_0. Therefore, given T_0, temperatures at different locations within the cylinder can be determined from Fig. 6-6c.

Figure 6-7 shows the dimensionless heat transferred Q/Q_0 as a function of dimensionless time for several different values of the Biot number for the cylinder problem given by Eqs. (6-15a) to (6-15d). Here Q_0 is as defined by Eq. (6-13), and Q represents the total amount of energy lost by the cylinder up to any time t during the transient heat transfer.

Transient-Temperature Chart for a Sphere

We now consider transient heat conduction for a solid sphere of radius b that is initially at a uniform temperature T_i. At time $t = 0$, the boundary surface at $r = b$ is suddenly subjected to convection with a heat transfer coefficient h into an ambient at temperature T_∞ and maintained so for times $t > 0$. Figure 6-8a illustrates the geometry, the coordinates, and the boundary conditions.

We introduce dimensionless variables as defined by Eqs. (6-14a) to (6-14d). Then the mathematical formulation of this heat conduction problem

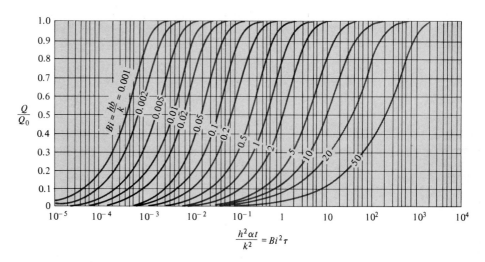

FIGURE 6-7
Dimensionless heat transferred Q/Q_0 for a long cylinder of radius b. (From Gröber, Erk and Grigull).

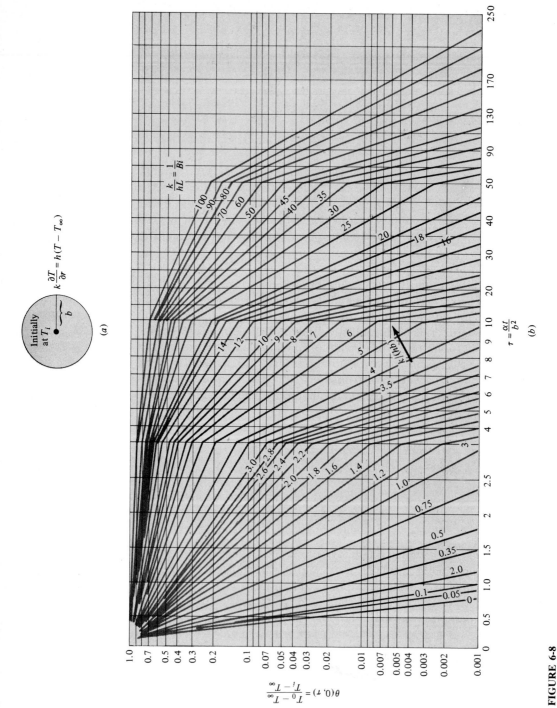

FIGURE 6-8

Transient temperature chart for a solid sphere of radius $r = b$ subjected to convection at the boundary surface $r = b$. (From Heisler). (a) Geometry, coordinates and boundary conditions; (b) Dimensionless temperature $\theta(0, \tau)$ at the center of the sphere;

154

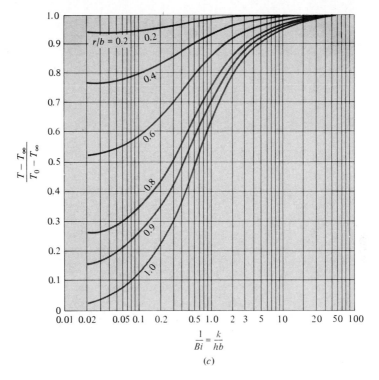

$$\frac{1}{Bi} = \frac{k}{hb}$$

(c)

FIGURE 6-8 (continued)
(c) Position correction for use with part (b).

in dimensionless form becomes

$$\frac{1}{R^2} \frac{\partial}{\partial R} \left(R^2 \frac{\partial \theta}{\partial R} \right) = \frac{\partial \theta}{\partial \tau} \qquad \text{in } 0 < R < 1, \text{ for } \tau > 0 \qquad (6\text{-}16a)$$

$$\frac{\partial \theta}{\partial R} = 0 \qquad \text{at } R = 0, \text{ for } \tau > 0 \qquad (6\text{-}16b)$$

$$\frac{\partial \theta}{\partial R} + \text{Bi } \theta = 0 \qquad \text{at } R = 1, \text{ for } \tau > 0 \qquad (6\text{-}16c)$$

$$\theta = 1 \qquad \text{in } 0 \le R \le 1, \text{ for } \tau = 0 \qquad (6\text{-}16d)$$

This problem has been solved, and Fig. 6-8b shows the dimensionless center temperature T_0 or $\theta(0, \tau)$ for the sphere as a function of dimensionless time τ for several different values of the parameter $1/\text{Bi}$. Figure 6-8c relates the temperatures at different locations within the sphere to the center temperature T_0.

Figure 6-9 shows the dimensionless heat transferred Q/Q_0 as a function of dimensionless time for several different values of the Biot number. Here, Q and Q_0 are as defined previously.

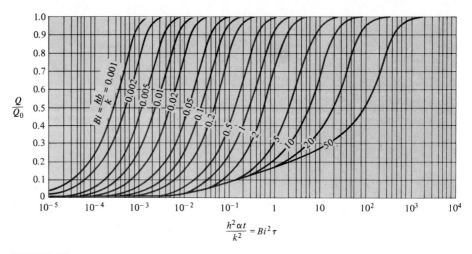

FIGURE 6-9
Dimensionless heat transfer Q/Q_0 for a sphere of radius b. (From Gröber, Erk and Grigull).

Example 6-4. Consider a slab of thickness 10 cm, a cylinder of diameter 10 cm, and a sphere of diameter 10 cm, each made of steel $[\alpha = 1.6 \times 10^{-5} \text{ m}^2/\text{s}$ and $k = 61 \text{ W/(m} \cdot ^\circ\text{C})]$ and initially at uniform temperature 300°C. Suddenly, they are all immersed into a well-stirred large bath at 50°C. The heat transfer coefficient between the surfaces and the fluid is 1000 W/(m² · °C). Calculate the time required for the centers of the slab, the cylinder, and the sphere to cool to 80°C.

Solution. Given $k = 61 \text{ W/(m} \cdot ^\circ\text{C})$, $h = 1000 \text{ W/(m}^2 \cdot ^\circ\text{C})$, $L = (r_0)_{\text{cyl}} = (r_0)_{\text{sph}} = 5 \text{ cm} = 0.05 \text{ m}$, $T_0 = 80^\circ\text{C}$, $T_i = 300^\circ\text{C}$, and $T_\infty = 50^\circ\text{C}$, we calculate

$$\frac{1}{\text{Bi}} = \frac{k}{hL} = \frac{61}{(1000)(0.05)} = 1.22$$

$$\theta(0, \tau) = \frac{T_0 - T_\infty}{T_i - T_\infty} = \frac{80 - 50}{300 - 50} = 0.12$$

Taking $\theta(0, \tau) = 0.12$ and $1/\text{Bi} = 1.22$, the dimensionless time for the slab is determined from Fig. 6-4b as $\tau = 3.5$. Then the time required t for the slab becomes

$$t = \frac{\tau L^2}{\alpha} = \frac{(3.5)(0.05)^2}{1.6 \times 10^{-5}} = 547 \text{ s}$$

Similarly, for the cylinder, the dimensionless time τ is determined from Fig. 6-6b as $\tau = 1.7$. Then t becomes

$$t = \frac{\tau r_0^2}{\alpha} = \frac{(1.7)(0.05)^2}{1.6 \times 10^{-5}} = 266 \text{ s}$$

For the sphere, the dimensionless time τ is determined from Fig. 6-8b to be $\tau = 1.2$. Therefore

$$t = \frac{\tau r_0^2}{\alpha} = \frac{(1.2)(0.05)^2}{1.6 \times 10^{-5}} = 188 \text{ s}$$

Example 6-5. A long steel shaft of radius 10 cm [$\alpha = 1.6 \times 10^{-5}$ m^2/s and $k = 61$ W/(m $\cdot$ °C)] is removed from an oven at a uniform temperature of 500°C and immersed in a well-stirred large bath of coolant maintained at 20°C. The heat transfer coefficient between the shaft surface and the coolant is 150 W/(m$^2 \cdot$ °C). Calculate the time required for the shaft surface to reach 300°C.

  Solution. Given $k = 61$ W/(m $\cdot$ °C), $h = 150$ W/(m$^2 \cdot$ °C),   and $r_0 = 10$ cm = 0.1 m, we calculate

$$\frac{1}{\text{Bi}} = \frac{k}{hr_0} = \frac{61}{150 \times 0.1} = 4.07$$

$$\frac{r}{r_0} = 1$$

Then, from Fig. 6-6c, for $1/\text{Bi} = 4.07$ and $r/r_0 = 1$, we have

$$\frac{T - T_\infty}{T_0 - T_\infty} \cong 0.87$$

where T is the surface temperature, T_∞ is the ambient temperature, and T_0 is the temperature at the axis of the shaft. Knowing the surface temperature $T = 300$°C and $T_\infty = 20$°C, the temperature at the axis, T_0, is determined as

$$T_0 = T_\infty + \frac{T - T_\infty}{0.87}$$

$$= 20 + \frac{300 - 20}{0.87} = 341.84°C$$

Next, given $T_i = 500$°C and $T_\infty = 20$°C, and taking $T_0 = 341.84$°C, we determine the dimensionless temperature $\theta(0, \tau)$ to be

$$\theta(0, \tau) = \frac{T_0 - T_\infty}{T_i - T_\infty} = \frac{341.84 - 20}{500 - 20} = 0.67$$

and then read the dimensionless time τ from Fig. 6-6b for $1/\text{Bi} = 4.07$ to be $\tau = 1$. We calculate the time required for the surface temperature to reach 300°C as follows:

$$\tau = \frac{\alpha t}{r_0^2} = 1$$

or

$$t = \frac{r_0^2}{\alpha} = \frac{(1.3)(0.1)^2}{1.6 \times 10^{-5}} = 625 \text{ s} \cong 10\tfrac{1}{2} \text{ min}$$

6-3 NUMERICAL ANALYSIS

The transient-temperature and heat flow charts presented in the previous section were constructed using the exact analytic solutions for the corresponding problem. A close scrutiny of these heat conduction problems reveals that they are restricted to bodies having simple shapes, such as a semi-infinite medium, a slab, a long solid cylinder, or a sphere. Furthermore, the thermal properties of the solid are assumed to be constant. When the body has an irregular shape, has temperature-dependent thermal properties, or is subjected to a nonlinear boundary condition, such as radiation to or from the boundary surface, exact analytic solution of the problem is not possible. The transient heat conduction problems for such situations can be solved using purely numerical approaches, such as finite-difference or finite-element methods.

The use of numerical methods in the solution of transient heat conduction in solids has been studied extensively, and there is a vast amount of literature on the subject. Here we present an introductory treatment of the use of the finite-difference method in the solution of one-dimensional transient heat conduction problems in the rectangular coordinate system. There are several different types of finite-difference schemes, each of which has advantages under certain conditions. To introduce the reader to the subject, we consider only the *explicit method* of finite-differencing, because it is a simple and straightforward approach, and the resulting system of algebraic equations is very easy to solve.

The Explicit Finite-Difference Scheme

The transient heat conduction problems considered in this chapter consist of a one-dimensional time-dependent heat conduction equation, two boundary conditions, and an initial condition. The basic idea of the finite-difference representation of such a problem is to transform the differential equation and its boundary conditions into a set of algebraic equations.

To illustrate the application of this concept, we consider a one-dimensional transient heat conduction problem for a slab with constant properties, confined to the region $0 \le x \le L$, that is subjected to some specified boundary conditions at the surfaces $x = 0$ and $x = L$, with an initial temperature distribution over the region.

The heat conduction equation for this problem is given by

$$\frac{\partial T(x, t)}{\partial t} = \alpha \frac{\partial^2 T}{\partial x^2} \quad \text{in } 0 < x < L \tag{6-17}$$

To obtain the finite-difference form of this differential equation, the x and t domains are divided into small steps Δx and Δt, as illustrated in Fig. 6-10. Suppose the region $0 \le x \le L$ is divided into M equal intervals, and a time step Δt is chosen. Then we have

$$x = (m - 1)\,\Delta x$$

or
$$x + \Delta x = m\,\Delta x \quad m = 1, 2, \ldots, M, M + 1 \tag{6-18a}$$

$$t = i\,\Delta t \quad i = 0, 1, 2, \ldots \tag{6-18b}$$

where M is the number of equal subintervals over the region $0 \le x \le L$,

$$\Delta x = \frac{L}{M} \tag{6-18c}$$

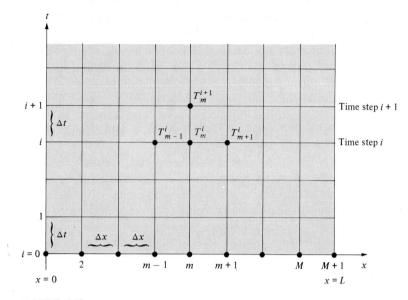

FIGURE 6-10
Subdivision of xt domain into intervals of Δx and Δt for finite-difference representation of the one-dimensional, time-dependent heat conduction equation.

and the temperature $T(x, t)$ at a location x and time t is denoted by the symbol T_m^i; that is,

$$T(x, t) = T[(m - 1)\, \Delta x,\, i\, \Delta t] \equiv T_m^i \qquad (6\text{-}18d)$$

Note that in the space domain there are $M + 1$ nodes (i.e., $m = 1, 2, 3, \ldots, M + 1$).

To develop the finite-difference equations for each of these nodes, it is desirable to consider them in two distinct groups:

1. The *interior nodes*, consisting of the nodes $m = 2, 3, \ldots, M$, for which the finite-difference equations are obtained by utilizing the heat conduction equation (6-17).

2. The *boundary nodes*, or the nodes $m = 1$ and $m = M + 1$, for which the finite-difference equations are determined by utilizing the boundary conditions for the problem.

We present below the development of the finite-difference equations for the *interior* and *boundary* nodes.

INTERIOR NODES. To develop the finite-difference form of the heat conduction equation (6-17) for an interior node m at time step i, we need to discretize the partial derivatives with respect to the space and time variables.

The second derivative of temperature with respect to x, at a node m (position $x + \Delta x = m\,\Delta x$) and at time step i (time $i\,\Delta t$), is represented in the finite-difference form as described previously in Chapter 5. We find

$$\left.\frac{\partial^2 T}{\partial x^2}\right|_{m,i} \cong \frac{T^i_{m-1} - 2T^i_m + T^i_{m+1}}{(\Delta x)^2} \tag{6-19a}$$

where T^i_{m-1} and T^i_{m+1} are the two neighboring points to the node T^i_m, and all of them are evaluated at the time step i.

The first derivative of temperature with respect to the time variable t at a position m and a time step i is represented by

$$\left.\frac{\partial T}{\partial t}\right|_{m,i} \cong \frac{T^{i+1}_m - T^i_m}{\Delta t} \tag{6-19b}$$

where T^{i+1}_m represents the temperature at the location $x + \Delta x = m\,\Delta x$ at the time $(i + 1)\,\Delta t$. The right-hand side of Eq. (6-19b) is a *forward finite-difference* representation with respect to time.

By introducing Eqs. (6-19) into Eq. (6-17), the finite-difference form of the one-dimensional time-dependent heat conduction equation at an internal node m becomes

$$\frac{T^{i+1}_m - T^i_m}{\Delta t} = \alpha \frac{T^i_{m-1} + T^i_{m+1} - 2T^i_m}{(\Delta x)^2} \tag{6-20}$$

This equation is rearranged in the form

$$\boxed{T^{i+1}_m = r(T^i_{m-1} + T^i_{m+1}) + (1 - 2r)T^i_m} \tag{6-21}$$

where

$$i = 0, 1, \ldots$$

$$m = 2, 3, \ldots, M$$

$$r = \frac{\alpha\,\Delta t}{(\Delta x)^2} \tag{6-22}$$

Equation (6-21) is called the *explicit finite-difference* form of the one-dimensional time-dependent heat conduction equation (6-17).

The method is called *explicit* because the temperature T^{i+1}_m at node m at a time step $i + 1$ is immediately determined from Eq. (6-21) if the temperatures of node m and its two neighboring points at the previous time step i are available and the value of the parameter r is specified.

The system of equations (6-21) provides $M - 1$ algebraic equations, but contains $M + 1$ unknown node temperatures (i.e., $T^i_1, T^i_2, \ldots, T^i_{M+1}$). The additional two relations are obtained by utilizing the boundary conditions at the nodes $m = 1$ and $m = M + 1$, as described below.

BOUNDARY NODES. The boundary conditions at the nodes $m = 1$ and $m = M + 1$ may be prescribed temperature, prescribed heat flux, or convection into an ambient at a specified temperature.

When the temperatures are specified at the boundaries, the node temperatures T_1^i and T_{M+1}^i are known for all times; this provides two additional relations, and so Eqs. (6-21) are a complete set for the solution of $M - 1$ unknown interior node temperatures for each time step.

In the case of insulated, prescribed heat flux, or convection boundary conditions at the nodes $m = 1$ and $m = M + 1$, the node temperatures T_1^i and T_{M+1}^i are unknown. Two additional equations are needed in order to make the number of equations equal to the number of unknowns in the system (6-21). These equations are developed by writing an energy balance equation for a differential volume ΔV of thickness $\Delta x / 2$ at node $m = 1$ and/or $m = M + 1$, as illustrated in Fig. 6-11. If the step size Δx is sufficiently small, the heat capacity associated with the differential volume element at the boundary node can be neglected. Then the finite-difference equations for the boundary nodes are developed by considering the following *steady-state energy balance equation.*

$$\left\{ \begin{array}{l} \text{Rate of heat entering } \Delta V \\ \text{from all its surfaces at} \\ \text{time step } i + 1 \end{array} \right\} + \left\{ \begin{array}{l} \text{rate of energy generation} \\ \text{in } \Delta V \text{ at time step } i + 1 \end{array} \right\} = 0 \qquad (6\text{-}23)$$

Suppose we have *convection* at the boundary surface $x = 0$, with a heat transfer coefficient h_0 into an ambient at temperature $T_{\infty 0}$, but *no energy generation* in the medium. The application of the energy equation (6-23) to the node $m = 1$ yields

$$h_0(T_{\infty 0} - T_1^{i+1}) + k \frac{T_2^{i+1} - T_1^{i+1}}{\Delta x} = 0 \qquad (6\text{-}24a)$$

Solving for T_1^{i+1}, we get

$$\boxed{T_1^{i+1} = \frac{1}{1 + \Delta x \, h_0 / k} \left(T_2^{i+1} + \frac{\Delta x \, h_0}{k} T_\infty \right)} \qquad \text{for } m = 0 \quad (6\text{-}24b)$$

where the parameter r is defined by Eq. (6-22).

FIGURE 6-11
Finite-difference equations for
nodes on the boundaries.

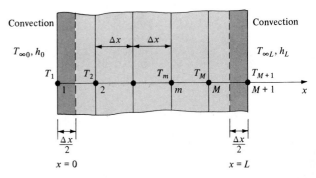

In the case of *convection* at the boundary surface $x = L$, with a heat transfer coefficient h_L into an ambient at temperature $T_{\infty L}$, but no energy generation in the medium, the application of the energy equation (6-23) to the node $m = M + 1$ gives

$$h_L(T_{\infty L} - T_{M+1}^{i+1}) + k \frac{T_M^{i+1} - T_{M+1}^{i+1}}{\Delta x} = 0 \qquad (6\text{-}25a)$$

Solving for T_M^{i+1}, we obtain

$$\boxed{T_{M+1}^{i+1} = \frac{1}{1 + \Delta x \, h_L/k} \left(T_M^{i+1} + \frac{\Delta x \, h_L}{k} T_{\infty L} \right)} \qquad \text{for } m = M \qquad (6\text{-}25b)$$

Equations (6-24b) and (6-25b) are the finite-difference equations for the boundary nodes $m = 1$ and $m = M + 1$, respectively, when these boundary surfaces are subjected to *convection*.

If the boundary surface at $x = 0$ is subjected to constant heat flux q_0, then the application of energy balance given by Eq. (6-23) to the node at $m = 1$ yields

$$q_0 + k \frac{T_2^{i+1} - T_1^{i+1}}{\Delta x} = 0 \qquad (6\text{-}26a)$$

or

$$T_1^{i+1} = q_0 \frac{\Delta x}{k} + T_2^{i+1} \qquad (6\text{-}26b)$$

where q_0 is a positive quantity when it is into the wall.

If the boundary surface at $x = L$ is subjected to constant heat flux q_L, then the application of energy balance given by Eq. (6-23) to the node at $m = M + 1$ yields

$$q_L + k \frac{T_M^{i+1} - T_{M+1}^{i+1}}{\Delta x} = 0 \qquad (6\text{-}27a)$$

or

$$T_{M+1}^{i+1} = q_L \frac{\Delta x}{k} + T_M^{i+1} \qquad (6\text{-}27b)$$

where q_L is a positive quantity when it is into the wall.

If the boundary surface at $m = 1$ or $m = M + 1$ is *insulated (adiabatic)*, the resulting finite-difference equation for the insulated boundary $m = 1$ or $m = M + 1$ is obtained from Eq. (6-24b) or (6-25b), respectively, by setting h_0 or h_L equal to zero.

RESTRICTION ON THE VALUE OF r. We note that the finite-difference equations (6-21) contain a parameter r, defined as

$$r = \frac{\alpha \, \Delta t}{(\Delta x)^2} \qquad (6\text{-}28)$$

The disadvantage of the simple explicit finite-difference scheme discussed above is that there is a restriction on the permissible value of r, imposed by a numerical stability condition. In finite-difference solution, the mesh size Δx is generally chosen first, and then the value of the thermal diffusivity α is established by the material property. If the value of r is fixed by the stability criterion, the permissible size of the time step Δt, according to Eq. (6-28), is restricted to

$$\Delta t \leq \frac{r(\Delta x)^2}{\alpha} \tag{6-29}$$

This condition implies that, if the mesh size Δx is reduced by a factor of n to improve the accuracy, the corresponding decrease in the time step Δt is by a factor of n^2. Then, a large number of calculations are needed to solve the problem over a given time interval.

There are various mathematical techniques for determining the stability criterion associated with the finite-difference representation of the time-dependent heat conduction equation. Here we avoid mathematical developments, and instead use a physical argument to establish such a stability criterion, as described below.

Let us examine the stability criterion associated with difference equation (6-21). Suppose that at any time step the temperatures T_{m-1}^i and T_{m+1}^i at nodes $m-1$ and $m+1$ are equal but less than T_m^i at node m between them. If the value of r exceeds $\frac{1}{2}$, the coefficient $(1-2r)$ becomes negative. Then, according to the finite-difference equation (6-21), for $(1-2r)$ negative, the temperature T_m^{i+1} at node m at the next time step should be less than that at the neighboring two nodes. This is not possible thermodynamically, since we assumed that T_m^i was higher than the temperature at the neighboring nodes. Therefore, to obtain meaningful solutions from Eq. (6-21), the coefficient $(1-2r)$ of T_m^i should not be negative; that is,

$$1 - 2r \geqq 0 \qquad \text{or} \qquad \boxed{r \leqq \tfrac{1}{2}} \tag{6-30}$$

where

$$r = \frac{\alpha \, \Delta t}{(\Delta x)^2}$$

CALCULATIONAL PROCEDURE. Having discussed the finite-difference form of the transient heat conduction equation at the internal nodes and that of the boundary conditions at the boundary nodes, we next summarize the mathematical formulation of the heat conduction problem and the corresponding finite-difference equations before discussing the calculational procedure.

Consider transient heat conduction in a slab $0 \leq x \leq L$ that is initially at a prescribed temperature $F(x)$ and for times $t > 0$ is subjected to convection from its boundaries into ambients at a constant temperature T_∞. The mathematical formulation of this heat conduction problem is given by

$$\frac{\partial T(x, t)}{\partial t} = \alpha \frac{\partial^2 T}{\partial x^2} \qquad \text{in } 0 < x < L, \ t > 0 \qquad (6\text{-}31a)$$

$$-k \frac{\partial T}{\partial x} + h_0 T = h_0 T_\infty \qquad \text{at } x = 0, \ t > 0 \qquad (6\text{-}31b)$$

$$k \frac{\partial T}{\partial x} + h_L T = h_L T_\infty \qquad \text{at } x = L, \ t > 0 \qquad (6\text{-}31c)$$

$$T(x, t) = F(x) \qquad \text{for } t = 0, \ 0 \le x \le L \qquad (6\text{-}31d)$$

To write the finite-difference form of this problem, the region $0 \le x \le L$ is divided into M equal subregions, each of thickness Δx, as illustrated in Fig. 6-11, and the time step Δt is chosen to satisfy the stability criterion.

The finite-difference form of the differential equation (6-31a) for the internal nodes $m = 2, 3, \ldots, M$, using the explicit scheme, is obtained from Eq. (6-21):

$$\boxed{T_m^{i+1} = r(T_{m-1}^i + T_{m+1}^i) + (1 - 2r)T_m^i \qquad \text{for } m = 2, 3, \ldots, M} \qquad (6\text{-}32a)$$

where

$$r = \frac{\alpha \, \Delta t}{(\Delta x)^2}$$

The finite-difference forms of the boundary conditions (6-31b) and (6-31c) at the nodes $m = 1$ and $m = M + 1$ are obtained, respectively, from Eqs. (6-24b) and (6-25b); with $T_{\infty 0} = T_{\infty L} = T_\infty$, we have

$$\boxed{\begin{aligned} T_1^{i+1} &= \frac{1}{1 + \Delta x \, h_0/k} \left(T_2^{i+1} + \frac{\Delta x \, h_0}{k} T_\infty \right) \qquad \text{for } m = 1 \qquad (6\text{-}32b) \\[2mm] T_{M+1}^{i+1} &= \frac{1}{1 + \Delta x \, h_L/k} \left(T_M^{i+1} + \frac{\Delta x \, h_L}{k} T_\infty \right) \qquad \text{for } m = M + 1 \qquad (6\text{-}32c) \end{aligned}}$$

Finally, the finite-difference form of the initial condition, Eq. (6-31d), is written as

$$T_m^0 = F(x + \Delta x) \equiv F_m \qquad \text{for } m = 1, 2, \ldots, M + 1 \text{ and } i = 0 \qquad (6\text{-}32d)$$

Thus Eqs. (6-32a) to (6-32c) provide $M + 1$ relations for the determination of $M + 1$ unknown node temperatures T_m^i ($m = 1, 2, \ldots, M + 1$) for each time step $i = 1, 2, 3, \ldots$.

The calculational procedure is very simple:

1. Choose the number of subdivisions M, hence determine the mesh size $\Delta x = L/M$.

2. Choose the value of the parameter r consistent with the stability criterion defined by Eq. (6-30), namely, set

$$r \le \frac{1}{2}$$

3. The material property establishes the value of the thermal diffusivity α. Knowing Δx, α, and r, establish the corresponding time step Δt according to the definition of r given by Eq. (6-28); that is,

$$\Delta t = \frac{r(\Delta x)^2}{\alpha}$$

4. Calculations are started by setting $i = 0$. Then Eq. (6-32a) gives the temperatures T_m^1 at all internal nodes $m = 2, 3, \ldots, M$ at the end of the first time step, $i = 1$ or $t = \Delta t$, since the node temperatures on the right-hand side of Eq. (6-32a) are available from the known initial temperatures.

5. Setting $i = 0$ in Eqs. (6-32b) and (6-32c), the temperatures T_1^1 and T_{M+1}^1 at the boundary nodes are determined, since T_2^1, T_M^1 and all the other quantities on the right-hand side of Eqs. (6-32b) and (6-32c) are known.

6. Given the node temperatures T_m^1 $(m = 1, 2, 3, \ldots, M + 1)$ at the end of the first time step, we repeat the procedure in steps 4 and 5 by setting $i = 1$ and calculating the node temperatures T_m^2 $(m = 1, 2, \ldots, M + 1)$ at the end of the second time step $i = 2$ or at the time $t = 2\Delta t$.

7. The node temperatures for subsequent times are found in a similar manner.

The application of the above calculational procedure is now illustrated with the following simple example.

Example 6-6. A large brick wall $(\alpha = 5 \times 10^{-7} \text{ m}^2/\text{s})$ of thickness $L = 20$ cm is initially at a uniform temperature $T_i = 100°\text{C}$. Suddenly both of its surfaces are lowered to $20°\text{C}$ and maintained at that temperature for times $t > 0$. Using four subdivisions (i.e., $M = 4$), corresponding to a mesh size of $\Delta x = 20/4 = 5$ cm, and an explicit finite-difference scheme, calculate the temperature distribution at the nodes for 10 consecutive time steps.

Solution. The mathematical formulation of this heat conduction problem is given by

$$\frac{\partial T}{\partial t} = \alpha \frac{\partial^2 T}{\partial x^2} \qquad \text{in } 0 < x < L, \ t > 0$$

subject to the boundary conditions

$$T = 20°\text{C} \qquad \text{at } x = 0, \ t > 0$$

$$T = 20°\text{C} \qquad \text{at } x = L, \ t > 0$$

and the initial condition

$$T = 100°\text{C} \qquad \text{for } t = 0, \text{ in } 0 \le x \le L$$

For illustration purposes and simplicity in the computations, we have chosen only $M = 4$ equal subdivisions of the region, which corresponds to a very coarse mesh size.

$$\Delta x = \frac{L}{M} = \frac{20}{4} = 5 \text{ cm}$$

Thus, over the entire region we have only five nodes, $m = 1, 2, \ldots, 5$, of which two are the boundary nodes and the remaining three are internal nodes.

The finite-difference equations for the interior nodes are obtained from Eq. (6-21) by setting the value of r such that it does not violate the stability criterion. For $r = \frac{1}{2}$, Eq. (6-21) becomes

$$T_m^{i+1} = \frac{1}{2} (T_{m-1}^i + T_{m+1}^i) \qquad m = 2, 3, 4 \qquad (a)$$

For the prescribed temperature boundary conditions considered in this problem, the equations for the boundary nodes $m = 1$ and $M + 1 = 5$ become

$$T_1^{i+1} = 20 \qquad m = 1 \qquad (b)$$

$$T_5^{i+1} = 50 \qquad M + 1 = 5 \qquad (c)$$

Finally, the initial condition for the problem is written as

$$T_m^0 = 200 \qquad i = 0, m = 1, 2, 3, 4, 5 \qquad (d)$$

and the corresponding time step Δt with $r = \frac{1}{2}$ becomes

$$\Delta t = \frac{r(\Delta x)^2}{\alpha} = \frac{1}{2} \frac{(0.05)^2}{5 \times 10^{-7}} = 2500 \text{ s}$$

By using the finite-difference equations (a) to (d) and following the computational procedure described previously, the unknown node temperatures T_m^i ($m = 2, 3, 4$) are readily determined. Table 6-1 shows the results of such calculations for 10 consecutive time steps. The step $i = 10$ corresponds to the time $t = 10 \times 2500 \text{ s} = 6.94 \text{ h}$.

TABLE 6-1
Numerical solution to Example 6-6

TIME STEP :	2500.	sec			
X =	0.000	0.050	0.100	0.150	0.200
Time(sec.)					
0.0000	100.000	100.000	100.000	100.000	100.000
2500.0000	20.000	100.000	100.000	100.000	20.000
5000.0000	20.000	60.000	100.000	60.000	20.000
7500.0000	20.000	60.000	60.000	60.000	20.000
10000.0000	20.000	40.000	60.000	40.000	20.000
12500.0000	20.000	40.000	40.000	40.000	20.000
15000.0000	20.000	30.000	40.000	30.000	20.000
17500.0000	20.000	30.000	30.000	30.000	20.000
20000.0000	20.000	25.000	30.000	25.000	20.000
22500.0000	20.000	25.000	25.000	25.000	20.000
25000.0000	20.000	22.500	25.000	22.500	20.000

In practice, a much finer mesh is needed to obtain sufficiently accurate results. This involves the use of both a large number of node points in the medium and a large number of time steps. In such situations, all the calculations are performed by using a digital computer.

In Table 6-2 we present a single computer program written in Fortran to solve transient heat conduction in a slab of thickness L, initially at a prescribed temperature, where for times $t > 0$ are boundary conditions at $x = 0$ and $x = L$ are maintained at constant temperature. The variable names used in this computer program are as follows:

Variable name	Quantity	Dimension
ALPHA	α, thermal diffusivity	m^2/s
DELX	Δx, spatial step size	m
IEND	maximum number of time steps	
M	$L/\Delta x$, number of subdivisions	
RL	L, slab thickness	m
R	$r = \dfrac{\alpha \, \Delta t}{(\Delta x)^2}$	dimensionless
T(J)	temperature at node J	°C
T(1)	surface temperature at $x = 0$, $m = 1$	°C
T(M + 1)	surface temperature at $x = L$, $m = M + 1$	°C

To illustrate the use of the computer, we applied the program given in Table 6-2 to the solution of the specific transient heat conduction problem considered in Example 6-6. The main features of this program and the significance of various statements are as follows:

Lines 12 to 22 specify the input data, which include M, α, L, Δx, r, and the maximum number of time steps required.

Lines 46 to 51 specify the types of boundary conditions at $x = 0$ and $x = L$ and the input data associated with them.

The boundary condition at $x = 0$ can be any one of the following three possibilities:

1. *Prescribed temperature*: Specify $T(1)$ in degrees Celsius to represent the surface temperature.
2. *Prescribed heat flux*: Specify q_0, to represent the prescribed heat flux in watts per square meter.
3. *Convection*: Specify h_0 in $W/(m^2 \cdot °C)$ and the ambient temperature $T_{\infty 0}$ in degrees Celsius associated with it.

The boundary condition at $x = L$ can be any of the following possibilities:

1. *Prescribed temperature*: Specify $T(M + 1)$ in degrees Celsius represent the surface temperature.
2. *Prescribed heat flux*: Specify q_L in watts per square meter to represent the prescribed heat flux at $x = L$.
3. *Convection*: Specify h_L in $W/(m^2 \cdot °C)$ and the ambient temperature $T_{\infty L}$ in degrees Celsius associated with it.

TABLE 6-2

Computer program for solving unsteady heat conduction in a slab applied for the solution of Example 6-6

```
 1   C****  COMPUTER PROGRAM FOR SOLVING UNSTEADY
 2   C         HEAT CONDUCTION IN A SLAB BY EXPLICIT
 3   C         FINITE DIFFERENCE SCHEME
 4             IMPLICIT REAL*8(A-H,O-Z)
 5             CHARACTER*72 TITLE
 6             DIMENSION T(100), TNEW(100), X(100)
 7   C
 8             TITLE = 'EXAMPLE  6-6'
 9             WRITE (6,810) TITLE
10     810 FORMAT(///A72)
11   C****  INPUT DATA
12             M = 4
13             ALPHA = 5.0D-7
14             RL = 0.2D0
15             DELX = RL / DBLE(M)
16             IEND = 10
17             R = 0.5D0
18   C
19   C
20   C
21   C
22   C
23             DO 400 I = 1,M+1
24                 X(I) = DBLE(I-1) * DELX
25     400 CONTINUE
26   C
27   C****  THE INITIAL TEMPERATURE DISTRIBUTION
28   C         IS GIVEN BY
29             DO 10 I = 1,M+1
30      10 T(I) = 100.0D0
31   C****  COMPUTE THE VALUE OF TIME STEP
32             DELT = R * (DELX)**2 / ALPHA
33             WRITE (6,200) DELT
34     200 FORMAT (/3X,'TIME STEP : ',G11.4, ' sec')
35   C****  COMPUTE INTERNAL TEMPERATURE FOR SUCCESSIVE
36   C         TIME STEPS
37             WRITE (6,60) (X(J),J=1,M+1)
38             WRITE (6,70)
39             ICOUNT = O
40             WRITE (6,50) ICOUNT, (T(J),J=1,M+1)
41      20 ICOUNT = ICOUNT + 1
42             DO 30 J = 2,M
43                 TNEW(J) = R * (T(J-1) + T(J+1)) + (1.0 - 2.0 * R) * T(J)
44      30 CONTINUE
45   C****  THE BOUNDARY CONDITIONS AT X=O AND X=L ARE
46   C
47                 T(1) = 20.0D0
48   C
49   C
50                 T(M+1) = 20.0D0
51   C
52   C****  SUBSTITUTE NEW TEMPERATURE
53             DO 40 J = 2,M
54      40 T(J) = TNEW(J)
55   C****  PRINT TEMPERATURES
56             WRITE (6,50) ICOUNT*DELT, (T(J),J=1,M+1)
57      50 FORMAT (1X,F11.4,2X,100(5F10.3/14X))
58             IF (ICOUNT .LT. IEND) GO TO 20
59      60 FORMAT (/6X,'X = ',4X,100(5F10.3/14X))
60      70 FORMAT(/4X,'Time(sec.)'/)
61             STOP
62             END
```

Lines 27 to 30 specify the initial temperature distribution. For the case of a uniform temperature, a "DO LOOP" is used to generate the initial temperatures at the nodes. For the case of nonuniform initial temperatures, the READ statement can be used to specify the initial temperatures at the nodes.

Lines 32 to 34 compute the permissible value of the time step Δt from the relation $\Delta t = r(\Delta x)^2/\alpha$.

Lines 42 to 44 compute the temperatures T_m^i at the internal nodes $m = 2$ to $m = M$ from the relation given by Eq. (6-32a),

$$T_m^{i+1} = r(T_{m-1}^i + T_{m+1}^i) + (1 - 2r)T_m^i$$

for the time step $i + 1$.

Lines 46 to 51 compute the temperatures at the boundary nodes $m = 1$ and $m = M + 1$ depending on the type of boundary conditions used. Note that in the computer program we used prescribed temperature boundary conditions.

Finally, lines 56 and 57 print the node temperatures at various times.

Example 6-7. A large brick wall $[\alpha = 5 \times 10^{-7} \text{ m}^2/\text{s}$ and $k = 1 \text{ W}/(\text{m} \cdot {}^\circ\text{C})]$ of thickness $L = 20 \text{ cm}$ is initially at a uniform temperature $T_i = 100{}^\circ\text{C}$. Suddenly one of its surfaces is subjected to convection with a heat transfer coefficient of $50 \text{ W}/(\text{m}^2 \cdot {}^\circ\text{C})$ into an ambient at $20{}^\circ\text{C}$. The other surface is subjected to a constant heat flux $q = 3 \times 10^3 \text{ W}/\text{m}^2$. Using 10 subdivisions (i.e., $M = 10$) and an explicit finite-difference scheme, calculate the temperature distribution at the nodes for 10 consecutive time steps.

Solution. The mathematical formulation of this heat conduction problem is given by

$$\frac{\partial T}{\partial t} = \alpha \frac{\partial^2 T}{\partial x^2} \quad \text{in } 0 < x < L, \, t > 0$$

subject to the boundary conditions

$$-(1) \frac{\partial T}{\partial x} + 50T = (50)(20) \quad \text{at } x = 0, \text{ for } t > 0$$

$$-(1) \frac{\partial T}{\partial x} = 3 \times 10^3 \quad \text{at } x = 0.2 \text{ m}, \text{ for } t > 0$$

$$T = 100{}^\circ\text{C} \quad \text{for } t = 0, \text{ in } 0 \le x \le 0.2 \text{ m}$$

Table 6-3 gives the modifications of the computer program in Table 6-2 needed to take into account the different number of subdivisions (i.e., $M = 10$) and the boundary conditions (convection at $x = 0$ and prescribed heat flux at $x = L$). The variable names used in this computer program are as follows:

Variable name	Quantity	Dimension
DXHKO	$\Delta x \cdot h_0/k$	Dimensionless
HO	h_0, heat transfer coefficient at $x = 0$	W/(m$^2 \cdot {}^\circ$C)
RK	k, thermal conductivity	W/(m$\cdot {}^\circ$C)
TINFO	$T_{\infty 0}$, ambient temperature at $x = 0$	${}^\circ$C
QL	q_L, heat flux at $x = L$	W/m^2

TABLE 6-3
Modification of the computer program in Table 6-2 for the solution of Example 6-7

```
Line 12  placed by
    12              M = 10
Lines 18  to 22 placed by
    18              RK = 1.0D0
    19              HO = 50.0D0
    20    C
    21              TINFO = 20.0D0
    22              QL = 3.0D0 + 3

-------- ---------------------------------------------------------

Lines 46  to 51 placed by

    46              DXHKO = DELX * HO / RK
    47              T(1) = 1.0D0/(1.0D0 + DXHKO)*(TNEW(2) + DXHKO*TINFO)
    48    C
    49    C
    50              T(M+1) = QL*DELX/RK + TNEW(M)
```

Table 6-4 gives the solution of the transient heat conduction problem considered in Example 6-7. This step $i = 10$ corresponds to the time $t = 4000$ s. The numerical solution for this problem is printed at each time step. The accuracy of numerical results could be increased by choosing smaller time steps, which in turn will require a large number of computations. In order to print output after a certain number of time steps, the given computer program will require minor modification.

The steady temperature distribution in Example 6-7 is given by

$$T(x) = \frac{q}{k} x + \frac{q}{h} + T_\infty$$

and for larger times the temperatures at $x = 0$ and $x = L = 0.2$ m approach the steady temperatures 40°C and 240°C, respectively.

TABLE 6-4
Numerical solution to Example 6-7

TIME STEP :	400.0	sec.			
X =	0.000	0.020	0.040	0.060	0.080
	0.100	·0.120	0.140	0.160	0.180
	0.200				
Time(sec.)					
0.0000	100.000	100.000	100.000	100.000	100.000
	100.000	100.000	100.000	100.000	100.000
	100.000				

TABLE 6-4 (continued)

400.0000	60.000	100.000	100.000	100.000	100.000
	100.000	100.000	100.000	100.000	100.000
	160.000				
800.0000	50.000	80.000	100.000	100.000	100.000
	100.000	100.000	100.000	100.000	130.000
	190.000				
1200.0000	47.500	75.000	90.000	100.000	100.000
	100.000	100.000	100.000	115.000	145.000
	205.000				
1600.0000	44.375	68.750	87.500	95.000	100.000
	100.000	100.000	107.500	122.500	160.000
	220.000				
2000.0000	42.969	65.938	81.875	93.750	97.500
	100.000	103.750	111.250	133.750	171.250
	231.250				
2400.0000	41.211	62.422	79.844	89.688	96.875
	100.625	105.625	118.750	141.250	182.500
	242.500				
2800.0000	40.264	60.527	76.055	88.359	95.156
	101.250	109.688	123.438	150.625	191.875
	251.875				
3200.0000	39.080	58.159	74.443	85.605	94.805
	102.422	112.344	130.156	157.656	201.250
	261.250				
3600.0000	38.381	56.761	71.882	84.624	94.014
	103.574	116.289	135.000	165.703	209.453
	269.453				
4000.0000	37.566	55.132	70.693	82.948	94.099
	105.151	119.287	140.996	172.227	217.578
	277.578				

6-4   INCLUSION OF HEAT CAPACITY IN FINITE-DIFFERENCE EQUATIONS FOR BOUNDARY NODES

The finite-difference expressions for the boundary nodes given by Eqs. (6-32b) and (6-32c) assume steady-state conditions, and hence do not include the effects of the heat capacity of the material in the control volume at the boundaries. If a sufficiently small mesh size Δx is used, the effect of heat capacity in the boundary node is negligible; hence Eqs. (6-24b) and (6-25b) are quite accurate to use for solving transient heat conduction problems. However, if the step size Δx is not sufficiently small, it is desirable to include the effects of heat capacitance in the difference equation for the boundary nodes by considering a time-dependent energy balance at the boundary node. The transient energy balance for a volume element ΔV at the boundaries can be stated as

$$\begin{pmatrix} \text{Rate of heat} \\ \text{entering } \Delta V \\ \text{from all its} \\ \text{surfaces at} \\ \text{time step } i \end{pmatrix} + \begin{pmatrix} \text{rate of energy} \\ \text{generation} \\ \text{in } \Delta V \text{ at} \\ \text{time step } i \end{pmatrix} = \begin{pmatrix} \text{rate of increase} \\ \text{of internal} \\ \text{energy of} \\ \Delta V \end{pmatrix} \qquad (6\text{-}33)$$

Suppose we have *convection* at the boundary surface $x = 0$, with a heat transfer coefficient h_0 into an ambient at temperature $T_{\infty 0}$, but no energy generation in the medium. The application of the energy equation (6-33) to the node $m = 1$ yields

$$h_0(T_{\infty 0} - T_1^i) + k \frac{T_2^i - T_1^i}{\Delta x} = \rho c_p \frac{\Delta x}{2} \frac{T_1^{i+1} - T_1^i}{\Delta t}$$

Solving for T_1^{i+1}, we get

$$\boxed{ T_1^{i+1} = 2r\left(T_2^i + \frac{\Delta x\, h_0}{k} T_{\infty 0} \right) + \left[1 - 2r\left(1 + \frac{\Delta x\, h_0}{k} \right) \right] T_1^i } \qquad \text{for } m = 1$$

$$(6\text{-}34a)$$

where

$$r = \frac{\alpha\, \Delta t}{(\Delta x)^2}$$

In the case of *convection* at the boundary surface $x = L$, with a heat transfer coefficient h_L into an ambient at temperature $T_{\infty L}$, but no energy generation in the medium, the application of the energy equation (6-33) to the node $m = M + 1$ give

$$h_L(T_{\infty L} - T_{M+1}^i) + k \frac{T_M^i - T_{M+1}^i}{\Delta x} = \rho c_p \frac{\Delta x}{2} \frac{T_{M+1}^{i+1} - T_{M+1}^i}{\Delta t}$$

Solving for T_{M+1}^{i+1}, we obtain

$$T_{M+1}^{i+1} = 2r\left(T_M^i + \frac{\Delta x\, h_L}{k}\, T_{\infty L}\right) + \left[1 - 2r\left(1 + \frac{\Delta x\, h_L}{k}\right)\right]T_{M+1}^i$$

for $m = M + 1$ (6-34b)

Equations (6-34a) and (6-34b) are the finite-difference equations for the boundary nodes $m = 1$ and $m = M + 1$, respectively, when these boundary surfaces are subjected to *convection*. These equations include the effects of the heat capacity of the volume element associated with the boundary nodes $m = 1$ and $m = M + 1$; hence they are different from the corresponding equations (6-24b) and (6-25b), which are developed by neglecting the transient effects at the boundaries.

If Eqs. (6-34a) and (6-34b) are used as the finite-difference expressions for the boundary nodes $m = 1$ and $m = M + 1$, an additional restriction on the permissible value of the parameter r should be recognized. By following an argument similar to that described previously for the derivation of the stability criterion given by Eq. (6-30), we conclude that the coefficients of T_1^i and T_{M+1}^i in Eqs. (6-34a) and (6-34b) should be positive; that is,

$$1 - 2r\left(1 + \frac{\Delta x\, h_0}{k}\right) \geq 0 \qquad \text{for } m = 1 \qquad (6\text{-}35a)$$

$$1 - 2r\left(1 + \frac{\Delta x\, h_L}{k}\right) \geq 0 \qquad \text{for } m = M + 1 \qquad (6\text{-}35b)$$

These conditions imply that the following restrictions should be imposed on the value of r in Eqs. (6-34a) and (6-34b), respectively:

$$0 < r \leq \frac{1}{2(1 + \Delta x\, h_0/k)} \qquad \text{for } m = 1 \qquad (6\text{-}36a)$$

$$0 < r \leq \frac{1}{2(1 + \Delta x\, h_L/k)} \qquad \text{for } m = M + 1 \qquad (6\text{-}36b)$$

Clearly, the stability criiteria imposed on the maximum permissible value of r that results from Eqs. (6-36a) and (6-36b) is more severe than that given by Eq. (6-30), which results from the interior modes. Among these three r's, only *the smallest value* should be considered as the stability criterion for the system when performing the numerical calculations.

If the boundary conditions at $x = 0$ and $x = L$ are subjected to constant heat flux q_0 and q_L, respectively, the application of the energy equation (6-33) to the nodes $m = 1$ and $m = M + 1$ yields, respectively, the following finite-difference expressions:

$$q_0 + k \frac{T_2^i - T_1^i}{\Delta x} = \rho c_p \frac{\Delta x}{2} \frac{T_1^{i+1} - T_1^i}{\Delta t} \qquad (6\text{-}37a)$$

or

$$\boxed{T_1^{i+1} = 2r\left(T_2^i + \frac{\Delta x}{k} q_0 \right) + (1 - 2r) T_1^i} \qquad (6\text{-}37b)$$

and

$$q_L + k \frac{T_M^i - T_{M+1}^i}{\Delta x} = \rho c_p \frac{\Delta x}{2} \frac{T_{M+1}^{i+1} - T_{M+1}^i}{\Delta t} \qquad (6\text{-}38a)$$

or

$$\boxed{T_{M+1}^{i+1} = 2r\left(T_M^i + \frac{\Delta x}{k} q_L \right) + (1 - 2r) T_{M+1}^i} \qquad (6\text{-}38b)$$

and the restriction on the permissible value of the parameter r for these boundary nodes becomes

$$0 < r \leq \frac{1}{2} \qquad \text{for } m = 1 \text{ and } m = M + 1 \qquad (6\text{-}39)$$

If the boundary surface at $m = 1$ or $m = M + 1$ is *insulated (adiabatic)*, the resulting finite-difference equation for the insulated boundary $m = 1$ or $m = M + 1$ is obtained from Eq. (6-34a) or (6-34b), respectively, by setting h_0 or h_L equal to zero. We find

$$\boxed{T_1^{i+1} = 2r T_2^i + (1 - 2r) T_1^i} \qquad \text{for } m = 1 \qquad (6\text{-}40a)$$

$$\boxed{T_{M+1}^{i+1} = 2r T_M^i + (1 - 2r) T_{M+1}^i} \qquad \text{for } m = M + 1 \qquad (6\text{-}40b)$$

The finite-difference form of the heat conduction equation for the internal nodes remains the same as that given by Eq. (6-21).

The computer program given in Table 6-2 can readily be adapted to this case by modifying the statements regarding the boundary conditions. That is, for convection boundary conditions, Eq. (6-32b) should be replaced by Eq. (6-34a), and Eq. (6-32c) by Eq. (6-34b). Similar replacements should be made for a prescribed heat flux boundary condition.

Table 6-5 gives the modifications of the computer program in Table 6-2 needed to take into account the effects of the heat capacitance of the boundary nodes. Lines 12 to 22 and 46 to 51 in Table 6-2 should be replaced with the statements given in Table 6-5.

TABLE 6-5
Modification of the computer program in Table 6-2 to take into account the heat capacitance at the boundary nodes, for the solution of Example 6-8

Lines 12 to 22 placed by:

```
12          M = 5
13          ALPHA = 1.2D-5
14          RL = 0.1D0
15          DELX = RL / DBLE(M)
16          IEND = 10
17          R = 0.4057D0
18          RK = 43.0D0
19          HO = 500.0D0
20          HL = 300.0D0
21          TINFO = 20.0D0
22          TINFL = 20.0D0
```

Lines 46 to 51 placed by:

```
46          DXHKO = DELX * HO/RK
47          T(1) = 2.0D0 * R * (T(2) + DXHKO * TINFO +
48                 (1.0D0 - 2.0D0 * R * (1.0D0 + DXHKO)) * T(1)
49          DXHKL = DELX * HL/RK
50          T(M+1) = 2.0D0 * R * (T(M) + DXHKL * TINFL) +
51                 (1.0D0 - 2.0D0*R*(1.0D0 + DXHKL)) * T(M+1)
```

Example 6-8. A steel plate [$\alpha = 1.2 \times 10^{-5}$ m^2/s and $k = 43$ W/(m $\cdot$ °C)] of thickness 10 cm, initially at a uniform temperature T_i, is suddenly immersed in an oil bath at $T_\infty = 20$°C. The convection heat transfer coefficient between the fluid and the top surface is $h_0 = 500$ W/(m$^2 \cdot$ °C), and that for the bottom surface is $h_L = 300$ W/(m$^2 \cdot$ °C). Using five equal subdivisions, or a mesh size $\Delta x = 2$ cm, determine the largest permissible value of time step Δt, if the heat capacity effects in the finite-difference equations for the boundary nodes are included.

Give the resulting finite-difference equations for all the nodes, using the largest permissible time step.

Solution. For $M = 5$, we have

$$\Delta x = \frac{L}{M} = \frac{10}{5} = 2 \text{ cm}$$

The nodes, $m = 1, 2, \ldots, 6$, are shown in the accompanying figure.

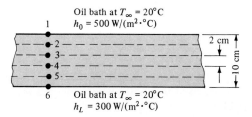

Oil bath at $T_\infty = 20$°C
$h_0 = 500$ W/(m$^2 \cdot$ °C)

2 cm

10 cm

Oil bath at $T_\infty = 20$°C
$h_L = 300$ W/(m$^2 \cdot$ °C)

FIGURE EXAMPLE 6-8

Nodes 2, 3, 4, and 5 are interior nodes, and 1 and 6 are the top and bottom boundary surface nodes. To ensure the stability of the finite-difference equations for the interior nodes, we should satisfy the inequality given by Eq. (6-30):

$$r \le 0.5$$

To ensure the stability of the finite-difference equations at the boundary nodes 1 and 6, we should satisfy the inequalities given by Eqs. (6-36a) and (6-36b), respectively:

$$r \le \frac{1}{2(1 + \Delta x \, h_0/k)} = 0.4057$$

$$r \le \frac{1}{2(1 + \Delta x \, h_L/k)} = 0.4388$$

where $\Delta x = 0.02$ m, $k = 43$ W/(m · °C), $h_0 = 500$ W/(m² · °C), and $h_L = 300$ W/(m² · °C). Clearly, the smallest value of r is that imposed by the top boundary node 0, which is $r \le 0.4057$. This gives the largest possible time increment Δt,

$$\Delta t = \frac{r(\Delta x)^2}{\alpha} = \frac{0.4057 \times (0.02)^2}{1.2 \times 10^{-5}} = 13.52 \text{ s}$$

The finite-difference equations for the interior nodes 2, 3, 4, and 5 are obtained from Eq. (6-32a):

$$T_2^{i+1} = r(T_1^i + T_3^i) + (1 - 2r) T_2^i$$

$$T_3^{i+1} = r(T_2^i + T_4^i) + (1 - 2r) T_3^i$$

$$T_4^{i+1} = r(T_3^i + T_5^i) + (1 - 2r) T_4^i$$

$$T_5^{i+1} = r(T_4^i + T_6^i) + (1 - 2r) T_5^i$$

The finite-difference equations for the boundary nodes 1 and 6 are given by Eqs. (6-34a) and (6-34b):

$$T_1^{i+1} = 2r\left(T_2^i + \frac{\Delta x \, h_0}{k} T_{\infty 0}\right) + \left[1 - 2r\left(1 + \frac{\Delta x \, h_0}{k}\right)\right] T_1^i$$

$$T_6^{i+1} = 2r\left(T_5^i + \frac{\Delta x \, h_L}{k} T_{\infty L}\right) + \left[1 - 2r\left(1 + \frac{\Delta x \, h_L}{k}\right)\right] T_6^i$$

where

$$\frac{\Delta x \, h_0}{k} = \frac{0.02 \times 500}{43} = 0.2326$$

$$\frac{\Delta x \, h_L}{k} = \frac{0.02 \times 300}{43} = 0.1395$$

and

$$r = 0.3057$$

Using the finite-difference equations and the computer program described in Table 6-5, the unknown node temperatures T_m^i are readily determined. Table 6-6 shows the results of these calculations for 10 consecutive time steps, for an initial temperature of 100°C.

TABLE 6-6
Numerical solution to Example 6-8

TIME STEP : 13.52   sec.

X =	0.000	0.020	0.040	0.060	0.080	0.100

Time(sec.)

Time(sec.)	0.000	0.020	0.040	0.060	0.080	0.100
0.0000	100.000	100.000	100.000	100.000	100.000	100.000
13.5200	84.908	100.000	100.000	100.000	100.000	90.945
27.0400	84.906	93.879	100.000	100.000	96.327	90.260
40.5600	79.940	92.722	97.517	98.510	95.356	87.229
54.0800	79.001	89.483	95.975	96.828	93.339	86.212
67.6000	76.373	87.865	93.688	95.067	91.863	84.499
81.1200	75.060	85.565	91.885	93.208	90.176	83.172
94.6400	73.195	83.868	89.859	91.442	88.565	81.703
108.1600	71.818	81.969	88.071	89.633	86.949	80.285
121.6800	70.277	80.326	86.229	87.911	85.335	78.867
135.2000	68.944	78.645	84.517	86.184	83.756	77.450

PROBLEMS

Transient Temperature and Heat Flow in a Semi-Infinite Solid

In Problems 6-1 to 6-12, treat the region as a semi-infinite medium and use transient charts for the solution.

6-1. A thick stainless-steel slab $[\alpha = 1.5 \times 10^{-5} \text{ m}^2/\text{s} \text{ and } k = 60 \text{ W/(m} \cdot \text{°C)}]$ is initially at a uniform temperature of 220°C. Suddenly one of its surfaces is lowered to 50°C and maintained at that temperature. By treating the slab as a semi-infinite solid, determine the temperature at a depth 1 cm from the surface and the heat flux at the surface 2 min after the surface temperature lowered.

6-2. A thick concrete slab $[\alpha = 7 \times 10^{-7} \text{ m}^2/\text{s} \text{ and } k = 1.4 \text{ W/(m} \cdot \text{°C)}]$ is initially at a uniform temperature of 400°C. Suddenly one of its surfaces is subjected to convective cooling with a heat transfer coefficient of 50 W/(m² · °C) into an ambient at 50°C. By treating the slab as a semi-infinite solid, calculate the temperature at a depth 5 cm from the surface and 0.5 h after the start of cooling.
Answer: 351°C

6-3. A fireclay brick slab $[\alpha = 5.4 \times 10^{-7} \text{ m}_2/\text{s} \text{ and } k = 1 \text{ W/(m} \cdot \text{°C)}]$ 10 cm thick is initially at a uniform temperature of 350°C. Suddenly one of its surfaces is subjected to convection with a heat transfer coefficient of 100 W/(m² · °C) into an ambient at 40°C. Calculate the temperature at a depth 1 cm from the surface 2 min after the start of cooling.
Answer: 309.7°C

6-4. A thick copper slab $[\alpha = 1.1 \times 10^{-4} \text{ m}^2/\text{s} \text{ and } k = 380 \text{ W/(m} \cdot \text{°C)}]$ is initially at a uniform temperature of 10°C. Suddenly one of its surfaces is raised to 100°C. Calculate the heat flux at the surface 1 min and 2 min after the raising of the surface temperature.

6-5. A thick aluminum slab $[\alpha = 8.4 \times 10^{-5} \text{ m}^2/\text{s} \text{ and } k = 200 \text{ W/(m} \cdot \text{°C)}]$ is initially at a uniform temperature of 20°C. Suddenly one of its surfaces is raised to 100°C. Calculate the time required for the temperature at a depth 5 cm from the surface to reach to 80°C.
Answer: 140.7 s

6-6. A thick concrete slab $[\alpha = 7 \times 10^{-7} \text{ m}^2/\text{s} \text{ and } k = 1.4 \text{ W/(m} \cdot \text{°C)}]$ is initially at a uniform temperature of 80°C. One of its surfaces is suddenly lowered to 20°C. Determine the temperatures at depths 2 and 4 cm from the surface 10 min after the surface temperature is lowered.

6-7. A thick stainless steel slab $[\alpha = 1.6 \times 10^{-5} \text{ m}^2/\text{s} \text{ and } k = 60 \text{ W/(m} \cdot \text{°C)}]$ is initially at a uniform temperature of 100°C. One of its surfaces is suddenly lowered to 30°C. Determine the time required for the temperature at a depth 2 cm from the surface to reach 50°C.

6-8. A thick bronze $[\alpha = 0.86 \times 10^{-5} \text{ m}^2/\text{s} \text{ and } k = 26 \text{ W/(m} \cdot \text{°C)}]$ is initially at a uniform temperature of 250°C. Suddenly one of its surface temperatures is lowered to 25°C. Determine the temperature at a location 5 cm from the surface 10 min after the surface temperature is lowered.
Answer: 110.4°C

6-9. A thick bronze $[\alpha = 0.86 \times 10^{-5} \text{ m}^2/\text{s} \text{ and } k = 26 \text{ W/(m} \cdot \text{°C)}]$ is initially at a uniform temperature of 250°C. Suddenly one of its surfaces is exposed to convective cooling by a fluid at 25°C. Assuming that the heat transfer coefficient for convection between the fluid and the surface is 150 W/(m² · °C), determine the temperature at a location 5 cm from the surface 10 min after the exposure.

6-10. A thick wood wall [$\alpha = 0.82 \times 10^{-7}$ m^2/s and $k = 0.15$ W/(m · °C)] is initially at a uniform temperature of 20°C. The wood may ignite at 400°C. If the surface is exposed to hot gases at $T_\infty = 500$°C and the heat transfer coefficient between the gas and the surface is 45 W/(m^2 · °C), how long will it take for the surface of the wood to reach 400°C?

Answer: ≈10 min

6-11. A thick wood wall [$\alpha = 0.82 \times 10^{-7}$ m^2/s and $k = 0.15$ W/(m · °C)] is initially at a uniform temperature of 20°C. Suddenly one of its surfaces is raised to 80°C. Calculate the temperature at a distance 2 cm from the surface 10 min after the exposure.

6-12. A water pipe is to be buried in soil ($\alpha = 0.2 \times 10^{-6}$ m^2/s) at sufficient depth to prevent freezing in winter. When the soil is at a uniform temperature of 5°C, the surface is subjected to a uniform temperature of −10°C continuously for 50 days. What minimum burial depth for the pipe is needed to prevent freezing?

Answer: 1.24 m

Transient Temperature and Heat Flow in a Slab, Cylinder, and Sphere

Slab. Use transient-temperature charts for the solution.

6-13. A steel plate [$\alpha = 1.2 \times 10^{-5}$ m^2/s, $k = 43$ W/(m · °C), $c_p = 465$ J/(kg · °C), and $\rho = 7833$ kg/m^3] thickness 5 cm, initially at a uniform temperature of 200°C, is suddenly immersed in an oil bath at 20°C. The convective heat transfer coefficient between the fluid and the surface is 500 W/(m^2 · °C). How long will it take for the center-plane to cool to 100°C?

6-14. A marble plate [$\alpha = 1.3 \times 10^{-6}$ m^2/s and $k = 3$ W/(m · °C)] of thickness 3 cm is initially at a uniform temperature of 130°C. The surfaces are suddenly lowered to 30°C. Determine the center-plane temperature 2 min after the lowering of the surface temperature.

Answer: 55°C

6-15. A marble plate [$\alpha = 1.3 \times 10^{-6}$ m^2/s and $k = 3$ W/(m · °C)] of thickness 5 cm is initially at a uniform temperature of 100°C. The surfaces are suddenly immersed in an oil bath at 20°C. The convective heat transfer coefficient between the oil and the marble surface is 400 W/(m^2 · °C). Calculate the temperature at a depth of 2 cm from one of the surfaces 5 min after the exposure. Determine the energy removed from the marble during this time for a marble plate 1 m × 1 m in size.

6-16. A copper plate [$\alpha = 1.1 \times 10^{-4}$ m^2/s and $k = 380$ W/(m · °C)] of thickness 2 cm is initially at a uniform temperature of 25°C. Suddenly both of its surfaces are raised to 50°C. Calculate the centerline temperature 10 min after the surface temperature is raised.

Answer: 50°C

6-17. A fireclay brick slab [$\alpha = 5.4 \times 10^{-7}$ m^2/s and $k = 1$ W/(m · °C)] of thickness 5 cm is initially at a uniform temperature of 350°C. Suddenly one of its surfaces is subjected to convection with a heat transfer coefficient of 100 W/(m^2 · °C) into an ambient at 40°C. The other surface is insulated. Calculate the centerline temperature 1 h after the start of cooling.

Answer: 240°C

Cylinder. Use transient-temperature charts for solution.

6-18. A long steel shaft of radius 20 cm [$\alpha = 1.6 \times 10^{-5}$ m^2/s and $k = 60$ W/(m · °C)] is taken out of an oven at a uniform temperature of 500°C and immersed in a well-stirred large bath of 20°C coolant. The heat transfer coefficient between the shaft surface and the coolant is 200 W/(m² · °C). Calculate the time required for the shaft surface to reach 100°C.

Answer: 2.4 h

6-19. A long steel bar [$\alpha = 1.6 \times 10^{-5}$ m²/s and $k = 60$ W/(m · °C)] of diameter 5 cm is initially at a uniform temperature of 200°C. Suddenly the surface of the bar is exposed to an ambient at 20°C with a heat transfer coefficient of 500 W/(m² · °C). Calculate the center temperature 2 min after the start of the cooling. Calculate the energy removed from the bar per meter length during this time period.

Answer: 77.6°C, 21.3 MJ/m²

6-20. A hot dog can be regarded as a solid having a shape in the form of a long solid cylinder. Consider a hot dog [$\alpha = 1.6 \times 10^{-7}$ m²/s and $k = 0.5$ W/(m · °C)] of diameter 2 cm, initially at a uniform temperature of 10°C, dropped into boiling water at 100°C. The heat transfer coefficient between the water and the surface is 150 W/(m² · °C). If the meat is considered cooked when its center temperature reaches 80°C, how long will it take for the centerline temperature to reach 80°C? Compare the result with that obtainable using lumped system analysis.

6-21. A long chrome-steel rod [$\alpha = 1.1 \times 10^{-5}$ m²/s and $k = 40$ W/(m · °C)] of diameter of 8 cm is initially at a uniform temperature of 225°C. It is suddenly exposed to a convective environment at 25°C with a surface heat transfer coefficient of 50 W/(m² · °C). Determine the center temperature at 0.1 and 1 h after exposure to the cooler ambient. Compare the result with that obtainable using lumped system analysis.

6-22. A long copper rod [$\alpha = 1.1 \times 10^{-4}$ m²/s and $k = 380$ W/(m · °C)] of diameter 5 cm is initially at a uniform temperature of 100°C. It is suddenly dropped into a coolant pool at 20°C. The heat transfer coefficient between the coolant and the rod is 500 W/(m² · °C). Determine the center temperature of the rod 100 s after exposure to the coolant. Calculate the energy removed from the rod per meter length during this time period.

Answer: 48°C, 12.1 MJ

6-23. A long aluminum wire [$\alpha = 8.4 \times 10^{-5}$ m²/s and $k = 200$ W/(m · °C)] of diameter 0.5 cm is initially at a uniform temperature of 20°C. Suddenly its surface temperature is increased to 30°C. Calculate the time required for the center temperature to reach 25°C.

6-24. A long bronze bar [$\alpha = 0.86 \times 10^{-5}$ m²/s and $k = 26$ W/(m · °C)] of radius 5 cm is initially at 30°C. Suddenly its surface temperature is increased to 80°C. Determine the time required for the temperature at a depth of 2 cm from the surface to reach 70°C. Calculate the heat transferred to the bar per meter length during this time.

Answer: 87.2 s

Sphere. Use transient-temperature charts for solution.

6-25. A solid aluminum sphere [$\alpha = 8.4 \times 10^{-5}$ m²/s and $k = 204$ W/(m · °C)] of diameter 10 cm is initially at 200°C. Suddenly it is immersed in a well-stirred bath at 100°C. The heat transfer coefficient between the fluid and the sphere surface is 500 W/(m² · °C). How long will it take for the center of the sphere to cool to 150°C?

6-26. A solid aluminum sphere [$\alpha = 8.4 \times 10^{-5}$ m²/s and $k = 204$ W/(m · °C)] of diameter 6 cm is initially at 100°C. Suddenly its surface temperature is lowered to 20°C. Determine the center temperature of the sphere 5 s after lowering the surface temperature.

6-27. An orange 12 cm in diameter is initially at a uniform temperature of 25°C. It is placed in a refrigerator in which the air temperature is 2°C. If the heat transfer coefficient between the air and the surface of the orange is 50 W/(m$^2 \cdot$°C), determine the time required for the center of the orange to reach 15°C. Assume the thermal properties of the orange to be $\alpha \cong 1.5 \times 10^{-7}$ m^2/s and $k \cong$ 6 W/(m $\cdot$ °C). Note that the thermal properties of the orange are very close to those of water.

Answer: 3.3 h

6-28. An 8-cm-diameter potato, initially at a uniform temperature of 25°C, is suddenly dropped into boiling water at 100°C. The heat transfer coefficient between the water and the surface of the potato is 5000 W/(m$^2 \cdot$°C). Assume the thermal properties of potato to be $\alpha \cong 1.5 \times 10^{-7}$ m^2/s and $k \cong 7$ W/(m $\cdot$ °C). Determine the time required for the center temperature of the potato to reach 80°C.

Answer: 22.2 min

6-29. A copper ball [$\alpha = 1.1 \times 10^{-4}$ m^2/s and $k = 380$ W/(m $\cdot$ °C)] 5 cm in diameter is initially at a uniform temperature of 100°C. It is suddenly dropped into a coolant pool at 20°C. The heat transfer coefficient between the coolant and the rod is 500 W/(m$^2 \cdot$°C). Determine the center temperature of the rod 100 s after exposure to the coolant. Calculate the energy removed from the ball during this time period.

Answer: 31.2°C, 11.6 MJ

6-30. A bronze sphere [$\alpha = 0.86 \times 10^{-5}$ m^2/s and $k = 26$ W/(m $\cdot$ °C)] 5 cm in radius is initially at 30°C. Suddenly its surface temperature is increased to 80°C. Determine the time required for the temperature at a depth of 2 cm from the surface to reach 70°C. Calculate the heat transferred to the sphere during this time.

6-31. A 5-cm-diameter wood sphere [$\alpha = 0.82 \times 10^{-7}$ m^2/s and $k = 0.15$ W/(m $\cdot$ °C)] is initially at a uniform temperature of 20°C. Suddenly its surface temperature is raised to 80°C. Calculate the temperature 2 cm from the surface of the sphere at 30 min after the exposure. Calculate the energy gained during this time.

6-32. A 10-cm-diameter marble sphere [$\alpha = 1.3 \times 10^{-6}$ m^2/s and $k = 3$ W/(m $\cdot$ °C)] is initially at a uniform temperature of 100°C. The sphere is suddenly immersed in an oil bath at 20°C. The convective heat transfer coefficient between the oil and the marble surface is 400 W/(m$^2 \cdot$°C). Calculate the temperature at a depth of 2 cm from the surface 20 min after the exposure. Determine the energy removed from the marble during this time.

Answer: 26.3°C, 8.4 MJ

6-33. Consider a slab 5 cm in thickness, a cylinder 5 cm in diameter, and a sphere 5 cm in diameter, each made of steel [$\alpha = 1.6 \times 10^{-5}$ m^2/s and $k = 60$ W/(m $\cdot$ °C)] and initially at a uniform temperature of 30°C. Suddenly, they are all immersed in a well-stirred large bath at 200°C. The heat transfer coefficient between the surfaces and the fluid is 1000 W/(m$^2 \cdot$°C). Calculate the time required for the centers of the slab, the cylinder, and the sphere to cool to 100°C.

Answer: 62.5 s, 35.2 s, 19.5 s

Numerical Analysis

Use finite-differences for solution.

6-34. An aluminum plate ($\alpha = 8 \times 10^{-5}$ m^2/s) 4 cm thick is initially at a uniform temperature of 20°C. Suddenly its surfaces are raised to 220°C. Using an explicit finite-difference scheme and a mesh size $\Delta x = 1$ cm, calculate the center temperature at 5, 10, and 15 s after the surfaces are exposed to high temperature.

6-35. An aluminum plate $(\alpha = 8 \times 10^{-5} \text{ m}^2/\text{s})$ 4 cm thick is initially at a uniform temperature of 20°C. Suddenly one of its surfaces is raised to 220°C while the other surface is kept insulated. Using an explicit finite-difference scheme and a mesh size $\Delta x = 1$ cm, calculate the temperature of the insulated surface 5, 10 and 15 s after the other surface is exposed to high temperature.

Answer: 102.8°C, 168.9°C, 197.8°C

6-36. An 8-cm-thick chrome-steel plate $[\alpha = 1.6 \times 10^{-5} \text{ m}^2/\text{s}$ and $k = 60 \text{ W/(m} \cdot °\text{C)}]$, initially at a uniform temperature of 325°C, is suddenly exposed to a cool airstream at 25°C at both of its surfaces. The heat transfer coefficient between the air and the surface is $400 \text{ W/(m}^2 \cdot °\text{C)}$. Using an explicit finite-difference scheme and a mesh size $\Delta x = 1$ cm, determine the center-plane temperature 5 and 15 min after the start of cooling. Neglect the heat capacity effects of the boundary elements.

Answer: 162.4°C, 51.5°C

6-37. A large brick wall $(\alpha = 5 \times 10^{-7} \text{ m}^2/\text{s})$ that is 20 cm thick is initially at a uniform temperature of 300°C. Suddenly its surfaces are lowered to 50°C and maintained at that temperature. Using an explicit finite-difference scheme and a mesh size $\Delta x = 5$ cm, determine the time required for the center temperature to reach 200°C after the start of cooling.

6-38. A large brick wall $(\alpha = 5 \times 10^{-7} \text{ m}^2/\text{s})$ that is 20 cm thick is initially at a uniform temperature of 150°C. Suddenly its surfaces are lowered to 30°C and maintained at that temperature. Using an explicit finite-difference scheme and a mesh size $\Delta x = 5$ cm, determine the center temperature 3 h after the start of cooling.

Answer: 74.34°C

6-39. A steel plate $[\alpha = 1.2 \times 10^{-5} \text{ m}^2/\text{s}$ and $k = 43 \text{ W/(m} \cdot °\text{C)}]$ 10 cm thick, initially at a uniform temperature T_0, is suddenly immersed in an oil bath at T_∞. The convection heat transfer coefficient between the fluid and the surfaces of the plate is $400 \text{ W/(m}^2 \cdot °\text{C)}$. Using a mesh size $\Delta x = 2$ cm, determine the largest permissible value of the time step Δt if the heat capacity effects in the finite-difference equations for the boundary nodes are included.

Answer: 14.05 s

6-40. A brick wall $[\alpha = 5 \times 10^{-7} \text{ m}^2/\text{s}$ and $k = 1 \text{ W/(m} \cdot °\text{C)}]$ 6 cm thick is initially at a uniform temperature of 100°C. Suddenly its surface temperatures are lowered by being exposed to a fluid at 20°C with a heat transfer coefficient of $150 \text{ W/(m} \cdot °\text{C)}$. Using an explicit finite-difference scheme and a mesh size $\Delta x = 2$ cm, determine the time required for the surface temperatures to drop to 80°C. Include the heat capacity effects of the boundary elements.

Answer: 40 s

6-41. Determine the center-plane temperature of the chrome-steel plate of Problem 6-36 5 min after the start of cooling by including the heat capacity effects of the boundaries.

REFERENCES

1. Schneider, P. J.: *Conduction Heat Transfer*, Addison-Wesley, Reading, Mass., 1955.
2. Heisler, M. P.: "Temperature Charts for Induction and Constant Temperature Heating," *Trans. ASME*, **69**: 227–236 (1947).
3. Gröber, H., S. Erk, and U. Grigull: *Fundamentals of Heat Transfer*, McGraw-Hill, New York, 1961.

CHAPTER
7

FORCED
CONVECTION
OVER
BODIES

A large proportion of heat transfer applications utilize *forced convection*, which involves transfer of heat between the outer surfaces of a solid body and a fluid that is forced to flow over it by some mechanism, such as a pump or a blower. For example, in a liquid-to-liquid heat exchanger, a pump is used to force the fluid to flow over the tube bundles. In a gas-cooled nuclear reactor, the hot gas from the reactor core is circulated over the tubes of the steam generator by special blowers. In the cooling of an automobile radiator, air is forced to flow over the hot radiator tubes as a result of the motion of the car.

The mechanism of heat transfer in forced convection is complicated. The transfer of heat between a solid surface and a fluid takes place through a combination of conduction and fluid motion. Consider a solid that is at a higher temperature than the fluid. Heat flows by conduction from the surface of the solid to the fluid adjacent to the wall. The heated fluid particles are then carried away into the cooler regions of the flow by fluid motion. Therefore, the characteristics of fluid motion govern the temperature distribution and heat transfer between the hot surface and the free stream. An understanding of velocity and temperature distribution in flow is most important in the study of heat transfer by forced convection. In this chapter we present a qualitative discussion of basic concepts pertaining to the distribution of velocity and temperature in the fluid and their consequences for heat transfer without getting involved in the study of equations of motion and energy. The significance of *dimensionless analysis* in the development of heat transfer correlations is illustrated, the velocity and thermal boundray-layer concepts basic to the study of forced convection over surfaces are introduced, and the approximate *integral method* is utilized to determine the velocity and temperature distributions in the flow. Some useful correlations are then presented for the determination of the drag force and heat transfer in both laminar and turbulent flow over solids with simple shapes, such as a flat plate, a circular or noncircular cylinder, and a sphere.

7-1   BASIC CONCEPTS

Heat transfer between the fluid and the surface of the solid, and the drag force exerted by the fluid on the body, are closely related to the distribution of temperature and velocity in the flow field. A good understanding of the factors affecting the temperature and velocity distribution is essential in the study of forced convection. Therefore, some qualitative discussion of the *velocity* and *thermal boundary-layer* concepts and the role of dimensionless parameters in the correlation of heat transfer in forced convection are given below.

Dimensionless Parameters

Heat transfer in flow over a body is affected by a variety of factors, including the flow pattern characteristics, the fluid properties, the flow passage geometry, and the surface conditions. The fluid flow pattern can be characterized as *laminar*, *transition*, or *turbulent*. The fluid properties that affect heat transfer include density, thermal conductivity, viscosity, and specific heat. The flow passage geometry may take such forms as flow over a flat plate, a cylinder, a sphere, tube banks or flow inside ducts and many others. Clearly there are too many factors that affect heat transfer between a solid surface and a fluid flowing over it.

Suppose the flow passage geometry and the character of the flow pattern are fixed; suppose, say, that a laminar flow along a flat plate and the local heat transfer coefficient h_x at a distance x from the leading edge of the plate (i.e., from the point where the flow first enters the plate) are considered. The parameters that affect heat transfer are x, k, u, ρ, μ, c_p, h_x where k, ρ, μ, and c_p are the thermal conductivity, density, viscosity, and specific heat of the fluid and u is the main flow velocity. For the specific case considered here, the local heat transfer coefficient h_x at a position x depends on six independent variables, and the determination of a correlation among these seven parameters is of practical concern.

For the specific case of flow over a flat plate considered here, it can be shown that the following three dimensionless groups can be established among these seven different parameters:

$$\mathrm{Nu}_x = \frac{h_x x}{k} = \text{Nusselt number (local)}$$

$$\mathrm{Pr} = \frac{c_p \mu}{k} = \text{Prandtl number}$$

$$\mathrm{Re}_x = \frac{\rho u x}{\mu} = \text{Reynolds number (local)}$$

Then, one envisions a correlation of heat transfer in the form

$$\mathrm{Nu}_x = f[\mathrm{Re}_x, \mathrm{Pr}] \qquad (7\text{-}1a)$$

or a specific functional form for the correlation may be chosen as

$$\mathrm{Nu}_x = C(\mathrm{Re}_x)^a (\mathrm{Pr})^b \qquad (7\text{-}1b)$$

where the numerical values for the constant C and the exponents a and b are to be determined by the "best fit" to experimental data. Later in this chapter, the values of a, b, and C for various flow configurations will be presented.

The principal advantages of correlating heat transfer in terms of dimensionless groups are now apparent. The experimental data is correlated in terms of three variables instead of the original seven variables, and the results of such correlations are applicable to a variety of fluids, flow velocities, and geometrical dimensions, provided that the flow characteristics and geometry considered remain the same.

Clearly, the success of dimensional analysis strongly depends upon the proper choice of the parameters involved, which requires a good insight into the physical nature of the problem. Suppose we had chosen six parameters instead of the seven; the resulting dimensionless groups obtained by dimensionless analysis would have been incorrect.

Velocity Boundary Layer

Consider the flow of a viscous fluid over a flat plate, as illustrated in Fig. 7-1. Let x be the coordinate axis measured along the plate in the direction of the flow, and let y be the coordinate axis normal to the surface of the plate. A fluid with a free-stream velocity u_∞ flows along the plate. Immediately before the leading edge of the plate (i.e., $x = 0$), the fluid velocity is uniform everywhere, but a significant change in the velocity profile occurs as fluid moves along the plate. The viscous forces tend to retard the flow in the regions near the surface. The fluid particles in contact with the surface assume zero velocity, whereas in the region sufficiently far away from the surface, the flow velocity u_∞ remains essentially unaffected. Such a situation implies that the flow velocity u parallel to the surface should increase from a value $u = 0$ at $y = 0$ to $u = u_\infty$ for y sufficiently large. Then, at each location x, one envisions a point $y = \delta(x)$ in the flow field where the axial velocity u is equal to 99 percent of the free-stream velocity u_∞. The locus of the points where $u = 0.99 u_\infty$ is called the *velocity*

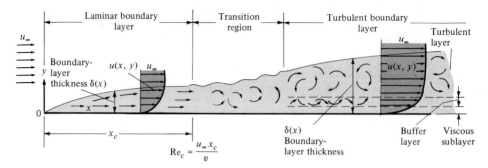

FIGURE 7-1
Boundary-layer concept for flow over a flat plate.

boundary layer, or simply the *boundary layer*. We use the notation $\delta(x)$ to denote the thickness of the boundary layer.

With these considerations, the flow field can be assumed to be separated into two distinct regions: (1) The *boundary-layer region*, where the flow velocity u increases rapidly with the distance from the wall surface, and (2) the *potential-flow region* outside the boundary layer, where the velocity gradients are negligible. It is the steep velocity gradients in the boundary layer that give rise to shear stress in the fluid, which in turn exerts drag on the plate surface. Therefore an understanding of the behavior of flow in the boundary-layer region is important in the study of flow over surfaces.

We now examine the boundary-layer flow in the x direction along the plate. The characteristics of this flow is governed by a dimensionless quantity called Reynolds number, Re_x, defined as

$$\text{Re}_x \equiv \frac{u_\infty x}{\nu} = \frac{\text{inertia force}}{\text{viscous force}} \qquad (7\text{-}2a)$$

where u_∞ = free-stream velocity
x = distance from leading edge
ν = kinematic viscosity of fluid

Depending on the value of the Reynolds number, the flow in the boundary layer can be *laminar*, *transitional*, or *turbulent*. When the flow is *laminar*, the fluid particles move along streamlines in an orderly manner. The term *turbulent* is used to indicate that the motion of the fluid is chaotic in nature and involves crosswise mixing or eddying superimposed on the motion of the mainstream. *Transitional*, as the name implies, indicates the region where the boundary-layer flow changes from laminar to turbulent behavior. Figure 7-1 schematically illustrates the laminar, transition, and turbulent boundary layers for flow along a flat plate. The boundary layer starts from the leading edge (i.e., $x = 0$) of the plate as laminar, and the flow stays laminar up to a critical distance $x = x_c$, where the transition from laminar to turbulent flow takes place. For most practical purposes, for flow along a flat plate the transition begins at a value of Reynolds number

$$\mathrm{Re}_x \equiv \frac{u_\infty x}{\nu} \cong 5 \times 10^5 \qquad (7\text{-}2b)$$

This critical value is strongly dependent on the surface roughness and the turbulence level of the free stream. For example, with very large disturbances in the free stream, the transition may begin at a Reynolds number as low as 10^5, whereas for flows which are free from disturbances, it may not start until a Reynolds number of 10^6 or more. For flow along a flat plate, the boundry layer is always turbulent for $\mathrm{Re}_x \geq 4 \times 10^6$.

In the *turbulent boundary layer*, next to the wall there is a very thin layer, called the *viscous sublayer*, where the flow retains its viscous-flow character. Adjacent to the viscous sublayer is a region called the *buffer layer*, in which there is fine-grained turbulence and the mean axial velocity rapidly increases with the distance from the wall. The buffer layer is followed by the turbulent layer, in which there is a larger-scale turbulence and the velocity changes relatively little with the distance from the wall. These concepts are also illustrated in Fig. 7-1.

The Reynolds number as defined above is based on three distinct parameters: velocity, kinematic viscosity, and a characteristic length. For the flow over a flat plate considered here, the characteristic length is chosen as the distance x measured along the plate. The choice of the characteristic length depends on the geometry of the body. For example, for flow across a sphere or a cylindrical tube, the diameter can be used as the characteristic dimension. Furthermore, the value of the critical Reynolds number at the transition from laminar to turbulent flow for bodies having different geometries is different from that given above.

DRAG COEFFICIENT. The presence of steep velocity gradients in the boundary layer gives rise to a drag force exerted on the body as a result of flow over it. The determination of this drag force is of interest in engineering applications. The drag force can be calculated by making use of the *drag coefficient* concept as now described.

The shear stress τ_x at the wall surface at any location x is related to the velocity gradient by

$$\tau(x) = \mu \left. \frac{\partial u(x, y)}{\partial y} \right|_{y=0} \qquad (7\text{-}3)$$

where μ is the viscosity of the fluid. However, in engineering practice calculating the shear stress from the knowledge of the velocity gradients at the wall surface is not practical. To alleviate this difficulty, the concept of *local drag coefficient* $c(x)$ is introduced, and the shear stress is related to the drag coefficient in the form

$$\tau(x) = c(x) \frac{\rho u_\infty^2}{2} \qquad \text{N/m}^2 \qquad (7\text{-}4)$$

where ρ is the density and u_∞ is the free-stream velocity of the flow. Clearly, knowing $c(x)$, the shear stress (or the drag force per unit area) at the location x is determined from Eq. (7-4).

By equating Eqs. (7-3) and (7-4), the following expression is obtained:

$$c(x) = \frac{2\nu}{u_\infty^2} \left. \frac{\partial u(x, y)}{\partial y} \right|_{y=0} \qquad (7\text{-}5)$$

where $\nu = \mu/\rho$ is the kinematic viscosity of fluid. Therefore, if the flow problem in the boundary layer can be solved and the velocity distribution u determined, then $c(x)$ is obtained from the definition in Eq. (7-5), and the resulting explicit expression can be used in practical applications.

The average value of $c(x)$ over the distance from $x = 0$ to $x = L$ along the plate is also of interest. Knowing the local values of $c(x)$, the *mean drag coefficient* c_m over the distance $x = 0$ to $x = L$ is determined from

$$c_m = \frac{1}{L} \int_{x=0}^{L} c(x)\, dx \qquad (7\text{-}6)$$

Once the *mean drag coefficient* c_m is available, the *drag force F* acting on the plate over the length from $x = 0$ to $x = L$ and for a width w is determined from

$$F = wLc_m \frac{\rho u_\infty^2}{2} \qquad \text{N} \qquad (7\text{-}7)$$

Thermal Boundary Layer

Analogous to the concept of velocity boundary layer, one can envision the development of a thermal boundary layer along the flat plate associated with the temperature profile in the fluid. To illustrate the concept, we consider the flow of a fluid at a uniform temperature T_∞ along a flat plate maintained at a constant temperature T_w. Let $T(x, y)$ be the temperature of the fluid at a location (x, y), where the coordinates x and y are measured along and perpendicular to the plate surface, respectively. Suppose the fluid is hotter than the wall, that is, $T_\infty > T_w$.

A dimensionless fluid temperature $\theta(x, y)$ is now defined as

$$\theta(x, y) = \frac{T(x, y) - T_w}{T_\infty - T_w} \qquad (7\text{-}8)$$

That is, we measure the fluid temperature $T(x, y)$ in excess of the wall temperature T_w and divide it by $T_\infty - T_w$. We consider two limiting cases of $\theta(x, y)$, one at the wall surface $y = 0$ and the other at distances sufficiently far away from the wall (i.e., y large). We have

$$
\begin{aligned}
\theta(x, y) &= 0 && \text{at } y = 0 \\
\theta(x, y) &\to 1 && \text{for large } y
\end{aligned}
\tag{7-9}
$$

The dimensionless fluid temperature $\theta(x, y)$ varies from a value of $\theta = 0$ at the wall surface to a value of $\theta = 1$ at sufficiently large distances from the wall. Then at each location x along the plate one envisions a point $y = \delta_t(x)$ in the flow such that $\theta(x, y)$ assumes a value of 0.99. The locus of such points where $\theta(x, y) = 0.99$ is called the *thermal boundary layer*. The concept is analogous to the velocity boundary layer discussed previously; it is illustrated in Fig. 7-2. In the thermal boundary-layer region the temperature gradients are very steep, and so heat transfer between the fluid and the wall is governed by the characteristics of the temperature profile in the thermal boundary layer. Therefore, the relative thickness of the thermal and velocity boundary layers is a factor that affects heat transfer between the fluid and the wall surface. Analytical studies have shown that the relative thickness of boundary layers depends on the Prandtl number of the fluid, defined as

$$
\text{Pr} = \frac{\mu c_p}{k} = \frac{\mu/\rho}{k/\rho c_p} = \frac{\nu}{\alpha} = \frac{\text{molecular diffusivity of momentum}}{\text{molecular diffusivity of heat}}
\tag{7-10}
$$

The dependence of the relative thickness of $\delta_t(x)$ and $\delta(x)$ can be stated as

$$
\delta_t(x) < \delta(x) \qquad \text{for Pr} > 1
\tag{7-11a}
$$

$$
\delta_t(x) = \delta(x) \qquad \text{for Pr} = 1
\tag{7-11b}
$$

$$
\delta_t(x) > \delta(x) \qquad \text{for Pr} < 1
\tag{7-11c}
$$

There is a wide difference in the values of the Prandtl number for various types of fluids. Figure 7-3 illustrates approximate ranges of Pr for different

FIGURE 7-2
Thermal boundary-layer concept for flow of a hot fluid over a cold wall.

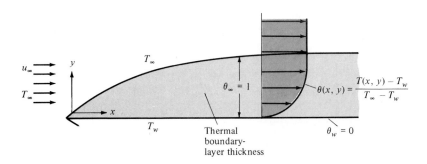

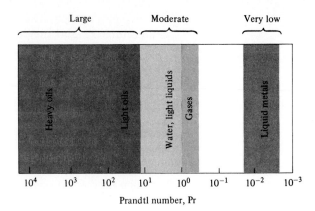

FIGURE 7-3
Typical ranges of Prandtl number for various types of fluids.

fluids. For practical purposes the fluids can be separated into three groups: fluids having very low, moderate, and large Prandtl numbers. Oils have large Prandtl numbers; fluids such as gas, water, and light liquids have moderate Prandtl numbers; and liquid metals have very low Prandtl numbers. Figure 7-3 shows that there is a gap between the very low and moderate values of the Prandtl number where no type of fluid appears to belong. A schematic illustration of the relative thickness of $\delta(x)$ and $\delta_t(x)$ for these three groups of fluids is shown in Fig. 7-4. For liquid metals the thermal boundary layer is much thicker than the velocity boundary layer, whereas for oils the thermal boundary layer is much smaller than the velocity layer.

FIGURE 7-4
Relative thickness of velocity and thermal boundary layers for different types of fluids.

HEAT TRANSFER COEFFICIENT. The concept of heat transfer coefficient is widely used in the determination of heat transfer between the wall surface and

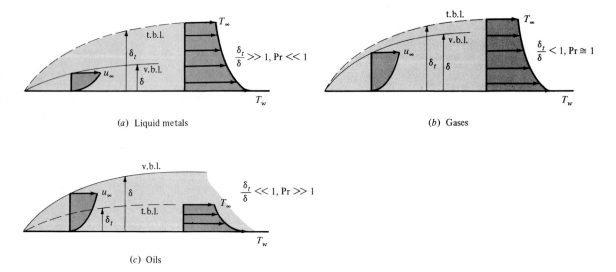

(a) Liquid metals

(b) Gases

(c) Oils

the fluid forced to flow over it. To illustrate the physical significance of the heat transfer coefficient in relation to the temperature distribution in the flow, we consider the flow of a hot fluid at a temperature T_∞ over a cold plate at a constant temperature T_w. Fluid particles in the immediate vicinity of the wall surface remain stationary. Therefore, heat transfer between the wall surface and the adjacent fluid particles is by pure conduction. Then the conduction heat flux $q(x)$ from the adjacent hot fluid to the wall surface at $y = 0$ (i.e., the plate surface) is given by

$$q(x) = k \left. \frac{\partial T(x, y)}{\partial y} \right|_{y=0} \qquad (7\text{-}12a)$$

where k is the thermal conductivity of the fluid. Here the negative sign is omitted from the conduction flux, because heat flow from the fluid to the wall, in the negative y direction, is considered.

In engineering applications it is not practical to calculate the derivative of temperature distribution in the fluid in order to determine the heat transfer rate as given by Eq. (7-12a). Instead, the heat transfer between the fluid and the wall surface is related to a local heat transfer coefficient $h(x)$, defined as

$$q(x) = h(x)(T_\infty - T_w) \qquad (7\text{-}12b)$$

where T_∞ and T_w are the free-stream and plate surface temperatures, respectively. The definition given by Eq. (7-12b) implies that heat transfer is from the fluid to the wall surface; hence it is consistent with the $q(x)$ defined by Eq. (7-12a).

The expression defining $h(x)$ is now determined by equating Eqs. (7-12a) and (7-12b).

$$\underbrace{h(x)(T_\infty - T_w)}_{\text{convection}} = \underbrace{k \left. \frac{\partial T(x, y)}{\partial y} \right|_{y=0}}_{\text{conduction}} \qquad (7\text{-}13a)$$

or

$$\boxed{h(x) = k \frac{[\partial T(x, y)/\partial y]_{y=0}}{T_\infty - T_w}} \qquad (7\text{-}13b)$$

Given the temperature distribution in the flow, the local heat transfer coefficient $h(x)$ can be calculated from Eq. (7-13b). This expression is also used in analytical studies to establish the relation for $h(x)$ by utilizing the solution of the temperature distribution in the thermal boundary layer.

In engineering applications, the mean value of h_x over a distance $x = 0$ to $x = L$ along the surface is of practical interest. The *mean heat transfer coefficient* h_m is determined from

$$\boxed{h_m = \frac{1}{L} \int_0^L h(x)\, dx} \qquad (7\text{-}14)$$

Knowing the mean heat transfer coefficient h_m, the heat transfer rate Q from the fluid to the wall over the distance from $x = 0$ to $x = L$ and width w is determined from

$$Q = wLh_m(T_\infty - T_w) \qquad (7\text{-}15)$$

Example 7-1. Air at atmospheric pressure and at 300 K flows with a velocity of 1 m/s over a flat plate. The transition from laminar to turbulent flow is assumed to take place at a Reynolds number of 5×10^5. Determine the distance from the leading edge of the plate at which transition occurs.

Solution. The kinematic viscosity ν of atmospheric air at $T_\infty = 300$ K is 0.168×10^{-4} m/s. The transition is assumed to occur at a distance $x = L$, where $\mathrm{Re}_L = 5 \times 10^5$. Then the distance L is determined from

$$\mathrm{Re}_L = \frac{u_\infty L}{\nu} = \frac{L}{0.168 \times 10^{-4}} = 5 \times 10^5$$

or

$$L = 8.4 \text{ m}$$

Example 7-2. Air at atmospheric pressure and at 300 K flows over a flat plate with a velocity of 2 m/s. The average drag coefficient c_m over a distance of 3 m from the leading edge is 2.22×10^{-3}. Calculate the drag force acting per 1-m width of the plate over the distance of 3 m from the leading edge.

Solution. The density of atmospheric air ρ at 300 K is 1.177 kg/m^3.
    The drag force acting on the plate is determined by Eq. (7-7) as

$$F = wLc_m \frac{\rho u_\infty^2}{2} = (1)(3)(2.22 \times 10^{-3}) \frac{(1.177)(2)^2}{2}$$

$$= 0.0157 \text{ N}$$

Example 7-3. The physical properties of engine oil at 100°C are given by $\rho = 840.01$ kg/m^3, $c_p = 2.219$ kJ/(kg·°C), $k = 0.137$ W/(m·K), and $\nu = 0.203 \times 10^{-4}$ m^2/s. Calculate the Prandtl number.

Solution.

$$\mathrm{Pr} = \frac{\nu}{\alpha} = \frac{\nu}{k/\rho c_p}$$

$$= \frac{0.203 \times 10^{-4}}{0.137/(840.01 \times 2.219 \times 10^3)}$$

$$= 276$$

7-2   LAMINAR FLOW OVER A FLAT PLATE

In the previous section we introduced the concepts of *drag* and *heat transfer* coefficients and showed that they are related to the velocity and temperature distributions in the flow, respectively. However, we have not shown how the velocity and temperature profiles in the flow can be determined by analytical means. In our discussion of dimensional analysis, it was shown that the functional form of the correlation of heat transfer coefficient (or the Nusselt number) is expected to be of the form given by Eq. (7-1*b*). The objective of this section is to illustrate, by approximate analytic approaches, the determination of the velocity and temperature profiles in the flow field for laminar flow over a flat plate. Then, by utilizing such information, we develop approximate analytic expressions for both the drag and heat transfer coefficients.

Velocity Distribution

The distribution of velocity for flow over bodies is governed by a system of coupled partial differential equations called the *continuity* and *momentum* equations. For the simple situation involving laminar flow along a plate, the flow velocity is characterized by two components. One is $u(x, y)$ in the flow direction parallel to the surface of the plate, and the other is $v(x, y)$ normal to the surface of the plate. The determination of these two velocity components requires two equations, the continuity and the x-direction momentum equations. Such equations can be solved exactly by numerical techniques and the velocity component $u(x, y)$ needed for the computation of the drag coefficient can be determined. There is also a relatively simple approximate method of analysis, called the *integral method*, available for solving the velocity problem in flow. To develop the integral form of the equations of motion, the continuity and momentum equations are manipulated by integrating them and then eliminating the normal velocity component $v(x, y)$ between them. Such a procedure results in the following equation, called the *momentum integral equation*:

$$\frac{d}{dx}\left[\int_0^{\delta(x)} u(u_\infty - u)\, dy\right] = \nu \left.\frac{\partial u}{\partial y}\right|_{y=0} \quad \text{in } 0 < y < \delta(x) \quad (7\text{-}16a)$$

where x and y are the coordinates measured, respectively, along the plate in the direction of flow and normal to the plate surface. In addition, u_∞ is the main stream velocity measured at distances away from the plate, and ν is the kinematic viscosity of the fluid, both of which are considered known. Then the velocity problem governed by Eq. (7-16*a*) involves two unknowns: the velocity component $u(x, y)$ and the boundary-layer thickness $\delta(x)$. We are concerned with the approximate method of solution of the momentum integral equation (7-16*a*). Basic steps in the analysis are as follows:

1. Assume a profile for the distribution of the axial velocity $u(x, y)$ in the boundary layer, $0 < y < \delta(x)$. Here we consider a second-degree polynomial of y in the form

$$u(x, y) = a_0(x) + a_1(x)y + a_2(x)y^2 \quad \text{in } 0 < y < \delta(x) \quad (7\text{-}16b)$$

where the coefficients $a_i(x)$ are, in general, functions of x and yet to be related to the velocity boundary-layer thickness $\delta(x)$.

2. The next step in the analysis is to relate $a_i(x)$ to $\delta(x)$ by utilizing the actual physical conditions for the variation of $u(x, y)$ within the boundary layer. We have only three such coefficients and consider the following three physically realistic requirements for $u(x, y)$:

$$u(x, y) = 0 \qquad \text{at } y = 0$$

$$u(x, y) = u_\infty \qquad \text{at } y = \delta(x)$$

$$\frac{\partial u(x, y)}{\partial y} = 0 \qquad \text{at } y = \delta(x)$$

By utilizing these three conditions, three equations are obtained for determining a_0, a_1, and a_2 in terms of the boundary-layer thickness $\delta(x)$. When these expressions are introduced into Eq. (7-16b), the following expression is obtained for the velocity profile:

$$u(x, y) = u_\infty \left[2\frac{y}{\delta} - \left(\frac{y}{\delta}\right)^2 \right] \qquad (7\text{-}16c)$$

where $\delta \equiv \delta(x)$. Clearly, if the boundary-layer thickness $\delta(x)$ is known, the velocity distribution $u(x, y)$ is determined from Eq. (7-16c).

3. The final step in the analysis is the determination of $\delta(x)$. The velocity profile Eq. (7-16c) is introduced into the momentum integral equation (7-16a),

$$\frac{d}{dx}\left(\int_0^{\delta(x)} u_\infty\left[2\frac{y}{\delta} - \left(\frac{y}{\delta}\right)^2\right]\left\{u_\infty - u_\infty\left[2\frac{y}{\delta} - \left(\frac{y}{\delta}\right)^2\right]\right\} dy \right) = \frac{2\nu u_\infty}{\delta(x)}$$

$$(7\text{-}16d)$$

where we made use of the following expressions:

$$\frac{\partial u}{\partial y} = u_\infty\left(\frac{2}{\delta} - \frac{2}{\delta^2}y\right) \qquad \text{and} \qquad \frac{\partial u}{\partial y}\bigg|_{y=0} = \frac{2u_\infty}{\delta}$$

When the integration with respect to y is performed, the following ordinary differential equation results for the determination of the velocity boundary-layer thickness $\delta(x)$:

$$\frac{d}{dx}\left[\frac{2}{15}u_\infty^2\delta(x)\right] = \frac{2\nu u_\infty}{\delta(x)}$$

or

$$\delta \, d\delta = \frac{15\nu}{u_\infty} dx \qquad (7\text{-}17a)$$

and the boundary condition needed to solve this differential equation is determined from the physical situation that the boundary-layer thickness vanishes at $x = 0$, as illustrated in Fig. 7-1. Then we have

$$\delta(x) = 0 \qquad \text{for } x = 0 \qquad (7\text{-}17b)$$

The solution of the ordinary differential equation in (7-17a) subject to the boundary condition of Eq. (7-17*b*) gives

$$\delta^2(x) = \frac{30\nu x}{u_\infty}$$

or

$$\delta(x) = \sqrt{\frac{30\nu x}{u_\infty}} \qquad (7\text{-}17c)$$

This expression for $\delta(x)$ is rearranged in dimensionless form as

$$\frac{\delta(x)}{x} = \sqrt{\frac{30}{\mathrm{Re}_x}} = \frac{5.48}{\mathrm{Re}_x^{1/2}} \qquad (7\text{-}17d)$$

where the local Reynolds number Re_x is defined by

$$\mathrm{Re}_x = \frac{u_\infty x}{\nu} \qquad (7\text{-}17e)$$

Thus, Eq. (7-16*c*) together with (7-17*d*) provides an expression for the velocity distribution $u(x, y)$ in the boundary-layer flow over a flat plate.

Drag Coefficient

Having established an approximate solution for the velocity distribution $u(x, y)$ in flow, we are now in a position to develop an approximate analytic expression for the drag coefficient c_x for laminar boundary-layer flow along a flat plate.
   From Eq. (7-5) we have

$$c_x = \frac{2\nu}{u_\infty^2} \left. \frac{\partial u(x, y)}{\partial y} \right|_{y=0} \qquad (7\text{-}18a)$$

where the velocity gradient is evaluated by utilizing the expressions given by Eqs. (7-16*c*) and (7-17*d*). That is,

$$\left. \frac{\partial u(x, y)}{\partial y} \right|_{y=0} = \frac{2u_\infty}{\delta(x)} = \frac{2u_\infty}{\sqrt{30\nu x/u_\infty}} \qquad (7\text{-}18b)$$

Introducing Eq. (7-18*b*) into Eq. (7-18*a*), the following approximate expression is determined for the local drag coefficient c_x:

$$c_x = \frac{4\nu}{u_\infty \sqrt{30\nu x/u_\infty}} = \frac{0.73}{\mathrm{Re}_x^{1/2}} \qquad (7\text{-}18c)$$

Our analysis for the determination of velocity profile being approximate, the resulting expression for the drag coefficient given above is also approximate. One of the shortcomings of approximate analysis is that the accuracy of the result cannot be assessed until the solution is compared with the exact result. For this particular problem, the velocity distribution in flow has been determined exactly by numerical solution of the continuity and momentum equations, and the resulting exact expression for the local drag coefficient is given by

$$c(x) = \frac{0.664}{\text{Re}_x^{1/2}} \qquad \text{for Re} \leq 10^5 \qquad\qquad (7\text{-}19a)$$

where $\text{Re}_x = u_\infty x / \nu$. We note that the local drag coefficient c_x decreases with increased distance from the leading in the form

$$c(x) \sim x^{-1/2} \qquad\qquad (7\text{-}19b)$$

Note that our approximate analysis for c_x is within 10 percent of the exact value. The average drag coefficient c_m, over the plate length from $x = 0$ to $x = L$, is determined from its definition as

$$c_m = \frac{1}{L} \int_{x=0}^{L} c(x)\, dx = 2c(x)|_{x=L} \qquad\qquad (7\text{-}20a)$$

Introducing Eq. (7-19a) into Eq. (7-20a), we obtain

$$c_m = \frac{1.328}{\text{Re}_L^{1/2}} \qquad \text{for Re} \leq 5 \times 10^5 \qquad\qquad (7\text{-}20b)$$

where

$$\text{Re}_L = \frac{u_\infty L}{\nu}$$

Given the average drag coefficient c_m, the drag force F acting on the plate over the length $x = 0$ to $x = L$ and for a width w is determined according to Eq. (7-7), that is,

$$F = wL c_m \frac{\rho u_\infty^2}{2} \qquad \text{N} \qquad\qquad (7\text{-}20c)$$

Temperature Distribution

Previously it was pointed out that given the temperature distribution in flow, the heat transfer coefficient can be determined from its definition, given by Eq. (7-13b). We are now concerned with the determination of the temperature distribution in the flow field. In general, the temperature distribution in flow is determined from the solution of the energy equation, which is a partial differential equation that contains the velocity components. The solution of such an equation is quite involved and beyond the scope of this work. However, an approximate solution for the temperature distribution $T(x, t)$ can be obtained by considering the integral form of the energy equation, called the *energy integral equation*, given by

$$\frac{d}{dx}\left[\int_0^{\delta_t} u(1 - \theta)\, dy\right] = \alpha \left.\frac{\partial \theta}{\partial y}\right|_{y=0} \qquad \text{in } 0 < y < \delta_t \qquad (7\text{-}21a)$$

where $\theta(x, y) = \dfrac{T(x, y) - T_w}{T_\infty - T_w} = $ dimensionless temperature

$T(x, y) = $ temperature distribution

$\delta_t \equiv \delta_t(x) = $ thermal boundary-layer thickness

This equation involves three unknowns, namely, $u(x, y)$, $\theta(x, y)$, and $\delta_t(x)$. If suitable profiles can be established for the velocity and temperature distributions, then the integral equation can be solved for the thermal boundary-layer thickness $\delta_t(x)$ by following a procedure similar to that used in the solution of the momentum integral equation. The basic steps are now described.

1. Suitable profiles are chosen for the velocity and temperature distributions. There are numerous possibilities. For simplicity in the analysis and for illustration purposes, we consider a parabolic profile for $u(x, y)$, given previously by Eq. (7-16c) as

$$u(x, y) = u_\infty\left[2\frac{y}{\delta} - \left(\frac{y}{\delta}\right)^2\right] \qquad \text{in } 0 < y < \delta \qquad (7\text{-}21b)$$

where

$$\delta = \sqrt{\frac{30\nu x}{u_\infty}}$$

A parabolic profile is also chosen for the dimensionless temperature $\theta(x, y)$:

$$\theta(x, y) = c_0(x) + c_1(x)y + c_2(x)y^2$$

where the coefficients $c_i(x)$ are functions of x and yet to be related to $\delta_t(x)$.

2. To determine the unknown coefficients c_0, c_1, and c_2, the following three physically realistic boundary conditions are imposed on $\theta(x, y)$:

$$\theta(x, y) = 0 \qquad \text{at } y = 0$$

$$\theta(x, y) = 1 \qquad \text{at } y = \delta_t(x)$$

$$\frac{\partial\theta(x, y)}{\partial y} = 0 \qquad \text{at } y = \delta_t(x)$$

Note that these conditions are similar to those imposed on the determination of the coefficients for the velocity profile. Therefore, the temperature profile takes the form

$$\theta(x, y) = 2\left(\frac{y}{\delta_t}\right) - \left(\frac{y}{\delta_t}\right)^2 \qquad \text{in } 0 < y < \delta_t \qquad (7\text{-}21c)$$

Note that this temperature profile is similar to the velocity profile given by Eq. (7-21b).

3. The final step in the analysis is the determination of the thermal boundary-layer thickness $\delta_t(x)$. The velocity profile, Eq. (7-21b), and the temperature profile, Eq. (7-21c), are introduced into the energy integral equation, Eq. (7-21a).

$$\frac{d}{dx}\left\{u_\infty \int_0^{\delta_t}\left[2\left(\frac{y}{\delta}\right)-\left(\frac{y}{\delta}\right)^2\right]\left[1-2\left(\frac{y}{\delta_t}\right)+\left(\frac{y}{\delta_t}\right)^3\right]dy\right\}=\frac{2\alpha}{\delta_t} \tag{7-21d}$$

The integration with respect to y is performed:

$$\frac{d}{dx}\left[\delta\left(\frac{1}{6}\Delta^2-\frac{1}{30}\Delta^3\right)\right]=\frac{2\alpha}{u_\infty\delta_t} \tag{7-22a}$$

where

$$\Delta=\frac{\delta_t(x)}{\delta(x)} \tag{7-22b}$$

Here we consider the situation in which the thermal boundary-layer thickness δ_t is smaller than the velocity boundary layer δ. Therefore, $\Delta<1$, and in Eq. (7-22a) the term $\frac{1}{30}\Delta^3$ can be neglected in comparison to $\frac{1}{6}\Delta^2$. Equation (7-22a) simplifies to

$$\frac{d}{dx}(\delta\Delta^2)=\frac{12\alpha}{u_\infty\delta\Delta} \tag{7-22c}$$

Differentiation with respect to x is performed:

$$\delta\frac{d\Delta^2}{dx}+\Delta^2\frac{d\delta}{dx}=\frac{12\alpha}{u_\infty(\delta\Delta)}$$

or

$$2\delta^2\Delta^2\frac{d\Delta}{dx}+\Delta^3\delta\frac{d\delta}{dx}=\frac{12\alpha}{u_\infty}$$

or

$$\frac{2}{3}\delta^2\frac{d\Delta^3}{dx}+\delta\frac{d\delta}{dx}\Delta^3=\frac{12\alpha}{u_\infty} \tag{7-22d}$$

since

$$\Delta^2\frac{d\Delta}{dx}=\frac{1}{3}\frac{d\Delta^3}{dx}$$

The velocity boundary-layer thickness δ is

$$\delta^2=\frac{30\nu x}{u_\infty} \tag{7-22e}$$

and differentiation, we obtain

$$\delta\frac{d\delta}{dx}=\frac{30\nu}{u_\infty} \tag{7-22f}$$

Substituting Eqs. (7-22f) and (7-22f) into Eq. (7-22d), we obtain

$$x\frac{d\Delta^3}{dx}+\frac{3}{4}\Delta^3=\frac{3}{5}\frac{\alpha}{\nu} \tag{7-22g}$$

This is an ordinary differential equation of first order in Δ^3, and its general solution is written as (see the note at the end of this chapter for an explanation)

$$\Delta^3 = Cx^{-3/4} + \frac{3/5}{3/4}\frac{\alpha}{\nu} \qquad (7\text{-}22h)$$

The integration constant C is determined by the application of the boundary condition

$$\delta_t = 0 \qquad \text{at } x = 0 \qquad (7\text{-}22i)$$

which implies that heat transfer starts at the leading edge of the plate. The application of this boundary condition gives $C = 0$. Therefore, the solution for Δ^3 becomes

$$\Delta^3 = \frac{4}{5}\frac{\alpha}{\nu} = \frac{4}{5\,\mathrm{Pr}} \qquad (7\text{-}22j)$$

or

$$\Delta = \frac{\delta_t(x)}{\delta(x)} = \left(\frac{4}{5}\right)^{1/3}\mathrm{Pr}^{-1/3} \qquad (7\text{-}22k)$$

where

$$\mathrm{Pr} = \frac{\nu}{\alpha}$$

The substitution of $\delta(x) = \sqrt{30\nu x/u_\infty}$ into Eq. $(7\text{-}22k)$ gives the thermal boundary-layer thickness $\delta_t(x)$ as

$$\delta_t(x) = \left(\frac{4}{5}\right)^{1/3}\mathrm{Pr}^{-1/3}\sqrt{\frac{30\nu x}{u_\infty}}$$

which can be rearranged as

$$\delta_t(x) = 5.085\,\frac{x}{\mathrm{Pr}^{1/3}\,\mathrm{Re}_x^{1/2}} \qquad (7\text{-}22l)$$

where

$$\mathrm{Re}_x = \frac{u_\infty x}{\nu}$$

Thus, Eq. $(7\text{-}21c)$, together with Eq. $(7\text{-}22l)$, provides an expression for the dimensionless temperature distribution $\theta(x, y)$ in the flow.

Heat Transfer Coefficient

Having established an approximate expression for the temperature distribution in the flow, we are now in a position to develop an approximate analytic expression for the local heat transfer coefficient $h(x)$ for laminar boundary-layer flow along a flat plate.

The local heat transfer coefficient $h(x)$ is defined by Eq. (7-13b) as

$$h(x) = k \frac{[\partial T(x, y)/\partial y]_{y=0}}{T_\infty - T_w} \tag{7-23a}$$

or it is related to $\theta(x, y)$ by

$$h(x) = k \frac{\partial \theta(x, y)}{\partial y} \bigg|_{y=0} \tag{7-23b}$$

Introducing Eq. (7-21c) into Eq. (7-23b), we obtain

$$h(x) = k \frac{2}{\delta_t(x)} \tag{7-23c}$$

For the thermal boundary-layer thickness $\delta_t(x)$ as given by Eq. (7-22l), the local heat transfer coefficient $h(x)$ becomes

$$h(x) = \frac{2}{5.085} \frac{k}{x} \operatorname{Re}_x^{1/2} \operatorname{Pr}^{1/3} \tag{7-23d}$$

or the local Nusselt number Nu_x is determined as

$$\operatorname{Nu}_x = \frac{xh(x)}{k} = 0.393 \operatorname{Re}_x^{1/2} \operatorname{Pr}^{1/3} \tag{7-23e}$$

which is valid for laminar flow, $\operatorname{Re}_x < 5 \times 10^5$. Note that the functional form of the local Nusselt number given by Eq. (7-23e) is consistent with the functional form established by dimensional analysis and given by Eq. (7-1b).

The Nusselt number given by Eq. (7-23e) is approximate because it is based on a temperature profile determined by an approximate method of analysis. Its accuracy cannot be established until it is compared with an exact analysis of the problem or verified experimentally. The exact analysis of heat transfer for laminar boundary-layer flow is available, and we present below some of these results. It has been shown that such solutions are dependent upon the range of the Prandtl number. Therefore, in presenting the results for the heat transfer coefficients determined through analytic solutions, we consider them in three groups: $h(x)$ applicable for fluids having moderate, very low, and high Prandtl numbers.

1. $h(x)$ for moderate Prandtl numbers. More than half a century ago Pohlhausen developed an analytic solution for the local heat transfer coefficient $h(x)$ for forced laminar convection over a flat plate maintained at a uniform temperature. His result, applicable for moderate Prandtl numbers, can be expressed in dimensionless form as

$$\boxed{\operatorname{Nu}_x = 0.332 \operatorname{Pr}^{1/3} \operatorname{Re}_x^{1/2}} \quad \begin{array}{l} \text{for } 0.6 < \operatorname{Pr} < 10 \\ \operatorname{Re}_x < 5 \times 10^5 \end{array} \tag{7-24a}$$

where Nu_x, Re_x, and Pr are defined as

$$\operatorname{Nu}_x = \frac{xh(x)}{k} \qquad \operatorname{Pr} = \frac{\nu}{\alpha} \qquad \operatorname{Re}_x = \frac{xu_\infty}{\nu} \tag{7-24b}$$

and x is the distance measured along the plate, starting from the leading edge. The physical properties of the fluid are to be calculated at the *film temperature* T_f, which is the arithmetic mean of the free-stream and wall surface temperatures; that is,

$$T_f = \tfrac{1}{2}(T_w + T_\infty) \tag{7-25}$$

The average value of the heat transfer coefficient over the distance from $x = 0$ to $x = L$ is determined from the definition given by Eq. (7-14). We find

$$h_m = \frac{1}{L} \int_0^L h(x)\, dx = 2h(x)\big|_{x=L}$$

which yields

$$\boxed{\mathrm{Nu}_m = 0.664\, \mathrm{Pr}^{1/3}\, \mathrm{Re}_L^{1/2}} \qquad \begin{array}{c} \text{for } 0.6 < \mathrm{Pr} < 10 \\ \mathrm{Re}_x < 5 \times 10^5 \end{array} \tag{7-26a}$$

where

$$\mathrm{Nu}_m = \frac{h_m L}{k} \qquad \text{and} \qquad \mathrm{Re}_L = \frac{u_\infty L}{\nu} \tag{7-26b}$$

and all the properties are evaluated at the *film temperature*.

These results are applicable for fluids such as gases, water, and light liquids.

2. $h(x)$ for large Prandtl numbers. The Prandtl number is very large for liquids such as oils. Pohlhausen's analysis for the limiting case of $\mathrm{Pr} \to \infty$ gives

$$\boxed{\mathrm{Nu}_x = 0.339\, \mathrm{Pr}^{1/3}\, \mathrm{Re}_x^{1/2}} \qquad \begin{array}{c} \text{for } \mathrm{Pr} \to \infty \\ \mathrm{Re}_x < 5 \times 10^5 \end{array} \tag{7-27a}$$

The average value of the Nusselt number over the distance $x = 0$ to $x = L$ is determined by Eq. (7-14) as

$$\boxed{\mathrm{Nu}_m = 0.678\, \mathrm{Pr}^{1/3}\, \mathrm{Re}_L^{1/2}} \qquad \text{for } \mathrm{Pr} \to \infty \tag{7-27b}$$

where Nu_m and Re_L are defined by Eq. (7-26b).

3. $h(x)$ for very small Prandtl numbers. The Prandtl number is very low for liquid metals. The limiting case of Pohlhausen's solution for $\mathrm{Pr} \to 0$ gives

$$\boxed{\mathrm{Nu}_x = 0.564\, \mathrm{Pe}_x^{1/2}} \qquad \begin{array}{c} \text{for } \mathrm{Pr} \to 0 \\ \mathrm{Re}_x < 5 \times 10^5 \end{array} \tag{7-28}$$

where the local Peclet number Pe_x is defined as

$$\mathrm{Pe}_x = \mathrm{Re}_x\, \mathrm{Pr} = \frac{u_\infty x}{\alpha}$$

The average value of the Nusselt number, Nu_m, is calculated by Eq. (7-14).

Example 7-4. An approximate expression for the local heat transfer coefficient $h(x)$ in the laminar boundary layer along a flat plate is given by

$$h(x) = 0.393 \frac{k}{x} \mathrm{Pr}^{1/3} \mathrm{Re}_x^{1/2}$$

Develop an expression for the average heat transfer coefficient h_m over a distance from the leading edge of the plate.

Solution. The average heat transfer coefficient h_m from $x = 0$ to $x = L$ is determined using Eq. (7-14) to be

$$h_m = \frac{1}{L} \int_0^L h(x)\, dx$$

Introducing the expression for $h(x)$, we find

$$h_m = \frac{1}{L} 0.393 \, \mathrm{Pr}^{1/3} \, (u_\infty \nu)^{1/2} \int_0^L \frac{1}{x^{1/2}}\, dx$$

$$= 2(0.393) \frac{k}{L} \mathrm{Pr}^{1/3} \mathrm{Re}_x^{1/2} = 2[h(x)]_{x=L}$$

That is, the average heat transfer coefficient h_m is twice the value of local heat transfer coefficient $h(x)$ evaluated at $x = L$.

Example 7-5. Consider laminar flow of ordinary fluid at temperature T_∞ over a flat plate maintained at a uniform temperature T_∞ in the region $0 \le x \le x_0$ and at a uniform temperature T_w in the region $x > x_0$. That is, heat transfer between the plate and the fluid does not start until the location $x = x_0$. Determine the thermal boundary-layer thickness by an approximate integral method.

Solution. The ratio $\delta_t(x)/\delta(x) = \Delta$ is given by Eq. (7-22h) as

$$\Delta^3 = Cx^{-3.4} + \frac{3/5}{3/4} \frac{\alpha}{\nu}$$

The integration constant C is determined by the application of the boundary condition $\delta_t = 0$ for $x = x_0$, which is $\Delta(x) = 0$ at $x = x_0$. We find

$$\Delta^3 = \frac{4}{5} \mathrm{Pr}^{-1} \left[1 - \left(\frac{x_0}{x} \right)^{3/4} \right]$$

where

$$\mathrm{Pr} = \frac{\nu}{\alpha}$$

Example 7-6. Air at atmospheric pressure and at 350 K flows over a flat plate with a velocity of 15 m/s. Determine the drag force per 1-m width of the plate acting over the distance 0.3 m from the leading edge.

Solution. The physical properties of atmospheric air at 350 K are

$$\rho = 0.998 \text{ kg/m}^3 \qquad \nu = 20.76 \times 10^{-6} \text{ m}^2/\text{s}$$

The Reynolds number at a distance $x = L = 0.3$ m is

$$\mathrm{Re}_L = \frac{u_\infty L}{\nu} = \frac{(15)(0.3)}{20.76 \times 10^{-6}} = 2.17 \times 10^5$$

This Reynolds number is less than the critical Reynolds number $\text{Re}_c \cong 5 \times 10^5$; therefore the flow is laminar. The average drag coefficient c_m over the length $x = 0$ to $x = L$ is given by Eq. (7-20b):

$$c_m = \frac{1.328}{\text{Re}_L^{1/2}} = \frac{1.328}{(2.17 \times 10^5)^{1/2}} = 2.85 \times 10^{-3}$$

Then the drag force acting on the plate is determined by Eq. (7-7):

$$F = wLc_m \frac{\rho u_\infty^2}{2} = (1)(0.3)(2.85 \times 10^{-3}) \frac{(0.988)(15)^2}{2}$$
$$= 0.095 \text{ N}$$

Example 7-7. Air at atmospheric pressure and at a mean temperature of 77°C flows over a flat plate with a velocity of 9 m/s. Plot the local and average heat transfer coefficients as a function of the distance from the leading edge of the plate for the laminar boundary layer.

Solution. The physical properties of atmospheric air taken at atmospheric pressure and at 77°C are as follows:

$$k = 0.03 \text{ W/(m} \cdot {}^\circ\text{C)} \qquad \nu = 2.076 \times 10^{-5} \text{ m}^2/\text{s} \qquad \text{Pr} = 0.697$$

For $\text{Re}_c = 5 \times 10^5$, the location x_c where the transition occurs is determined as

$$x_c = \frac{\nu \, \text{Re}_c}{u_\infty} = \frac{(2.076 \times 10^{-5})(5 \times 10^5)}{9} = 1.153 \text{ m}$$

The local heat transfer coefficient for the laminar boundary layer is determined by Eq. (7-24):

$$\text{Nu}_x = 0.332 \, \text{Pr}^{1/3} \, \text{Re}_x^{1/2}$$

$$\frac{xh(x)}{k} = 0.332 \, \text{Pr}^{1/3} \left(\frac{xu_\infty}{\nu} \right)^{1/2}$$

or

$$h(x) = 0.332 \, \text{Pr}^{1/3} \left(\frac{u_\infty}{\nu} \right)^{1/2} k \frac{1}{x^{1/2}}$$

$$= 0.332(0.697) \left(\frac{9}{2.076 \times 10^{-5}} \right)^{1/2} (0.03) \frac{1}{x^{1/2}}$$

$$= \frac{5.82}{x^{1/2}}$$

The average heat transfer coefficient over any distance from $x = 0$ to $x = l_x$ is determined by Eq. (7-14) to be

$$h_m(x = l_x) = \frac{1}{l_x} \int_0^{l_x} h(x) \, dx = 2h(x)|_{x=l_x}$$

or

$$h_m(x = l_x) = 2 \left(\frac{5.82}{x^{1/2}} \right) \Big|_{x=l} = \frac{11.64}{l^{1/2}}$$

A plot of h_x and $h_m(x = l_x)$ is given in the accompanying figure.

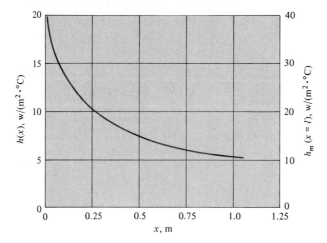

Example 7-8. Mercury at 70°C flows with a velocity of 0.1 m/s over a 0.4-m-long flat plate maintained at 130°C. Assuming that the transition from laminar to turbulent flow takes place at $\text{Re}_c = 5 \times 10^5$, determine the average heat transfer coefficient over the length of the plate.

Solution. The physical properties of the fluid are taken at the film temperature $T_f = (70 + 130)/2 = 100°C$ as follows:

$$k = 10.51 \text{ W/(m} \cdot °\text{C)} \qquad \nu = 0.0928 \times 10^{-6} \text{ m}^2/\text{s} \qquad \text{Pr} = 0.0162$$

For $\text{Re}_c = 5 \times 10^5$, the location x_c where the transition occurs is determined to be

$$x_c = \frac{\nu \, \text{Re}_c}{u_\infty} = \frac{(0.0928 \times 10^{-6})(5 \times 10^5)}{0.1} = 0.464 \text{ m}$$

Hence $L = 0.4 \text{ m} < x_c = 0.464 \text{ m}$ and the flow is laminar over the entire length of the plate.

The local Nusselt number for the laminar boundary layer is determined at $x = L$ by Eq. (7-28):

$$\text{Nu}|_{x=L} = 0.564 \, \text{Pe}^{1/2}$$

$$= 0.564 \, (\text{Re}_L \, \text{Pr})^{1/2}$$

$$= 0.564 \left[\frac{(0.4)(0.1)}{0.0928 \times 10^{-6}} (0.0162) \right]^{1/2} = 47.13$$

and the average Nusselt number is calculated from Eq. (7-14):

$$\text{Nu}_m = 2 \, \text{Nu}|_{x=L}$$

$$= 2(47.13) = 94.26$$

Then the average heat transfer coefficient over the length of the plate $L = 0.4 \text{ m}$ becomes

$$h_m = \frac{k}{L} \, \text{Nu}_m$$

$$= \left(\frac{10.51}{0.4} \right) 94.26 = 2476.64 \text{ W/(m}^2 \cdot °\text{C)}$$

7-3 TURBULENT FLOW OVER A FLAT PLATE

For forced flow over a flat plate, a transition from laminar to turbulent flow takes place in the range of Reynolds numbers from 2×10^5 to 5×10^5. The term *turbulent* is used to denote that the motion of the fluid is chaotic in nature and involves crosswise mixing and eddying superimposed on the motion of the mainstream.

The eddying or crosswise mixing is advantageous in that it assists greatly in improving heat transfer between the fluid and the surface of the solid body, but it has the disadvantage of increasing the drag generated by the flow.

The flow patterns in turbulent flow are so complex that most of the correlations of velocity distribution in flow are of a semiempirical nature. The local drag coefficient c_x and the local heat transfer coefficient $h(x)$ are quantities of practical interest in turbulent flow over a flat plate. Here we present some of the recommended expressions for the determination of the drag and heat transfer coefficients.

Drag Coefficient

Schlichting examined a vast amount of experimental data and recommended the following correlation for the local drag coefficient c_x for turbulent flow over a flat plate:

$$c(x) = 0.0592 \, \text{Re}_x^{-0.2} \qquad (7\text{-}29)$$

valid for $5 \times 10^5 < \text{Re}_x < 10^7$.

At higher Reynolds number, the following correlation is recommended by Schultz-Grunow:

$$c(x) = 0.370 \, (\log \text{Re}_x)^{-2.584} \qquad (7\text{-}30)$$

valid for $10^7 < \text{Re}_x < 10^9$.

In many applications, the flow is laminar over the part of the plate surface $0 < x \le x_c$, where x_c is the location of the transition from laminar to turbulent, and turbulent over the remaining region $x_c < x \le L$. The drag coefficient for the laminar region was given previously by Eq. (7-19a), and for the turbulent flow region Eq. (7-29) can used. Then the mean drag coefficient c_m over the entire region $0 < x < L$ can be determined by proper averaging of these two local drag coefficients. The resulting expression for the average drag coefficient c_m becomes

$$c_m = 0.074 \, \text{Re}_L^{-0.2} - \frac{B}{\text{Re}_L} \qquad \text{for } \text{Re}_c < \text{Re}_L < 10^7 \qquad (7\text{-}31)$$

which is valid over the entire region of laminar and turbulent flow. The value of B depends on the value of the critical Reynolds number Re_c chosen for the transition from laminar flow; corresponding values are listed in Table 7-1.

TABLE 7-1
Constant B of Eq. (7-31)

B	Re_c
700	2×10^5
1050	3×10^5
1740	3×10^5
3340	1×10^6

RELATION BETWEEN $c(x)$ AND $h(x)$. The drag and heat transfer coefficients for laminar flow over a flat plate can readily be determined by theoretical means; but this is not quite true for the case of turbulent flow. Therefore, the drag and heat transfer coefficients for turbulent flow are generally determined by experimental means. It is easier to measure the drag force than the heat transfer coefficient. Furthermore, correlations are generally available for the determination of the drag coefficient in turbulent flow. Therefore, efforts have been directed toward developing relations between the drag coefficient and the heat transfer coefficient so that $h(x)$ can be determined once $c(x)$ is known. One such relation is given in the form

$$St_x \, Pr^{2/3} = \tfrac{1}{2}c(x) \qquad (7\text{-}32a)$$

where St_x is the *local Stanton number*, defined as

$$St_x \equiv \frac{Nu_x}{Pr \, Re_x} \qquad (7\text{-}32b)$$

and Nu_x, Pr, and Re_x are as given by Eq. (7-24b). The relationship in Eq. (7-32a) is referred to as the *Reynolds-Colburn* analogy between the drag and heat transfer coefficients.

Heat Transfer Coefficient

We now make use of the Reynolds-Colburn analogy to develop expressions for the heat transfer coefficient $h(x)$ from the knowledge of the drag coefficient for turbulent flow over a flat plate.

Equations (7-29) and (7-30) give the local drag coefficient in turbulent flow. They can be used in Eqs. (7-32) to develop expressions for the local heat transfer coefficient for turbulent flow over a flat plate. The use of Eq. (7-29) gives

$$Nu_x = 0.0296 \, Re_x^{0.8} \, Pr^{1/3} \qquad \text{for } 5 \times 10^5 < Re_x < 10^7 \qquad (7\text{-}33)$$

and Eq. (7-30) yields

$$Nu_x = 0.185 \, Re_x (\log Re_x)^{-2.584} \, Pr^{1/3} \qquad \text{for } 10^7 < Re_x < 10^9 \qquad (7\text{-}34)$$

where

$$\mathrm{Re}_x = \frac{u_\infty x}{\nu} \qquad \mathrm{Pr} = \frac{\nu}{\alpha} \qquad \mathrm{Nu}_x = \frac{h(x)x}{k}$$

and all properties are evaluated at the *film temperature*.

The following expression for the local heat transfer coefficient in turbulent flow over a flat plate has been proposed [Whitaker, 1976]:

$$\boxed{\mathrm{Nu}_x = 0.029\,\mathrm{Re}_x^{0.8}\,\mathrm{Pr}^{0.43}} \qquad \text{for } 2 \times 10^5 < \mathrm{Re}_x < 5 \times 10^5 \qquad (7\text{-}35)$$

where all properties are evaluated at the film temperature. A large number of experimental data are found to correlate well with this expression.

In most practical applications the flow is *laminar* over a portion $0 < x < x_c$ of the plate and turbulent over the remaining portion $x_c < x < L$ of the plate, where x_c is the location of the transition from laminar to turbulent flow. The average value of the heat transfer coefficient h_m over the entire region $0 < x < L$ can be determined by properly averaging the local heat transfer coefficients for the laminar and turbulent flow regions. Taking the local heat transfer coefficients from Eqs. (7-24a) and (7-35) for the laminar- and turbulent-flow regions, respectively, averaging them properly, and assuming that the transition from laminar to turbulent flow takes place at a critical Reynolds number $\mathrm{Re}_c = 2 \times 10^5$, the following expression is obtained for the average Nusselt number applicable over the entire region $0 < x < L$:

$$\boxed{\mathrm{Nu}_m = 0.036\,\mathrm{Pr}^{0.43}(\mathrm{Re}_L^{0.8} - 17{,}400) + 297\,\mathrm{Pr}^{1/3}} \qquad (7\text{-}36)$$

The last term on the right-hand side can be approximated as

$$297\,\mathrm{Pr}^{1/3} \cong 297\,\mathrm{Pr}^{0.43}$$

and the viscosity correction can be introduced by multiplying the right-hand side of the resulting expression by $(\mu_\infty/\mu_w)^{0.25}$. Then the following equation is obtained:

$$\boxed{\mathrm{Nu}_m = 0.036\,\mathrm{Pr}^{0.43}(\mathrm{Re}_L^{0.8} - 9200)\left(\frac{\mu_\infty}{\mu_w}\right)^{0.25}} \qquad (7\text{-}37)$$

All physical properties are evaluated at the free-stream temperature except μ_w, which is evaluated at the wall temperature. The viscosity correction factor takes care of the effects of the property variation with temperature in liquids. For gases, the viscosity correction is neglected, and for such a case the physical properties are evaluated at the film temperature.

The relation given by Eq. (7-37) is plotted in Fig. 7-5 as a function of the Reynolds number Re_L for several different values of the Prandtl number. This figure also shows the average drag coefficient determined from Eq. (7-13).

Equation (7-37) gives the average Nusselt number over the laminar and

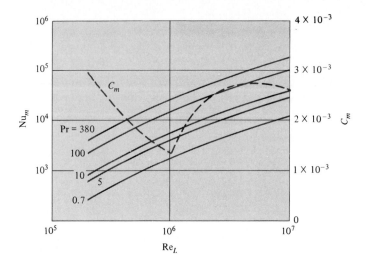

FIGURE 7-5
Average Nusselt number (Eq. 7-37) and the average drag coefficient (Eq. 7-31) for turbulent flow over a flat plate.

turbulent boundary layers over a flat plate for $\mathrm{Re}_L > 2 \times 10^5$. It has been proposed to correlate the experimental data for air, water, and oil covering the following ranges:

$$2 \times 10^5 < \mathrm{Re}_L < 5.5 \times 10^6$$

$$0.70 < \mathrm{Pr} < 380$$

$$0.26 < \frac{\mu_\infty}{\mu_w} < 3.5$$

Equation (7-37) correlates the experimental data reasonably well when the free-stream turbulence is small. If high-level turbulence is present in the free stream, Eq. (7-37) without the constant 9200 correlates the data reasonably well.

Example 7-9. Atmospheric air at $T_\infty = 40°C$ with a free-stream velocity $u_\infty = 8\,\mathrm{m/s}$ flows along a flat plate $L = 3\,\mathrm{m}$ long which is maintained at a uniform temperature of 100°C. Calculate the average heat transfer coefficient over the entire length of the plate. Assume that transition occurs at $\mathrm{Re}_c = 2 \times 10^5$.

Solution. The physical properties of atmospheric air at $T_f = (T_w + T_\infty)/2 = (100 + 40)/2 = 70°C$ are taken as follows:

$$k = 0.0295\ \mathrm{W/(m \cdot °C)}$$

$$\nu = 2.005 \times 10^{-5}\ \mathrm{m^2/s}$$

$$\mathrm{Pr} = 0.699$$

The Reynolds number at $L = 3\,\text{m}$ is

$$\text{Re}_L = \frac{u_\infty L}{\nu} = \frac{8 \times 3}{2.005 \times 10^{-5}} = 1.2 \times 10^6$$

The average heat transfer coefficient over $L = 3\,\text{m}$, without the viscosity correction, is determined from Eq. (7-37) as

$$\text{Nu}_m = 0.036\,\text{Pr}^{0.43}[\text{Re}_L^{0.8} - 9200]$$

$$= 0.036(0.699^{0.43}[(1.2 \times 10^6)^{0.8} - 9200]$$

$$= 1969$$

$$h_m = \frac{k}{L}\,\text{Nu}_m = \frac{0.0295}{3}\,1969$$

$$= 19.4\,\text{W/(m}^2 \cdot {}^\circ\text{C})$$

Here the physical properties are evaluated at the film temperature because the viscosity correction is neglected. We note that when the viscosity correction needs to be included, the physical properties are evaluated at the free-stream temperature.

Example 7-10. Calculate the total heat transfer rate Q from the plate of Example 7-9 to the air over the length $L = 3\,\text{m}$ and width $w = 1\,\text{m}$.

Solution. The heat transfer rate is

$$Q = wLh_m(T_w - T_\infty)$$

$$= (1)(3)(19.4)(100 - 40)$$

$$= 3492\,\text{W}$$

7-4 FLOW ACROSS A SINGLE CIRCULAR CYLINDER

Flow across a single circular cylinder is frequently encountered in practice, but the determination of the heat transfer coefficient is a very complicated matter because of the complexity of the flow patterns around the cylinder. Figure 7-6 illustrates with sketches the flow characteristics around a circular cylinder. The magnitude of the Reynolds number,

$$\text{Re} = \frac{u_\infty D}{\nu} \tag{7-38}$$

FIGURE 7-6
Flow around a circular cylinder at various Reynolds numbers.

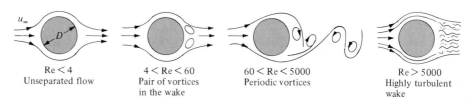

$\text{Re} < 4$	$4 < \text{Re} < 60$	$60 < \text{Re} < 5000$	$\text{Re} > 5000$
Unseparated flow	Pair of vortices in the wake	Periodic vortices	Highly turbulent wake

affects the formation of various types of flow patterns. Here the Reynolds number is based on the diameter D of the cylinder. For a Reynolds number less than about 4, the flow around the cylinder remains unseparated, and the flow patterns are relatively simple. At a Reynolds number of about 4, vortices start to form in the wake region, and for $\mathrm{Re} > 4$ the flow patterns become very complicated. The local drag and heat transfer coefficients vary around the circumference of the cylinder.

In most engineering applications the average values of the drag and heat transfer coefficients over the circumference are of interest. Therefore, we focus our attention on the determination of the average drag and heat transfer coefficients.

Average Drag Coefficient

Consider the flow of a fluid with a free-stream velocity u_∞ across a single circular tube of diameter D and length L. The drag force F acting on the tube is of interest in many engineering applications. To calculate this drag force F, an *average drag coefficient* c_D is defined as

$$\boxed{\frac{F}{LD} = c_D \frac{\rho u_\infty^2}{2}} \tag{7-39}$$

Here LD represents the area normal to the flow; the density ρ and the free-stream velocity u_∞ are known. Then, given c_D, the drag force F acting over the length L of the tube can be readily determined from Eq. (7-39).

Figure 7-7 shows the drag coefficient c_D for flow across a single cylinder. The physical significance of the variation of c_D with the Reynolds number can be better envisioned if we examine the results in Fig. 7-7 in relation to the sketches in Fig. 7-6. For $\mathrm{Re} < 4$, the drag is caused by viscous forces only, since the boundary layer remains attached to the cylinder. In the region $4 < \mathrm{Re} < 5000$, vortices are formed in the wake; therefore, the drag is due partly to the viscous forces and partly to the wake formation, that is, the low pressure caused by the flow separation. In the region $5 \times 10^3 < \mathrm{Re} < 3.5 \times 10^5$, the drag is caused predominantly by the highly turbulent eddies in the wake. The sudden reduction in drag at $\mathrm{Re} \cong 3.5 \times 10^5$ is caused by the boundary layer changing to turbulent; this causes the point of flow separation to move toward the rear of the cylinder, which in turn reduces the size of the wake, and hence the drag.

Average Heat Transfer Coefficient

The average heat transfer coefficient h_m for flow across a single cylinder has been studied experimentally by various investigators using a variety of fluids, such as air, water, paraffin, transformer oil, liquid sodium, and many others, covering a wide range of Prandtl numbers. Therefore there are several correlations available for the determination of h_m.

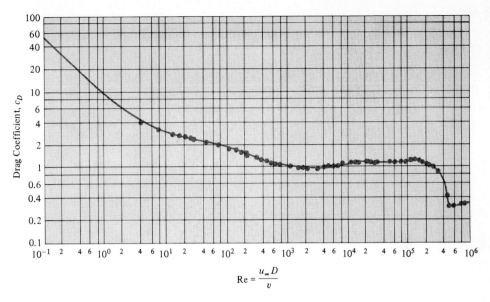

FIGURE 7-7
Average Drag coefficient for flow across a single circular cylinder. (From Schlichting).

A sufficienty general correlation for h_m is given in the form [Churchill and Bernstein]

$$\text{Nu}_m = 0.3 + \frac{0.62\,\text{Re}^{1/2}\,\text{Pr}^{1/3}}{[1 + (0.4/\text{Pr})^{2/3}]^{1/4}}\left[1 + \left(\frac{\text{Re}}{282{,}000}\right)^{5/8}\right]^{4/5}$$

$$\text{for } 10^2 < \text{Re} < 10^7 \quad \text{and} \quad \text{Pe} = \text{Re}\,\text{Pr} > 0.2 \qquad (7\text{-}40)$$

Equation (7-40) underpredicts most data by about 20 percent in the range $20{,}000 < \text{Re} < 400{,}000$. Therefore, for this particular range of Reynolds numbers, the following modified form of Eq. (7-40) is recommended.

$$\text{Nu}_m = 0.3 + \frac{0.62\,\text{Re}^{1/2}\,\text{Pr}^{1/3}}{[1 + (0.4/\text{Pr})^{2/3}]^{1/4}}\left[1 + \left(\frac{\text{Re}}{282{,}000}\right)^{1/2}\right]$$

$$\text{for } 20{,}000 < \text{Re} < 400{,}000 \qquad (7\text{-}41)$$

For ready reference, the expression given by Eq. (7-40) is plotted in Fig. 7-8 as a function of the Reynolds number Re for several different values of the Prandtl number. For the Peclet number less than 0.2, the following expression can be used [Nakai and Okazaki]:

$$\text{Nu}_m = (0.8237 - \ln \text{Pe}^{1/2})^{-1} \qquad \text{for Pe} < 0.2 \qquad (7\text{-}42)$$

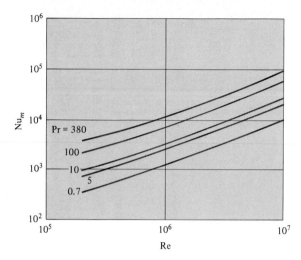

FIGURE 7-8
Average Nusselt number (Eq. 7-40) as a function of Reynolds number for flow across a single circular cylinder.

In the above relations all properties are to be evaluated at the *film temperature* $T_f = (T_w + T_\infty)/2$, and the Reynolds and Peclet numbers are defined as

$$Re = \frac{u_\infty D}{\nu} \qquad (7\text{-}43a)$$

$$Pe = Re\ Pr = \frac{u_\infty D}{\alpha} \qquad (7\text{-}43b)$$

where D is the tube diameter, α is the thermal diffusivity of the fluid, and u_∞ is the free stream velocity.

Variation of $h(\theta)$ around the Cylinder

In the above discussion we focused our attention on the determination of the average value of the heat transfer coefficient for the cylinder. Actually, the local value of the heat transfer coefficient $h(\theta)$ varies with the angle θ around the cylinder. It has a fairly high value at the *stagnation point* $\theta = 0$ and decreases around the cylinder as the boundary layer thickens. The decrease of the heat transfer coefficient is continuous until the boundary layer separates from the wall surface or the laminar boundary layer changes to turbulent; then an increase occurs with the distance around the cylinder. The variation of the local heat transfer coefficient $h(\theta)$ with θ around a circular cylinder has been investigated. In order to give some idea of the variation, $h(\theta)$ plotted against the angle θ measured from the stagnation point is presented in Fig. 7-9 for several different values of the Reynolds number. The flow mechanism around the cylinder must be complicated because $h(\theta)$ exhibits several maxima and minima at different locations which vary with the Reynolds number. For example, in the curve for $Re = 140,000$, the first minimum occurs at the transition from the laminar to the turbulent boundary layer at an angle $\theta \cong 80°$; the second minimum occurs at $\theta \cong 130°$, where flow separation takes place.

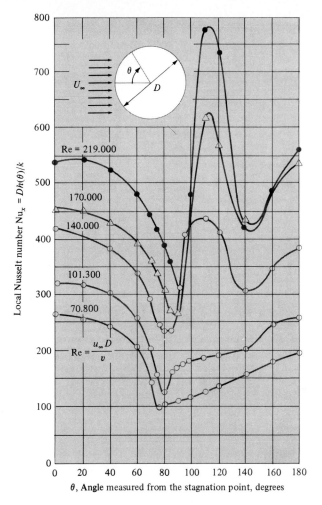

FIGURE 7-9
Variation of the local heat transfer coefficient $h(\theta)$ around a circular cylinder for flow of air. (From Giedt)

Example 7-11. Atmospheric air at $T_\infty = -5°C$ and a free-stream velocity $u_\infty = 6$ m/s flows across a single tube of outside diameter $D = 25$ cm. The tube's surface is kept at a uniform temperature $T_w = 180°C$. Determine the average heat transfer coefficient.

Solution. The physical properties of air at the film temperature $T_f = (T_w + T_\infty)/2 = (-5 + 180)/2 = 87.5°C$ are

$$k = 0.0308 \text{ W/(m·°C)} \qquad Pr = 0.695 \qquad \nu = 2.184 \times 10^{-5} \text{ m}^2/\text{s}$$

The Reynolds number becomes

$$Re = \frac{u_\infty D}{\nu} = \frac{6 \times 0.25}{2.184 \times 10^{-5}} = 6.868 \times 10^4$$

Equation (7-41) is applied to calculate h_m:

$$\text{Nu}_m = 0.3 + \frac{0.62\,\text{Re}^{1/2}\,\text{Pr}^{1/3}}{[1 + (0.4/\text{Pr})^{2/3}]^{1/4}}\left[1 + \left(\frac{\text{Re}}{282,000}\right)^{1/2}\right]$$

$$= 188.9$$

$$h_m = \frac{k}{D}\,\text{Nu}_m = \frac{0.0308}{0.25} \times 188.8$$

$$= 23.3\ \text{W}/(\text{m}^2 \cdot {}^\circ\text{C})$$

Example 7-12. Calculate the rate of heat loss from a 1-m length of the tube of Example 7-11 into the air.

Solution. The heat loss to the air becomes

$$Q = h_m(\pi DL)(T_w - T_\infty)$$

$$= 23.3(\pi)(-0.25)(1)[180 - (-5)]$$

$$= 3385\ \text{W}$$

7-5   FLOW ACROSS A SINGLE NONCIRCULAR CYLINDER

The results of experiments to determine the average heat transfer coefficient h_m for the flow of gases across a single, noncircular, long cylinder of various geometries have been correlated through the following simple relationship [Jacob]:

$$\text{Nu}_m = \frac{h_m D_e}{k} = c\left(\frac{u_\infty D_e}{\nu}\right)^n \tag{7-44}$$

where the constant c, the exponent n, and the characteristic dimension D_e for various geometries are presented in Table 7-2. The physical properties of the fluid are evaluated at the arithmetic mean of the free-stream and wall temperatures.

Example 7-13. Atmospheric air at $T_\infty = 27^\circ\text{C}$ with a free-stream velocity of $u_\infty = 10\ \text{m/s}$ flows across a square duct of 12.5 cm by 12.5 cm cross section oriented such that one of its lateral surfaces is perpendicular to the direction of flow. The duct's surface is kept at a uniform temperature $T_w = 100^\circ\text{C}$. Determine the average heat transfer coefficient.

Solution. The physical properties of air at the film temperature $T_f = (T_w + T_\infty)/2 = (100 + 27)/2 = 63.5^\circ\text{C}$ are

$$k = 0.029\ \text{W}/(\text{m} \cdot {}^\circ\text{C}) \qquad \text{Pr} = 0.7 \qquad \nu = 1.94 \times 10^{-5}\ \text{m}^2/\text{s}$$

The Reynolds number becomes

$$\text{Re} = \frac{u_\infty D}{\nu} = \frac{10 \times 0.125}{1.94 \times 10^{-5}}$$

$$= 6443$$

TABLE 7-2
Constants c and n of Eq. (7-44)

Flow direction and geometry	$Re = \dfrac{u_\infty D_e}{\nu}$	n	c
$u_\infty \rightarrow$ ◇ D_e	5,000–100,000	0.588	0.222
$u_\infty \rightarrow$ ⬭ D_e	2,500–15,000	0.612	0.224
$u_\infty \rightarrow$ ◇ D_e	2,500–7,500	0.624	0.261
$u_\infty \rightarrow$ ⬡ D_e	5,000–100,000	0.638	0.138
$u_\infty \rightarrow$ ⬡ D_e	5,000–19,500	0.638	0.144
$u_\infty \rightarrow$ ▢ D_e	5,000–100,000	0.675	0.092
$u_\infty \rightarrow$ ▢ D_e	2,500–8,000	0.699	0.160
$u_\infty \rightarrow$ │ D_e	4,000–15,000	0.731	0.205
$u_\infty \rightarrow$ ⬡ D_e	19,500–100,000	0.782	0.035
$u_\infty \rightarrow$ ⬯ D_e	3,000–15,000	0.804	0.085

Source: Jakob [29].

We use Eq. (7-44) with Table 7-2 to determine

$$\mathrm{Nu}_m = C\,\mathrm{Re}^n = 0.092(6443)^{0.675}$$

$$= 34.3$$

$$h_m = \frac{k}{D_e}\,\mathrm{Nu}_m = \frac{0.029}{0.125}(34.3)$$

$$= 7.96 \ \mathrm{W/(m^2 \cdot {}^\circ C)}$$

Example 7-14. Calculate the heat transfer rate from the duct to the air per meter length of the duct of Example 7-13.

Solution. The heat transfer rate becomes

$$Q = h_m(4D_e L)(T_w - T_\infty)$$

$$= 80.5(4 \times 0.012 \times 1)(100 - 27)$$

$$= 282 \ \mathrm{W}$$

7-6   FLOW ACROSS A SINGLE SPHERE

The basic characteristics of flow over a single sphere are somewhat similar to those for flow over a single circular tube, shown in Fig. 7-6. Therefore, the variation of the drag and heat transfer coefficients with the Reynolds number is expected to be similar to that for a single circular cylinder. The drag and heat transfer coefficients vary with the position around the sphere. Here we are concerned only with the average values of the drag and heat transfer coefficient over the entire sphere surface, because they are the quantities of interest in most engineering applications.

Average Drag Cofficient

Consider a fluid flowing with a free-stream velocity u_∞ over a single sphere of diameter D. The drag force F exerted by flow on the sphere is related to the *average drag coefficient* c_D in the form

$$\frac{F}{A} = c_D \frac{\rho u_\infty^2}{2} \qquad (7\text{-}45)$$

where A is the frontal area (that is, $A = \pi D^2/4$) and u_∞ is the free-stream velocity. We note that F/A is the drage force per unit frontal area of the sphere.

Figure 7-10 shows the average drag coefficient c_D plotted as a function of

FIGURE 7-10
Average Drag coefficient for flow over a single sphere. (From Schlichting).

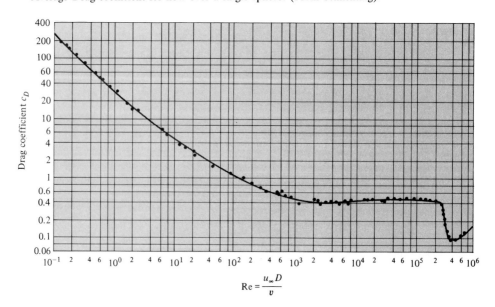

$$\text{Re} = \frac{u_\infty D}{v}$$

the Reynolds number for flow across a single sphere. A comparison of the drag coefficient curves for a single cylinder and a single sphere, shown in Figs. 7-7 and 7-10, respectively, reveals similar behavior on the part of these two curves.

Average Heat Transfer Coefficients

A sufficiently general correlation for the average heat transfer coefficient h_m for flow across a single sphere is given by [Whitaker, 1972]

$$\text{Nu}_m = 2 + (0.4\,\text{Re}^{0.5} + 0.06\,\text{Re}^{2/3})\,\text{Pr}^{0.4}\left(\frac{\mu_\infty}{\mu_w}\right)^{0.25} \qquad (7\text{-}46)$$

which is valid over the ranges

$$3.5 < \text{Re} < 8 \times 10^4$$

$$0.7 < \text{Pr} < 380$$

$$1 < \frac{\mu_\infty}{\mu_w} < 3.2$$

with the physical properties evaluated at the free-stream temperature, except for μ_w, which is evaluated at the wall temperature. For gases the viscosity correction is neglected, but the physical properties are evaluated at the film temperature. Here the Reynolds number is defined as

$$\text{Re} = \frac{u_\infty D}{\nu}$$

where D is the diameter of the sphere and u_∞ is the free-stream velocity.

For ready reference, the expression given by Eq. (7-46) is plotted in Fig. 7-11 as a function of the Reynolds number Re for several different values of the Prandtl number.

FIGURE 7-11
Average Nusselt number (Eq. 7-46) as a function of Reynolds number for flow across a single sphere.

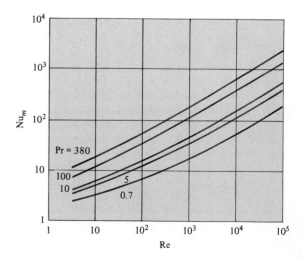

Example 7-14. Water at $T_\infty = 20°C$ flows with a free-stream velocity of $u_\infty = 1$ m/s over a 2.5-cm-diameter sphere whose surface is maintained at a uniform temperature of $T_w = 140°C$. Determine the average heat transfer coefficient.

Solution. The physical properties of water at the free-stream temperature $T_\infty = 20°C$ are

$$k = 0.597 \text{ W/(m} \cdot °C)\qquad Pr = 7.02 \qquad \nu = 1.006 \times 10^{-6} \text{ m}^2/\text{s}$$

$$\mu_\infty = 1.006 \times 10^{-3} \text{ kg/(m} \cdot \text{s})$$

and the viscosity of water at $T_w = 140°C$ is

$$\mu_w = 0.198 \times 10^{-3} \text{ kg/(m} \cdot \text{s})$$

The Reynolds number becomes

$$Re = \frac{u_\infty D}{\nu} = \frac{1 \times 0.025}{1.006 \times 10^{-6}}$$

$$= 24{,}850$$

We use Eq. (7-46) to determine Nu_m:

$$Nu_m = 2 + (0.4 \, Re^{0.5} + 0.06 \, Re^{2/3}) \, Pr^{0.4} \left(\frac{\mu_\infty}{\mu_w}\right)^{0.25}$$

$$= 375.1$$

$$h_m = \frac{k}{D} Nu_m = \frac{0.597}{0.025}(375.1)$$

$$= 8958 \text{ W/(m}^2 \cdot °C)$$

Example 7-15. Calculate the heat loss from the sphere of Example 7-14 to the air.

Solution. The heat loss becomes

$$Q = h_m(\pi D^2)(T_w - T_\infty)$$

$$= 8958(\pi \times 0.025^2)(140 - 20)$$

$$= 2111 \text{ W}$$

NOTE

The differential equation (7-22g) is of the form

$$x \frac{dy}{dx} + Ay = B$$

where A and B are constants. A particular solution y_p of this equation is given by

$$y_P = \frac{B}{A}$$

and the homogeneous solution y_H is

$$y_H = x^{-A}$$

Then the complete solution becomes

$$y = Cx^{-A} + \frac{B}{A}$$

where the integration constant C is determined from the boundary condition for the problem.

PROBLEMS

Basic Concepts

7-1. Air at atmospheric pressure and at 400 K flows over a flat plate with a velocity of 5 m/s. The transition from laminar to turbulent flow is assumed to take place at a Reynolds number of 5×10^5; determine the distance from the leading edge of the plate at which transition occurs.

7-2. Air at atmospheric pressure and at 350 K flows over a flat plate with a velocity of 5 m/s. The average drag coefficient c_m over a distance of 2 m from the leading edge is 0.0019. Calculate the drage force acting per 1-m width of the plate over the distance of 2 m from the leading edge.

   Answer: 0.0477 N

7-3. Hydrogen at atmospheric pressure and at 350 K flows over a flat plate with a velocity of 5 m/s. The transition from laminar to turbulent flow is assumed to take place at a Reynolds number of 5×10^5; determine the distance from the leading edge of the plate at which the transition occurs.

7-4. Hydrogen at atmospheric pressure and at 350 K flows over a flat plate with a velocity of 5 m/s. The average drag coefficient c_m over a distance of 2 m from the leading edge is 0.005. Calculate the drag force acting per 1-m width of the plate over a distance of 3 m from the leading edge.

7-5. Air at 1/20 atm and at 345 K has $\rho = 0.0508 \text{ kg/m}^3$, $c_p = 1.009 \text{ kJ/(kg} \cdot {}^\circ\text{C)}$, $k = 0.03 \text{ W/(m} \cdot {}^\circ\text{C)}$, and $\mu = 2.052 \times 10^{-5} \text{ kg/(m} \cdot \text{s)}$. Calculate the Prandtl number.

7-6. Mercury at 205°C has $\rho = 13{,}168 \text{ kg/m}^3$, $c_p = 0.1356 \text{ kJ/(kg} \cdot {}^\circ\text{C)}$, $k = 12.63 \text{ W/(m} \cdot {}^\circ\text{C)}$, and $\mu = 1.005 \times 10^{-3} \text{ kg/(m} \cdot \text{s)}$. Calculate the Prandtl number.

Laminar Flow over a Flat Plate

7-7. Air at atmospheric pressure and at 350 K flows over a flat plate with a velocity of 5 m/s. Calculate the average drag coefficient and the drag force acting per 1-m width of the plate over a distance of 2 m from the leading edge.

7-8. Atmospheric air at 27°C flows along a flat plate with a velocity of $U_\infty = 8$ m/s. The critical Reynolds number for transition from laminar to turbulent flow is $\text{Re}_c = 5 \times 10^5$. (*a*) Determine the distance from the leading edge of the plate at which the transition occurs; (*b*) determine the local drag coefficient at the location where the transition occurs; and (*c*) determine the average drag coefficient over the distance where the flow is laminar.

7-9. Hydrogen at atmospheric pressure and at 80°C flows over a flat plate with a velocity 2 m/s. Determine (*a*) the local drag coefficient at a distance of 2 m from the leading edge, and (*b*) the average drag coefficient over a distance of 2 m from the leading edge.

7-10. Engine oil at 80°C flows over a flat plate with a velocity of 1 m/s. Determine (*a*) the local drag coefficient at a distance of 2 m from the leading edge, and (*b*) the average drag coefficient and the drag force acting per 1-m width of the plate over the distance of 2 m from the leading edge.

7-11. Ethylene at 80°C flows over a flat plate with a velocity of 0.2 m/s. Determine (*a*) the local drag coefficient at a distance of 2 m from the leading edge, and (*b*) the average drag coefficient and the drag force acting per 1-m width of the plate over the distance of 2 m from the leading edge.

7-12. Air at 2 atm pressure and at 350 K flows over a flat plate with a velocity of 1.5 m/s. Determine the drag force acting per 1-m width of the plate over a distance of 0.3 m from the leading edge.

7-13. Air at 2 atm pressure and at 350 K flows over a flat plate with a velocity of 2 m/s. Determine the drag force acting per 1-m width of the plate over a distance of 0.5 m from the leading edge.

7-14. Air at 0.5 atm pressure and at 350 K flows over a flat plate with a velocity of 2 m/s. Determine the drag force acting per 1-m width of the plate over a distance of 0.5 m from the leading edge.

7-15. Determine the drag force exerted on a 2-m-long flat plate per meter width by the flow of the following fluids at atmospheric pressure and at 350 K with a velocity of 5 m/s: (*a*) air, (*b*) hydrogen, and (*c*) helium.
   Answers: (*a*) 0.0477 N, (*b*) 0.0088 N, (*c*) 0.0189 N

7-16. A fluid of 80°C flows with a free-stream velocity of 0.8 m/s along a 5-m-long flat plate. Compute the average drag coefficient and the drag force acting per 1-m width of the plate over a distance of 0.5 m from the leading edge for the following fluids: (*a*) air and (*b*) CO_2.
   Answers: (*a*) 0.0048 N, (*b*) 0.0054 N

7-17. A fluid at 80°C flows with a free-stream velocity of 0.5 m/s along a 5-m-long flat plate. Compute the average drag coefficient and the drag force acting per 1-m width of the plate over a distance of 0.5 m from the leading edge for the following fluids: (*a*) water and (*b*) ethylene glycol.
   Answers: (*a*) 52.8 N/m^2, (*b*) 79.7 N/m^2

7-18. Air at atmospheric pressure and at 40°C flows with a velocity of 5 m/s over a 2-m-long flat plate whose surface is kept at a uniform temperature of 120°C. Determine the average heat transfer coefficient over the 2-m length of the plate.
   Answer: 6.14 W/(m$^2 \cdot$ °C)

7-19. Engine oil at 40°C flows with a velocity of $u_\infty = 1$ m/s, over a 2-m-long flat plate whose surface is maintained at a uniform temperature of 80°C. Determine the average heat transfer coefficient over the 2-m length of the plate.
   Answer: 74.5 W/(m$^2 \cdot$ °C)

7-20. Air at atmospheric pressure and at 40°C flows with a velocity of 5 m/s over a 2-m-long flat plate whose surface is kept at a uniform temperature of 120°C. Determine the rate of heat transfer between the plate and the air per meter width of the plate.
   Answer: 982.4 W/m width

7-21. Air at atmospheric pressure and at 54°C flows with a velocity of 10 m/s over a 1-m-long flat plate maintained at 200°C. Calculate the average drag and heat transfer coefficients over the 1-m length of the plate. Determine the rate of heat transfer between the plate and the air per meter width of the plate.
   Answer: −1790 W

7-22. Mercury at 65°C flows with a velocity of 0.1 m/s over a flat plate maintined at 120°C. Assuming that the transition from laminar to turbulent flow takes place at $Re_c = 5 \times 10^5$, determine the average heat transfer coefficient over the length of the plate where the flow is laminar.

Answer: 2267 W/(m$^2 \cdot$°C)

7-23. Atmospheric air at 24°C flows with a velocity of 4 m/s along a 2-m-long flat plate maintained at a uniform temperature of 130°C.

(*a*) Determine the average heat transfer coefficient over the entire length of the plate.

(*b*) Determine the heat transfer rate from that plate to the air per meter width of the plate.

7-24. Consider the flow of air, hydrogen, and helium at atmospheric pressure and at 77°C with a velocity of 4 m/s along a 2-m-long flat plate. Determine the value of the local heat transfer coefficient at a distance of 1 m from the leading edge of the plate.

Answers: 3.88 W/(m$^2 \cdot$°C), 10.2 W/(m$^2 \cdot$°C), 7.64 W/(m$^2 \cdot$°C)

7-25. Determine the value of the local heat transfer coefficient h_x at a distance of 1 m from the leading edge of a flat plate for the flow of air at 77°C with a velocity of 4 m/s at pressures of 0.5, 1.0, and 2 atmospheres.

Answers: 2.74 W/(m$^2 \cdot$°C), 3.88 W/(m$^2 \cdot$°C), 5.48 W/(m$^2 \cdot$°C)

7-26. Ethylene glycol at 70°C flows over a 0.5-m-long flat plate at 90°C with a velocity of 2 m/s. Calculate the average heat transfer coefficient over the entire length of the plate.

Answer: 654 W/(m$^2 \cdot$°C)

7-27. Helium at atmospheric pressure and at 20°C flows with a velocity of 10 m/s over a $L = 2$ m long flat plate maintained at a uniform temperature of 140°C. Calculate the rate of heat loss from the plate per meter width of the plate.

Answer: 4.1 kW/m width

7-28. Atmospheric air at 20°C flows with a velocity of 2 m/s over the 3 m $\times$ 3 m surface of a wall that absorbs solar energy flux at a rate of 500 W/m^2 and dissipates heat by convection into the air stream. Assuming negligible heat loss at the back surface of the wall, determine the average temperature of the wall under equilibrium conditions.

Answer: 178°C

7-29. Atmospheric air at 27°C flows with a velocity of 4 m/s over a 0.5 m $\times$ 0.5 m flat plate which is uniformly heated with an electric heater at a rate of 2000 W/m^2. Calculate the average temperature of the plate.

Turbulent Flow over a Flat Plate

7-30. Air at atmospheric pressure and at 24°C flows with a velocity of $U_\infty = 10$ m/s along a $L = 4$ m long flat plate which is maintained at a uniform temperature of 130°C. Assuming $R_c = 2 \times 10^5$, calculate (*a*) the local heat transfer coefficient at the locations $x = 2$, 3, and 4 m from the leading edge of the plate; (*b*) the average heat transfer coefficient over the length $L = 4$ m; and (*c*) the heat transfer rate from the plate to the air per meter width of the plate.

7-31. Ethylene glycol at 40°C flows with a free-stream velocity of 8 m/s along a 3-m-long flat plate maintained at a uniform temperature of 100°C. Calculate the average heat transfer coefficient over the entire length of the plate. Assume transition occurs at $Re_c = 2 \times 10^5$.

7-32. Helium at atmospheric pressure and at 300 K flows with a velocity of 30 m/s over a 5-m-long and 1-m-wide plate which is maintained at a uniform temperature of 600 K. Calculate the average heat transfer coefficient and the total heat transfer rate. Assume that transition occurs at $Re_c = 2 \times 10^5$.
Answers: 39.5 W/(m$^2 \cdot$ °C), −59.3 kW

7-33. Ethylene glycol at 300 K flows with a velocity of 15 m/s over a 5-m-long and 1-m-wide plate which is maintained at a uniform temperature of 330 K. Calculate the average heat transfer coefficient and the total heat transfer rate. Assume that transition occurs at $Re_c = 2 \times 10^5$.

7-34. Air at 24°C flows along a 4-m-long flat plate with a velocity of 5 m/s. The plate is maintained at 130°C. Calculate the average heat transfer coefficient over the entire length of the plate and the heat transfer rate per meter width of the plate.
Answers: 9.73 W/(m$^2 \cdot$ °C), −4120 W/m width

7-35. For the plate in Problem 7-34, calculate the average heat transfer coefficient over the entire length of the plate and the heat transfer rate per meter width of the plate if the air velocity is increased at 10 m/s.
Answers: 22.6 W/(m$^2 \cdot$ °C), −9560 N/m width

Flow across a Single Circular Cylinder

7-36. Atmospheric air at 27°C with a velocity of 20 m/s flows across a single tube of outside diameter $D = 2.5$ cm. The surface of the cylinder is maintained at a uniform temperature of 127°C. Determine the average heat transfer coefficient and the heat transfer rate from the tube to the air per meter length of the tube.
Answers: 117 W/(m$^2 \cdot$ °C), 915 W

7-37. Determine the heat transfer rate from the tube to the air per meter length of the tube if the pressure of the air in Problem 7-36 is increased to 2 atm.
Answers: 150 W/(m$^2 \cdot$ °C), 1412 W

7-38. Engine oil at 20°C flows with a velocity of 1 m/s across a 2.5-cm-outside-diameter (OD) tube which is maintained at a uniform temperature of 100°C. Determine the average heat transfer coefficient and the heat transfer rate between the tube surface and the oil per meter length of the tube.

7-39. Water at 20°C with a free-stream velocity of 1.5 m/s flows across a single circular tube with an outside diameter of 2.5 cm. The tube surface is maintained at a uniform temperature of 80°C. Calculate the average heat transfer coefficient and the heat transfer rate per meter length of the tube.

7-40. Atmospheric air at $T_\infty = 300$ K and free-stream velocity of $U_\infty = 30$ m/s flows across a single cylinder with an outside diameter $D = 2.5$ cm. The cylinder surface is at a uniform temperature of $T_w = 400$ K. Calculate the mean heat transfer coefficient h_m and the heat transfer rate per meter length of the cylinder.
Answers: 150 W/(m$^2 \cdot$ °C), 1180 W

Flow across a Single Noncircular Cylinder

7-41. Atmospheric air at 300 K with a free-stream velocity of 30 m/s flows across a square duct 2 cm by 2 cm with its cross section oriented such that one of its lateral surfaces is perpendicular to the direction of flow. The duct's surface is kept at a uniform temperature of 400 K. Calculate the average heat transfer coefficient and the heat transfer rate per meter length of the duct.

7-42. Engine oil at 20°C flows with a velocity of 10 m/s normal to a 5-cm plate which is maintained at a uniform temperature of 100°C. Determine the average heat transfer coefficient and the heat transfer rate between the plate surface and the oil per meter length of the plate.

7-43. Water at 20°C with a free-stream velocity 1.5 m/s flows across a 2.5 cm by 2.5 cm duct which is maintained at a uniform temperature of 80°C and oriented such that one of its lateral surfaces is perpendicular to the direction of flow. Calculate the average heat transfer coefficient and the heat transfer rate per meter length of the tube.

Flow across a Sphere

7-44. Water at 80°C flows with a mass velocity of 50 kg/(m^2 · s) over a 5-cm-diameter sphere whose surface is maintained at a uniform temperature of 140°C. Determine the average heat transfer coefficient and the heat transfer rate from the sphere to the water.

7-45. Ethylene glycol at 40°C flows with a velocity of 2 m/s across a 2.5-cm-diameter sphere. The surface of the sphere is maintained at a uniform temperature of 80°C. Compute the average heat transfer coefficient and the rate of heat transfer from the sphere to the fluid.
Answers: 4110 W/(m^2 · °C), 323 W

7-46. A fluid at 40°C flows with a velocity of 2 m/s across a 2.5-cm-diameter sphere. The surface of the sphere is maintained at a uniform temperature of 100°C. Compute the average heat transfer coefficient for the following fluids: (*a*) CO_2 at 1 atm, and (*b*) water.

7-47. Atmospheric air at 20°C flows with a free-stream velocity of 0.5 m/s over a 2-m-diameter spherical tank which is maintained at 80°C. Compute the average heat transfer coefficient and the heat transfer rate from the sphere to the air.
Answers: 2.19 W/(m^2 · °C), 1650 W

7-48. Water at 20°C flows with a free-stream velocity of 1 m/s over a 2.5-cm-diameter sphere whose surface is maintained at a uniform temperature of 80°C. Determine the average heat transfer coefficient and the rate of heat loss from the sphere to the air.

REFERENCES

1. Churchill, S. W., and M. Bernstein: "A Correlating Equation for Forced Convection from Gases and Liquids to a Circular Cylinder in Cross Flow," *J. Heat Transfer*, **99**:300–306 (1977).
2. Giedt, W. H.: "Investigation of Variation of Point Unit-Heat-Transfer Coefficient around a Cylinder Normal to an Air Stream," *Trans. ASME*, **71**:375–381 (1949).
3. Jakob, Max: *Heat Transfer*, vol. 1, Wiley, New York, 1949.
4. Kreith, F.: *Principles of Heat Transfer*, Intext, New York, 1973.
5. Nakai, S., and T. Okazaki: "Heat Transfer from a Horizontal Circular Wire at Small Reynolds and Grashof Numbers—1. Pure Convection," *Int. J. Heat Mass Transfer*, **18**:387 (1975).
6. Pohlhausen, E.: *Angew. Math. Mech.*, **1**:115 (1921).
7. Schlichting, H.: *Boundary Layer Theory*, 7th ed., McGraw-Hill, New York, 1979.
8. Schutz-Grunow, F.: "Neues Widerstandsgesetz für glatte Platten," *Luftfahrtforschung*, **17**:239 (1940); also NACA *Tech. Memo* 986, 1941.
9. Whitaker, S.: *Elementary Heat Transfer Analysis*, Pergamon, New York, 1976.
10. Whitaker, S.: "Forced Convection Heat Transfer Calculations for Flow in Pipe, past Flat Plates Single Cylinders, and for Flow in Packed Beds and Tube Bundles," *AIChE J.*, **18**:361–371 (1972).

CHAPTER
8

FORCED CONVECTION INSIDE DUCTS

This chapter is devoted to heat transfer and pressure drop in forced convection inside ducts under both laminar and turbulent flow conditions, with emphasis on the understanding of the physical mechanism and the use of various correlations for predicting heat flow and pressure drop. Turbulent pipe flow is widely used in various industrial applications, and the available correlations of the heat transfer coefficient and the friction factor are mostly of an empirical or semiempirical nature. We present such correlations with particular emphasis on their ranges of validity.

Laminar pipe flow is encountered generally in compact heat exchangers, cryogenic coolant systems, the heating or cooling of heavy fluids such as oils, and many other applications. Numerous analytic expressions are available for predicting the friction factor and heat transfer coefficient in laminar tube flow. We present some of these results and discuss their range of validity.

Heat transfer correlations developed for ordinary fluids break down when they are applied to liquid metals, because the Prandtl number for liquid metals is very low. Therefore, heat transfer to liquid metals has been the subject of numerous investigations, but reliable correlations are still rather limited. We discuss some of the available correlations of heat transfer for the flow of liquid metals in tubes.

8-1 BASIC CONCEPTS

The basic concepts, discussed in the previous chapter, on the development of velocity and thermal boundary layers for flow along a flat plate also apply to flow inside ducts at the entrance regions, but some additional features need to be presented.

Hydrodynamic Entry Region

To illustrate the concept of the hydrodynamic entry region and its physical significance in heat transfer and pressure drop problems, we consider the flow entering a circular tube, shown in Fig. 8-1. The fluid has a uniform velocity u_0 at the tube inlet. As the fluid enters the tube, a velocity boundary layer starts to develop along the wall surface. The velocity of the fluid particles at the wall surface becomes zero, and that in the vicinity of the wall is retarded; as a result, the velocity in the central portion of the tube increases to satisfy the requirement of the continuity of flow. The thickness of the velocity boundary layer $\delta(z)$ continuously grows along the tube surface until it fills the entire tube. The region from the tube inlet to a little beyond the hypothetical location where the boundary layer reaches the tube center is called the *hydrodynamic entry region*. In this region the shape of the velocity profile changes in both the axial and radial directions. The region beyond the hydrodynamic entry length is called the *hydrodynamically developed region*, because in this region the velocity profile is invariant with distance along the tube.

If the boundary layer remains laminar until it fills the tube, fully developed laminar flow with a parabolic velocity profile prevails in the hydrodynamically developed region. However, if the boundary layer changes to turbulent before its thickness reaches the tube center, fully developed turbulent flow is experienced in the hydrodynamically developed region. When the flow is turbulent, the velocity profile is flatter than the parabolic velocity profile of laminar flow.

For flow inside a circular tube, the Reynolds number, defined as

$$\mathrm{Re} \equiv \frac{u_m D}{\nu} \tag{8-1}$$

FIGURE 8-1
Concept of hydrodynamic entry region.

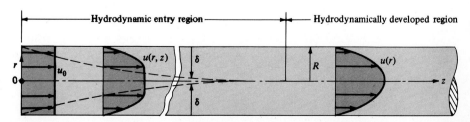

is used as a criterion for change from laminar to turbulent flow. In this definition, u_m is the mean flow velocity, D is the tube inside diameter, and ν is the kinematic viscosity of the fluid. For flow inside a circular tube, the turbulent flow is usually observed for

$$\boxed{\text{Re} = \frac{u_m D}{\nu} > 2300}$$ (8-2)

However, this critical value is strongly dependent on the surface roughness, the inlet conditions, and the fluctuations in the flow. In general, the transition from laminar to turbulent flow occurs in the range $2000 < \text{Re} < 4000$.

Thermal Entry Region

When a fluid enters a tube whose wall is maintained at a uniform temperature different from that of the fluid temperature or is subjected to heating at a uniform rate, heat transfer takes place between the fluid and the tube surface. The resulting temperature distribution in flow inside the tube is such that, starting from the tube inlet, a thermal boundary layer similar to the one for flow over a flat plate can be assumed to develop and grow with the distance along the tube, but it is different from that for flow over a flat plate.

Consider a laminar flow inside a circular tube subjected to uniform heat flux at the wall. Let r and z be the radial and axial coordinates, respectively. A dimensionless temperature $\theta(r, z)$ is defined as

$$\theta(r, z) = \frac{T(r, z) - T_w(z)}{T_m(z) - T_w(z)}$$ (8-3)

where $T_w(z)$ = tube wall temperature

$T_m(z)$ = bulk mean fluid temperature over cross-sectional area of tube at z

$T(r, z)$ = local fluid temperature

Clearly, $\theta(r, z)$ is zero at the tube wall surface and attains some finite value at the tube center. Then, one envisions the development of a thermal boundary layer along the wall surface. The thickness of the thermal boundary layer $\delta_t(z)$ continuously grows along the tube surface until it fills the entire tube. The region from the tube inlet to the hypothetical location where the thermal boundary-layer thickness reaches the tube center is called the *thermal entry region*. In this region, the shape of the dimensionless temperature profile $\theta(r, z)$ changes in both the axial and radial directions.

The region beyond the thermal entry region is called the *thermally developed region*, because in this region the dimensionless temperature profile remains invariant with the distance along the tube; that is,

$$\theta(r) = \frac{T(r, z) - T_w(z)}{T_m(z) - T_w(z)}$$ (8-4)

It is difficult to explain qualitatively why $\theta(r)$ should be independent of the z variable while the temperatures on the right-hand side of Eq. (8-4) depend on both r and z. However, it can be shown mathematically that for either constant temperature or constant heat flux at the wall, the dimensionless temperature $\theta(r)$ depends only on r for sufficiently large values of z.

Fully Developed Region

Consider a fluid at a uniform temperature T_0 and a uniform velocity u_0 entering a tube whose wall is maintained at a constant temperature T_w (i.e., $T_0 \neq T_w$) or subjected to heating at a constant rate. As discussed previously, starting from the inlet, the thermal and velocity boundary layers start to continuously grow along the tube surface until each fills the entire tube. The relative thickness of the velocity and thermal boundary layers during the growth depends on the magnitude of the Prandtl number for the fluid, that is, $\delta > \delta_t$ for $\Pr > 1$, $\delta = \delta_t$ for $\Pr = 1$, and $\delta < \delta_t$ for $\Pr < 1$. Figure 8-2 schematically illustrates, for the case $\Pr > 1$, the hypothetical locations L_h and L_t where δ and δ_t reach the tube center. The distances L_h and L_t are called the *hydrodynamic entrance length* and the *thermal entrance length*, respectively. In the region beyond L_h, the velocity profile does not vary with the axial position along the tube. Similarly, in the region beyond L_t, the dimensionless temperature profile $\theta(r)$ defined by Eq. (8-4) does not vary with the axial position. Then the region beyond L_h or L_t (whichever is greater), where both the velocity and temperature profiles do not vary in the axial direction, is called the *fully developed region* (i.e., the hydrodynamically and thermally developed region).

Bulk Mean Fluid Temperature

When a fluid flowing inside a duct is subjected to heating or cooling as a result of heat transfer between the fluid and the tube wall, the temperature of the

FIGURE 8-2
Concept of hydrodynamic and thermal entrance lengths and the fully developed region.

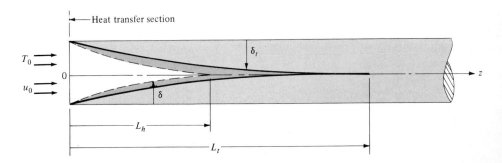

fluid varies in both the radial and axial directions. The average temperature of the fluid over the duct cross section at any axial location z is of practical interest. Such an averaging cannot be performed by taking a simple arithmetic average of the temperature over the cross section. The averaging must be based on the thermal energy transported with the bulk motion of the fluid as it passes through the cross section. If u is the axial velocity, the quantity

$$\rho c_p u T$$

represents energy flux per unit of cross-sectional area of the duct. With this consideration, the *mean fluid temperature* $T_m(z)$ for flow inside a circular tube of inside radius R is given by

$$T_m(z) = \frac{\int_0^R \rho c_p u(r) T(r, z)(2\pi r)\, dr}{\int_0^R \rho c_p u(r)(2\pi r)\, dr} \tag{8-5a}$$

For a constant-property fluid, the ρc_p term cancels out, and Eq. (8-5a) reduces to

$$T_m(z) = \frac{\int_0^R u(r) T(r, z)(2\pi r)\, dr}{\int_0^R u(r)(2\pi r)\, dr} = \frac{\int_0^R u(r) T(r, z)(2\pi r)\, dr}{u_m \pi R^2} \tag{8-5b}$$

In the case of flow between two parallel plates, the appropriate area term should be used in the integration.

Heat Transfer Coefficient

In engineering applications involving fluid flow inside a duct, the concept of *heat transfer coefficient* is frequently used in determining heat transfer between the fluid and the wall surface. Here we present the basic definition of the heat transfer coefficient.

Consider a fluid flowing inside a circular tube of inside radius R. Let $T(r, z)$ be the temperature distribution in the fluid, where r and z are the radial and axial coordinates, respectively. Since the fluid particles next to the tube wall are stationary, heat transfer between the fluid and the wall surface is by conduction. Then the heat flux from the fluid to the wall surface is determined from

$$q(z) = -k \left. \frac{\partial T(r, z)}{\partial r} \right|_{\text{wall}} \tag{8-6}$$

where k is the thermal conductivity of fluid.

In engineering applications, using this expression to determine the heat transfer between the fluid and the tube wall is not practical because it requires the evaluation of the derivative of temperature at the wall. To avoid this difficulty, a local heat transfer coefficient $h(z)$ is defined as

$$q(z) = h(z)[T_m(z) - T_w(z)] \tag{8-7}$$

where $T_m(z)$ = bulk mean fluid temperature over the tube cross-sectional area
at location z, defined previously
$T_w(z)$ = tube wall temperature at z

Equating Eqs. (8-6) and (8-7), we obtain

$$h(z) = - \frac{k}{T_m(z) - T_w(z)} \left. \frac{\partial T(r, z)}{\partial r} \right|_{r=R \text{ wall}} \tag{8-8}$$

Clearly, if the temperature distribution in the flow is known, the heat transfer
coefficient $h(z)$ is determined from the definition given by Eq. (8-8).

Equation (8-8) can be expressed in terms of the dimensionless temperature θ defined by Eq. (8-3) as

$$h(z) = -k \left. \frac{\partial \theta(r, z)}{\partial r} \right|_{r=R \text{ wall}} \tag{8-9}$$

An examination of Eq. (8-9) reveals that if $\theta(r, z)$ varies in the axial direction,
the heat transfer coefficient $h(z)$ also varies with the position along the tube.

In the region sufficiently far away from the tube inlet, the velocity and the
dimensionless temperature profiles do not vary with the axial position. Hence
in the *fully developed* region the dimensionless temperature $\theta(r)$ is independent
of z, and Eq. (8-9) reduces to

$$h = -k \left. \frac{d\theta(r)}{dr} \right|_{r=R \text{ wall}} \tag{8-10}$$

where $\theta(r)$ is defined by Eq. (8-4). This result implies that in the thermally
developed region the heat transfer coefficient does not vary with the distance
along the tube; and it is valid for heat transfer under conditions of constant
wall heat flux or constant wall temperature.

Logarithmic Mean Temperature Difference

Consider the flow of a fluid inside a duct whose walls are maintained at a
uniform temperature T_w. Let the fluid enter the tube at a temperature T_1 and
leave it at a temperature T_2. Figure 8-3 illustrates the axial variation of the
bulk mean temperature $T_m(z)$ along the tube for the case of a hot wall. In the
determination of the total heat transfer rate between the fluid and the wall, an
average temperature difference ΔT between the wall and the fluid is needed.
Let ΔT_1 and ΔT_2 be the inlet and outlet temperature differences between the
fluid and the wall at the inlet and outlet, respectively, as illustrated in Fig. 8-3.
The *logarithmic mean temperature difference* (LMTD) between the wall and

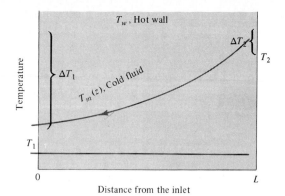

FIGURE 8-3
Nomenclature for the definition
of the Logarithmic Mean
Temperature Difference.

the fluid temperatures over the entire length L of the tube is defined as

$$\Delta T_{\ln} = \frac{\Delta T_1 - \Delta T_2}{\ln(\Delta T_1/\Delta T_2)} \tag{8-11}$$

whereas the arithmetic mean (AM) of ΔT_1 and ΔT_2 is defined as

$$\Delta T_{AM} = \tfrac{1}{2}(\Delta T_1 + \Delta T_2) \tag{8-12}$$

Note that $\Delta T_{\ln}$ is always less then ΔT_{AM}. If the quantity $\Delta T_1/\Delta T_2$ is not greater than 0.5, then $\Delta T_{\ln}$ can be approximated by the arithmetic mean difference within about 1.4 percent.

Friction Factor

In engineering applications the pressure drop ΔP associated with the flow of fluid inside a duct is a quantity of practical interest, because if the pressure drop is known, the pumping power required to get the fluid pumped through the fluid can be determined.

Consider a fluid flowing with a mean velocity u_m through a circular tube of inside diameter D and length L. The pressure drop ΔP over the length L of the tube is related to the *friction factor* f by the following expression:

$$\Delta P = f\,\frac{L}{D}\,\frac{\rho u_m^2}{2} \qquad \text{N/m}^2 \tag{8-13}$$

In the case of flow through a duct which is not circular, the tube diameter D in the above expression is replaced by the *hydraulic diameter* D_h, and we obtain

$$\Delta P = f\,\frac{L}{D_h}\,\frac{\rho u_m^2}{2} \tag{8-14a}$$

where the hydraulic diameter D_h is defined as

$$D_h = \frac{4A_c}{P}$$ (8-14b)

where A_c = the cross-sectional area for flow
P = the wetted perimeter

Clearly, for a circular tube $D_h = D$.

If M is the flow rate through the pipe in cubic meters per second, the pumping power required to get the fluid through the pipe against the pressure drop ΔP becomes

$$\text{Pumping power} = M \Delta P \qquad \text{N} \cdot \text{m/s or W}$$ (8-15)

Dimensionless Parameters

The dimensionless parameters, such as the Reynolds, Prandtl, Nusselt, and Stanton numbers, have been used in the study of forced convection over surfaces. They will also be used for forced convection through ducts. Therefore, it is instructive to discuss the physical significance of these parameters in relation to fluid flow and heat transfer.

We consider the Reynolds number based on a characteristic length L, rearranged in the form

$$\text{Re} = \frac{u_\infty L}{\nu} = \frac{u_\infty^2 / L}{\nu u_\infty / L^2} = \frac{\text{inertia force}}{\text{viscous force}}$$ (8-16a)

Then the Reynolds number represents the ratio of inertia force to viscous force. This result implies that viscous forces are dominant for small Reynolds numbers and inertia forces are dominant for large Reynolds numbers. Recall that the Reynolds number was used as the criterion for determining the change from laminar to turbulent flow. As the Reynolds number is increased, the inertia forces become dominant, and small disturbances in the fluid may be amplified to cause the transition from laminar to turbulent flow.

The Prandtl number can be arranged in the form

$$\text{Pr} = \frac{c_p \mu}{k} = \frac{\mu \rho}{k / (\rho c_p)} = \frac{\nu}{\alpha} = \frac{\text{molecular diffusivity of momentum}}{\text{molecular diffusivity of heat}}$$ (8-16b)

Thus it represents the relative importance of momentum and energy transport by the diffusion process. Hence for gases with $\text{Pr} \cong 1$, the transfer of momentum and energy by the diffusion process is comparable. For oils, $\text{Pr} \gg 1$, and hence the momentum diffusion is much greater than the energy diffusion; but for liquid metals, $\text{Pr} < 1$, and the situation is reversed. We recall that in discussing the development of velocity and thermal boundary layers for flow along a flat plate, the relative thickness of velocity and thermal boundary layers depended on the magnitude of the Prandtl number.

The Nusselt number, based on a characteristic length L, can be rearranged in the form

$$\text{Nu} = \frac{hL}{k} = \frac{h\,\Delta T}{k\,\Delta T/L} \tag{8-16c}$$

where ΔT is the reference temperature difference between the wall surface and the fluid temperature. Then the Nusselt number may be interpreted as the ratio of heat transfer by convection to that conduction across the fluid layer of thickness L. Based on this interpretation, a value of the Nusselt number equal to unity implies that there is no convection—the heat transfer is by pure conduction. A larger value of the Nusselt number implies enhanced heat transfer by convection.

The Stanton number can be rearranged as

$$\text{St} = \frac{h}{\rho c_p u_m} = \frac{h\,\Delta T}{\rho c_p u_m \,\Delta T} \tag{8-16d}$$

where ΔT is a reference temperature difference between the wall surface and the fluid. The numerator represents convection heat flux between the wall and the fluid, and the denominator represents the heat transfer capacity of the fluid flow.

Example 8-1. Engine oil ($\nu = 0.75 \times 10^{-4}\ \text{m}^2/\text{s}$) flows with a mean velocity of $u_m = 0.3\ \text{m/s}$ inside a circular tube that has an inside diameter $D = 2.5\ \text{cm}$. Determine whether the flow is laminar or turbulent.

Solution. We calculate the Reynolds number

$$\text{Re} = \frac{u_m D}{\nu} = \frac{0.3 \times 0.025}{0.75 \times 10^{-4}} = 100$$

Since $\text{Re} < 2300$, the flow is laminar.

Example 8-2. Consider a flow of a fluid in a channel whose walls are maintained at a uniform temperature $T_w = 100°\text{C}$. The fluid is heated from $T_1 = 30°\text{C}$ to $T_2 = 70°\text{C}$ over the length L of the channel. Determine the logarithmic mean temperature difference between the channel wall and the fluid.

Solution. From Eq. (8-11), the LMTD is calculated as

$$\Delta T_{\ln} = \frac{T_2 - T_1}{\ln[T_w - T_1)/(T_w - T_2)]} = \frac{70 - 30}{\ln[(100 - 30)/(100 - 70)]}$$

$$= 47.2°\text{C}$$

Example 8-3. A fluid flows through a square duct 2 cm by 2 cm. Calculate the hydraulic diameter.

Solution. From Eq. (8-14b) the hydraulic diameter of the duct is calculated as

$$D_h = \frac{4A_c}{P} = \frac{4(2 \times 2)}{4 \times 2} = 2\ \text{cm}$$

Example 8-4. Engine oil is pumped through a tube at a flow rate of $M = 2 \times 10^{-4}$ m^3/s. The pressure drop across the tube length L is $\Delta P = 1.5 \times 10^4$ N/m^2. Calculate the power required for pumping the oil through a tube bundle consisting of 500 such tubes.

Solution. The pumping power requirement is determined by Eq. (8-15):

$$W = M \cdot \Delta P = (2 \times 10^{-4})(1.5 \times 10^4)$$

$$= 3 \text{ N} \cdot \text{m/s} \quad (1 \text{ tube})$$

$$= (3 \text{ N} \cdot \text{m/s})\left[\frac{1}{745.7} \text{ (hp} \cdot \text{s)/(N} \cdot \text{m)}\right]$$

$$= 0.004 \text{ hp/tube}$$

Then

$$W(\text{for 500 tubes}) = 0.004 \times 500 = 2 \text{ hp}$$

8-2 FULLY DEVELOPED LAMINAR FLOW

We now present expressions for the determination of the friction factor and the heat transfer coefficient for laminar flow in ducts such as a circular tube, a parallel-plate channel, and many others. The use of these relations in the determination of pressure drop and heat transfer associated with the flow is illustrated with examples.

Flow Inside a Circular Tube

FRICTION FACTOR. Consider an incompressible, constant-property fluid in laminar forced convection inside a circular tube of radius R in the region where flow is hydrodynamically developed. The friction factor for flow inside a circular tube is related to the velocity gradient at the wall by

$$f = -\frac{8\mu}{\rho u_m^2} \frac{du}{dr}\bigg|_{r=R} \tag{8-17}$$

and the fully developed velocity profile for flow inside a circular tube is given by

$$\boxed{\frac{u(r)}{u_m} = 2\left[1 - \left(\frac{r}{R}\right)^2\right]} \tag{8-18a}$$

Then the velocity gradient at the wall surface is determined from this expression:

$$\frac{du(r)}{dr}\bigg|_{r=R} = -\frac{4u_m}{R} = -\frac{8u_m}{D} \tag{8-18b}$$

Introducing Eq. (8-18b) into Eq. (8-17), the following expression is obtained for the friction factor f for laminar flow inside a circular tube:

$$f = \frac{64\mu}{\rho u_m D} = \frac{64}{\text{Re}} \qquad (8\text{-}19a)$$

where D is the tube inside diameter and

$$\text{Re} = \frac{\rho u_m D}{\mu} = \frac{u_m D}{\nu} \qquad (8\text{-}19b)$$

is the Reynolds number.

In the literature, the friction factor also has been defined on the basis of hydraulic radius. If f_r denotes the friction factor based on the hydraulic radius, it is related to the friction factor f defined by Eq. (8-19a) by $f = 4f_r$.

Example 8-5. Compare the mean velocities, friction factors, and pressure drops for fully developed laminar flow at 20°C of Freon [$\mu = 2.65 \times 10^{-4}$ kg/(m·s), $\rho = 1330$ kg/m^3], ethylene glycol [$\mu = 2.24 \times 10^{-2}$ kg/(m·s), $\rho = 1117$ kg/m^3], and engine oil [$\mu = 0.8$ kg/(m·s), $\rho = 888$ kg/m^3] through a 2.5-cm-diameter, 50-m-long tube at a rate of 0.01 kg/s.

Solution. The mean velocities of the fluid are

$$\text{Freon: } u_m = \frac{m}{\rho A} = \frac{0.01}{1330 \times (\pi/4 \times 0.025^2)} = 0.0153 \text{ m/s}$$

$$\text{Ethylene glycol: } u_m = \frac{0.01}{1117 \times (\pi/4 \times 0.025^2)} = 0.0182 \text{ m/s}$$

$$\text{Engine oil: } u_m = \frac{0.01}{888 \times (\pi/4 \times 0.025^2)} = 0.0229 \text{ m/s}$$

The Reynolds number for the flows are

$$\text{Freon: } \text{Re} = \frac{u_m D}{(\mu/\rho)} = \frac{0.0153 \times 0.025}{2.63 \times 10^{-4}/1330} = 1934$$

$$\text{Ethylene glycol: } \text{Re} = \frac{0.0182 \times 0.025}{2.24 \times 10^{-2}/1117} = 22.7$$

$$\text{Engine oil: } \text{Re} = \frac{0.0229 \times 0.025}{0.8/888} = 0.635$$

These Reynolds numbers are less than 2300; therefore, the flows are laminar. The friction factors for laminar flows inside a circular tube, according to Eq. (8-19a), become

$$\text{Freon: } f = \frac{64}{\text{Re}} = \frac{64}{1934} = 0.0331$$

$$\text{Ethylene glycol: } f = \frac{64}{22.7} = 2.82$$

$$\text{Engine oil: } f = \frac{64}{0.635} = 100.8$$

The pressure drops across the tube from Eq. (8-14a) are calculated as

$$\text{Freon: } \Delta P = f \frac{L}{D} \frac{\rho u_m^2}{2} = 0.0331 \frac{100}{0.025} \frac{(1330)(0.0153)^2}{2} = 20.6 \text{ N/m}^2$$

$$\text{Ethylene: } \Delta P = 2.82 \frac{100}{0.025} \frac{(1117)(0.0182)^2}{2} = 2087 \text{ N/m}^2$$

$$\text{Engine oil: } \Delta P = 100.8 \frac{100}{0.025} \frac{(888)(0.0229)^2}{2} = 93,880 \text{ N/m}^2$$

HEAT TRANSFER COEFFICIENT. The heat transfer coefficient for flow inside a circular tube in the thermally developed region is related to the dimensionless temperature gradient at the wall by Eq. (8-10) as

$$h = -k \left. \frac{d\theta(r)}{dr} \right|_{r=R} \qquad (8\text{-}20a)$$

where $\theta(r)$ is defined by Eq. (8-4) as

$$\theta(r) = \frac{T(r, z) - T_w(z)}{T_m(z) - T_w(z)} \qquad (8\text{-}20b)$$

To determine h, the temperature distribution in the flow is needed. We consider below two specific cases: (1) constant heat flux at the wall, and (2) constant temperature at the wall.

1. Constant heat flux at the wall. Consider laminar flow inside a circular tube of radius R that is subjected to a uniform heat flux at the wall. In the region sufficiently away from the inlet where the velocity and temperature profiles are fully developed, the dimensionless temperature distribution $\theta(r)$ in the fluid is given by

$$\theta(r) = \frac{96}{11} \left[\frac{3}{16} + \frac{1}{16} \left(\frac{r}{R} \right)^4 - \frac{1}{4} \left(\frac{r}{R} \right)^2 \right] \qquad (8\text{-}21a)$$

and the derivative of this temperature at the wall $r = R$ becomes

$$\left. \frac{d\theta(r)}{dr} \right|_{r=R} = -\frac{48}{11} \frac{1}{2R} = -\frac{48}{11D} \qquad (8\text{-}21b)$$

Introducing Eq. (8-21b) into Eq. (8-20a), the expression for the heat transfer coefficient h becomes

$$\boxed{h = \frac{48}{11} \frac{k}{D}} \qquad (8\text{-}22a)$$

or

$$\boxed{\text{Nu} \equiv \frac{hD}{k} = \frac{48}{11} = 4.364} \qquad (8\text{-}22b)$$

where D is the tube's inside diameter and Nu is the Nusselt number.

The result given by Eqs. (8-22*a*) and (8-22*b*) represents the heat transfer coefficient for laminar forced convection inside a circular tube in the hydrodynamically and thermally developed region under the constant wall heat flux boundary condition.

2. Constant wall temperature. The heat transfer problem described above for the hydrodynamically and thermally developed region also can be solved under a constant wall temperature boundary condition; the resulting expression for the Nusselt number is

$$\text{Nu} \equiv \frac{hD}{k} = 3.66 \qquad (8\text{-}23)$$

which is valid for laminar forced convection inside a circular tube in the hydrodynamically and thermally developed region under the constant wall temperature boundary condition.

In the results given by Eqs. (8-22) and (8-23), the thermal conductivity of the fluid k depends on temperature. When the fluid temperature varies along the tube, k may be evaluated at the *fluid bulk mean temperature* T_b, defined as

$$T_b = \tfrac{1}{2}(T_i + T_o) \qquad (8\text{-}24)$$

where T_i = bulk fluid temperature at the inlet and T_o = bulk fluid temperature at the outlet.

Flow Inside Ducts of Various Cross Sections

The Nusselt number and the friction factor for laminar flow in ducts of various cross sections have been determined in the region where velocity and temperature profiles are fully developed. For ducts with a noncircular cross section, the heat transfer coefficient and friction factor, for many cases of practical interest, may be based on the hydraulic diameter D_h, defined as

$$D_h = \frac{4A_c}{P} \qquad (8\text{-}25)$$

where A_c = cross-sectional area for flow and P = the wetted perimeter. Then the Nusselt and Reynolds numbers for such cases are defined as

$$\text{Nu} = \frac{hD_h}{k} \qquad (8\text{-}26a)$$

$$\text{Re} = \frac{u_m D_h}{\nu} \qquad (8\text{-}26b)$$

The basis for choosing D_h as in Eq. (8-25) is that for a circular tube D_h becomes the tube diameter D, since $A_c = (\pi/4)D^2$ and $P = \pi D$. The hydraulic diameter D_h for parallel plates is twice the distance between the plates.

In Table 8-1 we present the Nusselt number and the friction factor for laminar flow inside ducts of various cross sections. Here the notation Nu_T and Nu_H refers to the Nusselt number under constant wall temperature and constant wall heat flux, respectively. The friction factor f is given as a product of f and Re. Both the Nusselt and the Reynolds number are based on the hydraulic diameter D_h as specified by Eqs. (8-25) and (8-26).

TABLE 8-1
Nusselt number and friction factor for hydrodynamically and thermally developed laminar flow in ducts of various cross sections

Geometry $(L/D_h > 100)$	Nu_T	Nu_H	f Re
◯	3.657	4.364	64.00
⬡	3.34	4.002	60.22
$2b$ △ $60°$ $\frac{2b}{2a} = \frac{\sqrt{3}}{2}$ $2a$	2.47	3.111	53.33
$2b$ ☐ $\frac{2b}{2a} = 1$ $2a$	2.976	3.608	56.91
$2b$ ▭ $\frac{2b}{2a} = \frac{1}{2}$ $2a$	3.391	4.123	62.20
$2b$ ▭ $\frac{2b}{2a} = \frac{1}{4}$ $2a$	3.66	5.099	74.8
$2b$ ▭ $\frac{2b}{2a} = \frac{1}{8}$ $2a$	5.597	6.490	82.34
$2b$ $\frac{2b}{2a} = 0$	7.541	8.235	96.00
$\frac{b}{a} = 0$ Insulated	4.861	5.385	96.00

* From Shah and London.

Example 8-6. Consider the flow of water with a velocity of 0.0379 m/s through a tube 2 cm in diameter whose wall is maintained at constant temperature. The flow is hydrodynamically and thermally developed. Calculate the heat transfer coefficient by taking the water properties at 50°C.

Solution. The properties of water at 50°C are

$$\nu = 0.568 \times 10^{-6} \, \text{m}^2/\text{s} \qquad k = 0.64 \, \text{W/(m} \cdot \text{°C)}$$

The Reynolds number for the flow is

$$\text{Re} = \frac{u_m D}{\nu} = \frac{0.0379 \times 0.02}{0.568 \times 10^{-6}} = 1335$$

Hence the flow is laminar. For the constant wall temperature at the boundary, the Nusselt number for laminar flow inside a circular tube in the hydrodynamically and thermally developed region is given by Eq. (8-23). Therefore the heat transfer coefficient is determined as

$$Nu = \frac{hD}{k} = 3.66$$

$$h = 3.66 \times \frac{0.64}{0.02} = 117.1 \, \text{W/(m}^2 \cdot \text{°C)}$$

Example 8-7. The wall of the tube in Example 8-6 is heated by electric resistance, thus maintaining a uniform surface heat flux. Calculate the heat transfer coefficient for this boundary condition.

Solution. The Nusselt number for a constant wall heat flux boundary condition for laminar flow inside a circular tube in the hydrodynamically and thermally developed region is given by Eqs. (8-22). Therefore the heat transfer coefficient is determined as

$$h = \frac{48}{11} \frac{k}{D}$$

$$= \frac{48}{11} \frac{0.64}{0.02} = 139.6 \, \text{W/(m}^2 \cdot \text{°C)}$$

Example 8-8. Consider the flow of water with a velocity of 0.0379 m/s through a square duct 2 cm by 2 cm whose walls are maintained at a uniform temperature $T_w = 100$°C. Assume that the flow is hydrodynamically and thermally developed. Determine the duct length required to heat the water from $T_1 = 30$°C to $T_2 = 70$°C.

Solution. The properties of water at the mean bulk temperature $T_b = (30 + 70)/2 = 50$°C are

$$\rho = 990 \, \text{kg/m}^3 \qquad \nu = 0.568 \times 10^{-6} \, \text{m}^2/\text{s}$$

$$k = 0.64 \, \text{W/(m} \cdot \text{°C)} \qquad c_p = 4184.4 \, \text{J(kg} \cdot \text{°C)}$$

The hydraulic diameter of the duct is

$$D_h = \frac{4A}{P} = \frac{4(2 \times 2)}{4 \times 2} = 2 \, \text{cm} = 0.02 \, \text{m}$$

The Reynolds number becomes

$$\text{Re} = \frac{u_m D_h}{\nu} = \frac{0.0379 \times 0.02}{0.568 \times 10^{-6}} = 1335$$

The Nusselt number for a square duct with uniform wall temperature is given in Table 8-1. The heat transfer coefficient is determined as follows:

$$\text{Nu} = 2.976$$

$$h = \text{Nu}\,\frac{k}{D_h} = 2.976\,\frac{0.64}{0.02} = 95.23 \text{ W/(m}^2 \cdot {}^\circ\text{C)}$$

The logarithmic mean temperature difference (LMTD) is

$$\text{LMTD} = \frac{T_2 - T_1}{\ln\left[(T_w - T_1)/(T_w - T_2)\right]} = 47.2^\circ\text{C}$$

Hence

$$Q = hA_s(\text{LMTD})$$

and

$$mc_p(T_2 - T_1) = hA_s(\text{LMTD})$$

where m is the mass flow rate of water and A_s is the surface area of the square duct.

$$m = \rho A u_m$$

$$= (990)(2 \times 2)(0.0379) = 0.015 \text{ kg/s}$$

and

$$(0.015 \text{ kg/s})\left(4181.4\,\frac{\text{J}}{\text{kg} \cdot {}^\circ\text{C}}\right)[(70 - 30)^\circ\text{C}] = [95.23 \text{ W/(m}^2 \cdot {}^\circ\text{C)}](A_s m^2)(47.2^\circ\text{C})$$

gives

$$A_s = 0.5582 \text{ m}^2 = (4 \times 0.02)L$$

and

$$L \cong 7 \text{ m}$$

Hydrodynamic and Thermal Entry Lengths

It is of practical interest to know the hydrodynamic entrance length L_h and the thermal entrance length L_t for flow inside ducts.

The hydrodynamic entrance length L_h is defined, somewhat arbitrarily, as the length from the duct inlet required to achieve a maximum velocity of 99 percent of the corresponding fully developed magnitude.

The thermal entrance length L_t is defined, somewhat arbitrarily, as the length from the beginning of the heat transfer section required to achieve a local Nusselt number Nu_x equal to 1.05 times the corresponding fully developed value.

In Table 8-2 we present the hydrodynamic entrance length L_h for laminar

TABLE 8-2
Hydrodynamic entrance length L_n and thermal entrance length L_t for laminar flow inside ducts

Geometry	$\dfrac{L_h/D_h}{\text{Re}}$	$\dfrac{L_t/D_h}{\text{Pe}}$ Constant wall temperature	$\dfrac{L_t/D_h}{\text{Pe}}$ Constant wall heat flux
circular, D	0.056	0.033	0.043
parallel plates, $2b$	0.011	0.008	0.012
rectangular $2a \times 2b$, $\dfrac{a}{b} = 0.25$	0.075	0.054	0.042
0.50	0.085	0.049	0.057
1.0	0.09	0.041	0.066

flow inside conduits of various cross sections based on the definition discussed previously. Included in this table are the thermal entrance lengths for constant wall temperature and constant wall heat flux boundary conditions for thermally developing, hydrodynamically developed flow. In this table, D_h is the hydraulic diameter, and the Reynolds number is based on the hydraulic diameter.

We note from Table 8-2 that for a given geometry, the hydrodynamic entry length L_h depends on the Reynolds number only, whereas the thermal entry length L_t depends on the Peclet number Pe, which is equal to the product of the Reynolds and Prandtl numbers. Therefore, for liquids that have a Prandtl number of the order of unity, L_h and L_t are of comparable magnitude. For fluids, such as oils, that have a large Prandtl number, $L_t \gg L_h$; and for liquid metals that have a small Prandtl number, $L_t \ll L_h$.

The thermal entry lengths given in Table 8-2 are for hydrodynamically developed, thermally developing laminar flow.

Example 8-9. Determine the hydrodynamic and thermal entrance lengths in terms of the tube inside diameter D for flow at a mean temperature $T_m = 60°C$ and Re = 200 inside a circular tube for mercury, air, water, ethylene glycol, and engine oil, under the constant wall heat flux boundary condition.

Solution. The hydrodynamic entrance length L_h, for laminar flow inside a circular tube, is obtained from Table 8-2 as

$$L_h = 0.056 \text{ Re } D$$

$$= (0.056)(200)D \cong 11D$$

Thus, L_h is approximately 11 diameters from the tube inlet for all the fluids considered here.

The thermal entrance length, given heat transfer under the constant wall heat flux boundary condition, is obtained from Table 8-2 as

$$L_t = 0.043 \text{ Re Pr } D$$

$$= (0.043)(200) \text{ Pr } D = 8.6 \text{ Pr } D$$

Here L_t depends on the Prandtl number, and for the fluids considered in this example it is determined as follows:

Fluid	Pr	L_t/d
Mercury	0.02	0.17
Air	0.7	6
Water	3	26
Ethylene glycol	50	430
Engine oil	1050	9030

We note that for flow at Re = 200, the thermal entrance length varies from a fraction of the tube diameter for mercury to about 9000 diameters for engine oil, while the hydrodynamic entrance length is about 11 diameters for all the fluids considered here.

8-3 DEVELOPING LAMINAR FLOW

The problems of heat transfer for flow inside ducts at the regions near the inlet where the velocity and temperature profiles are developing are of interest in numerous engineering applications. For fluids that have a large Prandtl number, such as oils, the thermal entry lengths are very long, as illustrated in the example in the previous section. Therefore, the heat transfer relations developed previously for the fully developed region are not applicable for the developing region.

In heat transfer problems, the following two different situations need to be taken into consideration:

1. There are situations, as illustrated in Fig. 8-2, in which the heat transfer starts as soon as fluid enters the duct. Then, both the velocity and the thermal boundary layers begin to develop simultaneously, and L_h and L_t are measured from the tube inlet. The region where the velocity and temperature boundary layers grow together is called the *simultaneously developing region*.

2. In some situations, heat transfer to the fluid begins after an isothermal calming section, as illustrated in Fig. 8-4. For such cases, L_h is measured

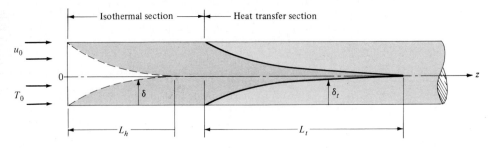

FIGURE 8-4
Hydrodynamically developed, thermal developing region concept.

from the duct inlet, because the velocity boundary layer begins to develop as soon as the fluid enters the duct, but L_t is measured from the location where the heat transfer starts, because the thermal boundary layer begins to develop in the heat transfer section. The region L_t is referred to as the *hydrodynamically developed, thermally developing region.*

In the study of heat transfer coefficients for flow through ducts, a distinction should be made between these two physically different situations. Therefore we examine below the determination of the heat transfer coefficient for these two cases separately.

Thermally Developing, Hydrodynamically Developed Laminar Flow

Consider the flow of a fluid inside a duct, as illustrated in Fig. 8-4, in which there is an isothermal section to allow for velocity development before the fluid enters the heat transfer zone. The physical situation is representative of that for fluids that have a large Prandtl number, such as oils, for which the hydrodynamic entrance length is very small in comparison with the thermal entrance length. A classic solution for laminar forced convection inside a circular tube subject to a uniform wall surface temperature with fully developed velocity profile and developing temperature profile was given by Graetz about 100 years ago. A vast amount of literature now exists on the extension of the Graetz problem for boundary conditions other than the uniform surface temperature and geometries other than a circular tube.

Figure 8-5 shows the local Nusselt number Nu_x and the average Nusselt number Nu_m for laminar flow inside a circular tube plotted against the dimensionless parameter $(x/D)/(\mathrm{Re}\,\mathrm{Pr})$, where x is the axial distance along the conduit measured from the beginning of the heated section. The inverse of this dimensionless parameter is called the Graetz number Gz:

$$(\mathrm{Gz})^{-1} = \frac{x/D}{\mathrm{Re}\,\mathrm{Pr}} \tag{8-27}$$

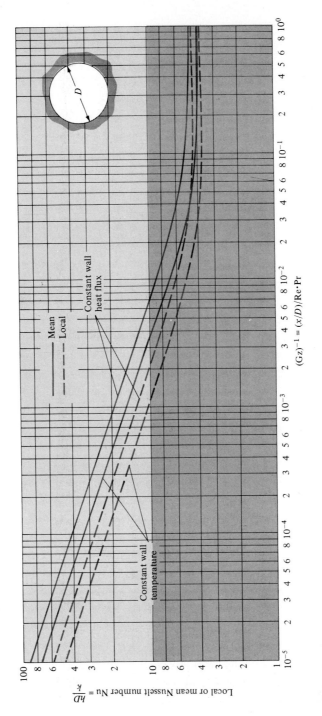

FIGURE 8-5
Mean and local Nusselt number for thermally developing, hydrodynamically developed laminar flow inside a circular tube.

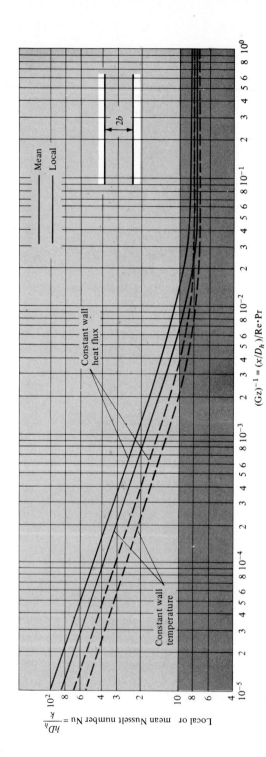

FIGURE 8-6
Mean and local Nusselt numbers for thermally developing, hydrodynamically developed laminar flow between parallel plates.

244

Here D is the inside diameter of the tube, and the Reynolds and Prandtl numbers are defined as

$$\text{Re} = \frac{u_m D}{\nu} \qquad \text{Pr} = \frac{\nu}{\alpha} \qquad (8\text{-}28)$$

In Fig. 8-5, the results are given for both the constant wall temperature and constant wall heat flux boundary conditions. The average Nusselt number represents the average value of the local Nusselt number over a distance from the inlet to the location considered. We note from Fig. 8-5 that at distances sufficiently far away from the tube inlet, the asymptotic values of the Nusselt number for the constant wall temperature and the constant wall heat flux are 3.66 and 4.364, respectively. These results are the same as those shown in Table 8-1 for the fully developed region.

Figure 8-6 gives the local and average Nusselt numbers for thermally developing, hydrodynamically developed laminar flow between parallel plates plotted against the dimensionless parameter $(x/D_h)/(\text{Re Pr})$, where D_h is the hydraulic diameter and x is the distance along the plate measured from the beginning of the heating section in the direction of flow. The Nusselt numbers are given for both constant wall heat flux and constant wall temperature. The asymptotic values of the Nusselt numbers 8.235 and 7.541 for constant wall heat flux and constant wall temperature, respectively, are the same as those given in Table 8-1 for the hydrodynamically and thermally developed region.

Figure 8-7 shows the mean Nusselt numbers for thermally developing,

FIGURE 8-7
Mean and local Nusselt numbers for thermally developing, hydrodynamically developed laminar flow inside a square duct.

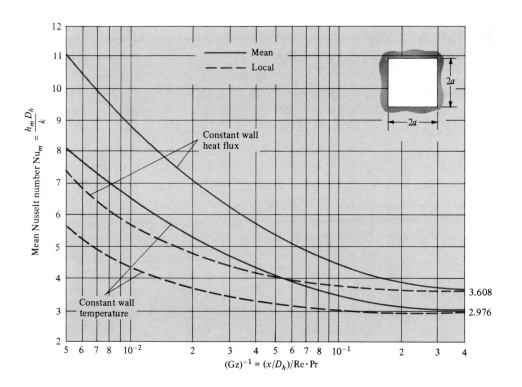

hydrodynamically developed laminar flow inside a square duct plotted against the dimensionless parameter $(x/D_h)/(\text{Re Pr})$. The asymptotic values 3.608 and 2.976 for the constant wall heat flux and constant wall temperature, respectively, are the same as those for Nu_H and Nu_T given in Table 8-1.

Simultaneously Developing Laminar Flow

When heat transfer starts as soon as the fluid enters the duct, as illustrated in Fig. 8-2, the velocity and temperature profiles start developing simultaneously. The analysis of temperature distribution in the flow, and hence of heat transfer between the fluid and the walls, for such situations is more involved because the velocity distribution varies in the axial direction as well as normal to it.

Figure 8-8 shows the mean Nusselt number for simultaneously developing laminar flow inside a circular tube subject to constant wall temperature. The results are given for the Prandtl numbers 0.7, 2, 5 and ∞ and are plotted against the dimensionless parameter $(x/D)/(\text{Re Pr})$. The case for $\text{Pr} = \infty$ corresponds to the thermally developing but hydrodynamically developed flow, discussed earlier. Clearly, the Nusselt number for simultaneously developing flow is higher than that for hydrodynamically developed flow. It is also apparent from this figure that for fluids with a large Prandtl number, the Nusselt number for simultaneously developing flow is very close to that for thermally developing, hydrodynamically developed flow. The asymptotic Nus-

FIGURE 8-8
Mean Nusselt numbers for simultaneously developing laminar flow inside a circular tube subjected to constant wall temperature. (Based on the results from Shah, and Hornbeck).

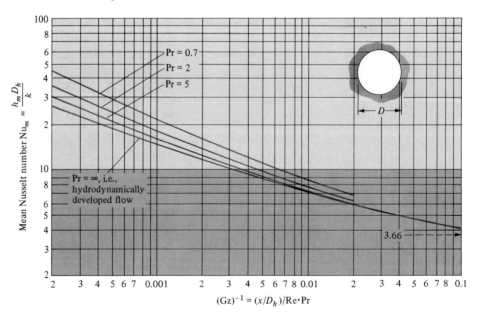

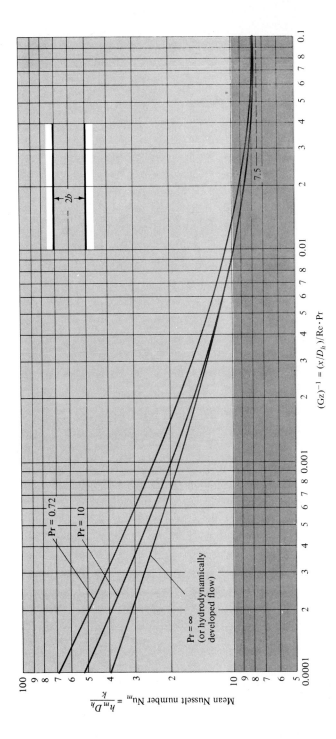

FIGURE 8-9
Mean Nusselt number for simultaneous developing laminar flow between parallel plates subjected to constant and the same temperature at both walls (Based on the results of C. L. Hwang given in Shah).

selt number for all the cases shown in this figure is equal to the fully developed value 3.66.

Figure 8-9 shows the mean Nusselt number for simultaneously developing flow inside a parallel-plate channel subjected to the same constant temperature at both walls. The results are given for the Prandtl numbers 0.72, 10, and ∞ and are plotted against the dimensionless parameter $(x/D_h)(\text{Re Pr})$. Here, the case for $\text{Pr} = \infty$ corresponds to thermally developing, but hydrodynamically developed flow.

Empirical Correlations

So far we have discussed heat transfer results based on theoretical predictions obtained from the solution of governing equations of motion and energy. In parallel to such studies, empirical relations for predicting the mean Nusselt number for laminar flow in the entrance region of a circular tube also have been developed. One such correlation for the mean Nusselt number for laminar flow in a circular tube at constant wall temperature is given by Hausen as

$$\text{Nu}_m = 3.66 + \frac{0.0668 \, \text{Gz}}{1 + 0.04(\text{Gz})^{2/3}} \qquad (8\text{-}29)$$

where

$$\text{Nu}_m = \frac{h_m D}{k} \qquad (8\text{-}30a)$$

$$\text{Gz} = \frac{\text{Re Pr}}{L/D} \qquad L = \text{distance from the inlet} \qquad (8\text{-}30b)$$

$$\text{Re} = \frac{u_m D}{\nu} \qquad (8\text{-}30c)$$

This relation is recommended for $\text{Gz} < 100$ (see Table 8-3), and all properties

TABLE 8-3
A comparison of empirical and theoretical correlations for mean Nusselt number for simultaneously developing laminar flow inside a circular tube

$\text{Gz}^{-1} = \dfrac{x/D}{\text{Re Pr}}$	Nu_m (Hausen), Eq. (8-29)	$\text{Nu}_m\left(\dfrac{\mu_w}{\mu_b}\right)^{0.14}$ (Sieder-Tate), Eq. (8-31)	Nu_m from Fig. 8-8		
			$\text{Pr} = 0.7$	$\text{Pr} = 5$	$\text{Pr} = \infty$
0.1	4.22	4.0			4.16
0.02	5.82	6.85	6.8	5.8	5.8
0.01	7.25	8.63	8.7	7.2	7.2
0.001	17.0	18.60	22.2	16.9	15.4
0.001	30.0	31.80	44.1	30.3	26.7

are evaluated at the fluid bulk mean temperature. Clearly, as the length L increases, the Nusselt number approaches the asymptotic value 3.66.

A rather simple empirical correlation has been proposed by Sieder and Tate to predict the mean Nusselt number for laminar flow in a circular tube at constant wall temperature:

$$\mathrm{Nu}_m = 1.86(\mathrm{Gz})^{1/3}\left(\frac{\mu_b}{\mu_w}\right)^{0.14} \tag{8-31}$$

And it is recommended for

$$0.48 < \mathrm{Pr} < 16{,}700 \tag{8-32a}$$

$$0.0044 < \frac{\mu_b}{\mu_w} < 9.75 \tag{8-32b}$$

$$(\mathrm{Gz})^{1/2}\left(\frac{\mu_b}{\mu_w}\right)^{0.14} > 2 \tag{8-32c}$$

All physical properties are evaluated at the fluid bulk mean temperature, except for μ_w, which is evaluated at the wall temperature. The Graetz number Gz is defined by Eq. (8-30b). The last restriction implies that this equation cannot be used for extremely long tubes, because, as $\mathrm{Gz} \to 0$ with increasing length, the equation has no provision to yield the correct asymptotic value, as is the case for the Hausen equation.

In Table 8-3, we compare the mean Nusselt numbers calculated from Eqs. (8-29) and (8-31) and obtained from Fig. 8-8. The Sieder and Tate equation underestimates Nu_m for $\mathrm{Gz} > 10$, which is consistent with the restriction of Eq. (8-32c). The results given in Table 8-3 show that the theoretical prediction of the Nusselt number for simultaneously developing flow given in Fig. 8-8 is reasonably accurate and includes the effects of the Prandtl number.

Example 8-10. Engine oil at 40°C with a flow rate of 0.3 kg/s enters a 2.5-cm-ID, 40-m-long tube maintained at a uniform temperature of 100°C. The flow is hydrodynamically developed. Determine (a) the mean heat transfer coefficient, and (b) the outlet temperature of the oil.

Solution. The Reynolds number should be determined to establish whether the flow is laminar or turbulent. The mean fluid temperature inside the tube cannot be calculated yet because the fluid exit temperature is not known. Therefore, we start by evaluating the physical properties of the fluid at an inlet temperature of, say, 40°C. (This assumption will be checked at the end of calculations.) We have

$$\rho = 876 \text{ kg/m}^3 \qquad c_p = 1964 \text{ J/(kg} \cdot °\text{C)}$$

$$\nu = 2.4 \times 10^{-4} \text{ m}^2/\text{s} \qquad h = 0.144 \text{ W/(m} \cdot °\text{C)} \qquad \mathrm{Pr} = 2870$$

Then

$$u_m = \frac{m}{\rho A} = \frac{0.3}{876(\pi/4)(0.025)^2} = 0.7 \text{ m/s}$$

$$\mathrm{Re} = \frac{u_m D}{\nu} = \frac{0.7(0.025)}{2.4 \times 10^{-4}} = 72.7$$

and the flow is laminar. For fluids with a high Prandtl number (e.g., the oils), the hydrodynamic entrance length is short compared with the thermodynamic entrance length. Therefore, the flow can be regarded as thermally developing and hydrodynamically developed.

We calculate the parameter

$$\frac{L/D}{\mathrm{Re\ Pr}} = \frac{40/0.025}{72.7 \times 2870} = 7.7 \times 10^{-3}$$

From Fig. 8-5, the mean Nusselt number for constant wall temperature is determined as

$$\mathrm{Nu}_m = \frac{h_m D}{k} \cong 7.5$$

$$h_m = 7.5 \frac{k}{D} = 7.5 \times \frac{0.144}{0.025} = 43.2 \ \mathrm{W/(m^2 \cdot {}^\circ C)}$$

To calculate the outlet temperature T_{out}, we consider an overall energy balance for the length L of the tube, as

Heat supplied to fluid from wall = energy removed by fluid by convection

$$h_m(\pi DL)\,\Delta T_m = \left(\frac{\pi}{4}\,D^2\right)(mc_p)(T_{\mathrm{out}} - T_{\mathrm{in}})$$

where ΔT_m is the logarithmic mean temperature difference. Introducing ΔT_m, we have

$$h_m(\pi DL)\,\frac{T_{\mathrm{out}} - T_{\mathrm{in}}}{\ln\left[(T_w - T_{\mathrm{in}})/(T_w - T_{\mathrm{out}})\right]} = \left(\frac{\pi}{4}\,D^2\right)(mc_p)(T_{\mathrm{out}} - T_{\mathrm{in}})$$

or

$$\ln \frac{T_w - T_{\mathrm{in}}}{T_w - T_{\mathrm{out}}} = \frac{(\pi DL)h_m}{mc_p}$$

$$\frac{T_w - T_{\mathrm{out}}}{T_w - T_{\mathrm{in}}} = \exp\left(-\frac{(\pi DL)h_m}{mc_p}\right)$$

$$\frac{100 - T_{\mathrm{out}}}{100 - 40} = \exp\left(-\frac{\pi \times 0.025 \times 40 \times 43.2}{0.3 \times 1964}\right) = 0.7943$$

$$T_{\mathrm{out}} = 52.34{}^\circ C$$

The results can be improved by repeating the above calculations for physical properties evaluated at the bulk fluid temperature $(T_{\mathrm{in}} + T_{\mathrm{out}})/2 = (40 + 52.34)/2 = 46.17{}^\circ C$.

Example 8-11. Water at an average temperature of 60°C with a flow rate of 0.015 kg/s flows through a 2.5-cm-ID tube which is maintained at a uniform temperature.

(*a*) Assuming a fully developed velocity profile, determine the local heat transfer coefficients at locations 1, 2, and 3 m from the inlet.

(*b*) Assuming simultaneously developing velocity and temperature, determine the mean heat transfer coefficients over distances of 1, 2, and 3 m from the inlet.

Solution. Various properties for water at $T_b = 60°C$ are taken as

$$\rho = 985.5 \text{ kg/m}^3 \qquad \mu_b = 4.71 \times 10^{-4} \text{ kg/(m} \cdot \text{s)} \qquad k = 0.651 \text{ W/(m} \cdot °C)$$

$$\text{Pr} = 3.02$$

Then

$$u_m = \frac{m}{\rho A} = \frac{0.015}{(985.5)(\pi/4)(0.025)^2} = 0.031 \text{ m/s}$$

and the Reynolds number becomes

$$\text{Re} = \frac{\rho u_m D}{\mu} = \frac{(985.5)(0.031)(0.025)}{4.71 \times 10^{-4}} = 1621$$

hence, the flow is laminar.

(*a*) For the hydrodynamically developed flow, we obtain the Nusselt number from Fig. 8-5. We have

$$\text{Gz}^{-1} = \frac{x/D}{\text{Re Pr}} = \frac{x/0.025}{1621 \times 3.02} = 8.2 \times 10^{-3} \times x$$

For the given tube lengths the Gz number can be calculated and the local heat transfer coefficient $h(x)$ can be determined from

$$h(x) = \frac{k}{D} \text{Nu}_x = \frac{0.651}{0.025} \text{Nu}_x = 26 \text{Nu}_x$$

x, m	Gz^{-1}	NU_x (From Fig. 8-8)	$h(x)$, W/(m$^2 \cdot$ °C)
1	8.2×10^{-3}	5.0	130
2	1.64×10^{-2}	4.4	114
3	2.46×10^{-2}	4.0	104

(*b*) For the simultaneously developing flow, we use Fig. 8-8.

x, m	Gz^{-1}	Nu_x (From Fig. 8-8)	$h(x)$, W/(m$^2 \cdot$ °C)
1	8.2×10^{-3}	7.8	203
2	1.64×10^{-2}	6.2	162
3	2.46×10^{-2}	5.6	146

8-4 FULLY DEVELOPED TURBULENT FLOW

Turbulent flow is important in engineering applications because it is involved in the vast majority of fluid flow and heat transfer problems encountered in engineering practice. Here we show how to determine the friction factor and heat transfer in turbulent flow inside conduits.

Friction Factor

For fully developed turbulent flow inside smooth pipes, an approximate analytic expression for the friction factor f is given in the form

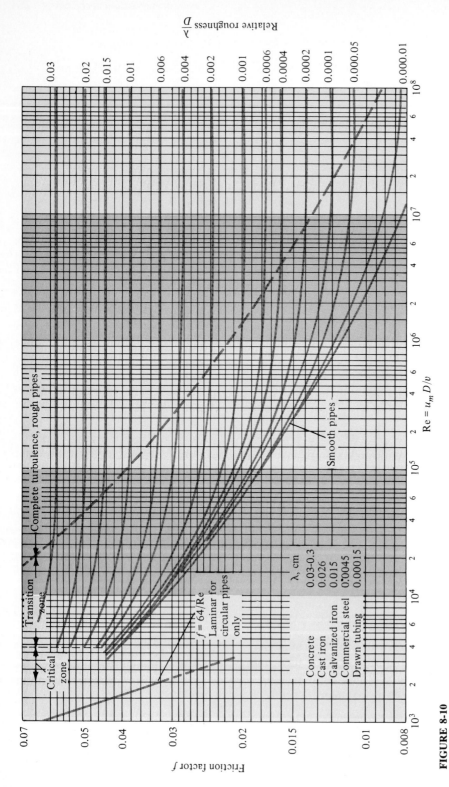

FIGURE 8-10
Friction factor for use in the relation $\Delta P = f(L/D)(\rho U_m^2/2)$ for pressure drop inside circular pipes (From Moody).

$$f = (1.82 \log \text{Re} - 1.64)^{-2} \tag{8-33}$$

The hydrodynamic development for turbulent flow occurs for x/D much shorter than that for laminar flow. For example, hydrodynamically developed flow conditions occur for x/D greater than about 10 to 20.

Somewhat simpler approximate expressions for f are also given in the form

$$f = 0.316 \, \text{Re}^{-0.25} \qquad \text{for } \text{Re} < 2 \times 10^4 \tag{8-34a}$$

$$f = 0.184 \, \text{Re}^{-0.2} \qquad \text{for } 2 \times 10^4 < \text{Re} < 3 \times 10^5 \tag{8-34b}$$

In laminar flow the surface roughness does not affect the friction factor, but in turbulent flow the friction factor is significantly affected by the surface roughness.

Nikuradse made extensive experiments with turbulent flow inside artificially roughened pipes over a wide range of relative roughnesses λ/D (i.e., protrusion height to diameter ratio), from about $\frac{1}{1000}$ to $\frac{1}{30}$. The sand-grain roughness used in these experiments has been adopted as a standard for the effects of roughness. A friction factor correlation has also been developed for turbulent flow inside rough pipes based on experiments performed with rough pipes.

Figure 8-10 shows the friction factor chart, originally presented by Moody, for turbulent flow inside smooth and rough pipes. Included on this figure is the friction factor $f = 64/\text{Re}$ for laminar flow inside circular pipes. It is apparent that for laminar flow the surface roughness has no effect on the friction factor; for turbulent flow, however, the friction factor is at a minimum for a smooth pipe. The laminar flow is confined to the region $\text{Re} < 2000$. The transitional turbulence occurs in the region $2000 < \text{Re} < 10{,}000$. The fully turbulent flow occurs in the region $\text{Re} > 10^4$.

Heat Transfer Coefficient

The analysis of heat transfer for turbulent flow is much more involved than that for laminar flow. Therefore, a large number of empirical correlations have been developed to determine the heat transfer coefficient.

DITTUS-BOELTER EQUATION. A relatively simple expression for the Nusselt number is given by Dittus and Boelter in the form

$$\text{Nu} = 0.023 \, \text{Re}^{0.8} \, \text{Pr}^n \tag{8-35}$$

where $n = 0.4$ for heating $(T_w > T_b)$ and $n = 0.3$ for cooling $(T_w < T_b)$ of the fluid, and $\mathrm{Nu} = hD/k$, $\mathrm{Re} = u_m D/\nu$, and $\mathrm{Pr} = \nu/\alpha$. Equation (8.35) is applicable for smooth pipes:

$$0.7 < \mathrm{Pr} < 160$$

$$\mathrm{Re} > 10{,}000$$

$$\frac{L}{D} > 60$$

and for small to moderate temperature differences. Fluid properties are evaluated at the bulk mean temperature T_b.

SIEDER AND TATE EQUATION. For situations involving a large property variation, the Sieder and Tate equation is recommended:

$$\mathrm{Nu} = 0.027\,\mathrm{Re}^{0.8}\,\mathrm{Pr}^{1/3}\left(\frac{\mu_b}{\mu_w}\right)^{0.14} \tag{8-36}$$

This equation is applicable for smooth pipes:

$$0.7 < \mathrm{Pr} < 16{,}700$$

$$\mathrm{Re} > 10{,}000$$

$$\frac{L}{D} > 60$$

All properties are evaluated at the bulk mean temperature T_b, except for μ_w, which is evaluated at the wall temperature.

PETUKHOV EQUATION. The previous relations are relatively simple, but they give maximum errors of ± 25 percent in the range of $0.67 < \mathrm{Pr} < 100$ and apply to turbulent flow in smooth ducts. A more accurate correlation, which is also applicable for rough ducts, has been developed by Petukhov and coworkers at the Moscow Institute for High Temperature:

$$\mathrm{Nu} = \frac{\mathrm{Re}\,\mathrm{Pr}}{X}\left(\frac{f}{8}\right)\left(\frac{\mu_b}{\mu_w}\right)^n \tag{8-37a}$$

where

$$X = 1.07 + 12.7(\mathrm{Pr}^{2/3} - 1)\left(\frac{f}{8}\right)^{1/2} \tag{8-37b}$$

for liquids

$$n = \begin{cases} 0.11 & \text{heating } (T_w > T_b) \\ 0.25 & \text{cooling } (T_w < T_b) \end{cases}$$

and for gases, although $n = 0$, the property variation will be explained later. Equations (8-37a) and (8-37b) are applicable for

$$10^4 < \text{Re} < 5 \times 10^6$$

$$2 < \text{Pr} < 140 \qquad \text{with 5 to 6 percent error}$$

$$0.5 < \text{Pr} < 2000 \qquad \text{with 10 percent error}$$

$$0.08 < \frac{\mu_w}{\mu_b} < 40$$

We note that $\mu_w/\mu_b < 1$ when a liquid is heated and $\mu_w/\mu_b > 1$ when the liquid is cooled. All physical properties, except for μ_w, are evaluated at the bulk temperature. The friction factor f can be evaluated from Eq. (8-33) for smooth pipes, or obtained from the Moody chart (Fig. 8-10) for both smooth and rough tubes.

Noncircular Ducts

So far, discussion of the friction factor and heat transfer coefficient for turbulent flow has been restricted to flow inside circular tubes. Numerous engineering applications involve turbulent forced convection inside ducts of noncircular cross section. The friction factor for a circular tube given by the Moody chart (Fig. 8-10) applies to turbulent flow inside noncircular ducts if the tube diameter D is replaced by the hydraulic diameter of the noncircular duct, defined by Eq. (8-25); that is,

$$D_h = \frac{4A_c}{P} \tag{8-38}$$

where A_c is the cross-sectional area for flow and P is the wetted perimeter.

For noncircular ducts, the turbulent flow also occurs for $\text{Re} > 2300$, where the Reynolds number is based on the hydraulic diameter.

With noncircular ducts, the heat transfer coefficient varies around the perimeter and approaches zero near the sharp corners. Therefore, for certain situations, difficulties may arise in applying the circular-tube results to a noncircular duct by using the hydraulic diameter concept.

Effects of Surface Roughness

The heat transfer coefficient for turbulent flow in rough-walled tubes is higher than that for smooth-walled tubes because roughness disturbs the viscous sublayer. The increased heat transfer due to roughness is achieved at the expense of increased friction to fluid flow. The correlation of heat transfer for turbulent flow in rough-walled tubes is very sparse in the literature. The Petukhov equation (8-37) can be recommended for predicting the heat transfer coefficient in hydrodynamically and thermally developed turbulent flow in rough pipes, because the friction factor f can be obtained from the Moody chart (Fig. 8-10) once the relative roughness of the pipe is known.

Effects of Property Variation

When heat transfer to or from a fluid flowing inside a duct takes place, the temperature varies over the flow cross section of the duct. For most liquids, although the specific heat and thermal conductivity are rather insensitive to temperature, the viscosity decreases significantly with temperature. For gases, the viscosity and thermal conductivity increase by approximately 0.8 power of the temperature. Therefore, the property variation affects both the heat transfer coefficient and the friction factor.

To compensate for the effects of nonisothermal conditions in the fluid, the Sieder and Tate equation (8-36) and the Petukhov equation (8-37) include a viscosity correction term in the form $(\mu_b/\mu_w)^n$.

For liquids, the variation of viscosity is responsible for the property effects. Therefore, viscosity corrections of the following power-law form are found to be sufficiently good approximations:

$$\frac{Nu}{Nu_{iso}} = \left(\frac{\mu_b}{\mu_w}\right)^n \qquad (8\text{-}39a)$$

$$\frac{f}{f_{iso}} = \left(\frac{\mu_b}{\mu_w}\right)^k \qquad (8\text{-}39b)$$

where

$$\mu_b = \text{viscosity evaluated at bulk mean temperature}$$
$$\mu_w = \text{viscosity evaluated at wall temperature}$$
$$Nu_{iso}, Nu = \text{Nusselt number under isothermal and nonisothermal}$$
conditions, respectively
$$f_{iso}, f = \text{friction factor under isothermal and nonisothermal}$$
conditions, respectively

In the case of gases, the viscosity, thermal conductivity, and density depend on the absolute temperature. Therefore, temperature corrections of the following form are found to be adequate for most practical applications:

$$\frac{Nu}{Nu_{iso}} = \left(\frac{T_b}{T_w}\right)^n \qquad (8\text{-}40a)$$

$$\frac{f}{f_{iso}} = \left(\frac{T_b}{T_w}\right)^k \qquad (8\text{-}40b)$$

where T_b and T_w are the absolute bulk mean and wall temperatures, respectively.

A number of experimental investigations and variable-property analyses to determine the values of the exponents n and k appearing in Eqs. (8-39) and (8-40) have been reported in the literature. In Table 8-4 we present recommended values of these exponents. Thus, by using the corrections given by Eqs. (8-39) and (8-40), we can adjust the Nusselt number and the friction factor for ideal isothermal conditions for the effects of property variations, if no viscosity correction is included in the equation.

TABLE 8-4
The exponents n and m associated with Eqs. (8-39) and (8-40)

Type of flow	Fluid	$T_w = $ constant condition	n	k
Laminar	Liquid	Cooling or heating	0.14	
	Gas	Cooling or heating	0	-1
Turbulent	Liquid	Cooling	0.25	
	Liquid	Heating	0.11	
	Liquid	Cooling or heating		-0.25
	Gas	Cooling	0	0.1
	Gas	Heating	0.5	0.1

Example 8-12. Water flows with a mean velocity of $u_m = 2$ m/s inside a circular smooth pipe of inside diameter $D = 5$ cm. The pipe is of commercial steel, and its wall is maintained at a uniform temperature $T_w = 100°C$ by condensing steam on its outer surface. At a location where the fluid is hydrodynamically and thermally developed, the bulk mean temperature of the water is $T_b = 60°C$. Calculate the heat transfer coefficient h by using

   (a) The Dittus and Boelter equation
   (b) The Sieder and Tate equation
   (c) The Petukhov equation

Solution. The physical properties at $T_b = 60°C$ are taken as

$$k = 0.651 \text{ W/(m} \cdot °C) \qquad \text{Pr} = 3.02 \qquad \text{Re} = 2.04 \times 10^5$$

$$\mu_b = 4.71 \times 10^{-4} \text{ kg/(m} \cdot \text{s)} \qquad \mu_w = 2.82 \times 10^{-4} \text{ kg/(m} \cdot \text{s)}$$

The friction factor for smooth pipe at $\text{Re} = 2.04 \times 10^5$ is obtained from Fig. 8-10 as

$$f = 0.0152$$

(a) The Dittus and Boelter equation gives

$$\text{Nu} = 0.023(2.04 \times 10^5)^{0.8}(3.02)^{0.4}$$

$$= 633$$

$$h = 633 \frac{0.651}{0.05} = 8242 \text{ W/(m}^2 \cdot °C)$$

(b) The Sider and Tate equation gives

$$\text{Nu} = 0.027(2.04 \times 10^5)^{0.8}(3.02)^{1/3}\left(\frac{4.71}{2.82}\right)^{0.14}$$

Then

$$\text{Nu} = 704$$

$$h = 704 \frac{0.651}{0.05} = 9166 \text{ W/(m}^2 \cdot °C)$$

(*c*) The Petukhov equation gives

$$Nu = \frac{(2.04 \times 10^5)(3.02)}{X} \left(\frac{0.0152}{8}\right)\left(\frac{4.71}{2.82}\right)^{0.11}$$

where

$$X = 1.07 + 12.7(3.02^{2/3} - 1)\left(\frac{0.0152}{8}\right)^{1/2}$$

Then

$$Nu = 741.65$$

$$h = 741.65 \frac{0.651}{0.05} = 9656 \ W/(m^2 \cdot {}^\circ C)$$

8-5   LIQUID METALS IN TURBULENT FLOW

The liquid metals are characterized by their very low Prandtl numbers, which vary from about 0.02 to 0.003. Therefore, the heat transfer correlations in previous sections do not apply to liquid metals: their range of validity does not extend to such low values of the Prandtl number.

Lithium, sodium, potassium, bismuth, and sodium–potassium are among the common low-melting metals that are suitable for heat transfer purposes as liquid metals. There has been interest in liquid-metal heat transfer in engineering applications because large amounts of heat can be transferred at high temperatures with a relatively low temperature difference between the fluid and the tube wall surface. The high heat transfer rates result from the high thermal conductivity of liquid metals compared with that of ordinary liquids and gases. Therefore, they are particularly attractive as heat transfer media in nuclear reactors and many other high-temperature, high-heat-flux applications. The major difficulty in their use lies in handling them. They are corrosive, and some may cause violent reactions when they come into contact with water or air.

As discussed previously, when $Pr \ll 1$, as in liquid metals, the thermal boundary layer is much thicker than the velocity boundary layer. This implies that the temperature profile, and hence the heat transfer for liquid metals, is not influenced by the velocity sublayer or viscosity. So in such cases one expects rather weak dependence of heat transfer on the Prandtl number. Thus most empirical correlations of liquid-metal heat transfer have been made by plotting the Nusselt number against the Peclet number, $Pe = Re \, Pr$.

We summarize now some empirical and theoretical correlations for heat transfer to liquid metals in fully developed turbulent flow inside a circular tube under the uniform wall heat flux and uniform wall temperature boundary conditions.

Uniform Wall Heat Flux

Lubarsky and Kaufman proposed the following empirical relation for calculating the Nusselt number in fully developed turbulent flow of liquid metals in smooth pipes:

$$\boxed{\mathrm{Nu} = 0.625\,\mathrm{Pe}^{0.4}} \qquad (8\text{-}41)$$

where

$$\text{Peclet number} \equiv \mathrm{Pe} = \mathrm{Re}\,\mathrm{Pr}$$

for $10^2 < \mathrm{Pe} < 10^4$, $L/D > 60$, and properties evaluated at the bulk mean fluid temperature.

Skupinski, Tortel, and Vautrey, basing their heat transfer experiments on sodium-potassium mixtures, recommended the following expression for liquid metals in fully developed turbulent flow in smooth pipes:

$$\boxed{\mathrm{Nu} = 4.82 + 0.0185\,\mathrm{Pe}^{0.827}} \qquad (8\text{-}42)$$

for $3.6 \times 10^3 < \mathrm{Re} < 9.05 \times 10^5$, $10^2 < \mathrm{Pe} < 10^4$, and $L/D > 60$. The physical properties are evaluated at the bulk mean fluid temperature.

Equation (8-41) predicts a lower Nusselt number than Eq. (8-42); therefore it is on the conservative side.

Uniform Wall Temperature

Expressions for the Nusselt number for fully developed turbulent flow of liquid metals in smooth pipes subject to uniform wall temperature boundary conditions have also been developed by empirical fits to the results of the theoretical solutions. We present now the results of such solutions.

Azer and Chao:

$$\boxed{\mathrm{Nu} = 5.0 + 0.05\,\mathrm{Pe}^{0.77}\,\mathrm{Pr}^{0.25}} \qquad \text{for } \mathrm{Pr} < 0.1,\ \mathrm{Pe} < 15{,}000 \qquad (8\text{-}43)$$

Notter and sleicher:

$$\boxed{\mathrm{Nu} = 4.8 + 0.0156\,\mathrm{Pe}^{0.85}\,\mathrm{Pr}^{0.08}} \qquad \text{for } 0.004 < \mathrm{Pr} < 0.1,\ \mathrm{Re} < 500{,}000 \qquad (8\text{-}44)$$

In calculating Nu, Pe, and Pr in these expressions, the physical properties are evaluated at the bulk mean fluid temperature; the expressions are applicable for $L/D > 60$.

Thermal Entry Region

The previous relations for liquid metals in turbulent flow are applicable in the fully developed region. Sleicher, Awad, and Notter examined the heat transfer calculations of Notter and Sleicher in the thermal entry region for both uniform wall heat flux and uniform wall temperature. They noted that the local Nusselt number for the thermal entrance region can be correlated within 20 percent with

$$\boxed{\text{Nu}_x = \text{Nu}\left(1 + \frac{2}{x/D}\right)} \qquad \text{for } \frac{x}{D} > 4 \qquad (8\text{-}45)$$

where

$$\text{Nu} = 6.3 + 0.0167\,\text{Pe}^{0.85}\,\text{Pr}^{0.08} \qquad \text{for uniform wall heat flux} \qquad (8\text{-}46)$$

$$\text{Nu} = 4.8 + 0.0156\,\text{Pe}^{0.85}\,\text{Pr}^{0.08} \qquad \text{for uniform wall temperature} \qquad (8\text{-}47)$$

and applies in the range $0.004 < \text{Pr} < 0.1$.

Example 8-13. Liquid sodium at 180°C with a mass flow rate of 3 kg/s enters a 2.5-cm-ID tube whose wall is maintained at a uniform temperature of 240°C. Calculate the tube length required to heat sodium to 230°C.

Solution. The physical properties of sodium at the bulk mean temperature T_b,

$$T_b = \frac{180 + 230}{2} = 205°C$$

are taken as

$$\rho = 907.5 \text{ kg/m}^3 \qquad c_p = 1339 \text{ J/(kg} \cdot °\text{C)}$$

$$\nu = 0.501 \times 10^{-6} \text{ m}^2/\text{s} \qquad k = 80.81 \text{ W/(m} \cdot °\text{C)} \qquad \text{Pr} = 0.0075$$

Then

$$u_m = \frac{m}{\rho A} = \frac{3}{(907.5)(\pi/4)(0.025)^2} = 6.73 \text{ m/s}$$

and

$$\text{Re} = \frac{u_m D}{\nu} = \frac{6.73 \times 0.025}{0.501 \times 10^{-6}} = 3.36 \times 10^5$$

$$\text{Pe} = \text{Re Pr} = (3.36 \times 10^5)(0.0075) = 2520$$

Using Eq. (8-44), we find

$$\text{Nu} = 4.8 + 0.0156\,\text{Pe}^{0.85}\,\text{Pr}^{0.08}$$

$$= 4.8 + 0.0156(2520)^{0.85}(0.0075)^{0.08} = 13$$

$$h = \frac{k}{D}\,\text{Nu} = \frac{80.81}{0.025} \times 13 = 42{,}021 \text{ W/(m}^2 \cdot °\text{C)}$$

The gross energy balance gives

$$mc_p(T_2 - T_1) = h(\pi DL)\,\frac{T_2 - T_1}{\ln\left[(T_w - T_1)/(T_w - T_2)\right]}$$

$$(3)(1339) = \frac{(42{,}021)(\pi)(0.025L)}{1.792}$$

gives

$$L = 2.18 \text{ m}$$

We check our assumption:

$$\frac{L}{D} = \frac{2.18}{0.025} = 87 > 60$$

hence the use of Eq. (8-44) is justified.

PROBLEMS

8-1. Air ($\nu = 20.76 \times 10^{-6}$ m^2/s) flows with a mean velocity of $u_m = 0.5$ m/s inside a circular tube that has an inside diameter $d = 2$ cm. Determine whether the flow is laminar or turbulent.

8-2. Air flows inside a duct of square cross section with sides of 2 cm. For experimental purposes a Reynolds number of 50,000 is required. Calculate the velocity of the air. Air properties are to be evaluated at 350 K.
Answer: 51.9 m/s

8-3. A fluid flows through a duct of rectangular cross section with sides 2 cm × 1.5 cm. Calculate the hydraulic diameter.
Answer: 1.71 cm

8-4. Engine oil is cooled from 120°C to 80°C while it is flowing with a mean velocity of 0.02 m/s through a circular tube of inside diameter 2 cm. Calculate the Reynolds number of the flow.
Answer: 19.7

Fully Developed Laminar Flow

8-5. Engine oil is pumped with a mean velocity of $u_m = 0.5$ m/s through a bundle of tubes, each of inside diameter 1.5 cm and length 5 m. The physical properties of the oil are $\nu = 0.75 \times 10^{-4}$ m^2/s and $\rho = 868$ kg/m^3. Calculate the pressure drop across each tube.

8-6. Determine the friction factor and the pressure drop for fully developed laminar flow of ethylene glycol at 40°C [$\mu = 0.96 \times 10^{-2}$ kg/(m · s), $\rho = 1101$ kg/m^3] through a tube with diameter 3 cm and length 10 m. The mass flow rate of ethylene glycol is 0.05 kg/s.
Answer: 219.3 N/m^2

8-7. Engine oil at 40°C [$\mu = 0.21$ kg/(m · s), $\rho = 875$ kg/m^3] flows inside a tube with a mean velocity of 0.5 m/s. The tube is 25 m long and has a 3-cm inside diameter. Determine the pressure drop for the flow.

8-8. Water at 21°C [$\mu = 9.8 \times 10^{-4}$ kg/(m · s), $\rho = 997.4$ kg/m^3] flows inside a 2-cm-diameter, 50-m-long tube at a rate of 0.01 kg/s. Determine the pressure drop for the flow.
Answer: 125 N/m^2

8-9. Mercury at 20°C [$\mu = 1.55 \times 10^{-3}$ kg/(m · s), $\rho = 13,579$ kg/m^3] flows inside a 2-cm-diameter, 50-m-long tube at a rate of 0.01 kg/s. Determine the pressure drop for the flow.
Answer: 14.5 N/m^2

8-10. Freon at 20°C $[\mu = 2.63 \times 10^{-4}\,\text{kg}/(\text{m} \cdot \text{s}),\ \rho = 1330\,\text{kg/m}^3]$ flows inside a 2-cm-diameter, 50-m-long tube at a rate of 0.005 kg/s. Determine the pressure drop for the flow.

Answer: 12.6 N/m²

8-11. Engine oil at 40°C $[\mu = 0.21\,\text{kg}/(\text{m} \cdot \text{s}),\ \rho = 875\,\text{kg/m}^3]$ with a mean velocity of 0.5 m/s flows through a bundle of 20 tubes, each of 2-cm inside diameter and 10 m long. Calculate the total power required to pump the oil through 20 tubes and overcome the fluid friction.

8-12. Determine the friction factor for the flow of engine oil at 40°C $[\mu = 0.21\,\text{kg}/(\text{m} \cdot \text{s}),\ \rho = 875\,\text{kg/m}^3]$ with a mean velocity of 0.5 m/s through a parallel-plate channel having a spacing of 2 cm between the plates.

Answer: 1.15

8-13. Determine the friction factor for the flow of Freon at 20°C $[\mu = 2.63 \times 10^{-4}\,\text{kg}/(\text{m} \cdot \text{s}),\ \rho = 1330\,\text{kg/m}^3]$ with a mean velocity of 0.01 m/s through a parallel-plate channel having a spacing of 3 cm between the plates.

8-14. Determine the friction factor for the flow of water at 50°C with a mean velocity of 0.05 m/s through a square duct 2 cm by 2 cm and through an equilateral triangular duct with sides of 2 cm, each having the same length of 20 m.

Answer: 0.0527

8-15. Consider the flow of water at a rate of 0.01 kg/s through an equilateral triangular duct with sides of 2 cm whose walls are kept at a uniform temperature of 100°C. Assume that the flow is hydrodynamically and thermally developed. Determine the duct length required to heat the water from 20°C to 70°C.

Answer: 5 m.

8-16. Engine oil at 50°C with a flow rate of 10^{-3} kg/s enters a 1-m-long equilateral triangular duct with sides of 0.5 cm whose walls are maintained at a uniform temperature of 120°C. Assuming hydrodynamically and thermally fully developed flow, determine the outlet temperature of the oil.

8-17. Oil at 50°C enters a 5-m-long conduit whose cross section can be approximated as two parallel plates with a spacing of 0.5 cm. The flow rate of the oil is $2\,\text{kg}/(\text{m}^2 \cdot \text{s})$. The walls are subjected to uniform heating at a rate of 100 W/m². Assuming hydrodynamically and thermally fully developed flow, determine the average heat transfer coefficient. [Properties of oil may be taken at the anticipated mean temperature of 80°C as $\mu = 0.032\,\text{kg}/(\text{m} \cdot \text{s})$, $c_p = 2130\,\text{W} \cdot \text{s}/(\text{kg} \cdot °\text{C})$, $k = 0.14\,\text{w}/(\text{m} \cdot °\text{C})$, $\rho = 850\,\text{kg/m}^3$, and Pr = 487.]

Answer: 115 W/(m² · °C)

8-18. Determine the friction factor f for the fully developed laminar flow of engine oil at 60°C with a flow rate of 0.1 kg/s through (*a*) a circular tube of 1 cm diameter, (*b*) a square duct 1 cm by 1 cm, and (*c*) an equilateral-triangular duct with sides of 1 cm and sharp corners. Also determine the pressure drop over the length $L = 10$ m for each of the duct flows.

8-19. Determine the friction factor f for the fully developed laminar flow of air at 350 K with a velocity of $u_m = 0.5$ m/s inside a square duct 2 cm by 2 cm and a rectangular duct 1 cm by 4 cm. Calculate the pressure drop over the length $L = 10$ m for each of the duct flows.

Answer: 7.36 N/m², 15.13 N/m²

8-20. Air at atmospheric pressure, 27°C and with a mean velocity $u_m = 1$ m/s flows inside a 10-m-long conduit whose cross section can be approximated as two parallel plates with a spacing of 8 cm. The bottom wall of the conduit is insulated, and the top wall is kept at a constant temperature of 80°C. Assuming hydrodynamically and thermally fully developed flow, determine the outlet temperature of the air.

8-21. Ethylene glycol at 60°C with a velocity of 0.2 m/s flows through a two-parallel-plate conduit with a spacing of 0.5 cm and length of 10 m. The bottom plate wall is well insulated, and the top plate wall is subjected to uniform heating at a rate of 1000 W/m². Assuming hydrodynamically and thermally fully developed flow, determine the heat transfer coefficient and the friction coefficient.

Hydrodynamic and Thermal Entry Lengths

8-22. Determine the hydrodynamic entry lengths for flow at 60°C and at a rate of 0.01 kg/s of water and engine oil through a circular tube with an inside diameter of 2 cm.

Answers: 1.51 m, 0.0098 m

8-23. Determine the thermal entry lengths for laminar flow at 60°C and at a rate of 0.01 kg/s of water and engine oil through a circular tube with an inside diameter of 2 cm subjected to uniform wall temperature.

8-24. Determine the thermal entry lengths for hydrodynamically developed laminar flow at 60°C at a rate of 0.01 kg/(m²·s) of water and engine oil through a parallel-plate channel with a spacing of 1 cm and subjected to uniform wall temperature.

8-25. Determine the thermal entry length for hydrodynamically developed laminar flow of water at 60°C flowing at 0.01 kg/s through a square duct 1 cm by 1 cm in cross section and subjected to uniform wall temperature.

Answer: 2.63 m

8-26. Determine the thermal entry length for hydrodynamically developed laminar flow of glycerin at 20°C flowing at 0.01 kg/s through a square duct 1 cm by 1 cm in cross section and subjected to uniform wall heat flux.

8-27. Glycerin at 20°C with a flow rate of 0.01 kg/s enters a 1-cm-ID tube which is maintained at a uniform temperature of 80°C. Determine the thermal entry length. Assuming hydrodynamically and thermally fully developed flow, determine the heat transfer coefficient and the tube length required to heat the glycerin to 50°C.

Answers: 104.6 W/(m²·°C), 5.23 m

Thermally Developing, Hydrodynamically Developed Laminar Flow

8-28. Engine oil at 100°C with a flow rate of 0.2 kg/s enters a 2-cm-ID, 50-m-long tube maintained at a uniform temperature of 120°C. Assume that the flow is hydrodynamically developed. Determine the mean heat transfer coefficient and the outlet temperature of the oil.

Answers: 44.5 W/(m²·°C), 105.4°C

8-29. Water at 20°C with a flow rate of 0.01 kg/s enters a 2-cm-ID, 5-m-long tube maintained at a uniform temperature of 100°C by condensing steam on the outer surface of the tube. Assume that the flow is hydrodynamically developed. Determine the thermal entry length, the average heat transfer coefficient, and the outlet temperature of the water.

8-30. Water at 20°C with a flow rate of 0.01 kg/s enters a 2-cm-diameter tube which is maintained at 100°C. Assuming hydrodynamically developed flow, determine the tube length required to heat the water to 70°C.
Answer: 5 m

8-31. Engine oil at 60°C with a flow rate of 0.2 kg/s enters a 2-cm-ID tube which is maintained at 120°C. Assuming hydrodynamically developed flow, determine the tube length required to heat the oil to 80°C.
Answer: 66.2 m

8-32. Water at an average temperature of 60°C with a flow rate of 0.01 kg/s flows through a square duct 2 cm by 2 cm in cross section which is maintained at a uniform temperature. Assuming a fully developed velocity profile, determine the average heat transfer coefficients over 1, 2, and 3 m from the inlet. Determine the average heat transfer coefficient for the hydrodynamically and thermally developed region.

8-33. Water at an average temperature of 50°C with a flow rate of 0.01 kg/s flows through a tube with a 2-cm ID which is subjected to uniform heat flux. Assuming a fully developed velocity profile, determine the average heat transfer coefficient for the hydrodynamically and thermally developed region.
Answer: 139.65 W/(m^2 · °C)

8-34. Water at an inlet temperature of 50°C with a flow rate of 0.01 kg/s flows through a square duct 2 cm by 2 cm in cross section which is maintained at a uniform temperature of 100°C. Assuming hydrodynamically developed flow, determine the length of the duct needed to heat the water to 70°C.
Answer: 1.06 m

8-35. Engine oil at 20°C with a flow rate of 0.1 kg/s enters a 100-m-long square duct 1 cm by 1 cm in cross section whose walls are maintained at a uniform temperature of 80°C. Assuming hydrodynamically developed flow, determine the duct length required to heat the oil to 60°C.

8-36. Water at 20°C with a flow rate of 0.02 kg/s enters a square duct 2 cm by 2 cm in cross section whose walls are maintained at a uniform temperature of 80°C. Assuming hydrodynamically developed flow, determine the thermal entry length.

8-37. Engine oil at 20°C with a flow rate of 0.001 kg/s enters a 3-cm-long duct 1 cm by 1 cm in cross section whose walls are maintained at a uniform temperature of 100°C. Assuming hydrodynamically developed flow, determine the outlet temperature of the oil.

8-38. Water at a mean temperature of 60°C with a flow rate of 0.01 kg/s flows through a 1-m-long square duct 2 cm by 2 cm in cross section maintained at a uniform temperature. Assuming hydrodynamically developed flow, determine the local heat transfer coefficient at a location 1 m from the inlet.

Simultaneously Developing Laminar Flow

8-39. Water at an average temperature of 80°C with a flow rate of 0.01 kg/s flows through a 2-cm-diameter tube which is maintained at a uniform temperature.
(*a*) Assuming a fully developed profile, determine the local heat transfer coefficients at locations 1 and 3 m from the inlet.
(*b*) Assuming simultaneously developing velocity and temperature, determine the mean heat transfer coefficients over distances of 1 and 3 m from the inlet.

8-40. Water at an average temperature of 60°C with a flow rate of 0.015 kg/s flows through a 2.5-cm-ID tube which is subjected to uniform wall temperature.
(*a*) Assuming a fully developed velocity profile, determine the local heat transfer coefficient at locations 1 and 3 m from the inlet.
(*b*) Assuming simultaneously developing velocity and temperature, determine the mean heat transfer coefficients over distances of 1 and 3 m from the inlet.
Answers: (*a*) 143 W/(m² · °C), 106.6 W/(m² · C)
(*b*) 203 W/(m² · °C), 145.8 W/(m² · °C)

8-41. Engine oil at an average temperature of 40°C with a flow rate of 0.3 kg/s enters a 2.5-cm-ID, 40-m-long tube maintained at constant wall temperature. The velocity and temperature profiles are simultaneously developing. Determine the heat transfer coefficient at the exit.

8-42. Water at 15°C with a flow rate of 0.01 kg/s enters a 2.5-cm-ID, 3-m-long tube maintained at a uniform temperature of 100°C by condensing steam on the outer surface of the tube. The velocity and temperature profiles are simultaneously developing. Determine the average outlet temperature of the water.

8-43. Engine oil at 30°C with a flow rate of 0.2 kg/s enters a 1-cm-ID tube which is maintained at a uniform temperature of 120°C. The velocity and temperature profiles are simultaneously developing. Determine the tube length needed to heat the oil to 40°C.

Fully Developed Turbulent Flow

8-44. Water at a mean temperature of 60°C flows inside a 4-cm-ID tube with a velocity $u = 5$ m/s. The tube wall temperature is 20°C. Determine the heat transfer coefficient for the fully developed turbulent flow.

8-45. Water at 20°C with a mass flow rate of 5 kg/s enters a 5-cm-ID, 10-m-long tube whose surface is maintained at a uniform temperature of 60°C. Calculate the outlet temperature of the water.
Answer: 41.5°C

8-46. Water at a mean temperature of 60°C flows inside a 2.5-cm-ID, 10-m-long tube with a velocity of 6 m/s. The tube wall is maintained at a uniform temperature of 100°C by condensing steam. Determine the heat transfer rate to the water. Assume an inlet temperature of 30°C.
Answer: 670 kW

8-47. A fluid at a mean temperature of 40°C flows with a velocity of 10 m/s inside a 5-cm-diameter tube. Assuming fully developed turbulent flow and $T_w < 40$°C, determine the average heat transfer coefficient for the flow of (*a*) helium at 1 atm, (*b*) air at 1 atm, (*c*) water, and (*d*) glycerin.

8-48. Water at 40°C with a mass flow rate of 2 kg/s enters a 2.5-cm-ID tube whose wall is maintained at a uniform temperature of 90°C. Calculate the tube length required to heat the water to 60°C.
Answer: 2.79 m

8-49. Air at atmospheric pressure and 27°C enters a 12-m-long, 1.5-cm-ID tube with a mass flow rate of 0.1 kg/s. The tube surface is maintained at a uniform temperature of 80°C. Calculate the average heat transfer coefficient and the rate of heat transfer to the air.
Answer: 5.3 kW

8-50. Air at 1 atm and 27°C mean temperature flows with a velocity of 10 m/s through a 5-cm-ID, 5-m-long tube. The flow is hydrodynamically developed and thermally developing. Determine the average heat transfer coefficient over the entire length of the tube for air to be heated.

8-51. Pressurized cooling water at a mean temperature of 200°C flows with a velocity of 2 m/s between parallel-plate fuel elements in a nuclear reactor. Assuming fully developed turbulent flow and a spacing of 0.25 cm between the plates, determine the average heat transfer coefficient.

8-52. Cooling water with a velocity of 2 m/s enters a condenser tube at 20°C and leaves the tube at 40°C. The inside diameter of the tube is $D = 2$ cm. Assuming fully developed turbulent flow, determine the average heat transfer coefficient.
  Answer: 7865 W/(m² · °C)

Liquid Metals in Turbulent Flow

8-53. Mercury at 20°C with a mass flow rate of 5 kg/s enters a 5-cm-ID tube whose surface is maintained at 110°C. Calculate the tube length required to raise the temperature of the mercury to 100°C.

8-54. Mercury at an average temperature of 100°C flows with a velocity of 0.5 m/s inside a 2.5-cm-diameter tube. The flow is hydrodynamically developed but thermally developing. Determine the entry region heat transfer coefficient under both uniform wall heat flux and uniform wall temperature conditions over the distances $x/D = 20$, 40, and 60.

8-55. Liquid bismuth at 500°C and with a velocity $u = 1$ m/s enters a 1.75-cm-ID tube which is maintained at a uniform temperature of 600°C. Determine the tube length required to raise the temperature of the liquid bismuth to 580°C.
  Answer: 1.1 m

8-56. Mercury at a temperature of 100°C and with a velocity of 1 m/s enters a 1.25-cm-ID tube which is maintained at a uniform temperature of 160°C. Determine the tube length required to raise the temperature of the mercury to 150°C.

REFERENCES

1. Azer, N. Z., and B. T. Chao: "Turbulent Heat Transfer in Liquid Metals—Fully Developed Pipe Flow with Constant Wall Temperature," *Int. J. Heat Mass Transfer* **3**:77–83 (1961).
2. Dittus, F. W., and L. M. K. Boelter: *Univ. Calif., Berkeley, Publ. Eng.* **2**:433 (1930).
3. Graetz, L.: "Uber die Warmeleitungsfahigkeit von Flussigkeiten (On the thermal conductivity of liquids) Part 1," *Ann. Phys. Chem.*, **18**: 79–94 (1883); Part 2, *Ann. Phys. Chem.* **25**:337–357 (1885).
4. Hausen, H.: "Darstellung des Warmeuberganges in Rohren durch verallgemeinerte Potenzbeziehungen, *VDIZ* **4**:91 (1943).
5. Hornbeck, R. W.: "An All-Numerical Method for Heat Transfer in the Inlet of a Tube," ASME paper No. 65-WA, HT-36 (1965).
6. Lubarsky, B., and S. J. Kaufman: "Review of Experimental Investigation of Liquid Metal Heat Transfer," NACA Tech. Note 3336, 1955.
7. Moody, L. F.: "Friction Factor for Pipe Flow," *Trans. ASME* **66**:671–684 (1944).
8. Nikuradse, J.: Forsch. Arb. Ing. Wes. no. 361 (1933); also,, "Laws of Flow in Rough Pipes" (trans), NACA Tech. Memo 1292, 1950.
9. Notter, R. H., C. A. Sleicher: "A Solution to the Turbulent Graetz Problem III. Fully Developed and Entry Heat Transfer Rates," *Chem. Eng. Sci.* **27**:2073–2093 (1972).

10. Petukhov, B. S.: "Heat Transfer and Friction in Turbulent Pipe Flow with Variable Physical Properties," in J. P. Hartnett and T. F. Irvine (eds.), *Advances in Heat Transfer*, Academic, New York, 1970, pp. 504–564.

11. Shah, R. K., and A. L. London: *Laminar Flow: Forced Convection in Ducts*, Academic, New York, 1978.

12. Shah, R. K.: "Thermal Entry Length Solutions for a Circular Tube and Parallel Plates," *Proc. Nat. Heat, Mass Transfer Conf.*, 3d, Indiana Inst. Technol., Bombay, vol. 1, Papar no. HMT-11-75 (1975).

13. Sieder, E. N., and G. E. Tate: "Heat Transfer and Pressure Drop of Liquids in Tubes," *Ind. Eng. Chem.* **28**: 1429–1435 (1936).

14. Skupinski, E. S., J. Torel, and L. Vautrey: "Determination des coefficients de convection d'un alliage sodium-potassium dans un tube circulaire," *Int. J. Heat Mass Transfer* **8**:937 (1965).

15. Sleicher, C. A., A. S. Awad, and R. H. Notter: "Temperature and Eddy Diffusivity Profiles in Nak," *Int. J. Heat Mass Transfer* **16**:1565–1575 (1973).

CHAPTER
9

FREE
CONVECTION

In the previous two chapters we examined heat transfer by forced convection, in which the motion of the fluid was imposed externally by a fan, a blower, or a pump. In some situations convective motion is set up within the fluid without a forced velocity. Consider, for example, a hot plate placed vertically in a body of fluid at rest which is at a uniform temperature lower than that of the plate. Heat transfer will take place first by pure conduction, and a temperature gradient will be established in the fluid. The temperature variation within the fluid will generate a density gradient, which, in a gravitational field, will give rise, in turn, to convective motion as a result of buoyancy forces. The fluid motion set up as a result of the buoyancy force is called free convection, or natural convection.

Energy transfer by free convection occurs in many engineering applications. Heat transfer from a hot radiator to heat a room, refrigeration coils, transmission lines, electric transformers, electric heating elements, and electronic equipment are typical examples. The seasonal thermal inversion of lakes is also caused by buoyancy-induced free-convection motions.

9-1 BASIC CONCEPTS

As an illustrative example of free convection, we consider a hot wall placed vertically in a large body of cold fluid at rest, as illustrated in Fig. 9-1a. The fluid near the wall is hotter than that away from it. The resulting density difference gives rise to a buoyancy force, which causes the hot fluid to rise along the plate and the velocity boundary layer to develop. Note that the velocity profile for free convection is different from that for forced convection over a flat plate. In free convection, the velocity at the wall is zero because of

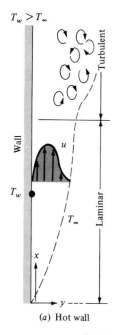

$T_w > T_\infty$

(a) Hot wall

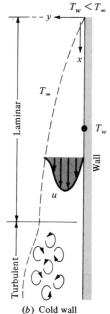

$T_w < T_\infty$

(b) Cold wall

FIGURE 9-1
Development of velocity boundary layer for free convection on a vertical plate for the case of hot wall in a cold fluid and cold wall in a hot fluid

the no-slip condition; it increases with distance from the wall but, after reaching some maximum value, decreases to zero because the fluid at distances sufficiently far from the wall is at rest. An examination of the boundary-layer development along the plate reveals that in the region near the leading edge of the plate, the boundary-layer development is *laminar*. However, after some distance from the leading edge of the plate, transition to a *turbulent boundary layer* begins; eventually, a fully developed *turbulent* boundary layer is established.

Figure 9-1*b* illustrates the development of the velocity boundary layer for the case of a cold vertical wall in a large body of hot fluid at rest. In this case, as expected, the direction of fluid motion is reversed; namely, the fluid in front of the cold plate moves vertically down. The velocity profile is similar to that discussed above for the hot plate, but the flow direction is reversed..

The flow velocity in free convection is much smaller than that encountered in forced convection. Therefore, the rate of heat transfer by free convection generally, is much smaller than that by forced convection.

We recall that in forced convection, the Reynolds number was an important dimensionless parameter that affected the flow regime; that is, depending on the magnitude of the Reynolds number, the flow regime could be identified as laminar or turbulent. In the case of free convection, the *Grashof number*, a dimensionless quantity, plays the same role as the Reynolds number in forced convection. Let X be the characteristic dimension of the body which is subjected to free convection, T_w its surface temperature, and T_∞ the temperature of the bulk fluid at rest. Then the Grashof number, Gr, is defined as

$$\text{Gr} = \frac{g\beta(T_w - T_\infty)X^3}{\nu^2} \qquad (9\text{-}1)$$

where g is the gravitational acceleration, β is the volume coefficient of thermal expansion, and ν is the kinematic viscosity of the fluid. The characteristic dimension X depends on the geometry of the body. In the case of the vertical plate example considered above, the distance x, measured from the leading edge of the plate, is taken as the characteristic length to define the local Grashof number at the location x. In the case of free convection from a horizontal tube immersed in a large body of fluid, the outside diameter D of the tube is taken as the characteristic dimension. The temperature difference is considered to be a positive quantity and hence it is taken as $|T_w - T_\infty|$. The volume expansion coefficient β may be determined from physical property tables, such as the one given in the Appendix B. In the case of *ideal gases* it may be calculated from

$$\beta = \frac{1}{T} \qquad \text{for ideal gases} \qquad (9\text{-}2)$$

where T is the absolute temperature.

We recall that in heat transfer correlations for forced convection, the Nusselt number generally depends on the Reynolds and Prandtl numbers, and the correlation can be expressed formally in the form

$$\mathrm{Nu} = f(\mathrm{Re}, \mathrm{Pr}) \qquad \text{for forced convection} \qquad (9\text{-}3a)$$

In the case of free convection, the Nusselt number generally depends on the Grashof and Prandtl numbers, and the heat transfer correlation can be expressed formally in the form

$$\mathrm{Nu} = f(\mathrm{Gr}, \mathrm{Pr}) \qquad \text{for free convection} \qquad (9\text{-}3b)$$

The physical situation discussed above on the formation of free convection implies that even in forced convection, the temperature gradients in the fluid may give rise to free convection. Therefore, it is useful to have some criteria for the relative importance of free convection in forced convection. It has been shown that the parameter

$$\frac{\mathrm{Gr}}{(\mathrm{Re})^2}$$

is a measure of the relative importance of free convection in relation to forced convection. If $\mathrm{Gr}/(\mathrm{Re})^2 \ll 1$, heat flow is considered to be primarily by forced convection, and the heat transfer correlation has the form given by Eq. (9-3a). Conversely, for $\mathrm{Gr}/(\mathrm{Re})^2 \gg 1$, free convection is dominant, and the heat transfer correlation has the form given by Eq. (9-3b).

When $\mathrm{Gr}/(\mathrm{Re})^2 \simeq 1$, free and forced convection are of the same order of magnitude, and hence both mechanisms must be considered in the heat transfer analysis. The Rayleigh number, Ra, defined as

$$\mathrm{Ra} = \mathrm{Gr}\,\mathrm{Pr} = \frac{g\beta(T_w - T_\infty)X^3}{\nu\alpha} \qquad (9\text{-}4)$$

has also been used instead of the Grashof number to correlate heat transfer in free convection.

For gases, $\mathrm{Pr} \cong 1$, and hence the Nusselt number of free convection is a function of the Grashof number only; that is,

$$\mathrm{Nu} = f(\mathrm{Gr}) \qquad \text{free convection for gases} \qquad (9\text{-}5)$$

The analysis of heat transfer in free convection is a complicated matter except for simple geometries. Therefore, experimental data are needed in order to develop reliable heat transfer correlations. Here we present some of the recommended empirical correlations for determining the free-convection heat transfer coefficient on geometries such as a flat plate, a single long cylinder, and a single sphere as well as for enclosures. The problems of simultaneous free and forced convection are far more complicated, and are not considered here.

9-2 FREE CONVECTION ON A FLAT PLATE

Heat transfer by free convection on a flat plate depends not only on the inclination and orientation of the heat transfer surface, but also on whether the heat transfer is taking place under uniform wall surface temperature or uniform wall heat flux conditions. Here we examine these cases and present some recommended heat transfer correlations.

Vertical Plate

1. UNIFORM WALL TEMPERATURE. McAdams proposed the following simple correlation for heat transfer by free convection on a vertical wall of height L maintained at a uniform temperature T_w in a fluid at constant temperature T_∞:

$$\mathrm{Nu}_m = c(\mathrm{Gr}_L\,\mathrm{Pr})^n = c\,\mathrm{Ra}_L^n \qquad (9\text{-}6)$$

where the Grashof number Gr_L and the mean Nusselt number Nu_m, based on the plate height L, are defined as

$$\mathrm{Nu}_m = \frac{h_m L}{k} \qquad \mathrm{Gr}_L = \frac{g\beta(T_w - T_\infty)L^3}{\nu^2} \qquad (9\text{-}7a,b)$$

and the Rayleigh number Ra_L becomes

$$\mathrm{Ra}_L = \mathrm{Gr}_L\,\mathrm{Pr} = \frac{g\beta(T_w - T_\infty)L^3}{\alpha\nu} \qquad (9\text{-}8)$$

The recommended values of the exponent n and the constant c of Eq. (9-6) are listed in Table 9-1 for both laminar and turbulent flow conditions. The physical properties of the fluid are evaluated at the *film temperature*

$$T_f = \tfrac{1}{2}(T_w + T_\infty) \qquad (9\text{-}9)$$

An examination of the values of the exponent n given in Table 9-1 reveals that in turbulent flow the mean heat transfer coefficient h_m is independent of the plate height L, since $\mathrm{Gr}_L \sim L^3$ and $h_m \sim (1/L)(\mathrm{Gr}_L)^{1/3}$.

A more elaborate but more accurate correlation has been proposed by Churchill and Chu for free convection from an isothermal vertical plate:

$$\mathrm{Nu}_m = \left\{0.825 + \frac{0.387\,\mathrm{Ra}_L^{1/6}}{[1 + (0.492/\mathrm{Pr})^{9/16}]^{8/27}}\right\}^2 \qquad \text{for } 10^{-1} < \mathrm{Ra}_L < 10^{12}$$

$$(9\text{-}10)$$

TABLE 9-1
Constant c and the exponent n of Eq. (9-6)

Type of flow	Range of $\mathrm{Gr}_L\,\mathrm{Pr}$	c	n
Laminar	10^4 to 10^9	0.59	1/4
Turbulent	10^9 to 10^{13}	0.10	1/3

which is applicable for both laminar and turbulent flow conditions. Again, the physical properties of the fluid are evaluated at the *film temperature*, defined by Eq. (9-9). The Rayleigh number Ra_L and the mean Nusselt number Nu_m were given previously by Eqs. (9-7) and (9-8). Figure 9-2 shows the mean Nusselt number Nu_m plotted as a function of the Grashof number Gr_L for several different values of the Prandtl number as calculated from Eq. (9-10).

Example 9-1. A vertical plate 0.3 m high and 1 m wide, maintained at a uniform temperature of 124°C, is exposed to quiescent atmospheric air at 30°C. Calculate the average heat transfer coefficient h_m over the entire height of the plate by using the correlations given by both Eqs. (9-6) and (9-10).

Solution. The physical properties of air at atmospheric pressure and the film temperature $T_f = (124 + 30)/2 = 77°C = 350\ K$ are

$$\nu = 2.076 \times 10^{-5}\ \text{m}^2/\text{s} \qquad Pr = 0.697$$

$$k = 0.03\ \text{W}/(\text{m} \cdot °\text{C}) \qquad \beta = \frac{1}{T_f} = 2.86 \times 10^{-3}\ \text{K}^{-1}$$

The Grashof number at $L = 0.3$ m becomes

$$Gr_{L=0.3} = \frac{g\beta(T_w - T_\infty)L^3}{\nu^2}$$

$$= \frac{(9.81)(2.86 \times 10^{-3})(124 - 30)(0.3)^3}{(2.076 \times 10^{-5})^2}$$

$$= 1.651 \times 10^8$$

and

$$Ra_{L=0.3} = Gr_{L=0.3}\ Pr$$

$$= (1.651 \times 10^8)(0.697) = 1.15 \times 10^8$$

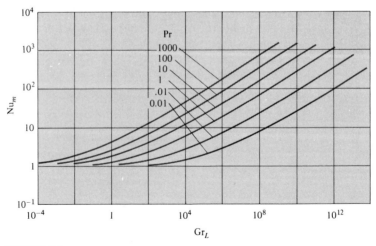

FIGURE 9-2
Average Nusselt number as a function of Grashof number for free convection on an isothermal plate calculated from Eq. (9-10).

Equation (9-6) with $c = 0.59$ and $n = \frac{1}{4}$ from Table 9-1 gives

$$\text{Nu}_m = \frac{h_m L}{k} = 0.59(\text{Ra}_L)^{1/4}$$

$$= 0.59(1.15 \times 10^8)^{1/4} = 61.1$$

then the average heat transfer coefficient becomes

$$h_m = \frac{k}{L} \text{Nu}_m$$

$$= \frac{0.03}{0.4} \, 61.1 = 4.58 \text{ W/(m}^2 \cdot {}^\circ\text{C)}$$

Equation (9-10) gives

$$\text{Nu}_m = \left\{ 0.825 + \frac{0.387 \, \text{Ra}_L^{1/6}}{[1 + (0.492/\text{Pr})^{9/16}]^{8/27}} \right\}^2$$

$$= \left\{ 0.825 + \frac{0.387(1.15 \times 10^8)^{1/6}}{[1 + (0.492/0.697)^{9/16}]^{8/27}} \right\}^2$$

$$= 63.51$$

and h_m becomes

$$h_m = 4.763$$

Example 9-2. Consider a rectangular plate 0.15 m by 0.3 m, maintained at a uniform temperature $T_w = 80°C$, placed vertically in quiescent atmospheric air at $T_\infty = 24°C$. Compare the heat transfer rates from the plate for the cases when the vertical height is (a) 0.15 and (b) 0.3 m.

Solution. The physical properties of air at atmospheric pressure and the film temperature $T_f = (80 + 24)/2 = 52°C = 325 \text{ K}$ are

$$\nu = 1.822 \times 10^{-5} \text{ m}^2/\text{s} \qquad \text{Pr} = 0.703$$

$$k = 0.02814 \text{ W/(m} \cdot {}^\circ\text{C)} \qquad \beta = \frac{1}{T_f} = 1/325 = 3.077 \times 10^{-3} \text{ K}^{-1}$$

(a) When the $L = 0.15$ m side is vertical, the Grashof and Rayleigh numbers are

$$\text{Gr}_{L=0.15} = \frac{g\beta(T_w - T_\infty)L^3}{\nu^2}$$

$$= \frac{(9.81)(3.077 \times 10^{-3})(80 - 24)(0.15)^3}{(1.822 \times 10^{-5})^2}$$

$$= 1.719 \times 10^7$$

and

$$\text{Ra}_{L=0.15} = \text{Gr}_{L=0.15} \, \text{Pr}$$

$$= (1.719 \times 10^7)(0.703) = 1.21 \times 10^7$$

Using Eq. (9-6) and Table 9-1, we have

$$\text{Nu}_m = \frac{h_m L}{k} = 0.59 \, \text{Ra}^{1/4}$$

$$= (0.59)(1.21 \times 10^7)^{1/4} = 34.8$$

then

$$h_m = \frac{k}{L} \text{Nu}_m$$

$$= \frac{0.02814}{0.15} 34.8 = 6.53 \text{ W/(m}^2 \cdot {}^\circ\text{C)}$$

and the heat transfer rate from one surface becomes

$$Q_{L=0.15} = A h_m (T_w - T_\infty)$$

$$= (0.15 \times 0.3)(6.53)(80 - 24)$$

$$= 16.5 \text{ W}$$

(b) When the $L = 0.3$ m side is vertical, the Rayleigh and Nusselt numbers are

$$\text{Ra}_{L=0.3} = \text{Ra}_{L=0.15} \times \left(\frac{0.3}{0.15}\right)^3$$

$$= (1.21 \times 10^7)(2)^3 = 0.968 \times 10^8$$

and

$$\text{Nu}_m = 0.59 \, \text{Ra}^{1/4}$$

$$= (0.59)(0.968 \times 10^8)^{1/4} = 58.5$$

Then

$$h_m = \frac{k}{L} \text{Nu}_m$$

$$= \frac{0.02814}{0.3} 58.5 = 5.49 \text{ W/(m}^2 \cdot {}^\circ\text{C)}$$

and the heat transfer rate from one surface becomes

$$Q_{L=0.3} = A h_m (T_w - T_\infty)$$

$$= (0.3 \times 0.15)(5.49)(80 - 24)$$

$$= 13.8 \text{ W}$$

which is lower than in the previous case.

2. UNIFORM WALL HEAT FLUX. Free convection on a vertical wall of height L subjected to a uniform heat flux q_w at the wall surface has been studied by several investigators. Based on the experiments with air and water by Vliet and Liu and by Vliet, the following correlations have been proposed for the mean Nusselt number for the laminar and turbulent flow regimes:

$$\text{Nu}_m = 0.75(\text{Gr}_L^* \, \text{Pr})^{1/5} \quad \text{for } 10^5 < \text{Gr}_L^* \, \text{Pr} < 10^{11} \text{ (laminar)} \tag{9-11}$$

$$\text{Nu}_m = 0.645(\text{Gr}_L^* \, \text{Pr})^{0.22} \quad \text{for } 2 \times 10^{13} < \text{Gr}_L^* \, \text{Pr} < 10^{16} \text{ (turbulent)} \tag{9-12}$$

where the modified Grashof number Gr_L^* is defined as

$$\text{Gr}_L^* = \frac{g\beta q_w L^4}{k\nu^2} \tag{9-13a}$$

and the average Nusselt number as

$$\mathrm{Nu}_m = \frac{h_m L}{k} \qquad (9\text{-}13b)$$

All physical properties are evaluated at the film temperature as defined by Eq. (9-9). However, in this case, since the wall heat flux is prescribed, the wall surface temperature T_w is *a priori* unknown, and hence the determination of T_w requires an initial guess and then iteration.

Horizontal Plate

The average Nusselt number for free convection on a horizontal plate depends, among other parameters, on the orientation of the heat transfer surface—that is, whether the hot surface is facing up or down, and conversely, whether the cold surface is facing down or up. We examine free convection on a horizontal plate for both uniform wall temperature and uniform wall heat flux.

1. UNIFORM WALL TEMPERATURE. The mean Nusselt number Nu_m for free convection on a horizontal plate maintained at a uniform temperature T_w in a large body of fluid at temperature T_∞ has been correlated by McAdams by the following simple expression:

$$\boxed{\mathrm{Nu}_m = c(\mathrm{Gr}\,\mathrm{Pr})^n} \qquad (9\text{-}14a)$$

where

$$\mathrm{Nu}_m = \frac{h_m L}{k} \qquad \mathrm{Gr}_L \equiv \frac{g\beta(T_w - T_\infty)L^3}{\nu^2} \qquad (9\text{-}14b)$$

The coefficient c and the exponent n are listed in Table 9.2. The physical properties of the fluid are evaluated at the film temperature T_f, defined by Eq. (9-9). Note that the values of c and n depend on the orientation of the heat transfer surface as well as on the flow regime. The *characteristic length L* of the plate is taken as:

TABLE 9-2
Constant c and exponent n of Eq. (9-14a) for free convection on a horizontal plate at uniform temperature

Orientation of plate	Range of $\mathrm{Gr}_L\,\mathrm{Pr}$	c	n	Flow regime
Hot surface facing up or cold surface facing down	10^5 to 2×10^9	0.54	1/4	Laminar
	2×10^7 to 3×10^{10}	0.14	1/3	Turbulent
Hot surface facing down or cold surface facing up	3×10^5 to 3×10^{10}	0.27	1/4	Laminar

Source: W. H. McAdams. *Heat Transmission*, 3rd ed., McGraw-Hill, New York, 1959.

- The length of a side for a square
- The arithmetic mean of the two dimensions for a rectangular square
- $0.9D$ for a circular disc of diameter D

Recent correlations of free convection suggest that improved accuracy may be obtained if the characteristic length L for the plate is defined as

$$L \equiv \frac{A}{P} \tag{9-15}$$

where A is the surface area of the plate and P is the perimeter that encompasses the area A.

It is apparent from Table 9-2 that for a turbulent flow regime the Nusselt number Nu_m is independent of the characteristic length.

Example 9-3. A circular disc heater 0.2 m in diameter is exposed to quiescent atmospheric air at $T_\infty = 25°C$. One surface of the disc is insulated, and the other surface is maintained at $T_w = 130°C$. Calculate the amount of heat transferred from the disc when it is (*a*) horizontal with the hot surface facing up, (*b*) horizontal with the hot surface facing down, and (*c*) vertical.

Solution. The physical properties of atmospheric air at $T_f = (T_w + T_\infty)/2 = (130 + 25)/2 = 77.5°C = 350.5$ K are taken as

$$\nu = 2.08 \times 10^{-5} \text{ m}^2/\text{s} \qquad \mathrm{Pr} = 0.697$$

$$k = 0.03 \text{ W/(m} \cdot °C) \qquad \beta = \frac{1}{T_f} = \frac{1}{350.5}$$

The area of the disc is

$$A = \frac{\pi}{4} D^2 = \frac{\pi}{4} (0.2)^2 = 0.0314 \text{ m}^2$$

The characteristic length $L = 0.9D$ for a circular disc of diameter D is suggested by McAdams. Then the Grashof number becomes

$$\mathrm{Gr} = \frac{g\beta(T_w - T_\infty)L^3}{\nu^2} = \frac{(9.81)(1/350.5)(130 - 25)(0.9 \times 0.2)^3}{(2.08 \times 10^{-5})^2}$$

$$= 3.96 \times 10^7$$

and

$$\mathrm{Gr} \, \mathrm{Pr} = (3.96 \times 10^7)(0.697) = 2.76 \times 10^7$$

(*a*) For the horizontal disc with the hot surface up, as illustrated in the accompanying figure, the average Nusselt number is determined from Eq. (9-14*a*) and Table 9-2. For the turbulent flow condition, we obtain

$$\mathrm{Nu}_m = \frac{h_m L}{k} = 0.14(\mathrm{Gr}_L \, \mathrm{Pr})^{1/3}$$

$$= 0.14(2.76 \times 10^7)^{1/3} = 42.3$$

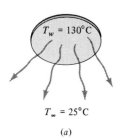

$T_w = 130°C$

$T_\infty = 25°C$

(a)

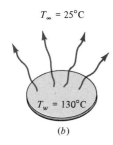

$T_\infty = 25°C$

$T_w = 130°C$

(b)

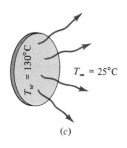

$T_w = 130°C$

$T_\infty = 25°C$

(c)

FIGURE EXAMPLE 9-3
A circular disc heater
(a) hot surface faces
up, (b) hot surface
faces down, (c)
vertical.

Then

$$h_m = \frac{k}{L} \text{Nu}_m = \frac{0.03}{0.9 \times 0.2} \quad (42.3)$$

$$= 7.05 \text{ W/(m}^2 \cdot °C)$$

and

$$Q = Ah(T_w - T_\infty) = (0.0314)(7.05)(130 - 25)$$

$$= 23.2 \text{ W}$$

(b) For the horizontal disc with the hot surface facing down, as illustrated in the accompanying figure, the average Nusselt number is determined from Eq. (9-14a) and Table 9-2:

$$\text{Nu}_m = \frac{h_m L}{k} = 0.27(\text{Gr}_L \text{ Pr})^{1/4}$$

$$= 0.27(2.76 \times 10^7)^{1/4} = 19.6$$

Then

$$h_m = \frac{k}{L} \text{Nu}_m = \frac{0.03}{0.9 \times 0.2} \quad (19.6)$$

$$= 3.27 \text{ W/(m}^2 \cdot °C)$$

and

$$Q = Ah(T_w - T_\infty) = 0.0314(3.27)(130 - 25)$$

$$= 10.8 \text{ W}$$

(c) For the vertical disc, as illustrated in the accompanying figure, the average Nusselt number is determined from Eq. (9-6) and Table 9.1:

$$\text{Nu}_m = \frac{h_m L}{k} = 0.59(\text{Gr}_L \text{ Pr})^{1/4}$$

$$= 0.59(2.76 \times 10^7)^{1/4} = 42.8$$

Then

$$h_m = \frac{k}{L} \text{Nu}_m = \frac{0.03}{0.9 \times 0.2} \quad (42.8)$$

$$= 7.1 \text{ W/(m}^2 \cdot °C)$$

and

$$Q = Ah(T_w - T_\infty) = 0.0314(7.1)(130 - 25)$$

$$= 23.4 \text{ W}$$

2. UNIFORM WALL HEAT FLUX. The average Nusselt number for free convection on a horizontal plate subjected to uniform surface heat flux q_w and exposed to an ambient at constant temperature T_∞ has been studied experimentally by Fujii and Imura, and the following correlations are proposed for the cases in which the heated surface is facing up and facing down.

Horizontal plate with the *heated surface facing upward*:

$$\text{Nu}_m = 0.13(\text{Gr}_L\,\text{Pr})^{1/3} \qquad \text{for Gr}_L\,\text{Pr} < 2 \times 10^8 \qquad (9\text{-}16a)$$

$$\text{Nu}_m = 0.16(\text{Gr}_L\,\text{Pr})^{1/3} \qquad \text{for } 5 \times 10^8 < \text{Gr}_L\,\text{Pr} < 10^{11} \qquad (9\text{-}16b)$$

Horizontal plate with the *heated surface facing downward*:

$$\text{Nu}_m = 0.58(\text{Gr}_L\,\text{Pr})^{1/5} \qquad \text{for } 10^6 < \text{Gr}_L\,\text{Pr} < 10^{11} \qquad (9\text{-}17)$$

The physical properties are to be evaluated at a mean temperature, defined as

$$T_m = T_w - 0.25(T_w - T_\infty) \qquad (9\text{-}18a)$$

and the thermal expansion coefficient β at $(T_w + T_\infty)/2$. In these expressions, the Grashof number is defined as

$$\text{Gr}_L = \frac{g\beta(T_w - T_\infty)L^3}{\nu^2} \qquad (9\text{-}18b)$$

and the mean Nusselt number over the length L as

$$\text{Nu}_m = \frac{h_m L}{k} = \frac{q_w L}{(T_w - T_\infty)k} \qquad (9\text{-}18c)$$

Inclined Plates

The heat transfer coefficient for free convection on an inclined plate can be predicted by the vertical plate formulas if the gravitational term g is replaced by $g \cos\theta$, where θ is the angle the plate makes with the vertical. However, the orientation of the heat transfer surface, whether the surface is facing upward or downward, is also a factor that affects the Nusselt number. In order to make a distinction between the two possible orientations of the surface, we designate the angle θ that the surface makes with the horizontal as *positive* if the hot surface is facing downward and *negative* if the hot surface is facing upward, as shown in Fig. 9-3.

Based on the extensive experimental studies by Fujii and Imura, the following correlations are recommended for free convection on an inclined surface subjected to a uniform surface heat flux q_w and exposed to an ambient at a uniform temperature T_∞. The notation for the angle of inclination is as defined in Fig. 9-3.

For an inclined plate with the heated surface facing downward,

$$\text{Nu}_m = 0.56(\text{Gr}_L\,\text{Pr}\cos\theta)^{1/4} \qquad \text{for } +\theta < 88, \quad 10^5 < \text{Gr}_L\,\text{Pr} < 10^{11}$$

$$(9\text{-}19)$$

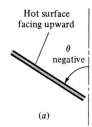

Hot surface
facing upward

θ
negative

(a)

θ
positive

Hot surface
facing downward

(b)

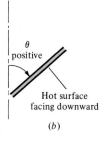

FIGURE 9-3
The concept of positive and negative inclination angles from the vertical to define the orientation of the hot surface.

When the plate is slightly inclined with the horizontal (that is, $88° < \theta < 90°$) and the heated surface is facing downward, Eq. (9-19) is applicable. The power of $\frac{1}{4}$ in Eq. (9-19) implies that the flow is always in the laminar regime.

For an inclined plate with the heated surface facing upward, the heat transfer correlation has been developed as

$$\text{Nu}_m = 0.145[(\text{Gr}_L \, \text{Pr})^{1/3} - (\text{Gr}_c \, \text{Pr})^{1/3}] + 0.56(\text{Gr}_c \, \text{Pr} \cos \theta)^{1/4} \qquad (9\text{-}20)$$

for $\text{Gr}_L \, \text{Pr} < 10^{11}$, $\text{Gr}_L > \text{Gr}_c$, and θ lying between $-15°$ and $-75°$. Here, the value of the transition Grashof number Gr_c depends on the angle of inclination θ, as listed in Table 9-3.

In Eqs. (9-19) and (9-20), all physical properties are evaluated at the mean temperature

$$T_m = T_w - 0.25(T_w - T_\infty)$$

and β is evaluated at $T_\infty + 0.25(T_w - T_\infty)$.

Example 9-4. Consider a 60 cm by 60 cm electrically heated plate with one of its surfaces thermally insulated and the other surface dissipating heat by free convection into atmospheric air at 30°C. The heat flux over the surface of the plate is uniform and results in a mean surface temperature of 50°C. The plate is inclined, making an angle of 50° from the vertical. Determine the heat loss from the plate for the following cases: (a) heated surface facing up; (b) heated surface facing down.

Solution. The physical properties of atmospheric air at a mean temperature $T_m = T_w - 0.25(T_w - T_\infty) = 50 - 0.25(50 - 30) = 45°C = 318$ K are taken as

$$\nu = 1.751 \times 10^{-5} \text{ m}^2/\text{s} \qquad \text{Pr} = 0.704$$

$$k = 0.0276 \text{ W}/(\text{m} \cdot °\text{C}) \qquad \beta = \frac{1}{T_\infty + 0.25(T_w - T_\infty)} = \frac{1}{308} \text{ K}^{-1}$$

$$\text{Gr}_L = \frac{g\beta(T_w - T_\infty)L^3}{\nu^2} = \frac{(9.81)\left(\dfrac{1}{308}\right)(50 - 30)(0.6)^3}{(1.751 \times 10^{-5})^2} = 4.487 \times 10^8$$

TABLE 9-3
Transition Grashof number (See Fig. 9-3 for the definition of θ)

θ, degrees	Gr_c
-15	5×10^9
-30	10^9
-60	10^8
-75	10^6

From the experimental data of T. Fujii and H. Imura, "Natural Convection Heat Transfer from a Plate with Arbitrary Inclination," *Int. J. Heat Mass Transfer* **15**:755 (1972).

Hot Surface

$-50°$

(a)

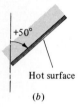

$+50°$

Hot surface

(b)

FIGURE EXAMPLE 9-4

(a) For the hot surface facing up, as illustrated in the accompanying figure, $\theta = -50°$, and we have

$$\text{Gr}_c \cong 3.33 \times 10^8$$

and $\text{Gr}_L > \text{Gr}_c$. Then, from Eq. (9-20) we obtain

$$\text{Nu}_m = 0.145[(\text{Gr}_L \, \text{Pr})^{1/3} - (\text{Gr}_c \, \text{Pr})^{1/3}] + 0.56(\text{Gr}_c \, \text{Pr} \cos \theta)^{1/4}$$

$$= 0.145[(4.487 \times 10^8 \times 0.704)^{1/3} - (3.33 \times 10^8 \times 0.704)^{1/3}]$$

$$+ 0.56(3.33 \times 10^8 \times 0.704 \times \cos(-50°))^{1/4}$$

$$= 71.39$$

Therefore,

$$h_m = \frac{k}{L} \text{Nu}_m = \frac{0.0276}{0.6} (71.39)$$

$$= 3.284 \text{ W/(m}^2 \cdot °\text{C)}$$

and

$$Q = Ah_m(T_w - T_\infty) = (0.6)^2(7.65)(50 - 30)$$

$$= 23.64 \text{ W}$$

(b) For the hot surface facing down, as illustrated in the accompanying figure, from Eq. (9-19) we have

$$\text{Nu}_m = 0.56(\text{Gr}_L \, \text{Pr} \cos \theta)^{1/4} = 0.56(3.246 \times 10^7 \times 0.704 \times \cos 50°)$$

$$= 34.7$$

$$h_m = \frac{k}{L} \text{Nu}_m = \frac{0.0276}{0.25} (34.7)$$

$$= 3.83 \text{ W/(m}^2 \cdot °\text{C)}$$

and

$$Q = Ah_w(T_w - T_\infty) = (0.6)^2(3.83)(50 - 30)$$

$$= 27.58 \text{ W}$$

9-3 FREE CONVECTION ON A LONG CYLINDER

We now examine free convection on a long cylinder in both the vertical and horizontal positions.

Vertical Cylinder

The average Nusselt number for free convection on a vertical cylinder is the same as that for a vertical plate if the curvature effects are negligible. Therefore, for an *isothermal vertical cylinder*, the average Nusselt number can be determined from the vertical plate correlation given by Eq. (9-6) or (9-10).

For fluids having a Prandtl number of 0.7 or higher, a vertical cylinder may be treated as a vertical flat plate when

$$\frac{LD}{(\mathrm{Gr}_D)^{1/4}} < 0.025 \qquad (9\text{-}21)$$

where D is the diameter of the cylinder and L is the height of the cylinder.

Horizontal Cylinder

Morgan presented a simple correlation for free convection from a *horizontal isothermal cylinder*, covering the range $10^{-10} < \mathrm{Ra}_D < 10^{12}$. It is given in the form

$$\boxed{\mathrm{Nu}_m = \frac{hD}{k} = c\,\mathrm{Ra}_D^n} \qquad (9\text{-}22)$$

where the constant c and the exponent n are listed in Table 9-4 and Nu_m and Ra_D are based on the cylinder diameter D:

$$\mathrm{Nu}_m = \frac{h_m D}{k} \qquad (9\text{-}23a)$$

$$\mathrm{Ra}_D = \mathrm{Gr}_D\,\mathrm{Pr} = \frac{g\beta(T_w - T_\infty)D^3}{\nu^2}\,\mathrm{Pr} \qquad (9\text{-}23b)$$

Figure 9-4 shows the isotherms for free convection on a uniformly heated horizontal circular cylinder 6 cm in diameter and 60 cm long at a temperature 9°C above the ambient air and with a Grashof number of 30,000.

TABLE 9-4
Constant c and exponent n of Eq. (9-22) for free convection on a horizontal cylinder

Ra_D	c	n
10^{-10}–10^{-2}	0.675	0.058
10^{-2}–10^{2}	1.02	0.148
10^{2}–10^{4}	0.850	0.188
10^{4}–10^{7}	0.480	0.250
10^{7}–10^{12}	0.125	0.333

Source: V. T. Morgan, "The Overall Convective Heat Transfer from Smooth Circular Cylinders," in T. F. Irvine and J. P. Hartnett (eds.), *Advances in Heat Transfer*, vol. 16, Academic, New York, 1975, pp. 199–269.

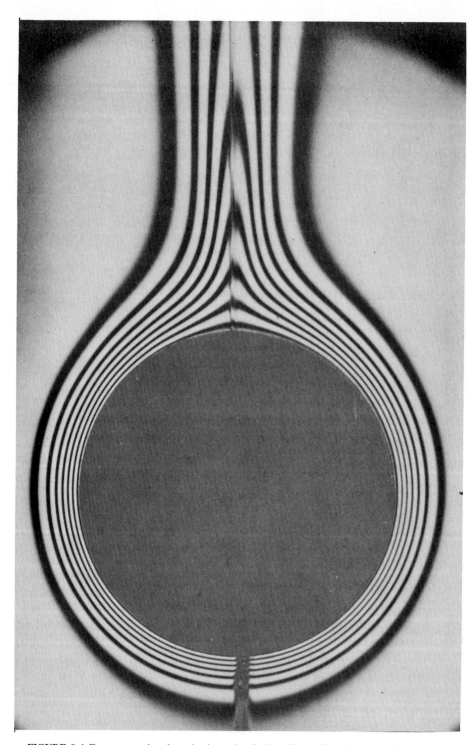

FIGURE 9-4 Free convection from horizontal cylinder. (From Grigull and Hauf).

Example 9-5. A 5-cm-outside-diameter (OD) horizontal tube at a uniform temperature $T_w = 400\,\text{K}$ is exposed to quiescent atmospheric air at $T_\infty = 300\,\text{K}$. Calculate the average heat transfer coefficient h_m.

Solution. The physical properties of the atmospheric air at $T_f = (T_w + T_\infty)/2 = (300 + 400)/2 = 350\,\text{K}$ are taken as

$$\nu = 2.076 \times 10^{-5}\,\text{m}^2/\text{s} \qquad \text{Pr} = 0.697$$

$$k = 0.03\,\text{W}/(\text{m}\cdot{}^\circ\text{C}) \qquad \beta = \frac{1}{350}\,\text{K}^{-1}$$

and

$$\text{Ra}_D = \text{Ga}_D\,\text{Pr} = \frac{g\beta(T_w - T_\infty)D^3}{\nu^2}\,\text{Pr} = 5.67 \times 10^5$$

From Eq. (9-22) and Table 9-4 we have

$$\text{Nu}_m = 0.48(\text{Ra}_D)^{0.25} = 0.48(5.67 \times 10^5)^{0.25}$$

$$= 13.17$$

Thus

$$h_m = \frac{k}{D}\,\text{Nu}_m = \frac{0.03}{0.05}\,(13.17)$$

$$= 7.9\,\text{W}/(\text{m}^2\cdot{}^\circ\text{C})$$

9-4 FREE CONVECTION ON A SPHERE

The average Nusselt number for free convection on a single *isothermal sphere*, for fluids with a Prandtl number close to unity, has been correlated by **Yuge** with the following expression:

$$\boxed{\text{Nu}_m = \frac{h_m D}{k} = 2 + 0.43\,\text{Ra}_D^{1/4}} \qquad (9\text{-}24)$$

for $1 < \text{Ra}_D < 10^5$ and $\text{Pr} \cong 1$.

Amato and Tien, based on experimental data for free convection on a single *isothermal sphere* in water, proposed the following correlation:

$$\boxed{\text{Nu}_m = 2 + 0.50\,\text{Ra}_D^{1/4}} \qquad (9\text{-}25)$$

for $3 \times 10^5 < \text{Ra}_D < 8 \times 10^8$ and $10 \le \text{Nu}_m \le 90$. Here the Rayleigh number, based on the sphere diameters, is defined as

$$\text{Ra}_D = \text{Gr}_D\,\text{Pr} = \frac{g\beta D^3(T_w - T_\infty)}{\nu^2}\,\text{Pr} \qquad (9\text{-}26)$$

The properties are evaluated at the film temperature $T_f = \frac{1}{2}(T_w + T_\infty)$.

9-5 SIMPLIFIED EQUATIONS FOR AIR

We present in Table 9-5 simplified expressions for rapid but approximate estimation of the average heat transfer coefficient from isothermal surfaces in air at atmospheric pressure and moderate temperatures. For more accurate results, the previously given, more exact expressions should be used. The correlations given in this table apply to air and to CO, CO_2, O_2, N_2, and flue gases for temperatures from 20 to 800°C. The Grashof number is defined as

$$\mathrm{Gr}_L \equiv \frac{g\beta \, \Delta\mathrm{T} \, \mathrm{L}^3}{\nu^2}$$

where L is the characteristic dimension of the body and ΔT is the temperature difference between the surface and the ambient air, that is, $\Delta T = T_w - T_\infty$.

The correlations given in Table 9-5 for air at atmospheric pressure can be extended to higher or lower pressures by multiplying by the following factors:

$$P^{1/2} \quad \text{for laminar regime}$$
$$P^{2/3} \quad \text{for turbulent regime}$$

where P is the pressure in atmospheres.

The expressions given in Table 9-5 are merely approximations; care must be exercised in their use.

Example 9-6. A vertical plate at a uniform temperature of 230°C is exposed to atmospheric air at 25°C. Determine the free-convection heat transfer coefficient, and compare the result with that obtainable from the simplified expressions for free convection to air at atmospheric pressure given in Table 9.5.

TABLE 9-5
Simplified equations for free convection to air at atmospheric pressure and moderate temperatures

Geometry	Characteristic dimension L	Type of flow	Range of Gr_L Pr	Heat transfer coefficient h_m, W/(m²·°C)
Vertical plates and Height cylinders	Height	Laminar	10^4–10^9	$h_m = 1.42(\Delta T/L)^{1/4}$
		Turbulent	10^9–10^{13}	$h_m = 1.31\,\Delta T^{1/3}$
Horizontal cylinders	Outside diameter	Laminar	10^4–10^9	$h_m = 1.32(\Delta T/D)^{1/4}$
		Turbulent	10^9–10^{12}	$h_m = 1.24\,\Delta T^{1/3}$
Horizontal plates (a) Upper surface hot or lower surface cold	As defined in the text	Laminar	10^5–2×10^7	$h_m = 1.32(\Delta T/L)^{1/4}$
		Turbulent	2×10^7–3×10^{10}	$h_m = 1.52\,\Delta T^{1/3}$
(b) Lower surface hot or upper surface cold	As defined in the text	Laminar	3×10^5–3×10^{10}	$h_m = 0.59(\Delta T/L)^{1/4}$

From W. H. McAdams, *Heat Transmission*, 3d ed., McGraw-Hill, New York, 1959.

Solution. The physical properties of atmospheric air at $T_f = (T_w + T_\infty)/2 = (230 + 25)/2 = 127.5°C = 450.5$ K are taken as

$$\nu = 2.59 \times 10^{-5} \text{ m}^2/\text{s} \qquad \text{Pr} = 0.689$$

$$k = 0.0337 \text{ W/(m} \cdot \text{°C)}$$

The Rayleigh number Ra is

$$\text{Ra} = \frac{g\beta(T_w - T_\infty)L^3}{\nu^2} \text{Pr}$$

$$= \frac{(9.81)(1/450.5)(230 - 25)(1)^3}{(2.59 \times 10^{-5})^2} \times 0.689$$

$$= 5.16 \times 10^9$$

therefore the flow is turbulent. From Eq. (9-10) we have

$$\text{Nu}_m = \left\{ 0.825 + \frac{0.387 \, \text{Ra}^{1/6}}{[1 + (0.492/\text{Pr})^{9/16}]^{8/27}} \right\}^2$$

$$= 204$$

and

$$h_m = \frac{k}{L} \text{Nu}_m = \frac{0.0337}{1}(172.8)$$

$$= 5.82 \text{ W/(m}^2 \cdot \text{°C)}$$

From the simplified expression given in Table 9-5 for air we have

$$h_m = 1.31 \, \Delta T^{1/3} = 1.31(230 - 25)^{1/3}$$

$$= 7.72 \text{ W/(m}^2 \cdot \text{°C)}$$

Example 9-7. A block $10 \text{ cm} \times 10 \text{ cm} \times 10 \text{ cm}$ in size, illustrated in the accompanying figure, is suspended in quiescent atmospheric air at 10°C with one of its surfaces in a horizontal position. All surfaces of the block are maintained at 150°C. Determine the free-convection heat transfer coefficient for all the surfaces of the block, and compare these results with those obtainable from the simplified expressions given in Table 9-5 for air at atmospheric pressure.

FIGURE EXAMPLE 9-7

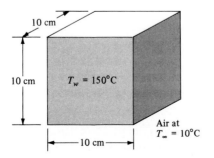

Solution. The physical properties of air at $T_f = (T_w + T_\infty)/2 = (150 + 10)/2 = 80°C = 353$ K are taken as

$$\nu = 2.107 \times 10^{-5} \text{ m}^2/\text{s} \qquad \text{Pr} = 0.697$$
$$k = 0.03 \text{ W}/(\text{m} \cdot °\text{C})$$

The Rayleigh number is

$$\text{Ra} = \frac{g\beta(T_w - T_\infty)L^3}{\nu^2} \text{Pr}$$
$$= \frac{(9.81)(1/353)(150 - 10)(0.1)^3}{(2.107 \times 10^{-5})^2} (0.697)$$
$$= 6.11 \times 10^6$$

Vertical surfaces of the cube: The flow is laminar; so from Eq. (9-6) and Table 9-1 we have

$$\text{Nu}_m = 0.59 \text{ Ra}^{1/4} = 29.33$$
$$h_m = \frac{k}{L} \text{Nu}_m = 8.8 \text{ W}/(\text{m}^2 \cdot °\text{C})$$

and the simplified expression of Table 9-5 for air gives

$$h_m = 1.42\left(\frac{\Delta T}{L}\right)^{1/4} = 1.42\left(\frac{150 - 10}{0.1}\right)^{1/4} = 8.69 \text{ W}/(\text{m}^2 \cdot °\text{C})$$

Top surface: The flow is laminar.
From Eq. (9-14a) and Table 9-2 we have

$$\text{Nu}_m = 0.54 \text{ Ra}^{1/4} = 26.85$$
$$h_m = \frac{k}{L} \text{Nu}_m = 8.05 \text{ W}/(\text{m}^2 \cdot °\text{C})$$

and the simplified expression of Table 9-5 for air gives

$$h_m = 1.32\left(\frac{\Delta T}{L}\right)^{1/4} = 1.32\left(\frac{150 - 10}{0.1}\right)^{1/4} = 8.07 \text{ W}/(\text{m}^2 \cdot °\text{C})$$

Bottom surface: The flow is laminar.
From Eq. (9-14a) and Table 9-2 we have

$$\text{Nu}_m = 0.27 \text{ Ra}^{1/4} = 13.42$$
$$h_m = \frac{k}{L} \text{Nu}_m = 4.03 \text{ W}/(\text{m}^2 \cdot °\text{C})$$

and the simplified expression of Table 9-5 for air gives

$$h_m = 0.59\left(\frac{\Delta T}{L}\right)^{1/4} = 0.59\left(\frac{150 - 10}{0.1}\right)^{1/4} = 3.61 \text{ W}/(\text{m}^2 \cdot °\text{C})$$

9-6 FREE CONVECTION IN ENCLOSED SPACES

Heat transfer by free convection in enclosed spaces has numerous engineering applications. Typical examples are free convection in wall cavities, between window glazing, in the annulus between concentric cylinders or spheres, and in flat-plate solar collectors.

Basic Concepts

The onset of free convection inside enclosed spaces involves an interesting but very complicated flow phenomenon. The subject has been studied experimentally, analytically, and numerically by various investigators, and correlations for predicting the free-convection heat transfer coefficient have been proposed. Before presenting such correlations, however, we shall provide some insight into the physical nature of the problem and give a qualitative discussion of the conditions leading to the onset of free convection in enclosed spaces.

As a classical example, we consider a fluid contained between two large horizontal plates separated by a distance δ, as illustrated in Fig. 9-5. The lower plate is maintained at a uniform temperature T_h which is higher than the temperature T_c of the upper colder plate. Then there is a heat flow through the fluid layer in the upward direction, and a temperature profile decreasing upward is established within the fluid. Clearly, the physical situation is such that the denser cold fluid lies above the lighter, warm layers. The fluid remains stationary and the heat transfer takes place by pure conduction so long as the viscous forces are stronger than the buoyancy forces.

Suppose that the temperature difference $T_h - T_c$ between the plate temperatures is increased to a value such that the buoyancy forces overcome the viscous forces. Then the fluid layer can no longer remain stationary, but gives rise to a convective motion under the influence of the buoyancy forces. Theoretical and experimental investigations have verified that, for such a horizontal enclosure, the fluid layer becomes unstable and convection currents are established if $T_h - T_c$ is increased beyond a value corresponding to the *critical Rayleigh number* Ra_c:

$$\mathrm{Ra}_c = \mathrm{Gr}\,\mathrm{Pr} = \frac{g\beta(T_h - T_c)\delta^3}{\nu^2}\,\mathrm{Pr} = 1708 \qquad (9\text{-}27)$$

FIGURE 9-5
A layer of fluid contained between two large horizontal plates.

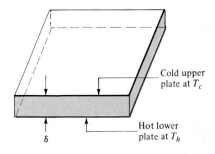

Cold upper plate at T_c

Hot lower plate at T_h

δ

Figure 9-6 shows the formation of convective flow patterns giving rise to hexagonal cells called *Bernard cells*. The criterion given by Eq. (9-27) is important in practical applications, because it signifies the transition in the mechanism of heat transfer from pure conduction to convection. Note that heat transfer across the fluid layer is higher with convection than with conduction.

If the lower plate temperature were less than the upper plate temperature, the fluid would always remain stationary, because there would be no buoyancy forces to initiate a convective motion.

These basic concepts concerning the onset of free convection inside horizontal layers can be extended to inclined layers, but the resulting convective flow patterns are more complicated. Consider a fluid contained between two large parallel plates of height H separated by a distance δ and inclined at an angle ϕ, as illustrated in Fig. 9-7. The lower plate is maintained at a uniform temperature T_h which is higher than the temperature T_c of the upper, colder plate. This situation gives rise to temperature gradients within the fluid that generate a buoyancy force opposing the stabilizing effects of the viscous forces. Under certain conditions the fluid layer remains virtually stagnant. However, in most practical cases the fluid circulates as a result of the buoyancy produced by the temperature gradients. For this particular situation, the dimensionless parameters that affect the onset of free convection in the fluid include the inclination ϕ of the enclosure from the horizontal plane in addition to the Grashof and Rayleigh numbers, defined as

$$Gr_\delta = \frac{g\beta(T_h - T_c)\delta^3}{\nu^2} \tag{9-28a}$$

$$\mathrm{Ra} = Gr_\delta\,\mathrm{Pr} = \frac{g\beta(T_h - T_c)\delta^3}{\nu\alpha} \tag{9-28b}$$

as well as the aspect ratio H/δ.

The determination of the criterion for the onset of free convection and the corresponding free-convection heat transfer coefficient is important in engineering applications.

We now present some of the correlations for heat transfer in free convection in enclosures.

FIGURE 9-6
The formation of Bernard cells in a horizontal layer of fluid for $Ra_c > 1708$.

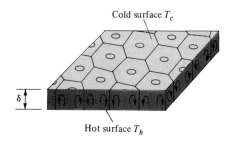

Cold surface T_c

Hot surface T_h

Vertical Layer, $\phi = 90°$

Consider a vertical layer of fluid between two parallel plates of height H separated by a distance δ, as illustrated in Fig. 9-7, but for the case of $\phi = 90°$. The hot and cold plates are maintained at uniform temperatures T_h and T_c, respectively. Let $a = H/\delta$ be the aspect ratio. The heat transfer across such a gap by free convection has been experimentally investigated by El Sherbiny, Raithby, and Hollands for air, and the following correlation is proposed for free convection across the air layer:

$$\boxed{\mathrm{Nu}_{90} = [\mathrm{Nu}_1, \mathrm{Nu}_2, \mathrm{Nu}_3]_{\max}} \qquad (9\text{-}29)$$

which implies that one should select the maximum of the three Nusselt numbers Nu_1, Nu_2, and Nu_3, which are defined as

$$\mathrm{Nu}_1 = 0.0605\,\mathrm{Ra}^{1/3} \qquad (9\text{-}30a)$$

$$\mathrm{Nu}_2 = \left\{1 + \left[\frac{0.104\,\mathrm{Ra}^{0.293}}{1 + (6310/\mathrm{Ra})^{1.36}}\right]^3\right\}^{1/3} \qquad (9\text{-}30b)$$

$$\mathrm{Nu}_3 = 0.242\left(\frac{\mathrm{Ra}}{a}\right)^{0.272} \qquad (9\text{-}30c)$$

for

$$\frac{H}{\delta} = 5 \text{ to } 110$$

$$10^2 < \mathrm{Ra}_\delta < 2 \times 10^7$$

where

$$a = \frac{H}{\delta} = \text{aspect ratio} \qquad (9\text{-}31a)$$

and

$$\mathrm{Ra} = \mathrm{Gr}_\delta\,\mathrm{Pr} = \frac{g\beta(T_h - T_c)\delta^3}{\nu^2}\,\mathrm{Pr} \qquad (9\text{-}31b)$$

FIGURE 9-7
Geometry and coordinates for free convection in an inclined layer.

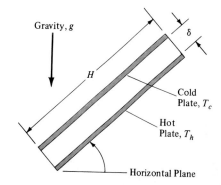

The Nusselt number can be defined as

$$Nu_{90} = \frac{q\delta}{k(T_h - T_c)} \qquad (9\text{-}32)$$

since $q = h(T_h - T_c)$, and all physical properties are evaluated at $(T_h + T_c)/2$. The experimental data for air agree with Eq. (9-29) to within 9 percent. Therefore, knowing the Nusselt number, the heat flux q can be determined.

Inclined Layers, $90° < \phi \le 60°$

Consider an inclined layer of fluid contained between two parallel plates separated by a distance δ, as illustrated in Fig. 9-7. The lower hot plate and the upper cold plate are at temperatures T_h and T_c, respectively. For the inclination $\phi = 60°$, the experimental data with air by El Sherbiny, Raithby, and Holland correlated the free-convection heat transfer across the gap by the following expression:

$$Nu_{60} = [Nu_1, Nu_2]_{max} \qquad (9\text{-}33)$$

where one should select the maximum of Nu_1 and Nu_2, defined as

$$Nu_1 = \left[1 + \left(\frac{0.093Ra^{0.314}}{1 + G}\right)^7\right]^{1/7} \qquad (9\text{-}34a)$$

$$Nu_2 = \left(0.104 + \frac{0.175}{a}\right)Ra^{0.283} \qquad (9\text{-}34b)$$

G is given by

$$G = \frac{0.5}{[1 + (Ra/3160)^{20.6}]^{0.1}} \qquad (9\text{-}34c)$$

and the Nusselt number is defined as

$$Nu_{60} = \frac{q\delta}{k(T_h - T_c)} \qquad (9\text{-}34d)$$

To determine the Nusselt number for inclinations where $60° \le \phi \le 90°$, a straight-line interpolation between Eqs. (9-29) and (9-33) is proposed; this yields

$$Nu_\phi = \frac{(90° - \phi°)Nu_{60°} + (\phi° - 60°)Nu_{90°}}{30°} \qquad \text{for } 60° \le \phi \le 90° \qquad (9\text{-}35)$$

The experimental data with air agree with Eq. (9-35) to within 6.5 percent.

Inclined Layers, $0° \le \phi \le 60°$

Based on the experimental investigations with air, the following correlation is proposed by Hollands, Unny, Raithby, and Konicek for free convection in inclined layers for inclinations $0° \le \phi \le 60°$:

$$\text{Nu}_\phi = 1 + 1.44\left[1 - \frac{1708}{\text{Ra} \cos \phi}\right]^*\left[1 - \frac{1708(\sin 1.8\phi)^{1.6}}{\text{Ra} \cos \phi}\right]$$
$$+ \left[\left(\frac{\text{Ra} \cos \phi}{5830}\right)^{1/3} - 1\right]^* \qquad (9\text{-}36)$$

for high aspect ratio (i.e., $H/\delta \rightarrow$ high) and in the range of $0 < \text{Ra} < 10^5$. Here the Nusselt number is defined as

$$\text{Nu}_\phi = \frac{q\delta}{k(T_h - T_c)}$$

$$q = \text{heat flux across layer, W/m}^2$$

and the notation [ ]* is used to denote that if the quantity in the bracket is negative, it should be set equal to zero. The experimental data with air agree with Eq. (9-36) to within 5 percent.

Figure 9-8 shows the Nusselt number Nu_ϕ computed from Eq. (9-36) plotted against $\text{Ra} \cos \phi$ for different inclinations between $\phi = 0°$ and $60°$.

Clearly, Eq. (9-36) includes the horizontal enclosure ($\phi = 0$) as a special case. Thus, by setting $\phi = 0$ in Eq. (9-36), the Nusselt number for free convection inside a horizontal enclosure becomes

$$\text{Nu}_{\phi=0} = 1 + 1.44\left[1 - \frac{1708}{\text{Ra}}\right]^* + \left[\left(\frac{\text{Ra}}{5830}\right)^{1/3} - 1\right]^* \qquad (9\text{-}37)$$

FIGURE 9-8
A plot of Nusselt number for free convection inside inclined enclosures for $0 \le \phi \le 60°$, computed from Eq. (9-36).

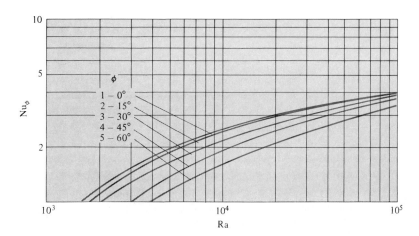

Note that for $Ra < 1708$, Eq. (9-37) reduces to

$$Nu_{\phi=0} = 1 \qquad \text{for } Ra < 1708$$

which is consistent with the fact that, as we have previously discussed, the free-convection flow between parallel plates sets up at a critical Rayleigh number 1708, and for $Ra < 1708$ the conduction regime prevails.

Example 9-8. A double-glass window consists of two vertical parallel sheets of glass, each 1.6 m by 1.6 m in size, separated by 1.5 cm of air space at atmospheric pressure. Calculate the free-convection heat transfer coefficient for the air space for a temperature difference of 35°C. Assume a mean temperature for the air of 27°C.

Solution. The physical properties of air at $T_f = 27°C = 300\,K$ are

$$\nu = 1.568 \times 10^{-5}\,m^2/s \qquad Pr = 0.708$$

$$k = 0.02624\,W/(m \cdot °C)$$

The Rayleigh number becomes

$$Ra = \frac{g\beta(T_h - T_c)\delta^3}{\nu^2}\,Pr$$

$$= \frac{(9.81)(1/300)(35)(0.015)^3}{(1.568 \times 10^{-5})^2}\,(0.708)$$

$$= 1.112 \times 10^4$$

The aspect ratio is

$$a = \frac{H}{\delta} = \frac{1.6}{0.015} = 106.67$$

We use Eqs. (9-29) and (9-30a) to (9-30c) to calculate the Nusselt number:

$$Nu = [Nu_1, Nu_2, Nu_3]_{max}$$

where Nu is the maximum of the following three Nusselt numbers:

$$Nu_1 = 0.0605\,Ra^{1/3} = 0.0605(1.112 \times 10^4)^{1/3} = 1.35$$

$$Nu_2 = \left\{1 + \left[\frac{0.104\,Ra^{0.293}}{1 + (6310/Ra)^{1.36}}\right]^3\right\}^{1/3} = 1.32$$

$$Nu_3 = 0.242\left(\frac{Ra}{a}\right)^{0.272} = 0.86$$

Therefore

$$Nu = Nu_{max} = 1.35$$

$$h = \frac{k}{\delta}\,Nu = \frac{0.02624}{0.015}\,1.35 = 2.36\,W/(m^2 \cdot °C)$$

$$q = h\,\Delta T = (2.36)(35) = 82.66\,W/m^2$$

Example 9-9. In a horizontal flat-plate solar collector the absorber plate and the glass cover are separated by an air gap at atmospheric pressure. Estimate the heat transfer coefficient for free convection across the air gap for a gap spacing 3.0 cm, assuming that the absorber plate is at 60°C and the glass cover at 30°C.

Solution. The physical properties of air at $T_f = (T_h + T_c)/2 = (60 + 30)/2 = 45°C = 318$ K are

$$\nu = 1.751 \times 10^{-5} \text{ m}^2/\text{s} \qquad \text{Pr} = 0.704$$

$$k = 0.0276 \text{ W}/(\text{m} \cdot °\text{C})$$

The Rayleigh number becomes

$$\text{Ra} = \frac{g\beta(T_h - T_c)\delta^3}{\nu^2} \text{ Pr}$$

$$= (9.81) \frac{(1/318)(60 - 30)(0.03)^3}{(1.751 \times 10^{-5})^2} (0.704)$$

$$= 57,356$$

For an horizontal enclosure with $\phi = 0°$, we use Eq. (9-37) to calculate the Nusselt number:

$$\text{Nu} = 1 + 1.44\left[1 - \frac{1708}{\text{Ra}}\right]^* + \left[\left(\frac{\text{Ra}}{5830}\right)^{1/3} - 1\right]^*$$

where [ ]* is used to denote that if the quantity in the bracket is negative, it should be set equal to zero. We have Ra = 57,356, and so both brackets in Eq. (9-37) are positive; therefore the Nusselt number becomes

$$\text{Nu} = 1 + 1.44\left[1 - \frac{1708}{57,356}\right] + \left[\left(\frac{57,356}{5830}\right)^{1/3} - 1\right]$$

$$= 3.54$$

Then,

$$h = \frac{k}{\delta} \text{ Nu} = \frac{0.0276}{0.03} (3.54) = 3.26 \text{ W}/(\text{m}^2 \cdot °\text{C})$$

and

$$q = h(T_h - T_c) = 3.26 \times 30 = 97.8 \text{ W}/\text{m}^2$$

Horizontal Cylindrical Annulus

Consider a fluid contained in a long, horizontal cylindrical annulus with a gap spacing $\delta = \frac{1}{2}(D_o - D_i)$, where D_o and D_i are, respectively, the diameters of the outer and inner cylinders, as illustrated in Fig. 9-9.

Raithby and Hollands proposed the following correlation for the heat transfer rate Q for the length H of a cylindrical annulus:

$$Q = \frac{2\pi k_{\text{eff}} H}{\ln (D_o/D_i)} (T_i - T_o) \qquad \text{W} \qquad (9\text{-}38)$$

where

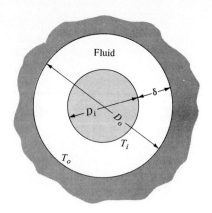

FIGURE 9-9
Free convection in horizontal
cylindrical annulus or spherical
annulus.

$$\frac{k_{eff}}{k} = 0.386\left(\frac{Pr}{0.861 + Pr}\right)^{1/4} (Ra_{cyl}^*)^{1/4}$$

$$(Ra_{cyl}^*)^{1/4} = \frac{\ln(D_o/D_i)}{\delta^{3/4}(D_i^{-3/5} + D_o^{-3/5})^{5/4}} \, Ra_\delta^{1/4}$$

$$H = \text{length of cylindrical annulus}$$

$$Ra_\delta = \frac{g\beta(T_i - T_o)\delta^3}{\nu^2} \, Pr$$

$$\delta = \tfrac{1}{2}(D_o - D_i)$$

This correlation is applicable over the range $10^2 < Ra_{cyl}^* < 10^7$. Figure 9-10 shows the flow path in free convection between two concentric horizontal cylinders for $D_o/D_i = 3$, $T_i - T_o = 14.5°C$, and Grashof number 120,000. Cigarette smoke shows the fully developed laminar flow pattern in air at atmospheric pressure.

Spherical Annulus

We now consider a fluid contained between two concentric spheres of inner and outer diameters D_i and D_o, respectively, as illustrated in Fig. 9-9. The surfaces are maintained at uniform temperatures T_i and T_o. Raithby and Hollands proposed the following correlation for the total heat transfer rate from the spheres:

$$\boxed{Q = k_{eff} \frac{\pi D_i D_o}{\delta} (T_i - T_o)} \qquad W \qquad (9\text{-}39)$$

where

$$\frac{k_{eff}}{k} = 0.74\left(\frac{Pr}{0.861 + Pr}\right)^{1/4} (Ra_{sph}^*)^{1/4}$$

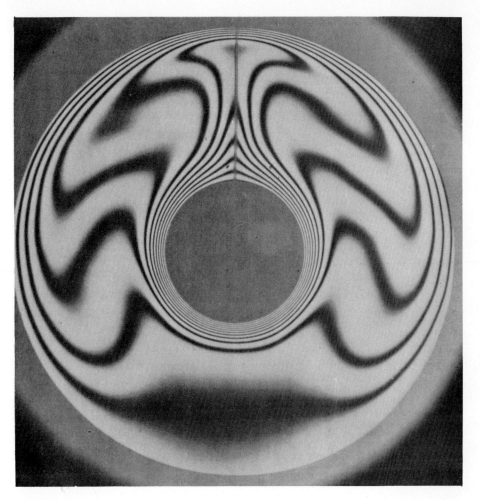

FIGURE 9-10 Isotherms in convection between concentric cylinders. (From Grigull and Hauf).

and

$$(\text{Ra}^*_{\text{sph}})^{1/4} = \frac{\delta^{1/4} \, \text{Ra}_\delta^{1/4}}{D_i D_o (D_i^{-7/5} + D_o^{-7/5})^{5/4}}$$

$$\text{Ra}_\delta = \frac{g\beta(T_i - T_o)\delta^3}{\nu^2} \, \text{Pr}$$

$$\delta = \tfrac{1}{2}(D_o - D_i)$$

This correlation is applicable over the range $10^2 < \text{Ra}^*_{\text{sph}} < 10^4$.

Example 9-10. The space between two concentric thin-walled spheres of diameters $D_1 = 10$ cm and $D_2 = 14$ cm contains air at atmospheric pressure. Calculate the heat transfer rate across the spheres by free convection for a temperature difference of 50°C and a mean air temperature of 27°C.

Solution. The physical properties of air at 300 K are

$$\nu = 1.568 \times 10^{-5} \text{ m}^2/\text{s} \qquad \text{Pr} = 0.708$$

$$k = 0.02624 \text{ W}/(\text{m} \cdot {}^\circ\text{C})$$

The Rayleigh number becomes

$$\text{Ra}_\delta = \frac{g\beta(T_i - T_o)\delta^3}{\nu^2} \text{ Pr}$$

$$= \frac{(9.81)(1/300)(50)(0.02)^3}{(1.568 \times 10^{-5})^2} (0.708)$$

$$= 37{,}666$$

Equation (9-39) is now used to calculate the total heat transfer rate as follows:

$$\text{Ra}_{\text{sph}}^* = \frac{\delta}{(D_i D_o)^4} \frac{\text{Ra}_\delta}{(D_i^{-7/5} + D_o^{-7/5})^5} = 173.4$$

$$k_{\text{eff}} = 0.74k \left(\frac{\text{Pr}}{0.861 + \text{Pr}}\right)^{1/4} (\text{Ra}_{\text{sph}}^*)^{1/4} = 0.0578$$

Then the total heat transfer rate Q becomes

$$Q = k_{\text{eff}} \frac{\pi D_i D_o}{\delta} (T_i - T_o)$$

$$= 0.0578 \frac{\pi(0.1)(0.14)}{0.02} (50)$$

$$= 6.35 \text{ W}$$

PROBLEMS

Free Convection on a Flat Plate

9-1. A vertical plate 0.4 m high and 2 m wide, maintained at a uniform temperature of 92°C, is exposed to quiescent atmospheric air at 32°C. Calculate the average heat transfer coefficient h_m over the entire height of the plate, using the correlations given by both Eqs. (9-6) and (9-10).

9-2. Calculate the heat transfer rates by free convection from a 0.3-m-high vertical plate maintained at a uniform temperature of 70°C to ambient air at 34°C at 1.0 and 2.0 atm.

9-3. A vertical plate 0.3 m high and 1 m wide, maintained at a uniform temperature of 134°C, is exposed to quiescent atmospheric air at 20°C. Calculate the total heat transfer rate from the plate to the air.

9-4. Calculate the heat transfer rate from a 0.3-m-high vertical plate at a uniform temperature of 80°C to quiescent ambient air at 24°C at 3 atm pressure.
Answer: 525 W/m^2

9-5. Calculate the total heat transfer rate by free convection from a vertical plate 0.3 m by 0.3 m with one surface insulated and the other surface maintained at 100°C and exposed to atmospheric air at 30°C.

Answer: 36 W

9-6. Compare the heat transfer rates by free convection from a 0.3 m by 0.3 m plate with one surface insulated and the other surface maintained at $T_w = 100°C$ and exposed to quiescent atmospheric air at $T_\infty = 30°C$ for the following conditions: (*a*) the plate is horizontal with the heated surface facing up; (*b*) the plate is horizontal with the heated surface facing down.

9-7. Consider a 0.15 m by 0.3 m rectangular plate maintained at a uniform temperature $T_w = 80°C$ placed vertically in quiescent atmospheric air at temperature $T_\infty = 30°C$. Compare the heat transfer rates from the plate when 0.15 m is the vertical height and when 0.3 m is the vertical height.

Answers: 16.5 W, 13.8 W

9-8. The south-facing wall of the building is 7 m high and 10 m wide. The wall absorbs the incident solar energy at a rate of 600 W/m². It is assumed that about 200 W/m² of the absorbed energy is conducted through the wall into the interior of the wall and that the remaining is dissipated by free convection into the surrounding quiescent atmospheric air at 30°C. Calculate the temperature of the outside surface of the wall.

9-9. An electrically heated 0.5 m by 0.1 m vertical plate is insulated on one side and dissipates heat from the other surface at a constant rate of 600 W/m² by free convection into atmospheric air at 30°C. Determine the surface temperature of the plate.

9-10. A cube 5 cm × 5 cm × 5 cm in size is suspended with one of its surfaces in a horizontal position in quiescent atmospheric air at 20°C. All surfaces of the cube are maintained at a uniform temperature of 100°C. Determine the heat loss by free convection from the cube into the atmosphere.

Answer: 9.4 W

9-11. A circular hot plate $D = 25$ cm in diameter with both of its surfaces maintained at a uniform temperature of 100°C is suspended in a horizontal position in atmospheric air at 30°C. Determine the heat transfer rate by free convection from the plate into the atmosphere.

Answer: 31.4 W

9-12. A thin electric strip heater 30 cm in width is placed with its width oriented vertically. It dissipates heat by free convection from both its surfaces into atmospheric air at $T_\infty = 20°C$. If the surface of the heater should not exceed 225°C, determine the length of the strip needed to dissipate 1500 W of energy into the room.

9-13. A hot iron block at 425°C 10 cm × 15 cm × 20 cm in size is placed on an asbestos sheet with its 10-cm side oriented vertically. There is negligible heat loss from the surface that is in contact with the asbestos sheet. Calculate the rate of heat loss from its five boundry surfaces by free convection into the surrounding quiescent air at atmospheric pressure and 25°C.

Answer: 528.5 W

9-14. A horizontal square plate heater 0.3 m by 0.3 m is thermally insulated on the bottom side and exposed to quiescent atmospheric air at $T_\infty = 30°C$ on the top side. The plate heater is electrically heated at a rate of 400 W/m², and heat dissipates by free convection into the air. Calculate the equilibrium temperature of the plate. [Assume $h = 6$ W/(m² · °C) as a first approximation.]

9-15. A circular hot plate $D = 50$ cm in diameter with both surfaces maintained at a uniform temperature of 300°C is suspended in a vertical position in water at 20°C. Determine the heat loss from the hot plate to the water.

9-16. A circular hot plate $D = 50$ cm in diameter with both surfaces maintained at a uniform temperature of 300°C is suspended in a vertical position in atmospheric air at 20°C. Determine the heat loss from the plate to the air.

Free Convection on a Long Cylinder

9-17. A bare 10-cm-diameter horizontal steam pipe is 100 m long and has a surface temperature of 150°C. Estimate the total heat loss by free convection to atmospheric air at 20°C.

9-18. An electric heater of outside diameter $D = 2.5$ cm and length $L = 0.5$ m is immersed horizontally in a large tank of engine oil at 20°C. If the surface temperature of the heater is 140°C, determine the rate of heat transfer to the oil.
    Answer: 575 W

9-19. A horizontal electric heater of outside diameter $D = 2.5$ cm and length $L = 0.5$ m dissipates heat by free convection into atmospheric air at $T_\infty = 30$°C. If the surface temperature of the heater is 250°C, calculate the rate of heat transfer from the heater into the air.
    Answer: 94.7 W

9-20. A cylindrical electric heater of outside diameter $D = 2.5$ cm and length $L = 2$ m is immersed horizontally in mercury at a temperature of 100°C. If the surface of the heater is maintained at an average temperature of 300°C, calculate the rate of heat transfer to the mercury.

9-21. Calculate the average heat transfer coefficient h from an isothermal horizontal cylinder of diameter $D = 5$ cm at temperature $T_w = 400$ K to quiescent air at atmospheric pressure and at temperature $T_\infty = 300$ K.
    Answer: 7.9 W/(m² · °C)

9-22. A hot gas at 220°C flows through a horizontal pipe of outside diameter $D = 15$ cm. The pipe has an uninsulated portion of length $L = 3$ m that is exposed to atmospheric air at temperature $T_\infty = 30$°C. Assuming that the outside surface of the pipe is at 220°C, determine the rate of heat loss into the atmosphere.

9-23. An uninsulated horizontal duct of diameter $D = 20$ cm carrying cold air at 10°C is exposed to quiescent atmospheric air at 35°C. Determine the heat gain by free convection per meter length of the duct.
    Answer: 21.3 W/m

9-24. A horizontal pipe of diameter $D = 2$ cm with the outer surface at 225°C is exposed to atmospheric air at 25°C. Calculate the heat transfer rate per meter length of the pipe by free convection.

9-25. A 1.5-m-long vertical cylinder of diameter $D = 2.5$ cm maintained at a uniform temperature of 140°C is exposed to atmospheric air at 15°C. Determine the free-convection heat transfer coefficient.
    Answer: 88.83 W/(m² · °C)

9-26. A $D = 5$ cm OD and $L = 40$ cm long tube, maintained at a uniform temperature $T_w = 400$ K, is placed vertically in quiesent air at atmospheric pressure and at temperature $T_\infty = 300$ K. Calculate the average free-convection heat transfer coefficient and the rate of heat loss from the tube to the air.

Free Convection on a Sphere

9-27. Calculate the heat transfer rate by free convection from a 2-cm-diameter sphere whose surface is maintained at 120°C to water at 20°C.

9-28. Calculate the heat transfer rate by free convection from a 2-cm-diameter sphere whose surface is maintained at 110°C to air at 20°C.

9-29. Calculate the heat transfer rate by free convection from a 20-cm-diameter sphere whose surface is maintained at 140°C to atmospheric air at 1 atm and at 20°C.

9-30. A sphere of diameter $D = 5$ cm whose surface is maintained at a uniform temperature of 120°C is submerged in quiescent water at 30°C. Determine the heat transfer rate by free convection from the sphere to the water.
Answer: 642.2 W

9-31. Calculate the heat transfer rate by free convection from a 20-cm-diameter sphere whose surface is maintained at 120°C to the surrounding air at 20°C and at 1 atm.
Answer: 83.4 W

9-32. A sphere of diameter $D = 5$ cm, maintained at a uniform temperature of 60°C, is immersed in water at 20°C. Calculate the rate of heat loss by free convection.
Answer: 3.72 W

9-33. Compare the heat loss by free convection from a spherical body of diameter $D = 0.5$ m, maintained at a uniform temperature 30°C, to ambient air at −10°C and 5 atm.

9-34. A 15-cm-diameter, electrically heated sphere is immersed in a quiescent body of air at 20°C. Calculate the amount of heat to be supplied by the electric heater in order to keep the surface temperature of the sphere at 100°C.

Simplified Equations for Air

9-35. A vertical plate 1 m high and 1 m wide at a uniform temperature of 40°C is exposed to atmospheric air at 20°C. Determine the free-convection heat transfer coefficient, and compare the result with that obtainable from the simplified expressions for free convection to air at atmospheric pressure.

9-36. A vertical plate 0.2 m by 0.2 m in size at a uniform temperature 140°C is exposed to atmospheric air at 15°C. Calculate the free-convection heat transfer coefficient, and compare the result with that obtainable from the simplified expressions for air at atmospheric pressure.

9-37. A horizontal plate 0.2 m by 0.2 m in size at a uniform temperature of 140°C is exposed to atmospheric air at 15°C. Calculate the free-convection heat transfer coefficient for (*a*) the heated surface facing up, and (*b*) the heated surface facing down, and compare the results with those obtainable from the simplified expressions for air at atmospheric pressure.
Answers: (*a*) 6.51 (or 7.6) W/(m$^2 \cdot$ °C), (*b*) 2.92 (or 2.95) W/(m$^2 \cdot$ °C)

9-38. A long horizontal cylinder 5 cm in diameter, maintained at a uniform temperature of 140°C, is exposed to atmospheric air at 10°C. Calculate the free-convection heat transfer coefficient, and compare it with that obtainable from the simplified expression for air at atmospheric pressure.

9-39. Calculate the heat transfer coefficient for free convection from a 2.5-cm-diameter horizontal cylinder maintained at a uniform temperature 325°C into atmospheric air at 30°C. Compare this result with that obtainable from the simplified expression for free convection in air.
Answers: 11.48 (or 13.76) W/(m$^2 \cdot$ °C)

9-40. A horizontal electric heater with an outside diameter of 2.5 cm and length of 0.5 m is exposed to atmospheric air at 15°C. If the surface of the heater is at 130°C, determine the heat transfer coefficient for free convection, and compare it with that obtainable from the simplified expression for air at atmospheric pressure.

9-41. A vertical plate 0.3 m by 0.3 m in size, maintained at a uniform temperature of 70°C, is exposed to cold atmospheric air at a temperature of 10°C. Calculate the free-convection heat transfer coefficient, and compare it with that obtainable from the simplified expressions for free convection to air at 4 atm.
Answers: 5.65 (or 5.34) W/(m$^2 \cdot$°C), 11.4 (or 12.9) W/(m$^2 \cdot$°C)

Free Convection in Enclosed Spaces

9-42. A double-glass window consists of two vertical parallel sheets of glass, each 1 m by 1 m in size, separated by a 1-cm air space at atmospheric pressure. Calculate the free-convection heat transfer coefficient for the air space for a temperature difference of 30°C. Assume the mean temperature for the air to be 27°C.

9-43. Two vertical parallel plates, each 1 m × 1 m in size, are separated by a distance of 3 cm filled with atmospheric air. One of the plates is maintained at a uniform temperature of 250°C and the other at 100°C. Calculate the heat transfer rate by free convection between the plates.

9-44. Two parallel vertical plates, each 2 m high, are separated by a 6-cm air space. One of the plates is maintained at a uniform temperature 130°C and the other at 25°C. Determine the free-convection heat transfer coefficient for free convection between the two plates.
Answer: 3.05 W/(m$^2 \cdot$°C)

9-45. Atmospheric air is contained between two large horizontal parallel plates separated by a distance of 5 cm. The lower plate is maintained at 100°C and the upper plate at 30°C. Determine the heat transfer rate between the plates by free convection per square meter of the plate surface.
Answer: 233.1 W/m^2

9-46. Two parallel horizontal plates are separated by a distance of 2 cm. The lower plate is at a uniform temperature of 220°C, and the upper plate is at 35°C. Determine the heat transfer rate between the plates per square meter of the plate surface.
Answer: 1006.5 W/m^2

9-47. Water is contained between two parallel horizontal plates separated by a distance of 2 cm. The lower plate is at a uniform temperature of 130°C, and the upper plate is at 30°C. Determine the heat transfer rate between the plates per square meter of the plate surface.
Answer: 58.1 kW/m^2

9-48. In a horizontal flat-plate solar collector the absorber plate and the glass cover are separated by an air gap at atmospheric pressure. Estimate the heat transfer coefficient for free convection across the air gap for a gap spacing of 2 cm and 4 cm, assuming that the absorber plate is at 80°C and the glass cover at 30°C.

9-49. A 50 cm by 50 cm horizontal double window glass is constructed of two glass plates separated by a 2-cm air gap at atmospheric pressure. If the upper and lower glass plates are at temperatures of −15°C and 20°C, respectively, calculate the heat transfer rate across the gap.
Answer: 23.2 W

9-50. The annular space between two horizontal thin-walled coaxial cylinders contains air at atmospheric pressure. The inner cylinder has a diameter $D_1 = 8$ cm and is maintained at a uniform temperature $T_1 = 100°C$, while the outer cylinder has a diameter $D_2 = 12$ cm and is maintained at a uniform temperature $T_2 = 50°C$. Calculate the heat transfer rate by free convection across the air space per meter length of the cylinders.

Answer: 42.9 W/m length

9-51. The annular space between two horizontal thin-walled coaxial cylinders contains air at atmospheric pressure. The diameters of the inner and outer cylinders are $D_1 = 8$ cm and $D_2 = 12$ cm, respectively. The temperature difference between the cylinders is 50°C, and the mean temperature of the air in the annular space is 27°C. Calculate the heat transfer rate by free convection across the air space per meter length of the cylinders.

Answer: 49.9 W/m length

9-52. The annular space between two horizontal thin-walled coaxial cylinders is filled with water. The inner and outer cylinders have diameters $D_1 = 10$ cm and $D_2 = 13$ cm, respectively. Calculate heat transfer by free convection across the water-filled space per meter length of the cylinder for a temperature difference of 50°C and a mean temperature of 40°C for the water.

9-53. The annular space between two thin-walled horizontal coaxial cylinders is filled with water. The inner and outer cylinders have diameters $D_1 = 7$ cm and $D_2 = 10$ cm, respectively. Calculate the heat transfer rate across the annulus by free convection when the inside and outside cylinders are maintained at $T_1 = 80°C$ and $T_2 = 40°C$, respectively.

9-54. The space between two concentric thin-walled spheres contains atmospheric air. The inner and outer spheres have diameters $D_1 = 4$ cm and $D_2 = 6$ cm, respectively, and the temperature difference between the surfaces is 20°C. Calculate the heat transfer rate across the space by free convection for air at a mean temperature of 27°C.

9-55. A spherical storage tank of diameter $D_1 = 2$ m contains a cold liquid at a temperature $T_1 = 10°C$. To reduce heat losses, this storage tank is enclosed inside another spherical shell, and the gap spacing is 3 cm. The temperature of the outer sphere is $T_2 = 20°C$. Determine the rate of heat loss by free convection across the gap filled with air at atmospheric pressure.

Answer: 125 W

REFERENCES

1. Amato, W. S., and C. Tien: "Free Convection Heat Transfer from Isothermal Spheres in Water," *Int. J. Heat Mass Transfer* **15**:327–339 (1972).
2. Churchill, S. W., and H. H. S. Chu: "Correlating Equations for Laminar and Turbulent Free Convection from a Vertical Plate," *Int. J. Heat Mass Transfer* **18**:1323 (1975).
3. El Sherbiny, S. M., G. D. Raithby, and K. G. T. Hollands: "Heat Transfer by Natural Convection across Vertical and Inclined Air Layers," *J. Heat Transfer* **104C**:96–102 (1982).
4. Fujii, T., and H. Imura: "Natural Convection Heat Transfer from a Plate with Arbitrary Inclination," *Int. J. Heat Mass Transfer* **15**:755 (1972).
5. Grigull, U., and W. Hauf: Proceedings of 3d International Heat Transfer Conference **2**:182–195.
6. Hollands, K. G. T., T. E. Unny, G. D. Raithby, and L. Konicek: "Free Convective Heat Transfer across Inclined Air Layers," *J. Heat Transfer* **98C**:189–193 (1976).
7. McAdams, W. H.: *Heat Transmission*, 3d ed., McGraw-Hill, New York, 1954.

8. Morgan, V. T.: "The Overall Convective Heat Transfer from Smooth Circular Cylinders," in T. F. Irvine and J. P. Hartnett (eds.), *Advances in Heat Transfer*, vol. 16, Academic, New York, 1975, pp. 199–264.

9. Raithby, G. D., and K. G. T. Hollands: "A General Method of Obtaining Approximate Solutions to Laminar and Turbulent Free Convection Problems," in T. F. Irvine and J. P. Hartnett (eds.), *Advances in Heat Transfer*, vol. 11, Academic, New York, 1975, pp. 265–315.

10. Rich, B. R.: "An Investigation of Heat Transfer from an Inclined Flat Plate in Free Convection," *Trans. ASME* **75**:489–499 (1953).

11. Sparrow, E. M., and J. L. Gregg: Laminar Free Convection from a Vertical Plate," *Trans. ASME*, **78**:435–440 (1956).

12. Vliet, G. C.: "Natural Convection Local Heat Transfer on Constant-Heat-Flux Inclined Surfaces," *J. Heat Transfer* **91C**:511–516 (1969).

13. Vliet, G. C., and C. K. Liu: "An Experimental Study of Natural Convection Boundary Layers," *J. Heat Transfer* **91C**:517–531 (1969).

14. Yuge, T.: "Experiments on Heat Transfer from Spheres Including Combined Natural and Free Convection," *J. Heat Transfer* **82C**:214–220 (1960).

CHAPTER
10

CONDENSATION AND BOILING

Condensers and boilers are important and widely used types of heat exchangers with unique characteristics of the heat transfer mechanism. If a vapor strikes a surface that is at a temperature below the corresponding saturation temperature, the vapor will immediately condense into the liquid phase. If the condensation takes place continuously over a surface which is kept cooled by some process, and if the condensed liquid is removed from the surface by the force of gravity, then the condensing surface is usually covered with a thin layer of liquid, and the situation is known as *filmwise condensation*. Under certain conditions, for example, if traces of oil are present during the condensation of steam on a highly polished surface, the film of condensate is broken into droplets, and the situation is known as *dropwise condensation*. The droplets grow, coalesce, and increase in size; eventually they run off the condensing surface under the influence of gravity, sweeping off other droplets in their path. Therefore, with dropwise condensation the condensing surface between the drops is exposed to condensing vapor; as a result, the thermal resistance to heat flow of the condensate layer is much less, and hence the heat transfer rates are several times higher with dropwise condensation than with filmwise condensation.

The presence of condensate acts as a barrier to heat transfer from the vapor to the metal surface, and dropwise condensation offers much less resistance to heat flow on the vapor side than does filmwise condensation. If the vapor contains some noncondensable gas, this gas will collect on the condensing side while condensation takes place, and the noncondensable gas will act as a resistance to heat flow on the condensing side. Therefore, an accurate prediction of the heat transfer coefficient for condensing vapors with and without the presence of noncondensable gas is important in the design of condensers.

When a liquid is in contact with a surface that is maintained at a temperature above the saturation temperature of the liquid, boiling may occur. The phenomenon of heat transfer in boiling is extremely complicated because a large number of variables are involved and because very complex hydrodynamic developments occur during the process. Therefore, considerable work has been directed toward gaining a better understanding of the boiling mechanism.

In this chapter we present some of the recommended correlations for predicting the heat transfer coefficient during condensation and boiling.

10-1   LAMINAR FILM CONDENSATION

When the temperature of a vapor is reduced below its saturation temperature, the vapor condenses. In engineering applications, the vapor is condensed by bringing it into contact with a cold surface. The steam condensers for power plants are typical examples of this application. If the liquid wets the surface, the condensation occurs in the form of a smooth film, which flows down the surface under the action of the gravity. The presence of a liquid film over the surface constitutes a thermal resistance to heat flow. The first fundamental analysis leading to the determination of the heat transfer coefficient during filmwise condensation of pure vapors (i.e., without the presence of noncondensable gas) on a flat plate and a circular tube was given by Nusselt in 1916. Over the years, improvements have been made on Nusselt's theory of film condensation. But except in the case of the condensation of liquid metals, Nusselt's original theory has been successful and still is widely used. Here we present the correlations of the heat transfer coefficient during film condensation of vapors based on Nusselt's theory of condensation.

Condensation on Vertical Surfaces

Consider a cold vertical plate at temperature T_w exposed to a large body of saturated vapor at temperature T_v ($>T_w$), as illustrated in Fig. 10-1. Here x is the axial coordinate measured downward along the plate, and y is the coordinate normal to the condensing surface. Note that the condensate film formed on the surface moves downward under the influence of gravity and that its thickness $\delta(x)$ increases with the distance along the surface. The velocity and temperature profiles are schematically illustrated in Fig. 10-1.

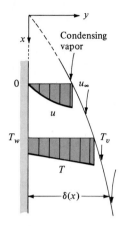

FIGURE 10-1
Schematic of filmwise
condensation on a
vertical surface.

This condensation problem was first analyzed by Nusselt under the following assumptions:

1. The plate is maintained at a uniform temperature T_w that is less than the saturation temperature T_v of the vapor.
2. The vapor is stationary or has low velocity, and so it exerts no drag on the motion of the condensate.
3. The downward flow of condensate under the influence of gravity is laminar.
4. The flow velocity associated with the condensate film is low; as a result, the flow acceleration in the condensate layer is negligible.
5. Fluid properties are constant.
6. Heat transfer across the condensate layer is by pure conduction; hence the liquid temperature distribution is linear.

With these considerations, Nusselt developed the following expression for the average value of the heat transfer coefficient h_m for filmwise condensation over a vertical plate of height L:

$$h_m = 0.943 \left[\frac{g\rho_l(\rho_l - \rho_v)h_{fg}k_l^3}{\mu_l(T_v - T_w)L} \right]^{1/4} \qquad \text{W/(m}^2 \cdot \text{°C)} \qquad (10\text{-}1a)$$

A comparison of this theoretical result with the results of experiments has shown that the measured heat transfer coefficient is about 20 percent higher than that predicted by the theory. Therefore, McAdams recommends that Nusselt's theoretical expression, Eq. (10-1a), should be multiplied by a factor of 1.2. Then, the recommended equation for the average heat transfer coefficient h_m for filmwise condensation over a vertical plate of height L becomes

$$h_m = 1.13 \left[\frac{g\rho_l(\rho_l - \rho_v)h_{fg}k_l^3}{\mu_l(T_v - T_w)L} \right]^{1/4} \qquad \text{for Re} < 1800 \qquad (10\text{-}1b)$$

where g = acceleration due to gravity, m/s^2
  h_{fg} = latent heat of condensation, J/kg
  k_l = thermal conductivity of liquid, W/(m · °C)
  L = length of vertical plate, m
  ρ_l, ρ_v = density of liquid and vapor, kg/m^3
  μ_l = viscosity of liquid, kg/(m.s)
  T_v, T_w = vapor and wall temperature, respectively, °C

and the physical properties, including h_{fg}, are evaluated at the film temperature

$$T_f = \tfrac{1}{2}(T_w + T_v) \qquad (10\text{-}2)$$

The additional energy needed to cool the condensate film below saturation temperature is accommodated approximately by evaluating h_{fg} at the film temperature instead of at the saturation temperature. The heat transfer coefficient derived above for a vertical plate is also applicable for condensation on the outside or inside surface of a vertical tube, provided that the tube radius is large compared with the thickness of the condensate film.

The *Reynolds number* for condensate flow is defined as

$$\text{Re} = \frac{4\dot{m}}{\mu_l P} \tag{10-3}$$

Here $\dot{m}$ is the total mass flow rate of condensate at the lowest part of the condensing surface (i.e., at $x = L$), in kilograms per second, and P is the wetted perimeter, defined as

$$P = \begin{cases} w & \text{for vertical plate of width } w \\ \pi D & \text{for vertical tube of outside diameter } D \end{cases} \tag{10-4}$$

Experiments have shown that the transition from laminar to turbulent condensate flow takes place at a Reynolds number of about 1800. Therefore, the correlation given by Eq. (10-1b) is valid only for the laminar flow of condensation over the surface. Generally, $\rho_v \ll \rho_l$; hence Eq. (10-1b) reduces to

$$h_m = 1.13 \left[\frac{g\rho_l^2 h_{fg} k_l^3}{\mu_l(T_v - T_w)L} \right]^{1/4} \qquad \text{for Re} < 1800 \tag{10-5}$$

which can be rearranged in the form

$$h_m \left(\frac{\mu_l^2}{k_l^3 \rho_l^2 g} \right)^{1/3} = 1.76 \, \text{Re}^{-1/3} \qquad \text{for Re} > 1800 \tag{10-6}$$

and the physical properties, including h_{fg}, are evaluated at the film temperature, defined by Eq. (10-2).

Condensation on Inclined Plates

Nusselt's analysis of filmwise condensation for a vertical surface, given above, can readily be extended to condensation on an inclined plane surface making an angle ϕ with the horizontal, as illustrated in Fig. 10-2, if the term g is replaced by $g \sin \phi$ in Eq. (10-1b). We find

$$h_m = 1.13 \left[\frac{g\rho_l(\rho_l - \rho_v) h_{fg} k_l^3}{\mu_l(T_v - T_w)L} \sin \phi \right]^{1/4} \qquad \text{for Re} < 1800 \tag{10-7}$$

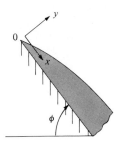

FIGURE 10-2
Nomenclature for filmwise condensation on an inclined plane surface.

For $\rho_v \ll \rho_1$, Eq. (10-7) reduces to

$$h_m = 1.13\left[\frac{g\rho_l^2 h_{fg}k_l^3}{\mu_1(T_v - T_w)L}\sin\phi\right]^{1/4} \qquad \text{for Re} > 1800 \qquad (10\text{-}8)$$

where ϕ is the angle between the plate and the horizontal. The physical properties are evaluated at the film temperature T_f, defined by Eq. (10-2).

Example 10-1. Air-free saturated steam at $T_v = 90°C$ and $P = 70.14$ kPa condenses on the outer surface of a $L = 1.5$ m long, $D = 2.5$ m outside diameter (OD) vertical tube maintained at a uniform temperature $T_w = 70°C$. Assuming filmwise condensation, calculate the average heat transfer coefficient over the entire length of the tube.

Solution. The physical properties of water at the film temperature $T_f = (70 + 90)/2 = 80°C$ are

$$k_l = 0.668 \text{ W/(m} \cdot °C) \qquad \mu_l = 0.355 \times 10^{-3} \text{ kg/(m} \cdot s)$$

$$\rho_l = 974 \text{ kg/m}^3 \qquad h_{fg} = 2309 \text{ kJ/kg}$$

and $g = 9.8 \text{ m}^2/\text{s}$. The average heat transfer coefficient h_m for laminar filmwise condensation on a vertical surface of the tube with $\rho_v \ll \rho_l$ is determined from Eq. (10-5):

$$h_m = 1.13\left[\frac{g\rho_l^2 h_{fg}k_l^3}{\mu_l(T_v - T_w)L}\right]^{1/4}$$

$$= 1.13\left[\frac{(9.81)(974)^2(2309 \times 10^3)(0.668)^3}{(0.355 \times 10^{-3})(90 - 70)(1.5)}\right]^{1/4} = 5596 \text{ W/(m}^2 \cdot °C)$$

Example 10-2. Air-free saturated steam at $T_v = 75°C$ and $P = 38.58$ kPa condenses on a 0.5-m-long plate inclined 45° from the horizontal and maintained at a uniform temperature of $T_w = 45°C$. Calculate the average film-condensation heat transfer coefficient h_m over the entire length of the plate.

Solution. The physical properties of water at the film temperature $T_f = (75 + 45)/2 = 60°C$ are

$$k_l = 0.651 \text{ W/(m} \cdot °C) \qquad \mu_l = 0.471 \times 10^{-3} \text{ kg/(m} \cdot s)$$

$$\rho_l = 985 \text{ kg/m}^3 \qquad h_{fg} = 2358.5 \text{ kJ/kg}$$

and $g = 9.8 \text{ m}^2/\text{s}$. The average heat transfer coefficient h_m for laminar filmwise condensation on an inclined plate with $\rho_v \ll \rho_l$ is determined from Eq. (10-8):

$$h_m = 1.13\left[\frac{g\rho_l^2 h_{fg}k_l^3}{\mu_l(T_v - T_w)L}\sin\phi\right]^{1/4}$$

$$= 1.13\left[\frac{(9.81)(985)^2(2358.5 \times 10^3)(0.651)^3}{(0.471 \times 10^{-3})(75 - 45)(0.5)}\sin 45°\right]^{1/4}$$

$$= 5639 \text{ W/(m}^2 \cdot °C)$$

Condensation on a Single Horizontal Tube

The analysis of heat transfer for condensation on the outside surface of a horizontal tube is more complicated than that for a vertical surface. Nusselt's analysis for laminar filmwise condensation on the surface of a horizontal tube gives the average heat transfer coefficient as

$$h_m = 0.725 \left[\frac{g\rho_l(\rho_l - \rho_v)h_{fg}k_l^3}{\mu_l(T_v - T_w)D} \right]^{1/4} \tag{10-9}$$

where D is the outside diameter of the tube.

For $\rho_v \ll \rho_l$, Eq. (10-9) reduces to

$$h_m = 0.725 \left[\frac{g\rho_l^2 h_{fg}k_l^2}{\mu_l(T_v - T_w)D} \right]^{1/4} \tag{10-10}$$

A comparison of Eqs. (10-1a) and (10-10), for filmwise condensation on a vertical tube of length L and a horizontal tube of diameter D, yields

$$\frac{h_{m,\text{vert}}}{h_{m,\text{horz}}} = 1.30 \left(\frac{D}{L} \right)^{1/4} \tag{10-11}$$

This result implies that for a given $T_v - T_w$, the average heat transfer coefficients for a vertical tube of length L and a horizontal tube of diameter D become equal when $L = 2.87D$. For example, when $L = 100D$, theoretically $h_{m,\text{horz}}$ would be 2.44 times $h_{m,\text{vert}}$. Given this consideration, horizontal tube arrangements are generally preferred to vertical tube arrangements in condenser design.

Condensation on a single horizontal tube hardly changes into turbulent flow. Therefore, Eq. (10-10) is valid for all practical purposes.

Condensation on Horizontal Tube Banks

Condenser design generally involves horizontal tubes arranged in vertical tiers, as illustrated in Fig. 10-3, in such a way that the condensate from one tube drains onto the tube just below. If it is assumed that the drainage from one tube flows smoothly onto the tube below, then for a vertical tier of N tubes, each of diameter D, the heat transfer coefficient h is obtained by replacing D by ND in Eq. (10-10). We find

$$[h_m]_{N \text{ tubes}} = 0.725 \left[\frac{g\rho_l(\rho_l - \rho_v)h_{fg}k_l^3}{\mu_l(T_v - T_w)ND} \right]^{1/4} = \frac{1}{N^{1/4}} [h_m]_{1 \text{ tube}} \qquad \text{for Re} < 1800$$

$$\tag{10-12}$$

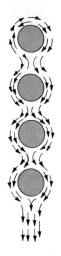

FIGURE 10-3
Film condensation on horizontal tubes arranged in a vertical tier.

This equation generally yields conservative results, since some turbulence and disturbance of the condensate film are unavoidable during drainage, and this increases the heat transfer coefficient.

Equation (10-12) is valid for laminar filmwise condensation, namely, for $\text{Re} < 1800$. However, for horizontal tubes arranged in a vertical tier, turbulent condensation may be possible at the bottom tubes. The Reynolds number for condensation was defined by Eq. (10-3) as

$$\boxed{\text{Re} = \frac{4\dot{m}}{\mu_l P}} \qquad (10\text{-}13a)$$

where $\dot{m}$ is the mass flow rate of condensate in Kilograms per second at the lowest part of the tube bank and P is the wetted perimeter. For horizontal tubes, each of length L, arranged in vertical tiers, P is given by

$$\boxed{P = 2L} \qquad (10\text{-}13b)$$

Calculation of Reynolds Number

To calculate the Reynolds number for condensate flow, defined previously, we need to know the condensation mass flow rate $\dot{m}$ at the lowest part of the system. Let A_t be the total condensing surface area, h_m the average condensation heat transfer coefficient, T_v the saturation temperature of the condensing vapor, T_w the average temperature of the cold condensing surface, and h_{fg} the latent heat of condensation. Then the total heat transfer rate Q is determined by

$$Q = A_t h_m (T_v - T_w) \qquad (10\text{-}14a)$$

and the total mass flow rate $\dot{m}$ at the bottom of the condenser becomes

$$\dot{m} = \frac{Q}{h_{fg}} \qquad (10\text{-}14b)$$

From Eqs. (10-14a) and (10-14b), we have

$$\dot{m} = \frac{A_t h_m (T_v - T_m)}{h_{fg}} \qquad (10\text{-}14c)$$

Introducing Eq. (10-14c) into the definition of the Reynolds number given by Eq. (10-13a), we obtain

$$\boxed{\text{Re} = \frac{4 A_t h_m (T_v - T_m)}{h_{fg} \mu_l P}} \qquad (10\text{-}15)$$

Thus, the Reynolds number can be calculated, since the wetted perimeter P was defined previously by Eqs. (10-4) and (10-13b); namely,

$$P = \begin{cases} \pi D & \text{for a vertical tube of outside diameter } D \\ w & \text{for a vertical or inclined plate of width } w \\ 2L & \text{for horizontal tubes, each of length } L, \text{ arranged} \\ & \text{in vertical tiers} \end{cases}$$

$$(10\text{-}16)$$

Example 10-3. Calculate the total mass flow rate of condensation $\dot{m}$, the total heat transfer rate Q over the entire surface, and the Reynolds number Re at the bottom of the surface in Example 10-1.

Solution. The total mass flow rate of condensate at the bottom of the tube of Example 10-1 is determined from Eq. (10-14c):

$$\dot{m} = \frac{(\pi DL)(h_m)(T_v - T_w)}{h_{fg}}$$

$$= \frac{(\pi \times 0.025 \times 1.5)(5596)(90 - 70)}{2309 \times 10^3} = 5.71 \times 10^{-3} \text{ kg/s}$$

where we set $A_t = \pi DL$ for the total condensing surface area.

The total heat transfer rate, from Eqs. (10-14a) and (10-14b), is

$$Q = A_t h_m (T_v - T_w)$$

$$= \dot{m} h_{fg}$$

$$= 5.71 \times 10^{-3} \times 2309 \times 10^3$$

$$= 13.184 \text{ kW}$$

The Reynolds number at the bottom of the tube, from Eqs. (10-3) and (10-16), is

$$\text{Re} = \frac{4\dot{m}}{\mu_l P} \qquad \text{where } P = \pi D$$

Then,

$$\text{Re} = \frac{(4)(5.71 \times 10^3)}{(0.355 \times 10^{-3})(0.025\pi)} = 819.2$$

which is less than 1800. Hence the condensate flow is in the laminar range and the decision made in Example 10-1 to use Eq. (10-5) which is valid for Re $<$ 1800, is justified.

Example 10-4. Calculate the Reynolds number at the bottom of the inclined plate of Example 10-2.

Solution. First we need to calculate the total mass rate of condensation $\dot{m}$ over the entire surface of the plate. In Example 10-2 the plate width w is not specified, but this is immaterial because the width cancels out in the calculation of the

Reynolds number. Therefore we determine $\dot{m}$ from Eq. (10-14c) for a unit width by setting $A_t = 0.5 \times 1$, and obtain

$$\dot{m} = \frac{A_t h_m (T_v - T_w)}{h_{fg}}$$

$$= \frac{(0.5 \times 1.0)(5639)(75 - 45)}{2358.5 \times 10^3} = 2.582 \times 10^{-2} \text{ kg/s}$$

Then the Reynolds number at the bottom of the inclined plate with the wetted perimeter $P = w = 0.5 \text{ m}$ becomes

$$\text{Re} = \frac{4\dot{m}}{\mu_l P}$$

$$= \frac{(4)(2.582 \times 10^{-2})}{(0.471 \times 10^{-3})(0.5)} = 438.6$$

which is less than 1800, and the laminar flow assumption in Example 10-2 is valid.

Example 10-5. Consider N horizontal tubes arranged in a vertical tier. What is the number of tubes that will produce an average condensation heat transfer coefficient for the vertical tier equal to one-half of that for the single horizontal tube at the top?

Solution. We assume laminar filmwise condensation and use Eq. (10-12) to calculate N as follows:

$$[h_m]_{N \text{ tubes}} = \frac{1}{N^{1/4}} [h_m]_{1 \text{ tube}}$$

$$\frac{1}{2} [h_m]_{1 \text{ tube}} = \frac{1}{N^{1/4}} [h_m]_{1 \text{ tube}}$$

$$\frac{1}{2} = \frac{1}{N^{1/4}}$$

which gives

$$N = 16$$

10-2 TURBULENT FILMWISE CONDENSATION

The previous results for filmwise condensation are applicable if the condensate flow is laminar. Kirkbride proposed the following empirical correlation for film condensation on a vertical plate after the start of turbulence:

$$h_m \left(\frac{\mu_l^2}{k_l^3 \rho_l^2 g} \right)^{1/3} = 0.0077 (\text{Re})^{0.4} \tag{10-17}$$

valid for $\text{Re} > 1800$; the physical properties of the condensate should be evaluated at $T_f = (T_w + T_v)/2$.

Figure 10-4 shows a plot of Eqs. (10-6) and (10-17) as a function of the

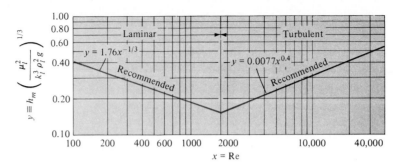

FIGURE 10-4
Average heat transfer coefficient for filmwise condensation on a vertical surface for laminar and turbulent flows.

Reynolds number for condensate flow in the laminar and turbulent regimes, respectively.

Example 10-6. Air-free steam at $T_v = 100°C$ and atmospheric pressure condenses on a $L = 2$ m long vertical plate. What is the temperature of the plate T_w below which the condensing film at the bottom of the plate will become turbulent?

Solution. To start the calculations, the physical properties of water at the mean film temperature are needed. As a first guess we take an average film temperature of 60°C. The properties of water at this temperature were given in Example 10-2, and therefore are not repeated here.

Let w be the width of the plate. The total condensate flow rate at the bottom of the plate at the transition Reynolds number Re = 1800 is determined from Eq. (10-3) by setting $P = w$:

$$\text{Re} = \frac{4\dot{m}}{\mu_l(w)}$$

$$1800 = \frac{4\dot{m}}{(0.471 \times 10^{-3})(w)}$$

which gives

$$\dot{m} = 0.212w$$

Using Eq. (10-17), the heat transfer coefficient at Re = 1800 is determined to be

$$h_m = 0.0077\,\text{Re}^{0.4}k_l\left(\frac{g\rho_l^2}{\mu_l^2}\right)^{1/3}$$

$$= (0.0077)(1800)^{0.4}(0.651)\left[\frac{(9.81)(985)^2}{(0.471 \times 10^{-3})^2}\right]^{1/3}$$

$$= 3518.4 \text{ W/(m}^2 \cdot °\text{C)}$$

The expressions for the total heat transfer rate Q are taken as

$$Q = \dot{m}h_{fg} = A_t h_m(T_v - T_w)$$

and the above values of $\dot{m}$, h_m, and A_t are introduced into this equation:

$$(0.212)(w)(2358.5 \times 10^3) = (w \times 2)(3518.4)(100 - T_w)$$

The plate width w cancels out, and the plate temperature T_w is determined to be

$$T_w = 28.9°C$$

To check the accuracy of the initial guess of 60°C for the mean condensate temperature, the film temperature T_f is computed:

$$T_f = \frac{28.9 + 100}{2} = 64.5°C$$

The initial guess is sufficiently close to this result; therefore there is no need for iteration.

Film Condensation inside Horizontal Tubes

The correlations of filmwise condensation given previously are based on the assumption that the vapor is stationary or has a negligible velocity. There are also applications, such as condensers for refrigeration and air conditioning systems, in which vapor condenses inside the tubes, and hence has a significant velocity. Consider, for example, the filmwise condensation on the inside surface of a long vertical tube. The upward flow of vapor retards the condensate flow and causes the thickening of the condensate layer, which in turn decreases the condensation heat transfer coefficient. Conversely, the downward flow of vapor decreases the thickness of the condensate film, and hence increases the heat transfer coefficient. The correlation of the effects of vapor velocity for condensation inside tubes is a complicated matter; therefore it is not considered here.

10-3 DROPWISE CONDENSATION

Since the original observation of dropwise condensation by Schmidt, Schurig, and Sellschopp, numerous investigations of dropwise condensation have been reported. If traces of oil are present in steam and the condensing surface is highly polished, the condensate film breaks into droplets, giving rise to so-called *dropwise condensation*. Figure 10-5 shows an ideal dropwise condensation of steam on a vertical surface. The droplets grow, coalesce, and run off the surface, leaving a greater portion of the condensing surface exposed to incoming steam. Since we never have the entire condensing surface covered with a continous layer of liquid film, the heat transfer coefficient for ideal dropwise condensation of steam is much higher than that for filmwise condensation of steam. The heat transfer coefficients may be 5 to 10 times greater, although the overall heat transfer coefficient between the steam and the coolant in a typical surface condenser may be about 2 to 3 times greater for dropwise than for filmwise condensation. If sustained dropwise condensation can be achieved in practice, the size of condensers can be reduced significantly, with considerable savings in the capital cost. Therefore, considerable research

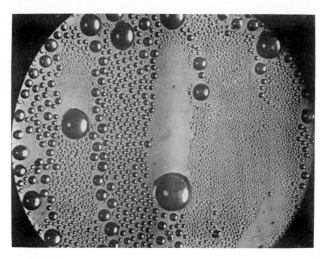

FIGURE 10-5
Dropwise condensation of steam under ideal conditions. (From Hampson and Ozisik).

has been done with the objective of producing long-lasting dropwise condensation. Various types of promoters—such as oleic, stearic, and linoleic acids; benzyl mercaptan; and many other chemicals—have been used to promote dropwise condensation. The periods for which continuous dropwise conditions are obtainable with different promoters vary between 100 and 300 h with pure steam and are shorter with industrial steam or intermittent operations. Failure occurs because of fouling or oxidation of the surface, because of the gradual removal of the promoter from the surface by the flow of condensate, or because of a combination of these effects.

To prevent failure of dropwise condensation as a result of oxidation, noble-metal coating of the condensing surface has been tried; coatings of gold, silver, rhodium, palladium, and platinum have been used. Although some of these coated surfaces could produce dropwise condensation under laboratory conditions for more than 10,000 h of continuous operation, the cost of coating the condensing surface with noble materials is so high that the economics of such an approach for industrial applications has yet to be proved.

It is unlikely that long-lasting dropwise condensation can be produced under practical conditions by a single treatment with any of the promoters currently available. Although it may be possible to produce dropwise condensation for up to a year by injecting a small quantity of promoter into the steam at regular intervals, the success of this operation depends on the amount and cost of the promoter and on the extent to which the cumulative effect of the injected promoter can be tolerated in the rest of the plant. Therefore, in the analysis of a heat exchanger involving the condensation of steam, it is recommended that filmwise condensation be assumed for the condensing surface.

10-4 CONDENSATION IN THE PRESENCE OF NONCONDENSABLE GAS

Earlier we considered the heat transfer coefficient for condensing vapors that did not contain any noncondensable gas. If noncondensable gas such as air is present in the vapor, even in very small amounts, the heat transfer coefficient for condensation is greatly reduced. The reason is that when a vapor containing noncondensable gas condenses, the noncondensable gas is left at the surface and the incoming condensable vapor must diffuse through this body of vapor–gas mixture collected in the vicinity of the condensate surface before it reaches the cold surface to condense. Therefore, the presence of noncondensable gas adjacent to the condensate surface acts as a thermal resistance to heat transfer. The resistance to this diffusion process causes a drop in the partial pressure of the condensing vapor, which in turn drops the saturation temperature; that is, the temperature of the outside surface of the condensate layer is lower than the saturation temperature of the bulk mixture.

Prediction of the condensation heat transfer coefficient in the presence of noncondensable gas has been the subject of numerous investigations. The results have shown that the heat transfer coefficient is very much dependent on the vapor flow patterns in the vicinity of the condensing surface. For example, high velocities over the condensing surface tend to reduce the accumulation of noncondensable gas and to alleviate the adverse effect of noncondensable gas on the heat transfer. If the noncondensable gas is allowed to accumulate over the condensing surface, a significant reduction in the heat transfer coefficient results. Depending on the vapor flow patterns in the vicinity of the condensing surface and the amount of noncondensable gas in the bulk mixture, the condensation heat transfer coefficient can be reduced substantially. For example, 0.5 percent by mass of air in the steam can reduce the filmwise condensation heat transfer coefficient by a factor of 2; or 5 percent by mass of air can easily cut the heat transfer coefficient by a factor of 5. Therefore, in practical applications, to alleviate the adverse effect of noncondensable gas accumulation on heat transfer, provisions are made in the design of a condenser to vent the noncondensable gas accumulating inside the condenser.

10-5 POOL BOILING

Pool boiling provides a convenient starting point for discussion of the mechanism of heat transfer in boiling systems. Despite the fact that this subject has been extensively studied and the mechanism of heat transfer is reasonably well understood, it is still not possible to predict theoretically the heat transfer characteristics of this apparently most simple boiling system.

Nukiyama was the first investigator to establish experimentally the characteristics of pool boiling phenomena. He immersed an electric resistance wire into a body of saturated water and initiated boiling on the surface of the wire by passing current through it. He determined both the heat flux and the

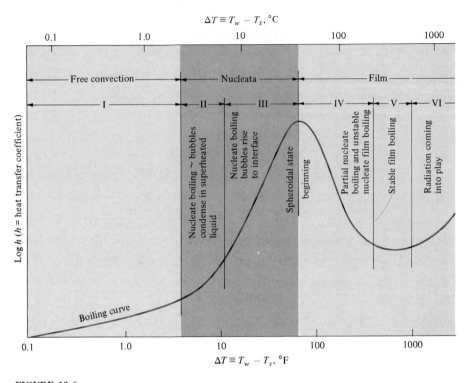

FIGURE 10-6
Principal boiling regimes in pool of boiling water at atmospheric pressure and saturation temperature T_s from an electrically heated platinum wire. (From Farber and Scorah).

temperature from measurements of current and voltage. Since Nukiyama's
original work, numerous investigations of the pool boiling phenomenon have
been reported. Figure 10-6 illustrates the characteristics of pool boiling for
water at atmospheric pressure. This boiling curve illustrates the variation of the
heat transfer coefficient or the heat flux as a function of the temperature
difference between the wire and water saturation temperatures. Scrutiny of this
boiling curve reveals that the mechanism of heat transfer can be divided into
three distinct regimes: the *free-convection*, *nucleate boiling*, and *film boiling*
regimes. We now examine the heat transfer characteristics of each of these
three regimes.

Free-Convection Regime

In this regime, the heat transfer from the heater surface to the saturated liquid
takes place by *free convection*. The heater surface is only a few degrees above
the saturation temperature of the liquid, but the flow produced by free
convection in the liquid is sufficient to remove the heat from the surface. The

heat transfer correlations, as given in Chap. 9, are in the form

$$\mathrm{Nu} = f(\mathrm{Gr}, \mathrm{Pr}) \qquad (10\text{-}18)$$

Once the heat transfer coefficient h is obtained, the heat flux for the free-convection regime is determined from

$$q = h(T_w - T_{\mathrm{sat}}) \qquad (10\text{-}19)$$

Various correlations are given in Chap. 9 for determining the free-convection heat transfer coefficient h for various geometries, such as vertical, inclined, and horizontal plates; vertical and horizontal cylinders; and many others.

Nucleate Boiling Regime

The nucleate boiling regime, in which bubbles are formed on the surface of the heater, can be separated into two distinct regions. In the region designated II, bubbles start to form at the favored sites on the heater surface, but as soon as the bubbles are detached from the surface, they are dissipated in the liquid. In region III, the nucleation sites are numerous and the bubble generation rate is so high that continuous columns of vapor appear. As a result, very high heat fluxes are obtainable in this region. In practical applications, the nucleate boiling regime is most desirable, because large heat fluxes are obtainable with small temperature differences. In the nucleate boiling regime, the heat flux increases rapidly with increasing temperature difference until the peak heat flux is reached. The location of this peak heat flux is called the burnout point, departure from nucleate boiling (DNB), or critical heat flux (CHF). The reason for calling the peak heat flux the burnout point is apparent from Fig. 10-6. As soon as the peak heat flux is exceeded, an extremely large temperature difference is needed to realize the resulting heat flux. Such a high temperature difference may cause the burning up, or melting away, of the heating element.

Clearly heat transfer in the nucleate boiling regime is affected by the nucleation process, the distribution of active nucleation sites on the surface, and the growth and departure of bubbles. If the number of active nucleation sites increases, the interaction between the bubbles may become important. In addition to these variables, the state of the fluid (i.e., the fluid properties) and the surface condition (i.e., the mechanical and material properties of the surface) are among the factors that affect heat transfer in the nucleate boiling regime.

Numerous experimental investigations have been reported, and various attempts have been made to correlate that portion of the boiling curve characterizing heat transfer in the nucleate boiling regime. The most successful and widely used correlation was developed by Rohsenow. By analyzing the significance of various parameters in relation to forced-convection effects, he proposed the following empirical relation to correlate the heat flux in the entire nucleate boiling regime:

$$\frac{c_{pl}\,\Delta T}{h_{fg}\,\mathrm{Pr}_l^n} = C_{sf}\left[\frac{q}{\mu_l h_{fg}}\sqrt{\frac{\sigma^*}{g(\rho_l - \rho_v)}}\right]^{0.33} \tag{10-20}$$

where c_{pl} = specific heat of saturated liquid, J/(kg·°C)
C_{sf} = constant, to be determined from experimental data, that depends on heating surface–fluid combination
h_{fg} = latent heat of vaporization, J/kg
g = gravitational acceleration, m/s^2
$\mathrm{Pr}_l = c_{pl}\mu_l/k_l$ = Prandtl number of saturated liquid
q = boiling heat flux, W/m^2
$\Delta T = T_w - T_{sat}$, temperature difference between wall and saturation temperature, °C
μ_l = viscosity of saturated liquid, kg/(m·s)
ρ_l, ρ_v = density of liquid and saturated vapor, respectively, kg/m^3
σ^* = surface tension of liquid–vapor interface, N/m

In Eq. (10-20) the exponent n and the coefficient C_{sf} are the two provisions for adjusting the correlation for the liquid–surface combination. Table 10-1 lists the experimentally determined values C_{sf} for water boiling on a variety of surfaces. The value of n for water should be taken as 1.

Table 10-2 gives the values of vapor–liquid surface tension for water at different saturation temperatures.

TABLE 10-1
Values of the coefficient C_{sf} of Eq. (10-20) for water surface combinations

Liquid–surface combination	C_{sf}
Water–copper	0.0130
Water–scored copper	0.0068
Water–emery-polished copper	0.0128
Water–emery-polished, paraffin-treated copper	0.0147
Water–chemically etched stainless steel	0.0133
Water–mechanically polished stainless steel	0.0132
Water–ground and polished stainless steel	0.0080
Water–Teflon pitted stainless steel	0.0058
Water–platinum	0.0130
Water–brass	0.0060

Source: R. I. Vahon, G. H. Nix, and G. E. Tanger,, "Evaluation of Constants for the Rohsenow Pool-Boiling Correlation," *J. Heat Transfer* **90C**:239–247 (1968).

TABLE 10-2
Values of liquid–vapor surface tension σ^* for water

Saturation temperature, °C	Surface tension $\sigma^* \times 10^3$, N/m
0	75.6
15.56	73.2
37.78	69.7
93.34	60.1
100	58.8
160	46.1
226.7	31.9
293.3	16.2
360	1.46
374.11	0

Example 10-7. During the boiling of saturated water at $T_s = 100°C$ with an electric heating element, a heat flux of $q = 7 \times 10^5$ W/m^2 is achieved with a temperature difference of $\Delta T = T_w - T_v = 10.4°C$. What is the value of the constant C_{sf} in Eq. (10-20)?

Solution. The physical properties of saturated water and vapor are taken as

$$c_{pl} = 4216 \text{ J/(kg} \cdot °C)} \qquad h_{fg} = 2257 \times 10^3 \text{ J/kg}$$

$$\rho_l = 960.6 \text{ kg/m}^3 \qquad \rho_v = 0.6 \text{ kg/m}^3$$

$$\text{Pr}_l = 1.74 \qquad \mu_l = 0.282 \times 10^{-3} \text{ kg/(m} \cdot \text{s)}$$

$$\sigma^* = 58.8 \times 10^{-3} \text{ N/m}$$

These numerical values are substituted into Eq. (10-20) with $n = 1$:

$$\frac{c_{pl} \, \Delta T}{h_{fg} \, \text{Pr}_l} = C_{sf} \left[\frac{q}{\mu_l h_{fg}} \sqrt{\frac{\sigma^*}{g(\rho_l - \rho_v)}} \right]^{0.333}$$

$$\frac{(4216)(10.4)}{(2257 \times 10^3)(1.74)} = C_{sf} \left[\frac{(7 \times 10^5)}{(0.282 \times 10^{-3})(2257 \times 10^3)} \sqrt{\frac{58.8 \times 10^{-3}}{9.81(960.6 - 0.6)}} \right]^{0.333}$$

Then the coefficient C_{sf} becomes

$$C_{sf} = 0.008$$

Example 10-8. A brass heating element of surface area $A = 0.04$ m^2, maintained at a uniform temperature $T_w = 112°C$, is immersed in saturated water at atmospheric pressure at temperature $T_s = 100°C$. Calculate the rate of evaporation.

Solution. The physical properties of saturated water and vapor at 100°C were given in Example 10-7. Introducing these properties into Eq. (10-20) with $n = 1$ and $\Delta T = T_w - T_s = 112 - 100 = 12°C$, and obtaining the coefficient C_{sf} for water-brass from Table 10-1 as $C_{sf.} = 0.006$, the heat flux becomes

$$q = \left(\frac{c_{p,l} \Delta T}{h_{fg} \, Pr_l \, C_{sf}}\right)^3 \mu_l h_{fg} \sqrt{\frac{g(\rho_l - \rho_w)}{\sigma^*}}$$

$$= \left[\frac{(4216)(12)}{(2257 \times 10^3)(1.74)(0.006)}\right]^3 (0.282 \times 10^{-3})(2257 \times 10^3) \sqrt{\frac{(9.81)(060.6 - 0.6)}{(58.8 \times 10^{-3})}}$$

$$= 2521.23 \text{ kW/m}^2$$

The total rate of heat transfer is

$$Q = \text{area} \times q$$
$$= (0.04)(2521.23) = 100.85 \text{ kW}$$

The rate of evaporation is

$$\dot{m} = \frac{Q}{h_{fg}}$$

$$= \frac{100.85 \times 10^3}{2257 \times 10^3} = 0.0447 \text{ kg/s} = 160.9 \text{ kg/h}$$

Peak Heat Flux

The correlation given by Eq. (10-20) provides information about the heat flux in nucleate boiling, but it cannot predict the peak heat flux. The determination of peak heat flux in nucleate boiling is of interest because of burnout considerations; that is, if the applied heat flux exceeds the peak heat flux, the transition takes place from the nucleate to the stable film boiling regime, in which, depending on the kind of fluid, boiling may occur at temperature differences well above the melting point of the heating surface.

Lienhard and coworkers modified the correlations originally proposed by Kutateladze and Zuber; they proposed the following expression for determination of the peak heat flux:

$$\boxed{q_{max} = F(L') \times 0.131 \rho_v^{1/2} h_{fg} [\sigma^* g(\rho_l - \rho_v)]^{1/4}} \qquad (10\text{-}21)$$

where $\sigma^* = $ surface tension of liquid–vapor interface, N/m
$\quad g = $ gravitational acceleration, m/s^2
$\rho_l, \rho_v = $ density of liquid and vapor, respectively, kg/m^3
$\quad h_{fg} = $ latent heat of vaporization, J/kg
$\quad q_{max} = $ peak heat flux, W/m^2

and $F(L')$ is a correction factor that depends on heat geometry; it is given in Table 10-3. The dimensionless characteristic length L' of the heater is defined as

$$L' = L\sqrt{\frac{g(\rho_l - \rho_v)}{\sigma^*}} \qquad (10\text{-}22)$$

where L is the characteristic dimension of the heater and the other quantities are as defined previously. In eq. (10-21) the physical properties of the vapor

TABLE 10-3
Correction factor $F(L')$ for use in Eq. (10-21)

Heater geometry	$F(L')$	Remarks
Infinite flat plate facing up	1.14	$L' \geq 2.7$; L is the heat width or diameter
Horizontal cylinder	$0.89 + 2.27e^{-3.44}\sqrt{L'}$	$L' \geq 0.15$; L is the cylinder radius
Large sphere	0.84	$L' \geq 4.26$; L is the sphere radius
Small sphere	$\dfrac{1.734}{(L')^{1/2}}$	$0.15 \leq L' \leq 4.26$; L is the sphere radius

Based on Lienhard and coworkers.

should be evaluated at

$$T_f = \tfrac{1}{2}(T_w + T_{sat})$$

The enthalpy of evaporation h_{fg} and the liquid properties should be evaluated at the saturated temperature of the liquid.

> **Example 10-9.** Water at saturation temperature and atmospheric pressure is boiled in the nucleate boiling regime with a large plate heating element facing up. Calculate the peak heat flux.
>
> **Solution.** The physical properties of saturated water and vapor at 100°C were given in Example 10-7. Introducing these properties into Eq. (10-21), with the correction factor $F(L') = 1.14$ obtained from Table 10-3 for a large plate heating element facing up, the peak heat flux q_{max} is determined to be
>
> $$q_{max} = 0.131 F(L') \rho_v^{1/2} h_{fg} [\sigma^* g(\rho_l - \rho_v)]^{1/4}$$
> $$= (0.131)(1.14)(0.6)^{1/2}(2257 \times 10^3)$$
> $$\times [(58.8 \times 10^{-3})(9.81)(960.6 - 0.6)]^{1/4}$$
> $$= 1.27 \, \text{MW/m}^2$$

Film Boiling Regime

The nucleate boiling region ends and the unstable film boiling region begins after the peak heat flux is reached. No correlations are available for the prediction of heat flux in this unstable region until the minimum point in the boiling curve is reached and the stable film boiling region starts. In the stable film boiling regions, V and VI, the heating surface is separated from the liquid by a vapor layer across which heat must be transferred. Since the thermal conductivity of the vapor is low, large temperature differences are needed for heat transfer in this region; therefore, heat transfer in this region is generally avoided when high temperatures are involved. However, stable film boiling has numerous applications in the boiling of cryogenic fluids. A theory for the prediction of the heat transfer coefficient for stable film boiling on the outside of a horizontal cylinder was developed by Bromley. The basic approach in the analysis is similar to Nusselt's theory for filmwise condensation on a horizontal tube. The resulting equation for the average heat transfer coefficient h_0 for stable film boiling on the outside of a horizontal cylinder, in the absence of radiation, is given by

$$h_0 = 0.62 \left[\frac{k_v^3 \rho_v (\rho_l - \rho_v) g h_{fg}}{\mu_v D_o \Delta T} \left(1 + \frac{0.4 c_{pv} \Delta T}{h_{fg}} \right) \right]^{1/4} \qquad (10\text{-}23)$$

where $h_0 =$ average boiling heat transfer coefficient in absence of radiation, $W/(m^2 \cdot °C)$

$c_{pv} =$ specific heat of saturated vapor, $J/(kg \cdot °C)$

$D_o =$ outside diameter of tube, m

$g =$ gravitational acceleration, m/s^2

$h_{fg} =$ latent heat of vaporization, J/kg

$k_v =$ thermal conductivity of saturated vapor, $W/(m \cdot °C)$

$\Delta T = T_w - T_{sat}$, temperature difference between wall and saturation temperatures, °C

In Eq. (10-23), the physical properties of vapor should be evaluated at $T_f = \frac{1}{2}(T_w + T_{sat})$, and the enthalpy of evaporation h_{fg} and the liquid density ρ_l should be evaluated at the saturation temperature T_{sat} of the liquid.

Example 10-10. Water at saturation temperature and atmospheric pressure is boiled with an electrically heated horizontal platinum wire of diameter $D = 0.2$ cm. Boiling takes place with a temperature difference of $T_w - T_s = 454°C$ in the stable film boiling range. Calculate the film boiling heat transfer coefficient and the heat flux, in the absence of radiation.

Solution. The physical properties of vapor are evaluated at $T_f = (T_w + T_s)/2 = (554 + 100)/2 = 327°C = 600$ K:

$$c_{p,v} = 2026 \text{ J}/(kg \cdot °C) \qquad k_v = 0.0422 \text{ W}/(m \cdot °C)$$

$$\mu_w = 2067 \times 10^{-5} \text{ kg}/(m \cdot s) \qquad \rho_v = 0.365 \text{ kg}/m^3$$

and the liquid density and h_{fg} are evaluated at the saturation temperature $T_s = 100°C$:

$$\rho_l = 960.6 \text{ kg}/m^3 \qquad h_{fg} = 2257 \times 10^3 \text{ J}/kg$$

The heat transfer coefficient h_0 for stable film boiling without the radiation effects is computed from Eq. (10-23):

$$h_0 = 0.62 \left[\frac{k_v^3 \rho_v (\rho_l - \rho_v) g h_{fg}}{\mu_v D_o \Delta T} \left(1 + \frac{0.4 c_{pv} \Delta T}{h_{fg}} \right) \right]^{1/4}$$

$$= 0.62 \left[\frac{(0.0422)^3 (0.365)(960.6 - 0.365)(9.81)(2257 \times 10^3)}{(2.067 \times 10^{-5})(0.002)(454)} \right.$$

$$\left. \times \left(1 + \frac{(0.4)(2026)(454)}{2257 \times 10^3} \right) \right]^{1/4}$$

$$= 270.3 \text{ W}/(m^2 \cdot °C)$$

PROBLEMS

Laminar Film Condensation

10-1. Air-free saturated steam at $T_v = 100°C$ condenses on the outer surface of a $L = 2$ m long, $D = 2$ cm OD vertical tube, maintained at a uniform temperature $T_w = 60°C$. Assuming filmwise condensation, calculate the average heat transfer coefficient over the entire length of the tube and the rate of condensate flow at the bottom of the tube.

10-2. Air-free saturated steam at $T_v = 65°C$ condenses on the outer surface of a 1 m by 1 m plate, maintained at a uniform temperature $T_w = 35°C$ by the flow of cooling water on one side. Assuming filmwise condensation, calculate the average heat transfer coefficient over the entire length of the plate and the rate of condensate flow at the bottom of the plate.

10-3. Determine the average heat transfer coefficient of Problem 10-1 when the tube is horizontal.

10-4. Air-free saturated steam at $T_v = 65°C$ condenses on the surface of a vertical tube with $D = 2$ cm OD which is maintained at a uniform temperature $T_w = 35°C$. Determine the tube length L for a condensate flow rate of 5×10^{-3} kg/s per tube.

10-5. Air-free saturated steam at $T_v = 50°C$ ($P = 12.35$ kPa) condenses on the outside surface of a $D = 2.5$ cm OD, $L = 2$ m long vertical tube maintained at a uniform temperature of $T_w = 30°C$ by the flow of cooling water through the tube. Assuming filmwise condensation, calculate: (a) the average condensation heat transfer coefficient over the entire length of the tube, and (b) the rate of condensate flow at the bottom of the tube.
Answers: (a) 4356 W/(m² · °C), (b) 5.685×10^{-3} kg/s

10-6. Calculate the average heat transfer coefficient h_m and the total condensation rate at the tube surface for Problem 10-5 when the tube is in the horizontal position.

10-7. Repeat Problem 10-5 for a tube length of 3 m. Calculate the average heat transfer coefficient h_m and the total condensation rate at the tube surface for the condensation problem when the tube is in the horizontal position.

10-8. Saturated, air-free steam at $T_v = 65°C$ ($P = 25.03$ kPa) condenses on the outer surface of a $D = 2.5$ cm OD vertical tube whose surface is maintained at a uniform temperature of $T_w = 35°C$. Determine the tube length needed to condense 30 kg/h of steam.
Answer: 2.06 m

10-9. Saturated, air-free steam at a temperature $T_v = 80°C$ ($P = 47.39$ kPa) condenses on the outer surface of a $L = 1.2$ m long, $D = 0.1$ m in diameter vertical tube which is maintained at a uniform temperature of $T_w = 40°C$. Calculate: (a) the average heat transfer coefficient h_m for filmwise condensation over the entire tube length, (b) the total rate of steam condensation at the tube surface, and (c) the condensate thickness at the bottom of the tube.
Answers: (a) 4597 W/(m² · °C), (b) 2.939×10^{-2} kg/s, (c) 0.189 mm

10-10. Saturated air-free steam at a temperature $T_v = 80°C$ ($P = 47.39$ kPa) condenses on the outer surface of a $L = 2$ m long, $D = 2.0$ cm OD vertical tube maintained at $T_w = 40°C$. Calculate the average heat transfer coefficient for filmwise condensation over the entire tube length and the total rate of condensation at the surface of the tube.

10-11. Saturated air-free steam at a temperature $T_v = 80°C$ ($P = 47.39$ kPa) condenses on the outer surface of a $D = 2.0$ cm OD vertical tube maintained at $T_w = 40°C$. Determine the tube length needed to condense 0.025 kg/s steam.

10-12. Saturated, air-free steam at a temperature $T_v = 50°C$ ($P = 12.35$ kPa) condenses on the outer surface of a $L = 2$ m long, $D = 2.0$ cm OD vertical tube maintained at a uniform temperature of $T_w = 10°C$. Assuming filmwise condensation, calculate the average condensation heat transfer coefficient h_m over the entire length of the tube and the total rate of condensation at the surface of the tube.

10-13. Air-free saturated steam at a temperature $T_v = 90°C$ ($P = 70.14$ kPa) condenses on the outer surface of a $D = 2.0$ cm OD vertical tube maintained at a uniform temperature $T_w = 70°C$. Determine the tube length needed to condense 20 kg/h steam.

10-14. Saturated Freon-12 vapor at a temperature $T_v = -5°C$ ($P = 0.26$ MPa) condenses on the outer surface of a $L = 1.2$ m long, $D = 1.27$ cm OD vertical tube maintained at a uniform temperature of $T_w = -15°C$. Calculate: (*a*) the average condensation heat transfer coefficient over the entire length of the tube, and (*b*) the total mass rate of condensation at the tube surface. (Take $h_{fg} = 154$ kJ/kg for Freon-12.)
Answers: (*a*) 847.6 W/(m² · °C), (*b*) 2.635×10^{-3} kg/s

10-15. Saturated ammonia vapor at a temperature $T_v = -5°C$ ($P = 0.3528$ MPa) condenses on the outer surface of a $L = 0.75$ m long, $D = 1.27$ cm OD vertical tube maintained at a uniform temperature $T_w = -15°C$. Calculate the average condensation heat transfer coefficient h_m and the total rate of condensation of ammonia over the entire length of the tube. (Take $h_{fg} = 1280$ kJ/kg for ammonia.)
Answers: 5243 W/(m² · °C), 1.226×10^{-3} kg/s

10-16. Saturated air-free steam at $T_v = 170°C$ ($P = 0.8$ MPa) condenses on the outer surface of a $D = 2$ cm OD $L = 1.5$ m long vertical tube maintained at a uniform temperature $T_w = 150°C$. Calculate: (*a*) the local film condensation heat transfer coefficient at the bottom of the tube, (*b*) the average condensation heat transfer coefficient over the entire length of the tube, and (*c*) the total condensation rate at the tube surface.

10-17. Air-free saturated steam at $T_v = 60°C$ ($P = 19.94$ kPa) condenses on the outer surface of $D = 2.5$ cm OD, $L = 2$ m long 100 horizontal tubes arranged in a 10 by 10 square array. The surface of the tubes is maintained at a uniform temperature $T_w = 40°C$. Calculate the average condensation heat transfer coefficient for the entire tube bundle and the total rate of condensation at the surface of the tubes in the bundle.
Answers: 4927 W/(m² · °C), 0.65 kg/s

10-18. A steam condenser consists of 625 $D = 1.25$ cm OD, $L = 3$ m long horizontal tubes arranged in a 25 by 25 square array. Saturated steam at $T_v = 50°C$ ($P = 12.35$ kPa) condenses on the outer surface of the tubes, which are maintained at a uniform temperature $T_w = 30°C$. Calculate the average heat transfer coefficient h_m and the total rate of condensation of steam in the condenser.
Answers: 4445 W/(m² · °C), 2.72 kg/s

10-19. Compare the average condensation heat transfer coefficient for filmwise condensation of air-free steam at atmospheric pressure on: (*a*) a single $L = 1$ m long, $D = 2.5$ cm OD vertical tube and (*b*) $D = 2.5$ cm OD, $N = 40$ horizontal tubes arranged in a vertical tier.

10-20. Saturated air-free steam at a temperature $T_v = 50°C$. ($P = 12.35$ kPa) condenses on the outer surface of a $D = 2.0$ cm OD horizontal tube maintained at a uniform temperature $T_w = 30°C$. Calculate the tube length required to condense $W = 50$ kg/h steam.

10-21. Air-free saturated steam at $T_v = 85°C$ ($P = 57.83$ kPa) condenses on the outer surface of 196 $D = 1.27$ cm OD, horizontal tubes arranged in a 14 by 14 square array. The tube surfaces are maintained at a uniform temperature $T_w = 75°C$. Calculate the length of the matrix needed to condense 0.7 kg/s of steam.

   Answer: 2.84 m

Turbulent Film Condensation

10-22. Air-free saturated steam at $T_v = 70°C$ ($P = 31.19$ kPa) condenses on the outer surface of a $D = 2.0$ cm OD vertical tube maintained at a uniform temperature $T_w = 50°C$. What length of the tube would produce turbulent film condensation?

10-23. Air-free saturated steam at $T_v = 90°C$ ($P = 70.14$ kPa) condenses on the outer surface of a $D = 2.5$ cm OD, $L = 6$ m long vertical tube maintained at a uniform temperature $T_w = 30°C$. Calculate the average heat transfer coefficient over the entire length of the tube and the total rate of condensation of steam at the tube surface.

   Answers: 6539 W/(m^2 · °C), 0.0784 kg/s

10-24. Air-free saturated steam at $T_v = 50°C$ ($P = 12.35$ kPa) condenses on the outer surface of a $D = 2.0$ cm OD vertical tube maintained at a uniform temperature $T_w = 30°C$. Calculate the tube length L that would produce turbulent film condensation at the bottom of the tube, and calculate the total condensation rate at the tube surface.

Pool Boiling

10-25. Saturated water at $T_s = 100°C$ is boiled inside a copper pan with a heating surface $A = 5 \times 10^{-2}$ m^2 that is maintained at a uniform temperature $T_w = 110°C$. Calculate the surface heat flux and the rate of evaporation.

   Answers: 7.173 kW, 11.4 kg/h

10-26. Saturated water at $T_v = 55°C$ ($P = 15.76$ kPa) is boiled with a copper heating element that has a heating surface $A = 5 \times 10^{-3}$ m^2, and maintained at a uniform temperature $T_w = 65°C$. Calculate the rate of evaporation.

   Answer: 0.235 kg/h

10-27. Saturated water at $T_s = 100°C$ is boiled in a $D = 20$ cm diameter Teflon pitted stainless steel pan with a temperature difference of $T_w - T_s = 10°C$. Calculate the rate of evaporation.

   Answer: 80.9 kg/h

10-28. Saturated water at $T_s = 100°C$ is boiled with a copper heating element. If the surface heat flux is $q = 300$ kW/m^2, calculate the surface temperature of the heating element.

10-29. Saturated water at $T_s = 100°C$ is boiled with a copper heating element. If the surface heat flux is $q = 400$ kw/m^2, calculate the surface temperature of the heating element.

   Answer: 114°C

10-30. An electrically heated copper plate with a heating surface $A = 0.2$ m^2, maintained at a uniform temperature $T_w = 108°C$, is immersed in a water tank at $T_s = 100°C$ and atmospheric pressure. Calculate the rate of evaporation.

   Answer: 23.4 kg/h

10-31. Water at a saturation temperature $T_s = 160°C$ $(P = 0.6178\,\text{MPa})$ is boiled using an electrically heated copper element with a temperature difference of $T_w - T_s = 10°C$. Calculate the surface heat flux.

10-32. An electrically heated copper kettle with a flat bottom of diameter $D = 25\,\text{cm}$ is to boil water at atmospheric pressure at a rate of 2.5 kg/h. What is the temperature of the bottom surface of the kettle?

    Answer: 106°C

10-33. Determine the peak heat flux obtainable with nucleate boiling of saturated water at 1 atm pressure in a gravitational field one-eighth that of the earth. Compare this result with that obtainable at the earth's gravitational field. Assume a large heating element.

    Answers: $0.753\,\text{MW/m}^2$; $1.27\,\text{MW/m}^2$

10-34. An electrically heated, copper spherical heating element of diameter $D = 10\,\text{cm}$ is immersed in water at atmospheric pressure and saturation temperature. The surface of the element is maintained at a uniform temperature $T_w = 115°C$. Calculate: (*a*) the surface heat flux, (*b*) the rate of evaporation, and (*c*) the peak heat flux.

10-35. Water at saturation temperature and atmospheric pressure is boiled with an electrically heated platinum wire of diameter $D_o = 0.2\,\text{cm}$ in the stable film boiling regime with a temperature difference of $T_w - T_s = 654°C$. In the absence of radiation, calculate the film-boiling heat transfer coefficient and the heat flux.

10-36. Water at saturation temperature and atmospheric pressure is boiled with an electrically heated platinum wire of diameter $D_o = 0.6\,\text{cm}$, in the stable film boiling regime with a temperature difference of $T_w - T_s = 654°C$. In the absence of radiation, calculate the film-boiling heat transfer coefficient and the heat flux.

REFERENCES

1. Bromley, L. A.: "Heat Transfer in Stable Film Boiling," *Chem. Eng. Prog.* **46**:221–227 (1950).
2. Ded, J.S., and J.H. Lienhard: "The Peak Pool Boiling Heat Flux from a Sphere," *AIChE J.* **18**:337–342 (1942).
3. Farber, E.A., and R.L. Scorah: "Heat Transfer to Water Boiling under Pressure," *Trans. ASHE*:369–384 (1948).
4. Hampson, H., and M.N. Ozisik: "An Investigation into the Condensation of Steam," *Proc. Inst. Mech. Engin.*, London **1B**:282–284 (1952).
5. Kirkbride, C. G.: "Heat Transfer by Condensing Vapors," *Trans. AIChE* **30**:524 (1951).
6. Kutateladze, S. S.: "A Hydrodynamic Theory of Changes in Boiling Process under Free Convection," *Iz. Akad. Nauk. SSSR*, Otd. Tekh. Nauk. (4):524 (1951).
7. Lienhard, J. H., and V. K. Dhir: "Hydrodynamic Prediction of Peak Pool-Boiling Heat Fluxes from Finite Bodies," *J. Heat Transfer* **95C**:152–158 (1973).
8. Nukiyama, S.: "The Maximum and Minimum Values of the Heat *Q* Transmitted from Metal to Boiling Water under Atmospheric Pressure," *J. Jap. Soc. Mech. Eng.* **37**:367–374 (1934).
9. Rohsenow, W. M.: "A Method of Correlating Heat Transfer Data for Surface Boiling Liquids," *Trans. ASME* **74**:969–975 (1952).
10. Schmidt, E., W. Schurig, and W. Sellschopp: "Versuche uber die Kondensation von Wasser-dampf in Film und Tropfenform," *Tech. Mech. u. Thermodynam.* **1**:53 (1930).
11. Sun, K. H., and J. H. Lienhard, "The Peak Pool Boiling Heat Flux on Horizontal Cylinders," *Int. J. Heat Mass Transfer* **13**:1425–1439 (1970).
12. Vahon, R. I., G. H. Nix, and G. E. Tanger: "Evaluation of Constants for the Rohsenow Pool-Boiling Correlation," *J. Heat Transfer* **90C**:239-247 (1968).
13. Zuber, N.: "On the Stability of Boiling Heat Transfer," *J. Heat Transfer* **80C**:711 (1958).

CHAPTER
11

RADIATION

11-1 BASIC RADIATION CONCEPTS

Thermal radiation refers to radiation energy that bodies emit because of their own temperature. All bodies at a temperature above absolute zero emit radiation. We recall that energy transfer by conduction and convecton requires some material carrier, but the transfer of energy by radiation can take place in a vacuum without a material carrier. A typical example of this phenomenon is the transfer of energy from the sun to the earth; that is, the thermal energy emitted from the sun travels through space and reaches the earth's surface. The actual mechanism of energy transport by radiation is not fully understood, but some theories based on certain concepts have been proposed to explain the propagation process.

One of the concepts, originally proposed by Maxwell, is the treatment of radiation as *electromagnetic waves*, just like radio or sound waves. This concept has been useful in studies on the prediction of the radiation properties of surfaces and materials.

Another concept, proposed by Max Planck, treats radiation as *photons* or *quanta of energy*. This concept has been employed to predict the magnitude of the radiation energy emitted by a body at a given temperature under idealized conditions.

Clearly, both concepts are useful in the study of radiation. We focus our attention on the wave nature of thermal radiation. A body at any given temperature emits thermal radiation at all wavelengths from $\lambda = 0$ to $\lambda = \infty$, but the distribution of the relative magnitude of energy emitted at each wavelength depends on the temperature. At temperatures encountered in most engineering applications, the bulk of the energy emitted by a body lies in the wavelengths, approximately, between $\lambda \cong 0.1\,\mu m$ and $\lambda \cong 100\,\mu m$. For this reason, this portion of the wavelength spectrum is generally referred to as the *thermal radiation*. For example, the Sun emits thermal radiation at an effective surface temperature of about 5760 K, and the bulk of this energy lies in wavelengths between $\lambda \cong 0.1\,\mu m$ and $\lambda \cong 3\,\mu m$; therefore, this portion of the spectrum is generally known as the solar radiation. The radiation emitted by the sun at wavelengths between $\lambda = 0.4\,\mu m$ and $\lambda = 0.7\,\mu m$ is visible to the eye; therefore, this portion of the spectrum is called visible radiation (i.e., light). Figure 11-1 illustrates such subdivisions on the electromagnetic wave spectrum.

Other types of radiation, such as x-rays, gamma rays, microwaves, etc., are well known and are utilized in various branches of science and engineering. X-rays are produced by the bombardment of a metal with high-frequency electrons, and the bulk of the energy is in the range between $\lambda \cong 10^{-4}\,\mu m$ and $\lambda \cong 10^{-2}\,\mu m$. Gamma rays are produced by the fission of nuclei or by radioactive disintegration, and the bulk of the energy is concentrated at wavelengths shorter than those of x-rays. In this book, we are not concerned with such radiation; our interest is only in *thermal radiation* as a mechanism of energy transport between objects at different temperatures.

The wave nature of thermal radiation implies that the wavelength λ should be associated with the frequency of radiation ν. The relation between λ and ν is given by

$$\lambda = \frac{c}{\nu} \tag{11-1}$$

where c is the speed of propagation in the medium. If the medium in which radiation travels is a vacuum, the speed of propagation is equal to the speed of light, that is,

$$c_0 = 2.9979 \times 10^8 \text{ m/s} \tag{11-2}$$

By utilizing this relationship between λ and ν, we included the corresponding frequency spectrum in Fig. 11-1.

FIGURE 11-1
Typical spectrum of electromagnetic radiation due to temperature of a body.

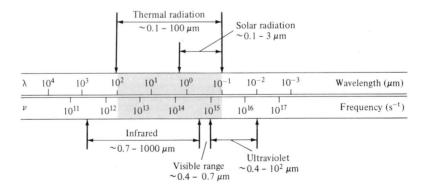

11-2 BLACKBODY RADIATION LAWS

A body at any temperature above absolute zero emits thermal radiation at all wavelengths and in all possible directions into space. A knowledge of the amount of radiation energy emitted by a body at any given temperature is most important in the study of thermal radiation. Therefore, the concept of a *blackbody* has been introduced as an idealized situation to serve as a reference in determining the emission and absorption of radiation by real bodies.

A blackbody is considered to absorb all incident radiation from all directions at all wavelengths without reflecting, transmitting, or scattering it. For a given temperature and wavelength, no other body can emit more radiation than a blackbody. The radiation emitted by a blackbody at any temperature T is the maximum possible emission at that temperature.

The term *black* should be distinguished from its common usage, where it indicates the blackness of a surface to visual observations. The human eye can detect blackness only in the visible range of the spectrum. For example, an object such as ice is bright to the eye but is almost black for long-wave thermal radiation. However, a blackbody is perfectly black to thermal radiation at all wavelengths from $\lambda = 0$ to $\lambda = \infty$.

Here we present the basic relations governing the emission of radiation by a blackbody.

Blackbody Emissive Power

Max Planck, at the beginning of this century, developed the theory of blackbody radiation in which the energy is assumed to be transported in the form of discrete photons. Based on the quantum arguments, Planck has shown that the radiation energy emitted by a blackbody into a vacuum is related to the absolute temperature T of the body and the wavelength (or frequency) of emission by the following expression:

$$E_{b\lambda}(T) = \frac{c_1}{\lambda^5 \{\exp\left[c_2/(\lambda T)\right] - 1\}} \qquad \text{W/(m}^2 \cdot \mu\text{m)} \qquad (11\text{-}3)$$

where $c_1 = 2\pi h c^2 = 3.743 \times 10^8 \text{ W} \cdot \mu\text{m}^4/\text{m}^2$
$c_2 = 1.4387 \times 10^4 \ \mu\text{m} \cdot \text{K}$
$T = $ absolute temperature, K
$\lambda = $ wavelength, μm

Here $E_{b\lambda}(T)$, called the *spectral blackbody emissive power*, represents the amount of radiation energy emitted by a blackbody at an absolute temperature T, per unit time, per unit area of the surface, per unit wavelength about the wavelength λ. Thus it has the units W/(m$^2 \cdot \mu$m). The validity of this formula has also been verified experimentally.

Figure 11-2 shows a plot of the function $E_{b\lambda}(T)$ against the wavelength of radiation for several different values of temperature. An examination of this figure reveals the following features of emission of radiation by a blackbody:

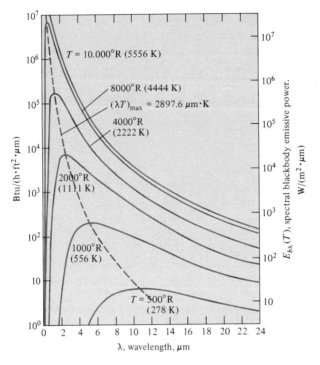

FIGURE 11-2
Spectral blackbody emissive power at different temperatures.

1. The emissive power increases with increasing temperature at all wavelengths.
2. Each curve exhibits a peak.
3. As temperature increases, the peaks tend to shift toward smaller wavelengths.
4. The top curve for $T = 5556$ K corresponds approximately to the emission of radiation by a body at the effective surface temperature of the sun ($T_{sun} = 5762$ K).

Wien's Displacement Law

In Fig. 11-2, the locus of the peaks of the curves is shown by a dashed line. Almost a century ago W. Wien developed an analytic formula for the locus of these peaks. The resulting expression, called *Wien's displacement law*, is given by

$$\boxed{(\lambda T)_{max} = 2897.6 \; \mu m \cdot K} \tag{11-4}$$

For example, for the solar radiation emitted at an effective temperature

$T = 5762$ K of the sun's surface, the application of Wien's law gives the wavelength at which the peak occurs as

$$\lambda_{\max} \cong 0.5 \ \mu\text{m} \tag{11-5}$$

which lies in the visible range of the spectrum.

Stefan-Boltzmann Law

The radiation energy emitted by a blackbody at an absolute temperature T over all wavelengths from $\lambda = 0$ to $\lambda = \infty$ is a quantity of practical interest. It is determined by integrating the special blackbody emissive power $E_{b\lambda}(T)$ from $\lambda = 0$ to $\lambda = \infty$:

$$E_b(T) = \int_{\lambda=0}^{\infty} E_{b\lambda}(T) \, d\lambda \qquad \text{W/m}^2 \tag{11-6}$$

For $E_{b\lambda}(T)$ given by Eq. (11-3), this integration can be performed, and the result is expressed as

$$\boxed{E_b(T) = \sigma T^4} \qquad \text{W/m}^2 \tag{11-7}$$

This relation is known as the *Stefan-Boltzmann law*. Here $E_b(T)$ is called the *blackbody emissive power*, T is the absolute temperature in K, and σ is the *Stefan-Boltzmann constant*, which has the numerical value

$$\boxed{\sigma = 5.67 \times 10^{-8} \ \text{W/(m}^2 \cdot \text{K}^4)} \tag{11-8}$$

Equation (11-7) shows that the radiation flux is proportional to the fourth power of the absolute temperature of the body. Therefore, at elevated temperatures radiation is an important mechanism of heat transfer; it can be more dominant than conduction or convection.

Blackbody Radiation Function

The radiation flux emitted by a blackbody at any given temperature T over all wavelengths from $\lambda = 0$ to $\lambda = \infty$ is given by the Stefan-Boltzmann law. The fraction of this total energy that is emitted over any given finite-wavelength band is also a quantity of interest in some applications.

The radiation energy emitted by a blackbody per unit area over a wavelength band from $\lambda = 0$ to λ is determined by

$$E_{b,0-\lambda}(T) = \int_0^{\lambda} E_{b\lambda}(T) \, d\lambda \tag{11-9}$$

By dividing this quantity by the total energy emission from $\lambda = 0$ to $\lambda = \infty$, we obtain

$$f_{0\lambda}(T) = \frac{\displaystyle\int_0^\lambda E_{b\lambda}(T)\, d\lambda}{\displaystyle\int_0^\infty E_{b\lambda}(T)\, d\lambda} = \frac{\displaystyle\int_0^\lambda E_{b\lambda}(T)\, d\lambda}{\sigma T^4} \tag{11-10}$$

Here, $f_{0\lambda}(T)$ is called the *blackbody radiation function*. It represents the emission of radiation by a blackbody over the wavelength band from $\lambda = 0$ to λ as a fraction of the total emission. In Table 11-1, $f_{0\lambda}(T)$ is listed as a function of λT. Here, λ is in micrometers and T is in Kelvins.

The radiation energy emitted by a blackbody at temperature T over a finite-wavelength band from $\lambda = \lambda_1$ to $\lambda = \lambda_2$ as a fraction of the total emission from $\lambda = 0$ to $\lambda = \lambda$ can be determined in the following manner.

We obtain from Table 11-1 the fractional functions $f_{0-\lambda_1}(T)$ and $f_{0-\lambda_2}(T)$, corresponding to $\lambda_1 T$ and $\lambda_2 T$, respectively. Then their difference gives

$$f_{\lambda_1 - \lambda_2}(T) = f_{0-\lambda_2}(T) - f_{0-\lambda_1}(T)$$

where $f_{\lambda_1 - \lambda_2}(T)$ represents the energy contained over the finite-wavelength band between λ_1 and λ_2 as a fraction of the total energy emission.

Example 11-1. At a wavelength of 4 μm, what is the temperature of the blackbody that will give an emissive power equal to 10^3 W/(m² · μm)?

Solution. Given $\lambda = 4$ μm and $E_{b\lambda}(T) = 10^3$ W/(m² · μm), and using the spectral blackbody emissive power expression given by Eq. (11-3), we solve for the temperature T as follows:

$$10^3 = \frac{3.743 \times 10^8}{4^5 \{\exp\left[1.4387 \times 10^4/(4T)\right] - 1\}}$$

or

$$\frac{1.4387 \times 10^4}{4T} = \ln 366.53$$

and

$$T = 609 \text{ K}$$

Example 11-2. Consider a blackbody emitting at 1500 K. Determine the wavelength at which the blackbody spectral emissive power $E_{b\lambda}$ is maximum.

Solution. Give $T = 1500$ K, from Wien's displacement law, Eq. (11-4), we have

$$(\lambda T)_{\max} = 2897.6 \text{ μm · K}$$

or

$$\lambda = \frac{2897.6}{1500} = 1.93 \ \mu m$$

Example 11-3. There is a hole of radius 0.2 cm in a large spherical enclosure whose inner surface is maintained at 800 K. Determine the rate of emission of radiative energy through this opening.

TABLE 11-1
Blackbody radiation functions

λT, $\mu m \cdot K$	$f_{0\lambda}(T)$	λT, $\mu m \cdot K$	$f_{0\lambda}(T)$
555.6	0.00000	5,777.8	0.71806
666.7	0.00000	5,888.9	0.72813
777.8	0.00000	6,000.0	0.73777
888.9	0.00007	6,111.1	0.74700
1,000.0	0.00032	6,222.2	0.75583
1,111.1	0.00101	6,333.3	0.76429
1,222.2	0.00252	6,444.4	0.77238
1,333.3	0.00531	6,555.6	0.78014
1,444.4	0.00983	6,666.7	0.78757
1,555.6	0.01643	6,777.8	0.79469
1,666.7	0.02537	6,888.9	0.80152
1,777.8	0.03677	7,000.0	0.80806
1,888.9	0.05059	7,111.1	0.81433
2,000.0	0.06672	7,222.2	0.82035
2,111.1	0.08496	7,333.3	0.82612
2,222.2	0.10503	7,444.4	0.83166
2,333.3	0.12665	7,555.6	0.83698
2,444.4	0.14953	7,666.7	0.84209
2,555.6	0.17337	7,777.8	0.84699
2,666.7	0.19789	7,888.9	0.85171
2,777.8	0.22285	8,000.0	0.85624
2,888.9	0.24803	8,111.1	0.86059
3,000.0	0.27322	8,222.2	0.86477
3,111.1	0.29825	8,333.3	0.86880
3,222.2	0.32300	8,888.9	0.88677
3,333.3	0.34734	9,444.4	0.90168
3,444.4	0.37118	10,000.0	0.91414
3,555.6	0.39445	10,555.6	0.92462
3,666.7	0.41708	11,111.1	0.93349
3,777.8	0.43905	11,666.7	0.94104
3,888.9	0.46031	12,222.2	0.94751
4,000.0	0.48085	12,777.8	0.95307
4,111.1	0.50066	13,333.3	0.95788
4,222.2	0.51974	13,888.9	0.96207
4,333.3	0.53809	14,444.4	0.96572
4,444.4	0.55573	15,000.0	0.96892
4.555.6	0.57267	15,555.6	0.97174
4,666.7	0.58891	16,111.1	0.97423
4,777.8	0.60449	16,666.7	0.97644
4,888.9	0.61941	22,222.2	0.98915
5,000.0	0.63371	27,777.8	0.99414
5,111.1	0.64740	33,333.3	0.99649
5,222.2	0.66051	38,888.9	0.99773
5,333.3	0.67305	44,444.4	0.99845
5,444.4	0.68506	50,000.0	0.99889
5,555.6	0.69655	55,555.6	0.99918
5,666.7	0.70754	∞	1.00000

Solution. Given $T = 800$ K, the blackbody emissive power can be calculated from the Stefan-Boltzmann law:

$$E_b(T) = \sigma T^4$$

$$= (5.67 \times 10^{-8})(800)^4 = 23{,}224.3 \text{ W/m}^2$$

The hole area is $A = \pi(0.002)^2$ m^2. Therefore the rate of emission of radiation energy Q through the hole becomes

$$Q = AE_b(T)$$

$$= \pi(0.002)^2(23{,}224.3) = 0.29 \text{ W}$$

Example 11-4. A tungsten filament is heated to 2500 K. What fraction of this energy is in the visible range?

Solution. The visible range of the spectrum is $\lambda_1 = 0.4$ μm to $\lambda_2 = 0.7$ μm. Given $T = 2500$ K, from Table 11-1, for $\lambda_1 T = 0.4(2500) = 1000$ μm · K^4, we obtain

$$f_{0-\lambda_1} = 0.00032$$

and for $\lambda_2 T = 0.7(2500) = 1750$ μm · K^4, we obtain

$$f_{0-\lambda_2} = 0.03392$$

Then

$$f_{0-\lambda_2} - f_{0-\lambda_1} = 0.03392 - 0.00032 = 0.0336$$

Therefore 3.36 percent of the energy is in the visible range.

11-3 RADIATION PROPERTIES OF SURFACES

Radiation emitted by a body because of its temperature, in general, originates in the interior of the body. In the case of materials such as metals, wood, stone, etc., which are opaque to thermal radiation, the emission is confined to within an extremely short distance from the surface. Similarly, radiation incident on an opaque body is absorbed within a very short distance from the surface. Therefore, for opaque materials the interaction of radiation with the matter is treated as a surface phenomenon, because radiation is absorbed or emitted within an extremely short distance from the surface.

In the case of solar radiation incident on a window glass or a body of water, radiation is not immediately absorbed at the surface, but penetrates into the depths of the material. Therefore, water and glass are semitransparent to solar radiation, which is considered to originate at an effective temperature $T = 5762$ K of the sun's surface. On the other hand, thermal radiation originating from a source at low temperatures, say 400 K or less, that is incident on a window glass or water is absorbed within an extremely short distance from the surface. These examples illustrate that the characteristics of a body for absorption and emission of radiation depend not only on the type of material but also on the temperature (or wavelength) of the source at which the radiation is generated.

Since most engineering materials are opaque to thermal radiation, the interaction of radiation with the matter in engineering applications is generally treated as a surface phenomenon. Given this concentration we examine below the radiation characteristics of surfaces.

Emissivity

Radiation emitted by a real body at a temperature T is always less than that of the blackbody. Therefore, the blackbody emission is chosen as a reference. The *emissivity* ε of a surface is defined as the ratio of the energy emitted by a real surface to that emitted by a blackbody at the same temperature; it has a value between zero and unity. Clearly, these are numerous possibilities for making such a comparison: for example, the comparison can be made at a given wavelength, over all wavelengths, for the energy emitted in a specified direction, or for the energy emitted into the hemispherical space. Most emissivity data reported in the literature are in the form of *hemispherical emissivity* averaged over all wavelengths. By definition, the hemispherical emissivity is the ratio of the radiation flux $q(T)$ emitted by a real surface at temperature T over all wavelengths into the hemispherical space, to that which would have been emitted by a blackbody, $E_b(T)$, at the same temperature T. Then the hemispherical emissivity $\varepsilon(T)$ is defined as

$$\varepsilon(T) = \frac{q(T)}{E_b(T)} \tag{11-11}$$

Given the emissivity of a real surface, the radiation energy $q(T)$ emitted by the surface per unit area at temperature T is determined by

$$q(T) = \varepsilon(T)E_b(T) = \varepsilon(T)\sigma T^4 \qquad \text{W/m}^2 \tag{11-12}$$

The hemispherical emissivity of typical engineering materials is listed in Appendix C. Figure 11-3 illustrates the effects of temperature and oxidation on emissivity. Clearly, oxidation increases the emissivity. We can also speak of the *spectral hemispherical emissivity* ε_λ if the emissivity is considered at a specific wavelength λ. The spectral hemispherical emissivity ε_λ is determined by laboratory tests performed at different wavelengths. Figure 11-4 shows the variation of ε_λ with wavelength for typical engineering materials. For example, for window glass there is a sharp change in spectral emissivity at wavelengths of about 3 μm.

Spectral information on ε_λ, such as that shown in Fig. 11-4, can be used to calculate the average value of emissivity ε over all wavelength from the following expression:

$$\varepsilon = \frac{\int_0^\infty \varepsilon_\lambda E_{b\lambda}(T)\,d\lambda}{\int_0^\infty E_{b\lambda}(T)\,d\lambda} = \frac{\int_0^\infty \varepsilon_\lambda E_{b\lambda}(T)\,d\lambda}{E_b(T)} = \frac{\int_0^\infty \varepsilon_\lambda E_{b\lambda}(T)\,d\lambda}{\sigma T^4} \tag{11-13}$$

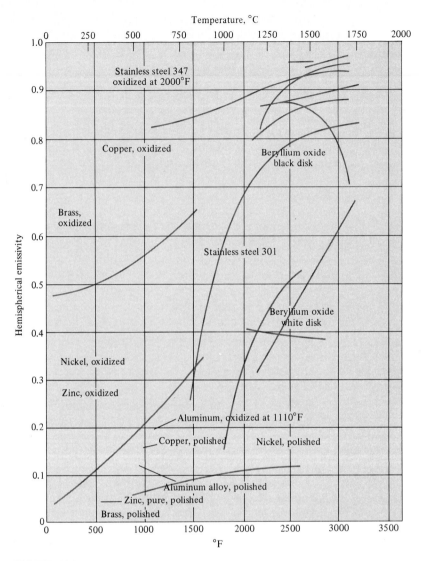

FIGURE 11-3
Effects of temperature and oxidation of hemispherical emissivity of metals. (Based on data from Gubareff, Janssen and Torborg.)

Note that, in this averaging process, the spectral blackbody emissive power $E_{b\lambda}(T)$ is used as the weight factor, because the energy emission is governed by the magnitude of $E_{b\lambda}(T)$.

Generally it is quite laborious to perform the integration in Eq. (11-13) when experimental data are to be used for ε_{λ}. To alleviate this difficulty, the integration can be transformed into a summation involving several wavelength bands, over each of which ε_{λ} can be regarded as uniform and taken out of the

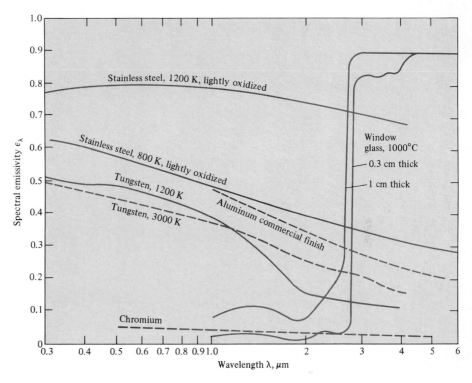

FIGURE 11-4
Spectral emissivity of materials.

integral sign. Then the resulting integrals are recast as the blackbody radiation functions, which are readily available from Table 11-1.

To illustrate this matter, we consider the entire wavelength spectrum to be divided into three wavelength bands:

From 0 to λ_1 : $\varepsilon_\lambda = \varepsilon_1 = $ constant

From λ_1 to λ_2: $\varepsilon_\lambda = \varepsilon_2 = $ constant

From λ_2 to ∞ : $\varepsilon_\lambda = \varepsilon_3 = $ constant

Then the integral in Eq. (11-13) is separated into three parts, and the definition of the blackbody radiation function given by Eq. (11-10) is utilized.

$$\varepsilon = \varepsilon_1 \frac{\int_0^{\lambda_1} E_{b\lambda}(T)\, d\lambda}{\sigma T^4} + \varepsilon_2 \frac{\int_{\lambda_1}^{\lambda_2} E_{b\lambda}(T)\, d\lambda}{\sigma T^4} + \varepsilon_3 \frac{\int_{\lambda_2}^{\infty} E_{b\lambda}(T)\, d\lambda}{\sigma T^4}$$

$$\varepsilon = \varepsilon_1 f_{0-\lambda_1}(T) + \varepsilon_2[f_{0-\lambda_2}(T) - f_{0-\lambda_1}(T)] + \varepsilon_3[1 - f_{0-\lambda_2}] \qquad (11\text{-}14)$$

Clearly, knowing ε_1, ε_2, and ε_3, the evaluation of emissivity ε from Eq. (11-14) by using Table 11-1 is a relatively easy matter.

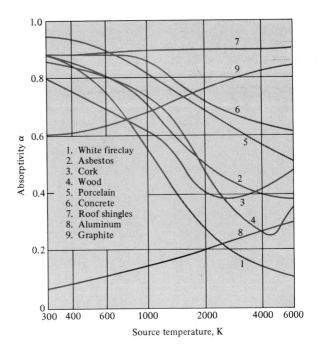

FIGURE 11-5
Variation of absorptivity with source temperature of incident radiation for various common materials at room temperature. (From Sieber.)

Absorptivity

A blackbody absorbs all radiation incident upon it, whereas a real surface absorbs only part of it. The *hemispherical absorptivity* α of a real surface is defined as the fraction of the radiation energy incident upon a surface from all directions over the hemispherical space and over all wavelengths that is absorbed by the surface. We can also speak of *spectral hemispherical absorptivity* α_λ if the absorptivity refers to radiation incident at a specific wavelength λ.

Figure 11-5 shows the absorptivity of various materials at room temperature as a function of the source temperature at which the incident radiation is originating. For example, the absorptivity of white fireclay for incident radiation originating from a source at 300 K is about 0.9, whereas for an incident radiation from a source at 6000 K it is about 0.1. On the other hand, the absorptivity of aluminum and graphite increases with an increase in the source temperature.

Relatively few data are available in the literature on the effect of source temperature on absorptivity of surfaces, except for the case of solar radiation.

Graybody Approximation

The radiation properties of a surface, such as absorptivity and emissivity, in general, depend on the wavelength of radiation. However, in radiation calculations, it is not practical to consider heat transfer at each wavelength. To alleviate such difficulties, it is common to assume a uniform emissivity ε over the entire wavelength spectrum. This is called the *graybody approximation*; it is frequently used in engineering applications.

Kirchhoff's Law

Under certain conditions, the absorptivity and emissivity of a body are related to each other by Kirchhoff's law. According to this law, the spectral emissivity $\varepsilon_\lambda(T)$ for the emission of radiation by a body at temperature T is equal to the spectral absorptivity $\alpha_\lambda(T)$ of the body for radiation originating from a blackbody source at the same temperature T. Hence, by Kirchhoff's law we write

$$\varepsilon_\lambda(T) = \alpha_\lambda(T) \qquad (11\text{-}15a)$$

Care must be exercised in generalizing this result to the average values of emissivity and absorptivity over the entire wavelength spectrum. That is, Eq. (11-15a) is always valid, but the equality

$$\varepsilon(T) = \alpha(T) \qquad (11\text{-}15b)$$

is applicable only where the radiation properties of the body are independent of the wavelength (i.e., a graybody) or when the incident and the emitted radiation have the same spectral distribution.

Reflectivity

When radiation is incident on a real surface, a fraction of it is reflected by the surface. If the surface is perfectly smooth, the incident and the reflected rays are symmetric with respect to the normal at the point of incidence, as illustrated in Fig. 11-6a. This mirrorlike reflection is called *specular* reflection. If the surface has some roughness, the incident radiation is scattered in all directions. An idealized reflection law assumes that in such a situation the intensity of the reflected radiation is constant for all angles of reflection and

FIGURE 11-6
Reflection from surfaces. (a) Specular reflection, (b) diffuse reflection, (c) irregular reflection.

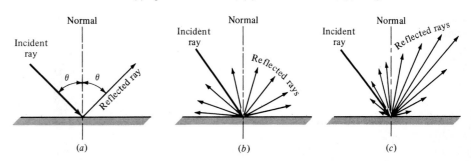

independent of the direction of the incident radiation. Such a reflection is called *diffuse* reflection. Figure 11-6b illustrates diffuse reflection from a surface. Reflection from real surfaces encountered in engineering applications is neither perfectly diffuse nor perfectly specular, but exhibits characteristics that are a combination of the diffuse and specular behavior, as sketched in Figs. 11-6c. However, the concept is useful for studying the effects of the two limiting cases on radiation transfer. The reflectivity of a surface is defined as the fraction of the incident radiation reflected by the surface.

The *hemispherical reflectivity* ρ of a surface is defined as the fraction of the radiation energy incident on a surface from all directions in the hemispherical space and over all wavelengths that is reflected back into the hemispherical space. For an *opaque surface*, the hemispherical reflectivity and the hemispherical absorptivity α are related by

$$\alpha = 1 - \rho \qquad (11\text{-}16)$$

The *spectral hemispherical reflectivity* ρ_λ is the reflection at a specific wavelength λ. For an opaque surface, the spectral hemispherical reflectivity ρ_λ and the spectral hemispherical absorptivity α_λ are related by

$$\alpha_\lambda = 1 - \rho_\lambda \qquad (11\text{-}17)$$

Transmissivity

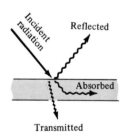

If the body is *semitransparent* to radiation, as glass is to solar radiation, part of the incident radiation is reflected by the surface, part is absorbed by the medium, and the remainder is transmitted, as schematically shown in Fig. 11-7. Then the sum of absorptivity and reflectivity is less than unity, and the difference is called the *transmissivity* of the body. With this consideration we write

$$\alpha_\lambda + \rho_\lambda + \tau_\lambda = 1 \qquad (11\text{-}18a)$$

$$\alpha + \rho + \tau = 1 \qquad (11\text{-}18b)$$

FIGURE 11-7
Reflection, absorption, and transmission of incident radiation by a semitransparent material.

where we define τ_λ as the *spectral transmissivity* and τ as the *transmissivity*.

Example 11-5. The spectral emissivity of an opaque surface at 1000 K is given by

$$\varepsilon_1 = 0.1 \qquad \text{for } \lambda_0 = 0 \text{ to } \lambda_1 = 0.5 \ \mu\text{m}$$

$$\varepsilon_2 = 0.5 \qquad \text{for } \lambda_1 = 0.5 \text{ to } \lambda_2 = 6 \ \mu\text{m}$$

$$\varepsilon_3 = 0.7 \qquad \text{for } \lambda_2 = 6 \text{ to } \lambda_3 = 15 \ \mu\text{m}$$

$$\varepsilon_4 = 0.8 \qquad \text{for } \lambda_3 > 15 \ \mu\text{m}$$

Determine the average emissivity over the entire range of wavelengths and the radiation flux emitted by the material at 1000 K.

Solution. The spectral distribution of emissivity is given in stepwise variations. Therefore, in applying Eq. (11-13) we break the integral into parts:

$$\varepsilon = \frac{\int_0^\infty \varepsilon_\lambda E_{b\lambda}(T) \, d\lambda}{e_b(T)}$$

$$= \varepsilon_1 \int_0^{\lambda_1} \frac{E_{b\lambda}(T)}{E_b(T)} \, d\lambda + \varepsilon_2 \int_{\lambda_1}^{\lambda_2} \frac{E_{b\lambda}(T)}{E_b(T)} \, d\lambda + \varepsilon_3 \int_{\lambda_2}^{\lambda_3} \frac{E_{b\lambda}(T)}{E_b(T)} \, d\lambda$$

$$+ \varepsilon_4 \int_{\lambda_3}^\infty \frac{E_{b\lambda}(T)}{E_b(T)} \, d\lambda$$

$$= \varepsilon_1 f_{0-\lambda_1} + \varepsilon_2 (f_{0-\lambda_2} - f_{0-\lambda_1}) + \varepsilon_3 (f_{0-\lambda_3} - f_{0-\lambda_2}) + \varepsilon_4 (f_{0-\infty} - f_{0-\lambda_3})$$

where $f_{0-\lambda}$ is given in Table 11-1. We have

$$\lambda_1 = 0.5 \qquad \lambda_1 T = 0.5(1000) = 500 \ \mu\text{m} \cdot \text{K} \qquad f_{0-\lambda_1} \cong 0$$

$$\lambda_2 = 6, \qquad \lambda_2 T = 6(1000) = 6000 \ \mu\text{m} \cdot \text{K} \qquad f_{0-\lambda_2} = 0.73777$$

$$\lambda_3 = 15, \qquad \lambda_3 T = 15(1000) = 15{,}000 \ \mu\text{m} \cdot \text{K} \qquad f_{0-\lambda_3} = 0.96892$$

Then

$$\varepsilon = 0.1(0) + 0.5(0.73777 - 0) + 0.7(0.96892 - 0.73777) + 0.8(1 - 0.96892)$$

$$= 0.5556$$

Knowing the average emissivity $\varepsilon(T)$, the radiation energy $q(T)$ emitted by the surface per unit area at a temperature $T = 1000$ K is determined from Eq. (11-12) as follows:

$$q(T) = \varepsilon(T)\sigma T^4$$

$$= 0.5556(5.67 \times 10^{-8})(1000)^4$$

$$= 31{,}502.5 \ \text{W/m}^2$$

11-4 VIEW FACTOR CONCEPT

In the study of radiation heat exchange among surfaces that are separated by a nonparticipating medium, such as a vacuum or air, the orientation and size of the surfaces relative to each other are factors that strongly affect the radiation exchange. To formalize the effects of orientation in the analysis of radiation heat exchange among the surfaces, the concept of *view factor* has been adopted. The terms *shape factor*, *angle factor*, and *configuration factor* also have been used in the literature. A distinction should be made between a diffuse view factor and a specular view factor. The former refers to the situation in which the surfaces are diffuse reflectors and diffuse emitters, whereas the latter refers to the situation in which the surfaces are diffuse emitters and specular reflectors. In this book we consider only the case in which the surfaces are *diffuse emitters and diffuse reflectors*; therefore, we do not need such a distinction. We simply use the term view factor, and it will imply the diffuse view factor.

The physical significance of the view factor between two surfaces is that it represents the fraction of the radiative energy leaving one surface that strikes the other surface directly. Therefore, the determination of the view factor between two surfaces is merely a geometrical problem.

View Factor between Two Elementary Surfaces

The determination of the view factor between two elementary surfaces is a very straightforward matter, and the results are useful in generalizing the concept to the determination of the view factor between two arbitrary finite surfaces.

Consider two elemental surfaces dA_1 and dA_2, as illustrated in Fig. 11-8. Let r be the distance between these two surfaces, θ_1 the polar angle between the normal $\hat{n}_1$ to the surface dA_1 and the line joining dA_1 to dA_2, and θ_2 the polar angle between the normal $\hat{n}_2$ to the surface element dA_2 and the line r.

The *elemental view factor* $dF_{dA_1-dA_2}$, by definition, is the ratio of the radiative energy leaving dA_1 that strikes dA_2 directly to the radiative energy leaving dA_1 in all directions into the hemispherical space. It is given by

$$dF_{dA_1-dA_2} = \frac{\cos\theta_1 \cos\theta_2 \, dA_2}{\pi r^2} \qquad (11\text{-}19)$$

The elemental view factor $dF_{dA_2-dA_1}$ from dA_2 to dA_1 is now immediately obtained from Eq. (11-19) by interchanging subscripts 1 and 2. We find

$$dF_{dA_2-dA_1} = \frac{\cos\theta_1 \cos\theta_2 \, dA_1}{\pi r^2} \qquad (11\text{-}20)$$

The *reciprocity relation* between the view factors $dF_{dA_1-dA_2}$ and $dF_{dA_2-dA_1}$

FIGURE 11-8
Coordinates for the definition of view factor.

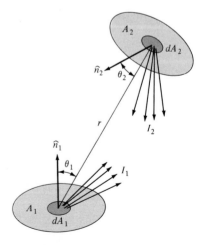

follows from Eqs. (11-19) and (11-20):

$$dA_1 dF_{dA_1 - dA_2} = dA_2 dF_{dA_2 - dA_1} \qquad (11\text{-}21)$$

This relation implies that for two elemental surfaces dA_1 and dA_2, when one of the view factors is known, the other can be readily computed by the reciprocity relation.

View Factor between Two Finite Surfaces

Although developing formal expressions for the view factor between two finite surfaces from the definition of elemental view factors given above is a relatively easy matter, computing the resulting expressions is very difficult except for very simple geometrical arrangements. To illustrate this, we consider the view factor $F_{A_1 - A_2}$ from a finite surface A_1 to another finite surface A_2, as shown in Fig. 11-8. By integrating the elemental view factor $dF_{dA_1 - dA_2}$ over the areas A_2 and A_1 and dividing the resulting expression by A_1, we obtain the following formal expression for $F_{A_1 - A_2}$:

$$F_{A_1 - A_2} = \frac{1}{A_1} \int_{A_1} \int_{A_2} \frac{\cos \theta_1 \cos \theta_2}{\pi r^2} \, dA_2 \, dA_1 \qquad (11\text{-}22)$$

Then the view factor $F_{A_2 - A_1}$ from area A_2 to A_1 is immediately obtained by interchanging the subscripts 1 and 2 in Eq. (11-22). We find

$$F_{A_2 - A_1} = \frac{1}{A_2} \int_{A_2} \int_{A_1} \frac{\cos \theta_1 \cos \theta_2}{\pi r^2} \, dA_1 \, dA_2 \qquad (11\text{-}23)$$

From Eqs. (11-22) and (11-23), the reciprocity relation between the view factors $F_{A_1 - A_2}$ and $F_{A_2 - A_1}$ is obtained as

$$A_1 F_{A_1 - A_2} = A_2 F_{A_2 - A_1} \qquad (11\text{-}24)$$

The reciprocity relations given above are useful for determining one of the view factors, given knowledge of the other.

The view factor $F_{A_1 - A_2}$ represents the fraction of the radiation energy leaving surface A_1 diffusely that strikes surface A_2 directly. The evaluation of the double surface integrals that appear in the resulting formal expressions is extremely difficult.

Properties of View Factors

We consider an *enclosure* consisting of N *zones*, each of surface area A_i ($i = 1, 2, \ldots, N$), as illustrated in Fig. 11-9. It is assumed that each zone is

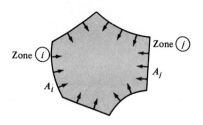

FIGURE 11-9
An N-zone enclosure.

isothermal, a diffuse emitter, and a diffuse reflector. The surface of each zone may be plane, convex, or concave. The view factors between surfaces A_i and A_j of the enclosure obey the following reciprocity relation:

$$A_i F_{A_i - A_j} = A_j F_{A_j - A_i} \tag{11-25}$$

The view factors from one surface, say A_i, of the enclosure to all surfaces of the enclosure, including itself, when summed, should be equal to unity by the definition of the view factor. This is called *the summation relation* among the view factors for an enclosure, and it is written as

$$\sum_{k=1}^{N} F_{A_i - A_k} = 1 \tag{11-26}$$

where N is the number of zones in the enclosure. In this summation the term $F_{A_i - A_i}$ is the view factor from the surface A_i to itself; it represents the fraction of radiative energy leaving the surface A_i that strikes itself directly. Clearly, $F_{A_i - A_i}$ vanishes if A_i is flat or convex, and it is nonzero if A_i is concave; this is stated as

$$F_{A_i - A_i} = 0 \qquad \text{if } A_i \text{ plane or convex} \tag{11-27a}$$

$$F_{A_i - A_i} \neq 0 \qquad \text{if } A_i \text{ concave} \tag{11-27b}$$

Example 11-6. Two small surfaces $dA_1 = 5\,\text{cm}^2$ and $dA_2 = 10\,\text{cm}^2$ are separated by $r = 100\,\text{cm}$ and oriented as illustrated in the accompanying figure. Calculate the view factors between the surfaces.

Solution. Both surfaces can be approximated as differential surfaces because $dA_i / r^2 \ll 1$. From Eq. (11-19) we have

$$dF_{dA_1 - dA_2} = \frac{\cos \theta_1 \cos \theta_2 \, dA_2}{\pi r^2}$$

$$= \frac{(\cos 60°)(\cos 30°)(10)}{\pi(100)^2} = 1.378 \times 10^{-4}$$

By the reciprocity relation, Eq. (11-21), we write

$$dF_{dA_2 - dA_1} = \frac{dA_1}{dA_2} dF_{dA_1 - dA_2}$$

$$= \tfrac{5}{10}(1.378 \times 10^{-4}) = 6.89 \times 10^{-5}$$

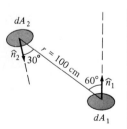

FIGURE EXAMPLE 11-6

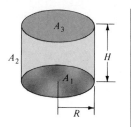

FIGURE EXAMPLE 11-7

Example 11-7. The view factor F_{1-3} between the base and the top surface of the cylinder shown in the accompanying figure is available from the charts. Develop a relation for the view factors F_{1-2} and F_{2-1} between the base and the lateral cylindrical surface in terms of F_{1-3}.

Solution. From the summation rule, Eq. (11-26),

$$F_{1-1} + F_{1-2} + F_{1-3} = 1$$

Since $F_{1-1} = 0$, we have

$$F_{1-2} = 1 - F_{1-3}$$

From the reciprocity relation, Eq. (11-25),

$$A_1 F_{1-2} = A_2 F_{2-1}$$

Therefore

$$F_{2-1} = \frac{A_1}{A_2} F_{1-2} = \frac{R}{2H} F_{1-2}$$

11-5 VIEW-FACTOR DETERMINATION

The computation of the elemental view factors defined by Eqs. (11-19) and (11-20) poses no problem, but the calculation of the view factor between two finite surfaces defined by Eqs. (11-22) or (11-23) is very difficult. Considerable effort has been devoted in the literature to the computation of view factors between surfaces, and the results are well documented. In Figs. 11-10 to 11-13 we present in graphical form some of the view factors for simple configurations. Comprehensive compilations of view factors are available in radiation heat transfer books.

View factor Algebra

The standard view-factor charts are available for only a limited number of simple configurations. However, it may be possible to split up the configuration of a complicated geometric arrangement into a number of simple configurations in such a manner that the view factors can be determined from the standard view-factor charts. Then it may be possible to determine the view factor for the original, complicated configuration by the algebraic sum of the view factors for the separate, simpler configurations. Such an approach is known as *view-factor algebra*. It provides a powerful method for determining the view factors for many complicated configurations.

No standard set of rules can be stated for this method, but appropriate use of the reciprocity relations and the summation rules is the key to the success of this technique.

To illustrate how the summation rule and the reciprocity relation can be applied, we consider the view factor from an area A_1 to an area A_2, which is divided into two areas A_3 and A_4:

$$A_2 = A_3 + A_4 \tag{11-28}$$

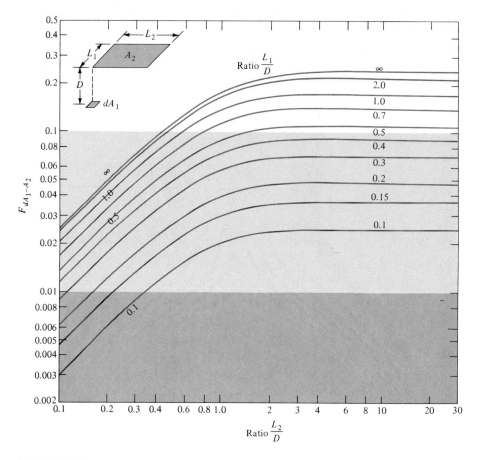

FIGURE 11-10

View factor $F_{dA_1-A_2}$ from an element surface dA_1 to a rectangular surface A_2. (From Mackey et al.)

Figure 11-14 shows the surfaces and their subdivision. Then the view factor from A_1 to A_2 can be written as

$$F_{1-2} = F_{1-3} + F_{1-4} \qquad (11\text{-}29)$$

which is consistent with the definition of the view factor. That is, the fraction of the total energy leaving A_1 that strikes A_3 and A_4 is equal to the fraction that strikes A_2.

Additional relationships among these view factors can be obtained. For example, suppose both sides of Eq. (11-29) are multiplied by A_1:

$$A_1 F_{1-2} = A_1 F_{1-3} + A_1 F_{1-4}$$

Then the reciprocity relation is applied to each term:

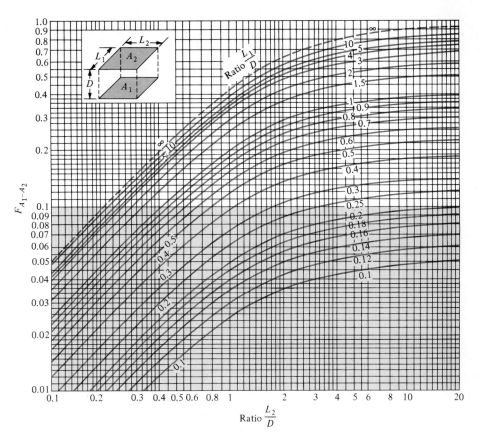

FIGURE 11-11

View factor $F_{A_1-A_2}$ from a rectangular surface A_1 to a rectangular surface A_2 which are adjacent and in perpendicular planes. (From Mackey et al.)

$$A_2 F_{2-1} = A_3 F_{3-1} + A_4 F_{4-1}$$

or

$$F_{2-1} = \frac{A_3 F_{3-1} + A_4 F_{4-1}}{A_2} = \frac{A_3 F_{3-1} + A_4 F_{4-1}}{A_3 + A_4} \qquad (11\text{-}30)$$

Suppose the area A_2 is divided into more parts:

$$A_2 = A_3 + A_4 + \cdots + A_N \qquad (11\text{-}31)$$

Then the corresponding form of Eq. (11-30) becomes

$$F_{2-1} = \frac{A_3 F_{3-1} + A_4 F_{4-1} + \cdots + A_N F_{N-1}}{A_3 + A_4 + \cdots + A_N} \qquad (11\text{-}32)$$

Clearly, similar manipulations can be applied to Eq. (11-30), and other relations among the view factors can be obtained.

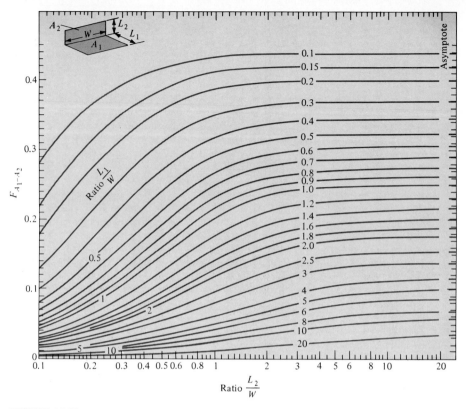

FIGURE 11-12

View factor $F_{A_1-A_2}$ from a rectangular surface A_1 to a rectangular surface A_2 which are parallel to and directly opposite each other. (From Mackey et al.)

FIGURE 11-13

View factor $F_{A_1-A_2}$ between two coaxial parallel disks.

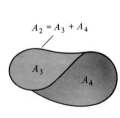

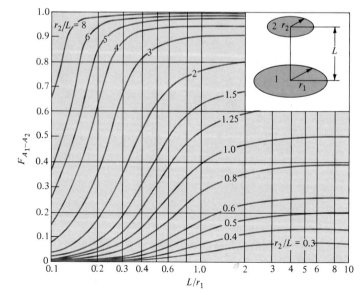

FIGURE 11-14

View factor algebra.

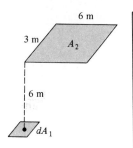

FIGURE EXAMPLE 11-8

Example 11-8. Determine the view factor F_{1-2} between an elemental surface dA_1 and the finite rectangular surface A_2 for the geometric arrangements shown in the accompanying figure.

Solution. Given $L_1 = 3\,\text{m}$, $L_2 = 6\,\text{m}$, and $D = 6\,\text{m}$, we have

$$\frac{L_1}{D} = \frac{3}{6} = 0.5$$

$$\frac{L_2}{D} = \frac{6}{6} = 1$$

Then we read the view factor $F_{dA_1-A_2}$ from Fig. 11-10 as

$$F_{1-2} = 0.09$$

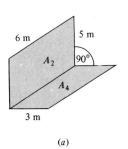

(a)

Example 11-9. Determine the view factors F_{4-2} and F_{3-2} between two rectangular surfaces for the geometric arrangement shown in the accompanying figure.

Solution. (a) Given $L_1 = 3\,\text{m}$, $L_2 = 5\,\text{m}$, and $W = 5\,\text{m}$, we have

$$\frac{L_1}{W} = \frac{3}{6} = 0.5$$

$$\frac{L_2}{W} = \frac{5}{6} = 0.833$$

Then we read the view factor $F_{A_1-A_2}$ from Fig. 11-11 as

$$F_{4-2} = 0.285$$

(b) Given $L_1 = 1\,\text{m}$, $L_2 = 5\,\text{m}$, and $W = 6\,\text{m}$, we have

$$\frac{L_1}{W} = \frac{1}{6} = 0.167$$

$$\frac{L_2}{W} = \frac{5}{6} = 0.833$$

Then we read the view factor $F_{A_1-A_2}$ from Fig. 11-11 as

$$F_{3-2} = 0.4$$

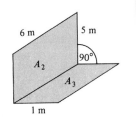

(b)

FIGURE EXAMPLE 11-9

Example 11-10. Determine the view factor F_{1-2} between two rectangular surfaces A_1 and A_2 for the geometric arrangement shown in the accompanying figure.

Solution. We split up this arrangement into the parts similar to those considered in Example 11-9. Then the view factors for the split-up arrangements are available in Fig. 11-11. From the laws of view-factor algebra applied to this configuration, we write

$$A_4 F_{4-2} = A_1 F_{1-2} + A_3 F_{3-2}$$

$$(3 \times 6)(0.285) = (2 \times 6)F_{1-2} + (1 \times 6)(0.4)$$

Therefore

$$F_{1-2} = 0.2275$$

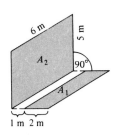

FIGURE EXAMPLE 11-10

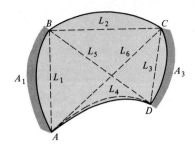

FIGURE 11-15
Determination of view factor
between the surfaces A_1 and A_3
of a long enclosure.

Crossed-String Method

Consider an enclosure, as shown in Fig. 11-15, consisting of four surfaces that
are very long in the direction perpendicular to the plane of the figure. The
surfaces can be flat, convex, or concave. Suppose we wish to find the view
factor F_{1-3} between surfaces A_1 and A_3. We assume that imaginary strings,
shown by dashed lines in Fig. 11-15, are tightly stretched among the four
corners A, B, C, and D of the enclosure. Let L_i ($i = 1, 2, 3, 4, 5, 6$) denote the
lengths of the strings joining the corners, as illustrated in the figure. Hottel has
shown that the view factor F_{1-3} can be expressed as

$$L_1 F_{1-3} = \frac{(L_5 + L_6) - (L_2 + L_4)}{2} \qquad (11\text{-}33)$$

Here we note that the term $L_5 + L_6$ is the sum of the lengths of the *cross
strings*, and $L_2 + L_4$ is the sum of the lengths of the uncrossed strings. Equation
(11-33) is useful for determining the view factor between the surfaces of a long
enclosure, such as a groove, which can be characterized as a two-dimensional
geometry of the form illustrated in Fig. 11-15.

Example 11-11. An infinitely long semicylindrical surface A_1 of radius b and an
infinitely long plate A_3 of half-width c are located a distance d apart, as illustrated
in the accompanying figure. Determine the view factor F_{1-3} between surfaces A_1
and A_3.

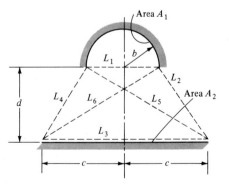

FIGURE EXAMPLE 11-11

Solution. The dashed lines in the figure are the imaginary strings. The view factor F_{1-3} between surfaces A_1 and A_3 is given by Eq. (11-33). Because of the symmetry, we have $L_5 = L_6$ and $L_2 = L_4$, and so Eq. (11-33) gives

$$F_{1-3} = \frac{L_6 - L_4}{L_1}$$

from the geometry we have

$$L_4 = [(c - b)^2 + d^2]^{1/2}$$

$$L_6 = [(c + b)^2 + d^2]^{1/2}$$

$$L_1 = 2b$$

Substituting the expressions for L_4, L_6, and L_1 into the above expression, the view factor F_{1-3} is determined as

$$F_{1-3} = \frac{[(c + b)^2 + d^2]^{1/2} - [(c - b)^2 + d^2]^{1/2}}{2b}$$

11-6 NETWORK METHOD FOR RADIATION EXCHANGE

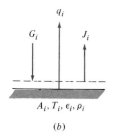

$A_i, T_i, \rho_i, \epsilon_i, \alpha_i$

(a)

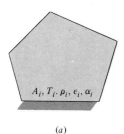

$A_i, T_i, \epsilon_i, \rho_i$

(b)

FIGURE 11-16
(a) An enclosure filled with a nonparticipating medium; (b) an energy balance per unit area of zone i.

The analysis of radiation exchange among the surfaces of an enclosure is complicated by the fact that when the surfaces are not black, radiation leaving a surface may be reflected back and forth among the surfaces several times, with partial absorption occurring at each reflection. Therefore, a proper analysis of the problem must include the effects of these multiple reflections. To simplify the analysis, we assume that a given enclosure can be divided into several zones, as illustrated in Fig. 11-16, in such a manner that the following conditions are assumed to hold for each of the zones $i = 1, 2, \ldots, N$:

1. Radiative properties (i.e., reflectivity, emissivity, and absorptivity) are uniform and independent of direction and frequency.
2. The surfaces are diffuse emitters and diffuse reflectors.
3. The radiative heat flux leaving the surface is uniform over the surface of each zone.
4. The irradiation is uniform over the surface of each zone.
5. Surfaces are opaque (that is, $\alpha + \rho = 1$).
6. Either a uniform temperature or a uniform heat flux is prescribed over the surface of each zone.
7. The enclosure is filled with a nonparticipating medium. Assumptions 3 and 4 are generally not correct, but the analysis becomes very complicated without them.

The objective of the analysis of radiation heat exchange in an enclosure is the *determination of net radiation heat flux* at the zones for which the temperature is prescribed. Conversely, the temperature can be determined at the zones for which the net heat flux is prescribed.

The electric network analogy heat exchange between surfaces. Before presenting the network method, we introduce the concepts leading to the definition of various radiation resistances in the path of the heat flow.

Radiosity Concept

We focus our attention on any one of the zones of the enclosure, say zone i. By assumption 3, the radiation energy leaving the surface is assumed to be uniform. We introduce the concept of *radiosity* J_i at zone i to represent the radiation energy leaving this surface per unit area per unit time, as viewed by an observer immediately above the surface of zone i. Figure 11-16b illustrates this hypothetical location symbolically by the dashed line. From physical considerations we conclude that the radiosity J_i is composed of two components:

$$J_i = \begin{pmatrix} \text{radiation} \\ \text{emitted by} \\ \text{surface } A_i \\ \text{per unit area} \end{pmatrix} + \begin{pmatrix} \text{radiation} \\ \text{reflected by} \\ \text{surface } A_i \\ \text{per unit area} \end{pmatrix} \quad \text{W/m}^2 \qquad (11\text{-}34a)$$

Each of these two components is now evaluated.

$$\begin{pmatrix} \text{Radiation emitted} \\ \text{by surface } A_i \text{ per} \\ \text{unit area} \end{pmatrix} = \varepsilon_i E_{bi} \qquad (11\text{-}34b)$$

where ε_i is the emissivity and $E_{bi} = \sigma T_i^4$ is the blackbody emissive power at temperature T_i of the zone A_i.

$$\begin{pmatrix} \text{Radiation reflected} \\ \text{by surface } A_i \text{ per} \\ \text{unit area} \end{pmatrix} = \rho_i G_i = (1 - \varepsilon_i) G_i \qquad (11\text{-}34c)$$

where G_i is the *irradiation* or the radiation flux incident on the surface of zone i, in watts per square meter, which is considered uniform over the zone by assumption 4. In addition, we assumed an opaque surface and set $\rho_i = 1 - \varepsilon_i$. By introducing Eqs. (11-34b) and (11-34c) into (11-34a), we obtain the following expression for the radiosity:

$$J_i = \varepsilon_i E_{bi} + (1 - \varepsilon_i) G_i \quad \text{W/m}^2 \qquad (11\text{-}35a)$$

and solving for G_i we have

$$G_i = \frac{J_i - \varepsilon_i E_{bi}}{1 - \varepsilon_i} \quad \text{W/m}^2 \qquad (11\text{-}35b)$$

This expression is needed in the relations that are now developed.

Radiation Resistance at a Surface

When a surface is not black, and hence has an emissivity $\varepsilon_i < 1$, the radiation heat transfer at the surface can be presented in a form that represents the *radiation resistance* at the surface as described below.

Let q_i be the net radiation heat flux, in watts per square meter, leaving the surface at zone i. By referring to the illustration in Fig. 11-16b, we conclude that q_i is equal to the difference between J_i and G_i, and write

$$q_i = J_i - G_i \qquad \text{W/m}^2 \tag{11-36}$$

Equation (11-35b) is introduced into Eq. (11-36) to eliminate G_i:

$$q_i = \frac{\varepsilon_i}{1 - \varepsilon_i} (E_{bi} - J_i) \qquad \text{W/m}^2 \tag{11-37}$$

The total net radiation heat flow Q_i leaving surface A_i becomes

$$Q_i = A_i q_i = A_i \frac{\varepsilon_i}{1 - \varepsilon_i} (E_{bi} - J_i) \tag{11-38}$$

which is rearranged as

$$\boxed{Q_i = \frac{E_{bi} - J_i}{R_i}} \qquad \text{W} \tag{11-39a}$$

where

$$\boxed{R_i = \frac{1 - \varepsilon_i}{A_i \varepsilon_i}} \tag{11-39b}$$

Clearly, Eq. (11-39a) is analogous to Ohm's law, where R_i represents the surface resistance to radiation. Equation (11-39a) is also analogous to the concept of *thermal resistance* (or the skin resistance), which we discussed in connection with convective heat transfer over a surface. That is, the total heat transfer rate is equal to the potential difference across the surface divided by the thermal resistance to heat flow over the surface.

When the surface is *black*, we have $\varepsilon_i = 1$, which implies that $R_i = 0$. Then Eq. (11-39a) reduces to

$$\boxed{J_i = E_{bi} = \sigma T_i^4} \qquad \text{for } \varepsilon_i = 1 \text{ or black surface} \tag{11-40}$$

Thus, for a black surface, the radiosity is equal to the blackbody emissive power of the surface.

Figure 11-17 illustrates the concept of surface thermal resistance to radiation between the potentials E_{bi} and J_i.

When the net radiation heat flow Q_i at the surface A_i vanishes, the zone A_i is called a *reradiating* or *adiabatic zone*. For such a case, we set Q_i equal to

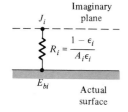

FIGURE 11-17
Surface resistance to radiation.

zero in Eq. (11-39a) and obtain

$$E_{bi} = J_i \qquad (11\text{-}41)$$

This result implies that at a reradiating or adiabatic zone, the radiosity J_i is equal to $E_{bi} = \sigma T_i^4$, and that the emissivity ε_i does not enter the analysis.

Radiation Resistance across Two Surfaces

The radiation resistance concept can also be developed for radiation exchange across two surfaces. Figure 11-18 illustrates two surfaces A_i and A_j, having emissivities ε_i and ε_j, and maintained at uniform but different temperatures T_i and T_j, respectively. Let Q_{i-j} denote the net radiation heat transfer from zero i to zone j. Then the energy balance for radiation heat exchange between the two zones can be stated as

$$Q_{i-j} = \begin{pmatrix} \text{radiation energy} \\ \text{leaving } A_i \text{ that} \\ \text{strikes } A_j \end{pmatrix} - \begin{pmatrix} \text{radiation energy} \\ \text{leaving } A_j \text{ that} \\ \text{strikes } A_i \end{pmatrix} \qquad (11\text{-}42)$$

To evaluate the two terms on the right-hand side of this equation, we consider J_i and J_j, the *radiosities* at the surfaces of zone i and zone j, respectively, and the view factors F_{i-j} and F_{j-i} between the two zones. Then, the mathematical expressions for each term on the right-hand side are written as

$$Q_{i-j} = J_i A_i F_{i-j} - J_j A_j F_{j-i} \qquad (11\text{-}43)$$

We utilize the reciprocity relation $A_i F_{i-j} = A_j F_{j-i}$ between the view factors and rewrite Eq. (11-43) as

$$Q_{i-j} = J_i A_i F_{i-j} - J_j A_i F_{i-j} = A_i F_{i-j}(J_i - J_j) \qquad (11\text{-}44)$$

Equation (11-44) is now rearranged in the form

$$Q_{i-j} = \frac{J_i - J_j}{R_{i-j}} \qquad (11\text{-}45a)$$

FIGURE 11-18
Various resistances to radiation in the path of heat flow between surfaces A_i and A_j.

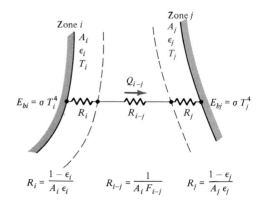

where

$$R_{i-j} = \frac{1}{A_i F_{i-j}}$$

(11-45b)

Clearly, Eq. (11-45a) is analogous to Ohm's law, where $R_{i-j} = 1/A_i F_{i-j}$ represents the *radiation resistance* to heat flow across the fictitious potentials J_i and J_j, which are considered to exist just above the surfaces A_i and A_j, respectively.

Radiation Network for Two Surfaces

Having established the necessary framework by defining various thermal resistances to radiation in the path of heat flow, we are now in a position to construct an analogous *electrical network* for radiation exchange between two surfaces.

Figure 11-18 illustrates various resistances to radiation in the path of heat flow Q_{i-j} by radiation from zone i to the zone j. The surface resistances to radiation for surfaces A_i and A_j are given, respectively, by

$$R_i = \frac{1 - \varepsilon_i}{A_i \varepsilon_i} \quad \text{and} \quad R_j = \frac{1 - \varepsilon_j}{A_j \varepsilon_j}$$

(11-46a,b)

and the resistance across the fictitious surfaces is given by

$$R_{i-j} = \frac{1}{A_i F_{i-j}}$$

(11-46c)

Then the total heat flow Q_{i-j} from zone i to zone j across the potential difference $E_{bi} - E_{bj}$ is determined by Ohm's law as

$$Q_{i-j} = \frac{E_{bi} - E_{bj}}{R_i + R_{i-j} + R_j} = \frac{\sigma(T_i^4 - T_j^4)}{R_i + R_{i-j} + R_j}$$

(11-47)

where various resistances are defined by Eqs. (11-46a) to (11-46c).

Radiation Transfer between Two Parallel Plates

Consider an enclosure consisting of two parallel, opaque, large plates, as illustrated in Fig. 11-19. Surfaces 1 and 2 are kept at uniform temperatures T_1 and T_2 and have emissivities ε_1 and ε_2, respectively. The configuration shown in this figure is a special case of that considered in Fig. 11-18. Therefore, the network analysis of heat flow given by Eq. (11-47) is applicable for this problem, with appropriate simplifications. We write

$$Q_{1-2} = \frac{\sigma(T_1^4 - T_2^4)}{R_1 + R_{1-2} + R_2}$$

(11-48a)

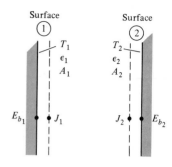

FIGURE 11-19
Radiation transfer between two
opaque large parallel plates.

where

$$R_1 = \frac{1 - \varepsilon_1}{A\varepsilon_1} \qquad R_{1-2} = \frac{1}{AF_{1-2}} \qquad R_2 = \frac{1 - \varepsilon_2}{A\varepsilon_2} \qquad (11\text{-}48b,c,d)$$

For two large parallel plates, $F_{1-2} = 1$; then Eq. (11-48a) reduces to the following expression for heat flow per unit area from surface 1 to surface 2:

$$q_{1-2} \equiv \frac{Q_{1-2}}{A} = \frac{\sigma(T_1^4 - T_2^4)}{1/\varepsilon_1 + 1/\varepsilon_2 - 1} \qquad \text{W/m}^2 \qquad (11\text{-}49)$$

Radiation Exchange in Three or More Zone Enclosures

The radiation network approach described above can readily be generalized to determine radiation exchange among the surfaces of a three or more zone enclosure by utilizing the resistances defined by Eqs. (11-39a), (11-39b), (11-45a), and (11-45b). That is, the radiation resistance at surface A_i is given by

$$R_i = \frac{1 - \varepsilon_i}{A_i \varepsilon_i} \qquad (11\text{-}50)$$

and the radiation resistance across the fictitious surfaces above A_i and A_j is given by

$$R_{i-j} = \frac{1}{A_i F_{i-j}} = \frac{1}{A_j F_{j-i}} \qquad (11\text{-}51)$$

To illustrate the application, we consider a three-zone enclosure, as shown in Fig. 11-20a. Zones 1, 2, and 3 have surface areas A_1, A_2, and A_3, emissivities ε_1, ε_2, and ε_3, and temperatures T_1, T_2, and T_3, respectively. Figure 11-20b shows the corresponding radiation network. To solve this problem, the algebraic sum of the currents at the nodes J_1, J_2, and J_3 is set equal to zero. This leads to three algebraic equations for the determination of three unknown radiosities J_1, J_2, and J_3. Knowing the radiosities, the heat flow rates Q_1, Q_2,

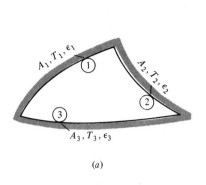

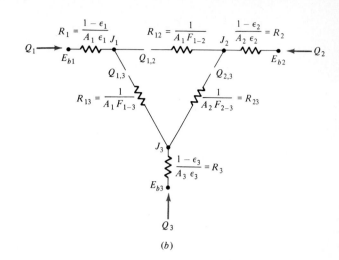

(b)

(a)

FIGURE 11-20
(a) A three-zone enclosure and (b) the corresponding radiation network.

and Q_3 at each zone are determined by the application of Ohm's law as

$$Q_1 = \frac{\sigma T_1^4 - J_1}{R_1} \quad \text{where } R_1 = \frac{1 - \varepsilon_1}{A_1 \varepsilon_1} \qquad (11\text{-}52a)$$

$$Q_2 = \frac{\sigma T_2^4 - J_2}{R_2} \quad \text{where } R_2 = \frac{1 - \varepsilon_2}{A_2 \varepsilon_2} \qquad (11\text{-}52b)$$

and

$$Q_3 = \frac{\sigma T_3^4 - J_3}{R_3} \quad \text{where } R_3 = \frac{1 - \varepsilon_3}{A_3 \varepsilon_3} \qquad (11\text{-}52c)$$

Example 11-12. Consider an enclosure consisting of two parallel, opaque large plates, as illustrated in Fig. 11-19. Calculate the net radiation flux q_{1-2} leaving plate 1 for $T_1 = 1000\,\text{K}$, $T_2 = 500\,\text{K}$, $\varepsilon_1 = 0.6$, and $\varepsilon_2 = 0.8$.

Solution. The net radiation flux q_{1-2} in this two-zone enclosure is given by Eq. (11-49) as

$$q_{1-2} = \frac{\sigma(T_1^4 - T_2^4)}{1/\varepsilon_1 + 1/\varepsilon_2 - 1} \quad \text{W/m}^2$$

Therefore, for the specified problem, q_{1-2} becomes

$$q_{1-2} = \frac{(5.67 \times 10^{-8})(1000^4 - 500^4)}{1/0.6 + 1/0.8 - 1} = 27{,}733.7 \text{ W/m}^2$$

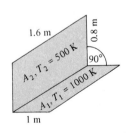

FIGURE EXAMPLE 11-13

Example 11-13. Two black rectangular surfaces A_1 and A_2, arranged as shown in the accompanying figure, are located in a large room whose walls are black and kept at 300 K. Determine the net radiative-heat exchange between these two surfaces when A_1 is kept at 1000 K and A_2 at 500 K. Neglect the radiation from the room.

Solution. Given: $L_1 = 1\,\text{m}$, $L_2 = 0.8\,\text{m}$, and $W = 1.6\,\text{m}$. The view factor F_{1-2} is obtained from Fig. 11-11 as

$$\frac{L_1}{W} = \frac{1}{1.6} = 0.625$$

$$\frac{L_2}{W} = \frac{0.8}{1.6} = 0.5$$

$$F_{1-2} = 0.2$$

The corresponding radiation network is illustrated in the accompanying figure. The heat transfer rate becomes

$$Q = \frac{E_{b1} - E_{b2}}{R_{12}} = \frac{E_{b1} - E_{b2}}{1/(A_1 F_{1-2})}$$

$$= A_1 F_{1-2} \sigma (T_1^4 - T_2^4)$$

where

$$E_{b1} = \sigma T_1^4 \quad \text{and} \quad E_{b2} = \sigma T_2^4$$

Introducing the numerical values, we have

$$Q = (1.0 \times 1.6)(0.2)(5.67 \times 10^{-8})(1000^4 - 500^4)$$

$$= 17\,\text{kW}$$

Example 11-14. A cubical room 3 m by 3 m by 3 m is heated through the ceiling by maintaining it at a uniform temperature $T_1 = 343\,\text{K}$ while the walls and the floor are at 283 K. Assuming that all surfaces have an emissivity of 0.8, determine the rate of heat loss from the ceiling by radiation.

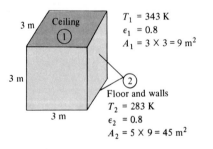

3 m   Ceiling
(1)

3 m

(2)
Floor and walls
$T_2 = 283\,\text{K}$
$\epsilon_2 = 0.8$
$A_2 = 5 \times 9 = 45\,\text{m}^2$

3 m

$T_1 = 343\,\text{K}$
$\epsilon_1 = 0.8$
$A_1 = 3 \times 3 = 9\,\text{m}^2$

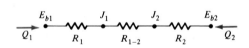

FIGURE EXAMPLE 11-14

Solution. The problem is illustrated in the accompanying figure, which shows the various given values. The floor and walls are considered to be a single surface. The problem is like a two-zone enclosure, and the equivalent network is illustrated in the accompanying figure. Various potentials and resistances are

$$E_{b1} = \sigma T_1^4 = (5.67 \times 10^{-8})(343)^4 = 784.8\,\text{W/m}^2$$

$$E_{b2} = \sigma T_2^4 = (5.67 \times 10^{-8})(283)^4 = 363.7\,\text{W/m}^2$$

$$R_1 = \frac{1 - \epsilon_1}{A_1 \epsilon_1} = \frac{1 - 0.8}{(9)(0.8)} = 0.02778$$

$$R_2 = \frac{1 - \varepsilon_2}{A_2 \varepsilon_2} = \frac{1 - 0.8}{(45)(0.8)} = 0.00556$$

$$R_{1-2} = \frac{1}{A_1 F_{1-2}} = \frac{1}{(9)(1)} = 0.11111$$

The radiation heat exchange is determined from Eq. (11-47) as

$$Q_{1-2} = \frac{E_{b1} - E_{b2}}{R_1 + R_{1-2} + R_2}$$

$$= \frac{784.8 - 363.7}{0.02778 + 0.11111 + 0.00556} = 2915.3 \text{ W}$$

Example 11-15. A cubical room 3 m by 3 m by 3 m is heated through the floor by maintaining it at a uniform temperature of 310 K. Since the side walls are well insulated, the heat loss through them can be considered negligible. The heat loss takes place through the ceiling, which is maintained at 280 K. All surfaces have an emissivity $\varepsilon = 0.85$. Determine the rate of heat loss by radiation through the ceiling.

Solution. The problem is illustrated in the accompanying figure, which shows various values. The view factors F_{1-2} and F_{1-3} are determined from Figs. 11-12 and 11-11 as

$$F_{1-2} = 0.2 \qquad F_{1-3} = 0.8$$

The equivalent network is illustrated in the accompanying figure.

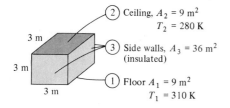

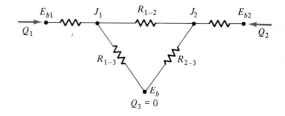

FIGURE EXAMPLE 11-15

Various potentials and resistances are

$$E_{b1} = \sigma T_1^4 = (5.67 \times 10^{-8})(310)^4 = 523.64 \text{ W/m}^2$$

$$E_{b2} = \sigma T_2^4 = (5.67 \times 10^{-8})(280) = 348.51 \text{ W/m}^2$$

$$R_{1-2} = \frac{1}{A_1 F_{1-2}} = \frac{1}{(9)(0.2)} = \frac{1}{1.8}$$

$$R_{2-3} = R_{1-3} = \frac{1}{A_1 F_{1-3}} = \frac{1}{(9)(0.8)} = \frac{1}{7.2}$$

$$R_1 = R_2 = \frac{1 - \varepsilon_1}{A_1 \varepsilon_1} = \frac{1 - 0.85}{(9)(0.85)} = \frac{1}{51}$$

The equivalent resistance between E_{b1} and E_{b2} is

$$R_{\text{equivalent}} = R_1 + \frac{1}{1/R_{1-2} + 1/(R_{1-3} + R_{2-3})} + R_2$$

$$= \frac{1}{51} + \frac{1}{1.8 + 1/[(1/7.2) + (1/7.2)]} + \frac{1}{51} = 0.2244$$

The radiation heat exchange between the floor and the ceiling is determined as

$$Q_{1-2} = \frac{E_{b1} - E_{b2}}{R_{\text{equivalent}}}$$

$$= \frac{523.64 - 348.51}{0.2244} = 780.43 \text{ W}$$

11-7 RADIATION SHIELDS

The radiation heat transfer between two surfaces can be reduced significantly if a radiation shield made of a low-emissivity material is placed between them. This fact has been utilized extensively in reducing heat gain from the surrounding atmosphere to a cryogenic tank filled with cryogenic fluid at very low temperature. The role of the radiation shield is to increase the thermal resistance to radiation in the path of heat flow, and hence reduce the heat transfer rate.

To illustrate this matter, we first consider radiation heat transfer between two large, opaque parallel plates. Let T_1 and T_2 be the temperatures and ε_1 and ε_2 the emissivities of the surfaces. Then the heat transfer rate Q_0 across an area A through the plates is determined from Eq. (11-49) as

$$Q_0 = \frac{A\sigma(T_1^4 - T_2^4)}{1/\varepsilon_1 + 1/\varepsilon_2 - 1} \qquad \text{W} \qquad (11\text{-}53)$$

We now consider a radiation shield placed between the plates. Let $\varepsilon_{3,1}$ and $\varepsilon_{3,2}$ be the emissivities of the shield at the surfaces facing plates 1 and 2, respectively. The radiation network for the assembly with one shield can be constructed by utilizing Eqs. (11-39b) and (11-45b) for the radiation resistances at the surface and across the surfaces, respectively. Figure 11-21 shows the resulting radiation network. By utilizing this network and noting that $F_{1,3} = F_{3,2} = 1$ for large parallel plates, the heat transfer rate Q_1 across the system with one shield becomes

$$Q_1 = \frac{A\sigma(T_1^4 - T_2^4)}{1/\varepsilon_1 + (1 - \varepsilon_{3,1})/\varepsilon_{3,1} + (1 - \varepsilon_{3,2})/\varepsilon_{3,2} + 1/\varepsilon_2}$$

which is rearranged as

$$Q_1 = \frac{A\sigma(T_1^4 - T_2^4)}{(1/\varepsilon_1 + 1/\varepsilon_2 - 1) + (1/\varepsilon_{3,1} + 1/\varepsilon_{3,2} - 1)} \qquad \text{W} \qquad (11\text{-}54)$$

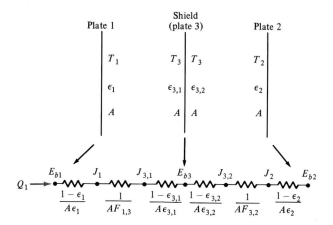

FIGURE 11-21
Plate-shield assembly and the corresponding radiation network.

A comparison of Eq. (11-54) with Eq. (11-53) shows that the effect of the radiation shield is to increase the thermal resistance of the system.

If the emissivities of all surfaces are equal, Eq. (11-54) reduces to

$$Q_1 = \frac{A\sigma(T_1^4 - T_2^4)}{2(2/\varepsilon - 1)} \quad \text{W} \tag{11-55}$$

For a parallel-plate system containing N shields between two surfaces with the emissivities of all surfaces equal, Eq. (11-55) is generalized as

$$\boxed{Q_N = \frac{A\sigma(T_1^4 - T_2^4)}{(N+1)(2/\varepsilon - 1)}} \quad \text{W} \tag{11-56}$$

The ratio of the heat transfer rates for parallel-plate systems with N shields and with no shield, when all emissivities are equal, is determined from Eq. (11-56) as

$$\frac{Q_N}{Q_0} = \frac{1}{N+1} \tag{11-57}$$

Example 11-16. Consider two large parallel plates, one at T_1 K with emissivity $\varepsilon_1 = 0.8$ and the other at T_2 K with emissivity $\varepsilon_2 = 0.4$. An aluminum radiation shield with emissivity $\varepsilon_2 = 0.05$ is placed between the plates. Calculate the percentage reduction in the heat transfer rate resulting from the radiation shield.

Solution. The heat transfer rate without the radiation shield is determined from Eq. (11-53):

$$Q_0 = \frac{A\sigma(T_1^4 - T_2^4)}{1/\varepsilon_1 + 1/\varepsilon_2 - 1}$$

$$= \frac{A\sigma(T_1^4 - T_2^4)}{1/0.8 + 1/0.4 - 1} = 0.36 A\sigma(T_1^4 - T_2^4)$$

The heat transfer rate with the radiation shield is determined from Eq. (11-54):

$$Q_1 = \frac{A\sigma(T_1^4 - T_2^4)}{(1/\varepsilon_1 + 1/\varepsilon_2 - 1) + (1/\varepsilon_{3,1} + 1/\varepsilon_{3,2} - 1)}$$

$$= \frac{A\sigma(T_1^4 - T_2^4)}{(1/0.8 + 1/0.4 - 1) + (1/0.05 + 1/0.05 - 1)}$$

$$= 0.024 A\sigma(T_1^4 - T_2^4)$$

The reduction in the heat transfer rate resulting from the shield is

$$\frac{Q_0 - Q_1}{Q_0} = \frac{0.36 - 0.024}{0.36} = 0.934$$

Therefore the radiation shield reduces the heat loss by 93.4 percent.

PROBLEMS

Blackbody Radiation Laws

11-1. At a wavelength of 0.7 μm, what is the temperature of the blackbody that will give an emissive power equal to 10^2 W/(m$^2 \cdot$ μm)?

11-2. Calculate the spectral blackbody emissive power $E_{b\lambda}(T)$ at $\lambda = 2$, 3, and 5 μm for a surface at 1000 K.

11-3. Determine the wavelength at which the blackbody spectral emissive power $E_{b\lambda}$ is maximum for a blackbody at 1600 K.

11-4. Calculate the wavelength at which the emission of radiation by a blackbody at the following temperatures is maximum:

(a) The effective surface temperature of the Sun, 5762 K
(b) A tungsten filament at 2300 K
(c) A body at room temperature, 300 K
    Answer: (a) 0.5 μm, (b) 1.26 μm, (c) 9.66 μm

11-5. Consider a blackbody emitting at 1500 K. Calculate the maximum spectral emissive power at this temperature.
    Answer: 97,766 W/(m$^2 \cdot$ μm)

11-6. Calculate the blackbody emissive power $E_b(T)$ at 1000 K and at 5762 K.

11-7. A 1 m × 1 m blackbody surface is at 800 K. Determine the rate of emission of radiative energy through this surface.

11-8. A blackbody filament is heated to 2300 K. What is the maximum radiative heat flux from the filament?
    Answer: 1586.7 kW/m^2

11-9. A large empty box (assumed to be black) has its inside wall at 700 K. If a 1-cm-diameter hole is drilled in the side, how much energy will be emitted from the hole?

11-10. A large blackbody enclosure has a small opening area of 1 cm^2. The radiant energy emitted by the opening is 5.67 W. Determine the temperature of the blackbody enclosure.
    Answer: $T = 1000$ K

11-11. The sun radiates as a blackbody at an effective surface temperature of 5762 K. What fraction of the total emitted energy is in the visible range?

11-12. The sun radiates as a blackbody at an effective surface temperature of 5762 K. What fraction of the total emitted energy is in the infrared range (i.e., $\lambda = 0.7$ to 1000 μm)?

Answer: 46.2 percent

11-13. A tungsten filament (assumed to be black) is heated to 2300 K. What fraction of the total energy is emitted in the wavelength range 0.6 to 1 μm?

11-14. A light bulb filament (assumed to be black) is at 3500 K. What percentage of its radiation is in (a) the visible range, and (b) the infrared range?

11-15. A blackbody at 556 K is emitting into air. Calculate (a) the wavelength at which the blackbody emissive power is maximum, and (b) the energy emitted over the wavelengths $\lambda = 1$ to 8 μm and $\lambda = 8$ to 18 μm.

11-16. What is the temperature of a blackbody such that 50 percent of the energy emitted should lie in the wavelength spectrum $\lambda = 0$ to 10 μm?

Answer: 411 K

Radiation Properties of Surfaces

11-17. The spectral emissivity of an opaque surface at 1500 K is given by

$$\varepsilon_1 = 0.1 \quad \text{for } \lambda_0 = 0 \text{ to } \lambda_1 = 0.5 \, \mu\text{m}$$

$$\varepsilon_2 = 0.5 \text{ for } \lambda_1 = 0.5 \text{ to } \lambda_2 = 6 \, \mu\text{m}$$

$$\varepsilon_3 = 0.8 \text{ for } \lambda_3 > 6 \, \mu\text{m}$$

Determine the average emissivity over the entire range of wavelengths and the radiative flux emitted by the material at 1500 K.

11-18. The spectral emissivity of a filament at 3000 K is given by

$$\varepsilon_1 = 0.5 \quad \text{for } \lambda_0 = 0 \text{ to } \lambda_1 = 0.5 \, \mu\text{m}$$

$$\varepsilon_2 = 0.1 \text{ for } \lambda_1 = 0.5 \, \mu\text{m to } \lambda_2 \rightarrow \infty$$

Determine the average emissivity of the filament over the entire range of wavelengths.

Answer: 0.105

11-19. The spectral emissivity of an opaque surface at 2000 K is given by

$$\varepsilon_1 = 0.2 \quad \text{for } \lambda_0 = 0 \text{ to } \lambda_1 = 0.3 \, \mu\text{m}$$

$$\varepsilon_2 = 0.4 \quad \text{for } \lambda_1 = 0.3 \text{ to } \lambda_2 = 3 \, \mu\text{m}$$

$$\varepsilon_3 = 0.6 \quad \text{for } \lambda_2 = 3 \text{ to } \lambda_3 = 4 \, \mu\text{m}$$

$$\varepsilon_4 = 0.8 \quad \text{for } \lambda_3 = 4 \, \mu\text{m to } \lambda_4 \rightarrow \infty$$

Determine the average emissivity over the entire range of wavelengths of the radiative flux emitted by the material at 2000 K.

11-20. Spectral emissivity of an opaque surface at 1500 K is given by

$$\varepsilon_1 = 0.3 \quad \text{for } \lambda_0 = 0 \text{ to } \lambda_1 = 1.0 \, \mu\text{m}$$

$$\varepsilon_2 = 0.8 \quad \text{for } \lambda_1 = 1.0 \text{ to } \lambda_2 \rightarrow \infty$$

Determine the average emissivity over the entire range of wavelengths and the emissive power at 1500 K.

Answers: 0.7935, 227.77 kW/m^2

View-Factor Concept

11-21. Two small surfaces $dA_1 = 4 \text{ cm}^2$ and $dA_2 = 6 \text{ cm}^2$ are separated by $r = 50$ cm and oriented as shown in the accompanying figure. Calculate the view factors between the surfaces.

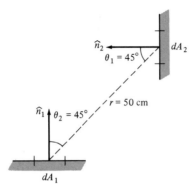

FIGURE PROBLEM 11-21

11-22. Determine the view factors F_{1-2}, F_{2-1}, between the surfaces of a long semicylindrical duct, shown in the accompanying figure, by view-factor algebra.

Answers: $F_{1-2} = 1$, $F_{2-1} = 0.637$, $F_{2-2} = 0.363$

FIGURE PROBLEM 11-22

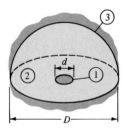

FIGURE PROBLEM 11-23

11-23. A small circular disk of diameter d is placed centrally at the base of a hemisphere of diameter D, as illustrated in diameter D, as illustrated in the accompanying figure. Determine the view factors F_{3-1} and F_{3-2} by view-factor algebra.

Answers: $F_{3-1} = d^2/2D^2$, $F_{3-2} = (D^2 - d^2)/2D^2$

11-24. Consider two concentric spheres. The inner sphere has a radius $R_1 = 5$ cm. Determine the radius R_2 of the outer sphere such that $F_{2-1} = 0.6$.

Answer: 6.45 cm

11-25. Consider two very long coaxial cylinders. The outer cylinder has a radius $R_2 = 10$ cm. Determine the radius R_1 of the inner cylinder such that $F_{2-1} = 0.7$.

11-26. Consider two very long coaxial cylinders. The inner cylinder has a radius $R_1 = 5$ cm. Determine the radius R_2 of the outer cylinder such that $F_{2-1} = 0.6$.

Answer: 8.333 cm

11-27. Consider two very long coaxial cylinders. The inner cylinder has a radius R_1 and the outer cylinder has a radius $R_2 = 3R_1$. Determine the shape factors F_{1-2}, F_{2-1}, and F_{2-2}, where the subscript 1 refers to the inner cylinder.

11-28. Determine the view factor F_{1-2} between two rectangular surfaces A_1 and A_2 for the geometric arrangement shown in the accompanying figure.

Answer: 0.165

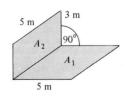

FIGURE PROBLEM 11-28

11-29. Determine the view factors between the surfaces shown in the accompanying figure.

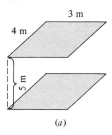

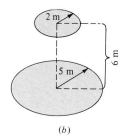

(a)

(b)

FIGURE PROBLEM 11-29

11-30. For the grooves shown in the accompanying figure (an equilateral triangle and a semicircle), determine the view factor of each groove with respect to the surroundings outside the groove.

11-31. For the rectangular groove with $H = 2W$ shown in the accompanying figure, determine the view factor F_{1-2}.
Answer: 0.618

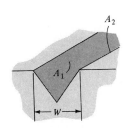

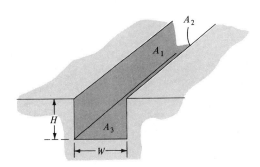

FIGURE PROBLEM 11-30

FIGURE PROBLEM 11-31

11-32. Two aligned, parallel square plates 0.5 m by 0.5 m are separated by 0.7 m, as illustrated in the accompanying figure. Determine the view factor F_{1-2}.
Answer: 0.12

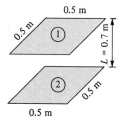

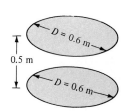

FIGURE PROBLEM 11-32

FIGURE PROBLEM 11-33

11-33. Two parallel circular disks of equal diameter 0.6 m separated by 0.5 m have a common central normal, as illustrated in the accompanying figure. Determine the view factor F_{1-2}.
Answer: 0.22

11-34. Determine the view factors F_{1-2}, F_{2-1}, and F_{1-3} of the infinitely long surfaces in the accompanying figure.

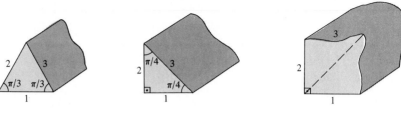

FIGURE PROBLEM 11-34

11-35. Determine the view factor F_{1-3} between surfaces A_1 and A_3 of Example 11-11 for $b = 3$, $c = 5$, and $d = 10$ cm.

Network Method for Radiation Exchange

11-36. Consider an enclosure consisting of two parallel, opaque, large plates that are kept at 500 K and 200 K, respectively, and that have emissivities of 0.8. Calculate the net radiation flux.
   Answer: 2302 W/m^2

11-37. Calculate the heat dissipation by radiation through a 0.2-m^2 opening of a furnace at 1100 K into an ambient at 300 K. Assume that both the furnace and the ambient are blackbodies.

11-38. Consider two parallel, opaque, large plates that are kept at uniform temperatures T_1 and T_2 and that have emissivities $\varepsilon_1 = \varepsilon_2 = 0.6$. Calculate the temperature T_2 of plate 2 for $T_1 = 800$ K and $q_{1-2} = 1000$ W/m^2.
   Answer: 779.1 K

11-39. A 15-cm-outside-diameter (OD) and 50-cm-long steam pipe whose surface is at 100°C passes through a room with a wall at 10°C. Assuming the emissivity of the pipe to be 0.9, determine the rate of heat loss from the pipe by radiation.

11-40. Two black surfaces A_1 and A_2, arranged as shown in the accompanying figure, are located in a large room whose walls are black and kept at 250 K. Determine the net radiative heat exchange for the given surface conditions. Neglect the radiation from the room.
   Answer: 331.7 W

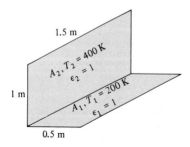

FIGURE PROBLEM 11-40

11-41. A cubical room 2 m by 2 m by 2 m is heated through the floor by maintaining it at a uniform temperature $T_1 = 250$ K, while the walls and the ceiling are at

200 K. Assume that the floor has an emissivity $\varepsilon_1 = 0.9$ and the walls and the ceiling have an emissivity $\varepsilon_2 = 0.6$. Determine the rate of heat loss from the floor.

Answer: 420.3 W

11-42. A cubical room 2 m by 2 m by 2 m is heated through the floor by maintaining it at a uniform temperature $T_1 = 250$ K. The side walls are well insulated. The heat loss takes place through the ceiling, which is maintained at 200 K. The floor has an emissivity $\varepsilon_1 = 0.9$, and the ceiling has an emissivity of 0.6. Determine the rate of heat loss from the floor.

Answer: 214 W

11-43. Two square plates, each 2 m by 2 m, are parallel to and directly opposite each other at a distance of 3 m apart. The hot plate is at $T_1 = 1000$ K and has an emissivity $\varepsilon_1 = 0.7$. The colder plate is at $T_2 = 400$ K and has an emissivity $\varepsilon_2 = 0.6$. The radiation heat exchange takes place between the plates as well as with a large ambient at $T_3 = 300$ K through the opening between the plates. Calculate the net heat transfer rate by radiation at each plate and to the ambient.

Radiation Shields

11-44. Consider two large parallel plates, one at $T_1 = 1000$ K with emissivity $\varepsilon_1 = 0.8$ and the other at $T_2 = 500$ K with emissivity $\varepsilon_2 = 0.4$. An aluminium radiation shield with an emissivity (on both sides) $\varepsilon_3 = 0.05$ is placed between the plates. Sketch the radiation network for the system with and without the radiation shield. Calculate the percentage reduction in the heat transfer rate as a result of the radiation shield.

Answer: 93.4 percent

11-45. Two large parallel plates are at temperatures T_1 and T_2 and have emissivities $\varepsilon_1 = 0.9$ and $\varepsilon_2 = 0.6$. A radiation shield having an emissivity ε_3 (on both sides) is placed between the plates. Calculate the emissivity ε_3 of the shield needed to reduce the radiation loss from the system to one-tenth of that without the shield.

11-46. Consider two large parallel plates, one at $T_1 = 1000$ K with emissivity $\varepsilon_1 = 0.8$ and the other at $T_2 = 300$ K with emissivity $\varepsilon_1 = 0.6$. A radiation shield is placed between them. The shield has an emissivity $\varepsilon_{3,1} = 0.1$ on the side facing the hot plate and emissivity $\varepsilon_{3,2} = 0.3$ on the side facing the cold plate. Sketch the radiation network. Calculate the reduction in the heat transfer rate between the hot and the cold plates as a result of the radiation shield.

Answer: 86.6 percent

11-47. Consider two large parallel plates, one at $T_1 = 800$ K with emissivity $\varepsilon_1 = 0.6$ and the other at $T_2 = 300$ K with emissivity $\varepsilon_2 = 0.5$. A radiation shield is placed between them. The shield has emissivity $\varepsilon_3 = 0.2$ (on both sides). Sketch the radiation network. Calculate the reduction in the heat transfer rate between the plates as a result of the radiation shield, and the final equilibrium temperature of the radiation shield.

11-48. Consider two large parallel plates, one at $T_1 = 800$ K with emissivity $\varepsilon_1 = 0.9$ and the other at $T_2 = 300$ K with emissivity $\varepsilon_2 = 0.5$. A radiation shield with an emissivity ε_3 (on both sides) is placed between the plates. Calculate the emissivity of the radiation shield needed to reduce the heat transfer rate to 10 percent of that without the shield.

Answer: 0.1

REFERENCES

1. Gubareff, G. G., J. E. Janssen, and R. H. Torborg: *Thermal Radiation Properties Survey*, Honeywell Research Center, Honeywell Regulator Company, Minneapolis, 1960.
2. Hottel, H. C.: "Radiant Heat Transmission," in W. H. McAdams (ed.), *Heat Transmission*, 3d ed., McGraw-Hill, New York, 1954, chap 4.
3. Mackey, C. O., L. T. Wright, R. E. Clark, and N. R. Gray: "Radiant Heating and Cooling, Part I," *Cornell Univ., Eng. Exp. Sta. Bull.* **22** (1943).
4. Sieber, W.: Zusammensetzung Der Von Werk-und Baustoffen Zurückgeworfenen Wärmestrahlung," *Z. Tech. Phys.* **22**:130–135 (1941).

CHAPTER
12

HEAT EXCHANGERS

Heat exchangers are commonly used for the transfer of heat between a hot and a cold fluid. For example, in car radiators, hot water from the engine is cooled by atmospheric air. In household refrigerators, hot refrigerant from the compressor is cooled by natural convection into the atmosphere by passing it through finned tubes. In space radiators, the waste heat carried by the coolant fluid is dissipated by thermal radiation into the atmosphere-free space. In steam condensers, the latent heat of condensation is removed by the coolant water passing through the condenser tubes. Clearly, different types of heat-exchanging devices are needed for different applications. There are also situations in which heat exchangers become among the most costly components of the system. Therefore, the design of heat exchangers for new applications is a very complicated matter. The design involves not only heat transfer and pressure drop analysis, but also optimization and cost estimation.

In this chapter we discuss the classification of heat exchangers, the determination of the overall heat transfer coefficient, the logarithmic mean temperature difference and ε-NTU methods for sizing, and heat transfer and estimation in heat exchangers. The problems of optimization and cost estimation are beyond the scope of this work.

12-1 CLASSIFICATION

Heat exchangers are designed in so many sizes, types, configurations, and flow arrangements and used for so many different purposes that some kind of classification, even though arbitrary, is needed in order to study heat exchangers. Here we briefly discuss the basic concepts leading to classification.

Heat exchangers can be classified according to *compactness*. The ratio of the heat transfer surface area on one side of the heat exchanger to the volume can be used as a measure of the compactness of a heat exchanger. A heat exchanger with a surface area density on any one side greater than about $700 \, \text{m}^2/\text{m}^3$ is referred to, quite arbitrarily, as a *compact heat exchanger* regardless of its structural design. For example, automobile radiators with an area density on the order of $1100 \, \text{m}^2/\text{m}^3$ and the glass ceramic heat exchangers for some vehicular gas-turbine engines that have an area density on the order of $6600 \, \text{m}^2/\text{m}^3$ are compact heat exchangers. By this classification, the commonly used plane tubular and shell-and-tube-type heat exchangers, which have an area density in the range of 70 to $500 \, \text{m}^2/\text{m}^3$, are not considered compact.

Heat exchangers can be classified according to their *construction* features—for example, *tubular*, *tube-fin*, *plate*, *plate-fin*, *extended surfaces*, and *regenerative* exchangers. The tubular exchangers are widely used; they are manufactured in many sizes, flow arrangements, and types. They can accommodate a wide range of operating pressures and temperatures. Tubular exchangers can be designed for high pressures relative to the environment and for high-pressures differences between the fluids. Their ease of manufacturing and relatively low cost have been the principal reason for their widespread use in engineering applications. A commonly used design, called the *shell-and-tube exchanger*, consists of round tubes mounted on a cylindrical shell with their axes parallel to that of the shell. Figure 12-1 illustrates the main features of a shell-and-tube exchanger that has one fluid flowing inside the tubes and the other flowing outside the tubes. The principal components of this type of heat exchanger are the tube bundle, shell, front- and rear-end headers, and baffles. A plate-fin heat exchanger and a round tube-fin heat exchanger are shown in Fig. 12-2.

The most common method of classifying heat exchangers is by *flow arrangement*. There are numerous possibilities; we summarize the principal ones.

Parallel-flow The hot and cold fluids enter at the same end of the heat exchanger, flow through in the same direction, and leave together at the other end, as illustrated in Fig. 12-3*a*.

FIGURE 12-1
A shell-and-tube heat exchanger; one shell pass and one tube pass.

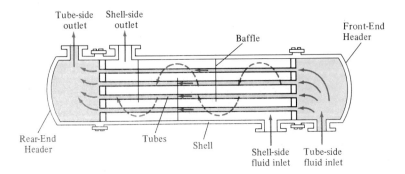

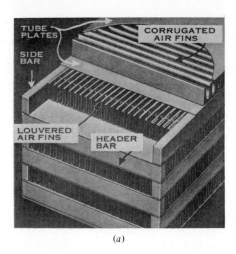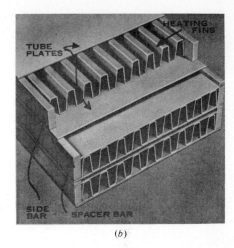

(a)   (b)

FIGURE 12-2
A plate-fin and a round tube-fin heat exchanger. (Courtesy of Harrison Division of General Motors Corporation).

FIGURE 12-3
(a) parallel-flow, (b) counter flow, and (c) cross-flow arrangements.

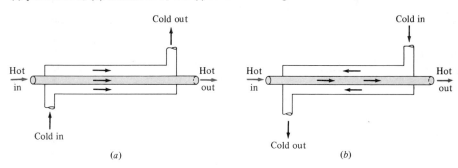

(a)   (b)

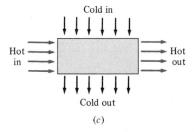

(c)

Counter-flow The hot and cold fluids enter at the opposite ends of the heat exchanger and flow through in opposite directions, as illustrated in Fig. 12-3*b*.

Cross-flow In the cross-flow exchanger, the two fluids usually flow at right angles to each other, as illustrated in Fig. 12-3*c*. In a cross-flow arrangement, the flow may be what we call *mixed* or *unmixed*, depending on the design. To illustrate the physical significance of the concept of *mixed* and *unmixed*, we refer to the illustration in Fig. 12-4. Here, Fig. 12-4*a* shows an arrangement in which both hot and cold fluids flow through individual channels formed by corrugation; therefore, the fluids are not free to move in the transverse direction. In this case each fluid stream is said to be unmixed. In the flow arrangements shown in Fig. 12-4*b*, the cold fluid flows inside the tubes and so is not free to move in the transverse direction. Therefore, the cold fluid is said to be unmixed. However, the hot fluid flows over the tubes and is free to move in the transverse direction. Therefore, the hot fluid stream is said to be mixed. The mixing tends to make the fluid temperature uniform in the transverse direction; therefore, the exit temperature of a mixed steam exhibits negligible variation in the crosswise direction. In a shell-and-tube heat exchanger, the presence of a large number of baffles serves to "mix" the shell-side fluid in the sense discussed above; that is, its temperature tends to be uniform at any cross section.

Multipass flow Multipass flow arrangements are frequently used in heat exchanger design, because multipassing increases the overall effectiveness. A wide variety of multipass flow arrangements are possible. Figure 12-5 illustrates typical arrangements.

Heat exchangers can also be classified according to the *heat transfer mechanism*—for example, condensers, boilers, and radiators. Condensers are used in such varied applications as steam power plants, chemical processing plants, and nuclear electric plants for space vehicles. Steam boilers are one of

FIGURE 12-4
Cross-flow arrangements; (a) both fluids unmixed, (b) cold fluid unmixed, hot fluid mixed.

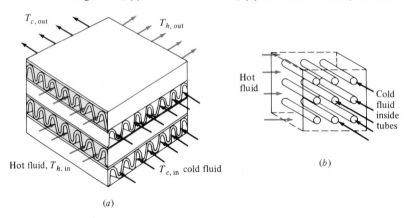

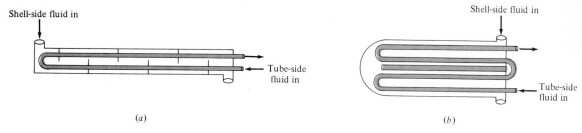

Shell-side fluid in

Tube-side fluid in

(a)

Shell-side fluid in

Tube-side fluid in

(b)

FIGURE 12-5
Multipass flow arrangements. (a) one shell pass, two tube pass; (b) two shell pass, four tube pass.

the earliest types of heat exchangers. Radiators for space applications are used in space vehicles to dissipate the waste heat by thermal radiation into the space by taking advantage of the fourth-power relationship between the absolute temperature of the surface and the radiation heat flux.

12-2 TEMPERATURE DISTRIBUTION

The heat transfer from the hot to the cold fluid causes a change in the temperature of one or both of the fluids flowing through the heat exchanger. Figure 12-6 illustrates how the temperature of the fluid varies along the path of the heat exchanger for a number of typical single-pass heat transfer matrices. In each instance, the temperature distribution is plotted as a function of the distance from the cold-fluid inlet end. Figure 12-6*a*, for example, is characteristic of a pure counter-flow heat exchanger in which the temperature rise in the cold fluid is equal to the temperature drop in the hot fluid; thus the temperature difference ΔT between the hot and cold fluids is constant throughout. However, in all other cases (Fig. 12-6*b* to *e*), the temperature difference ΔT between the hot and cold fluids varies with the position along the path of flow. Figure 12-6*b* corresponds to a situation in which the hot fluid condenses and heat is transferred to the cold fluid, causing its temperature to rise along the path of flow. In Fig. 12-6*c*, cold liquid is evaporating and cooling the hot fluid along its path of flow. Figure 12-6*d* shows a parallel-flow arrangement in which both fluids flow in the same direction, with the cold fluid experiencing a temperature rise and the hot fluid a temperature drop. The outlet temperature of the cold fluid cannot exceed that of the hot fluid. Therefore, the temperature effectiveness of parallel-flow exchangers is limited. Because of this limitation, generally they are not considered for heat recovery. However, since the metal temperature lies approximately midway between the hot and cold fluid temperatures, the wall is almost at a uniform temperature. Figure 12-6*e* shows a counter-flow arrangement, in which fluids flow in opposite directions. The exit temperature of the cold fluid can be higher than that of the hot fluid.

In multipass and cross-flow arrangements, the temperature distributions in the heat exchanger exhibit more complicated patterns. For example, Fig. 12-7 shows the temperature distribution in a one-shell-pass, two-tube-pass heat exchanger.

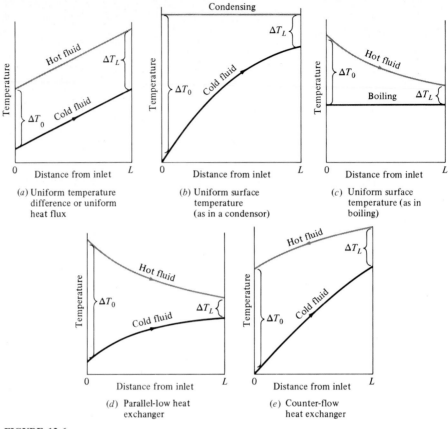

FIGURE 12-6
Axial temperature distribution in typical single-pass transfer matrices.

(a) Uniform temperature difference or uniform heat flux

(b) Uniform surface temperature (as in a condensor)

(c) Uniform surface temperature (as in boiling)

(d) Parallel-low heat exchanger

(e) Counter-flow heat exchanger

FIGURE 12-7
Axial temperature distribution in a one shell pass, two tube pass heat exchanger.

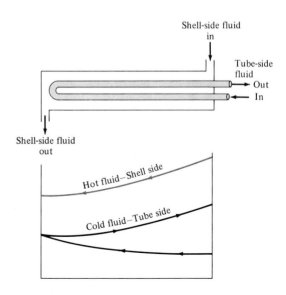

12-3 OVERALL HEAT TRANSFER COEFFICIENT

In a heat exchanger, the thermal resistances in the path of heat flow from the hot to the cold fluid include: (1) the skin resistances associated with the flow, (2) the scale resistances resulting from the formation of deposits on the walls, and (3) the thermal resistance of the wall material itself. Therefore, the total thermal resistance to heat flow needs to be determined in order to calculate heat transfer through a heat exchanger. Here we present an overview of the determination of the *total thermal resistance R* in heat exchangers, then relate this quantity to the commonly used concept of the *overall heat transfer coefficient U*.

Consider a tubular heat exchanger with heat transfer taking place between the fluids flowing inside and outside the tubes. The thermal resistance R to heat flow across the tubes, between the inside and outside flow, is composed of the following resistances:

$$R = \begin{pmatrix} \text{thermal} \\ \text{resistance} \\ \text{of inside} \\ \text{flow} \end{pmatrix} + \begin{pmatrix} \text{thermal} \\ \text{resistance} \\ \text{of tube} \\ \text{material} \end{pmatrix} + \begin{pmatrix} \text{thermal} \\ \text{resistance} \\ \text{of outside} \\ \text{flow} \end{pmatrix} \tag{12-1}$$

and the various terms are given by

$$R = \frac{1}{A_i h_i} + \frac{t}{k A_m} + \frac{1}{A_o h_o} \tag{12-2}$$

where A_o, A_i = outside and inside surface areas of tube, respectively, m^2

$A_m = \dfrac{A_o - A_i}{\ln(A_o/A_i)}$ = logarithmic mean area, m^2

h_i h_o = heat transfer coefficients for inside and outside flow, respectively, $W/(m^2 \cdot {}^\circ C)$

k = thermal conductivity of tube material, $W/(m \cdot {}^\circ C)$

R = total thermal resistance from inside to outside flow, $^\circ C/W$

t = thickness of tube, m

The thermal resistance R given by Eq. (12-2) can be expressed as an *overall heat transfer coefficient* based on either the inside or the outside surface of the tube. For example, the overall heat transfer coefficient U_o based on the *outside surface* of the tube is defined as

$$U_o = \frac{1}{A_o R} = \frac{1}{(A_o/A_i)(1/h_i) + (A_o/A_m)(t/k) + 1/h_o}$$

$$= \frac{1}{(D_o/D_i)(1/h_i) + (1/2k)D_o \ln(D_o/D_i) + 1/h_o} \tag{12-3}$$

since

$$\frac{A_o}{A_m} = \frac{D_o}{2t} \ln\left(\frac{D_o}{D_i}\right) \qquad D_o - D_i = 2t \tag{12-4}$$

and D_i and D_o are the inside and outside diameters of the tube, respectively.

Similarly, the overall heat transfer coefficient U_i based on the *inside surface* of the tube is defined as

$$U_i = \frac{1}{A_i R} = \frac{1}{1/h_i + (A_i/A_m)(t/k) + (A_i/A_o)(1/h_o)}$$

$$= \frac{1}{1/h_i + (1/2k)D_i \ln (D_o/D_i) + (D_i/D_o)(1/h_o)} \qquad (12\text{-}5a)$$

When the wall thickness is small and its thermal conductivity is high, the tube resistance can be neglected and Eq. (12-5a) reduces to

$$U_i = \frac{1}{1/h_i + 1/h_o} \qquad (12\text{-}5b)$$

In heat exchanger applications, the heat transfer surface is fouled with accumulated deposits, which introduce additional thermal resistance in the path of heat flow. The effect of fouling is generally introduced in the form of a fouling factor F, which has the dimensions $(m^2 \cdot {}^\circ C)/W$.

Based on the experience of manufacturers and users, the Tubular Equipment Manufacturers Association (TEMA) prepared tables of fouling factors as a guide in heat transfer calculations. We present some of their results in Table 12-1. Fouling is a very complicated matter, and representing it by such a simple listing is highly questionable. But in the absence of anything better, this remains the only reference for estimating the effects of fouling in reducing heat transfer.

We now consider heat transfer across a tube which is fouled by deposit formation on both the inside and outside surfaces. The thermal resistance R in the path of heat flow for this case is given by

$$R = \frac{1}{A_i h_i} + \frac{F_i}{A_i} + \frac{t}{k A_m} + \frac{F_o}{A_o} + \frac{1}{A_o h_o} \qquad (12\text{-}6)$$

where F_i and F_o are the fouling factors (i.e., the unit fouling resistances) at the inside and outside surfaces of the tube, respectively, and the other quantities are as defined previously.

In heat exchanger applications, the overall heat transfer coefficient is usually based on the outer tube surface. Then Eq. (12-6) can be represented in terms of the overall heat transfer coefficient based on the outside surface of the tube as

$$U_o = \frac{1}{(D_o/D_i)(1/h_i) + (D_o/D_i)F_i + (D_o/2k) \ln (D_o/D_i) + F_o + 1/h_o}$$
$$(12\text{-}7)$$

The values of overall heat transfer coefficients for different types of applications vary widely. Typical ranges of U_o are as follows:

Water-to-oil exchangers:	60 to	350 W/(m² · °C)
Gas-to-gas exchangers:	60 to	600 W/(m² · °C)
Air condensers:	350 to	800 W/(m² · °C)
Ammonia condensers:	800 to	1400 W/(m² · °C)
Steam condensers:	1500 to	5000 W/(m² · °C)

TABLE 12-1
Unit fouling resistance (or fouling factor) F

	F, m$^2 \cdot$ °C/W
Water (52°C or less; velocity 1 m/s or less)	
Seawater	0.000088
Distilled	0.000088
Engine jacket	0.00018
Great Lakes	0.00018
Boiler blowdown	0.00035
Brackish water	0.00035
River water	
Minimum	0.00036
Mississippi	0.00053
Delaware, Schuylkill	0.00053
East River & New York Bay	0.00053
Chicago Sanitary Canal	0.00141
Muddy or Silty	0.00053
Industrial fluids	
Clean recirculating oil	0.00018
Vegetable oils	0.00053
Quenching oil	0.00070
Fuel oil	0.00088
Organic vapors	0.000088
Steam (non-oil-bearing)	0.000088
Steam, exhaust	0.00018
Refrigerating vapors	0.00035
Air	0.00035
Organic liquids	0.00018
Refrigerating liquids	0.00018
Brine (cooling)	0.00018

Source: Tubular Exchanger Manufacturers Association.

It is apparent that U_o is generally low for fluids that have low thermal conductivity, such as gases or oils.

Example 12-1. Determine the overall *heat transfer coefficient* U_o based on the outer surface of a $D_i = 2.5$ cm, $D_o = 3.34$ cm brass $[k = 110$ W/(m $\cdot$ °C)] tube for the following conditions: The inside and outside heat transfer coefficients are, respectively, $h_i = 1200$ W/(m$^2 \cdot$ °C) and $h_o = 2000$ W/(m$^2 \cdot$ °C); the fouling factors for the inside and outside surfaces are $F_i = F_o = 0.00018$ (m$^2 \cdot$ °C)/W.

Solution. Equation (12-7) can be used to determine U_o:

$$U_o = \frac{1}{(D_o/D_i)(1/h_i) + (D_o/D_i)F_i + (D_o/2k)\ln(D_o/D_i) + F_o + 1/h_o}$$

Introducing the numerical values, we obtain

$$U_o = \frac{1}{(3.34/2.5)(1/1200) + (3.34/2.5)(0.00018) + 0.0334/(2 \times 110)\ln(3.34/2.5) + 0.00018 + 1/2000}$$

$$= 481.3 \text{ W/(m}^2 \cdot \text{°C)}$$

Example 12-2. Hot water at a mean temperature $T_m = 80°C$ and with a mean velocity $u_m = 0.4$ m/s flows inside a 3.8-cm-ID, 4.8-cm-OD steel tube [$k = 50$ W/(m · °C)]. The flow is considered hydrodynamically and thermally developed. The outside surface is exposed to atmospheric air at $T_\infty = 20°C$, flowing with a velocity of $u_\infty = 3$ m/s normal to the tube. Calculate the overall *heat transfer coefficient* U_o based on the outer surface of the tube, and the heat loss per meter length of the tube.

Solution. The physical properties of water at $T_m = 80°C$ are

$$\nu = 0.364 \times 10^{-6} \text{ m}^2/\text{s}$$

$$k = 0.668 \text{ W/(m · °C)} \qquad \text{Pr} = 2.22$$

The Reynolds number for water flow is

$$\text{Re} = \frac{u_m D}{\nu} = \frac{0.4 \times 0.038}{0.364 \times 10^{-6}} = 41,760$$

We use the Dittus-Boelter equation to determine h_i for water flow:

$$\text{Nu} = 0.023 \text{ Re}^{0.8} \text{ Pr}^{0.3}$$

$$= 0.023(41,760)^{0.8}(2.22)^{0.3}$$

$$= 145.3$$

$$h_i = \text{Nu} \frac{k}{D_i} = 145.3 \frac{0.668}{0.038} = 2554 \text{ W/(m}^2 \cdot °\text{C})$$

To evaluate the physical properties of air at the film temperature, the closest approximation for the film temperature is taken as $T_f \cong (80 + 20)/2 = 50°C$. Then

$$\nu = 18.22 \times 10^{-6} \text{ m}^2/\text{s} \qquad k = 0.0281 \text{ W/m · °C}$$

$$\text{Pr} = 0.703$$

The Reynolds number for the air flow becomes

$$\text{Re} = \frac{u_\infty D}{\nu} = \frac{(3)(0.048)}{18.22 \times 10^{-6}} = 7990$$

The Nusselt number is determined from Eq. (8-81):

$$\text{Nu} = (0.4 \text{ Re}^{0.5} + 0.06 \text{ Re}^{2/3}) \text{ Pr}^{0.4}$$

$$= [0.4(7990)^{0.5} + 0.06(7990)^{2/3}](0.703)^{0.4}$$

$$= 51.9$$

$$h_o = \text{Nu} \frac{k}{D_o} = 51.9 \frac{0.0281}{0.048} = 30.3 \text{ W/(m}^2 \cdot °\text{C})$$

Neglecting the tube wall resistance to heat flow, the overall heat transfer coefficient becomes

$$U_o = \frac{1}{(D_o/D_i)(1/h_i) + (D_o/2k) \ln (D_o/D_i) + 1/h_o}$$

$$= \frac{1}{(4.8/3.8)(1/2554) + 0.048/(2 \times 50) \ln (4.8/3.8) + 1/30.3}$$

$$= 29.8 \text{ W/(m}^2 \cdot °\text{C})$$

The heat loss per meter length of the tube is determined as

$$Q = A_o U_o \, \Delta T$$
$$= \pi D_o U_o (T_i - T_o)$$
$$= (\pi)(0.048)(29.8)(80 - 20)$$
$$= 269.6 \text{ W/m}$$

Example 12-3. Engine oil at a mean temperature $T_i = 100°C$ flows inside a $D = 3$ cm ID, thin-walled copper tube with a heat transfer coefficient $h_i = 20$ W/(m² · °C). The outer surface of the tube dissipates heat by free convection into atmospheric air at temperature $T_\infty = 20°C$ with a heat transfer coefficient $h_o = 8$ W/(m² · °C). Calculate the overall heat transfer coefficient and the heat loss per meter length of the tube.

Solution. We assume that the thermal resistance and the curvature effect of a thin-walled copper tube are negligible. The overall heat transfer coefficient becomes

$$U = \frac{1}{1/h_i + 1/h_o}$$
$$= \frac{1}{1/20 + 1/8} = 5.71 \text{ W/(m}^2 \cdot °C)$$

The heat loss per meter length of the tube is

$$Q = \pi DU(T_i - T_\infty)$$
$$= \pi \times 0.03 \times 5.71(100 - 20) = 43.1 \text{ W/m}$$

12-4 LOGARITHMIC MEAN TEMPERATURE DIFFERENCE (LMTD) METHOD

In the thermal analysis of heat exchangers, the total heat transfer rate Q through the heat exchanger is a quantity of primary interest. Here we turn our attention to single-pass heat exchangers that have flow arrangements of the type illustrated in Fig. 12-6. It is apparent from this illustration that the temperature difference ΔT between the hot and cold fluids, in general, is not constant; it varies with distance along the heat exchanger.

In the heat transfer analysis of heat exchangers, it is convenient to establish a mean temperature difference ΔT_m between the hot and cold fluids such that the total heat transfer rate Q between the fluids can be determined from the following simple expression:

$$Q = A_t U_m \, \Delta T_m \tag{12-8}$$

where A_t is the total heat transfer area and U_m is the average overall heat transfer coefficient based on that area.

An explicit expression for ΔT_m can be established by considering an energy balance over a differential length along the heat exchanger, then integrating it over the entire path of the flow. It can be shown that this temperature difference is a logarithmic mean of the temperature differences between the hot and cold fluids at both ends of the heat exchanger, and is given by

$$\Delta T_m \equiv \Delta T_{\ln} = \frac{\Delta T_0 - \Delta T_L}{\ln\left(\Delta T_0/\Delta T_L\right)} \tag{12-9}$$

where $\Delta T_{\ln}$ is called the *logarithmic mean temperature difference*, and

$$\Delta T_0, \Delta T_L = \left\{ \begin{array}{l} \text{temperature differences between the hot and} \\ \text{cold fluids at the two ends of the heat} \\ \text{exchanger } x = 0 \text{ and } x = L, \text{ respectively} \end{array} \right\}$$

The formula for $\Delta T_{\ln}$ given in Eq. (12-9) is applicable for the single-pass heat transfer matrices illustrated in Fig. 12-6. We note that for the special case of $\Delta T_0 = \Delta T_L$, Eq. (12-9) leads to $\Delta T_{\ln} = 0/0 =$ indeterminate. But by the application of L'Hospital's rule [i.e., by differentiating the numerator and denominator of Eq. (12-9) with respect to ΔT_0], it can be shown that for this particular case $\Delta T_{\ln} = \Delta T_0 = \Delta T_L$.

It is of interest to compare the LMTD of ΔT_0 and ΔT_L with their arithmetic mean:

$$\Delta T_a = \frac{\Delta T_0 + \Delta T_L}{2} \tag{12-10}$$

We present in Table 12-2 a comparison of the logarithmic and arithmetic means of the two quantities ΔT_0 and ΔT_L. We note that the arithmetic and logarithmic means are equal for $\Delta T_0 = \Delta T_L$. When $\Delta T_0 \neq \Delta T_L$, the LMTD is always less than the arithmetic mean; if ΔT_0 is not more than 50 percent greater than ΔT_L, the LMTD can be approximated by the arithmetic mean within about 1.4 percent.

Thus, the total heat transfer rate between the cold and hot fluids for the single-pass arrangements shown in Fig. 12-6 can be calculated from the following simple expression:

$$Q = A_t U_m \Delta T_{\ln} \tag{12-11}$$

where $A_t =$ total heat transfer area
$\quad\quad\;\; U_m =$ average overall heat transfer coefficient between the hot and cold fluids

TABLE 12-2
Comparison of logarithmic $\Delta T_{\ln}$ and arithmetic ΔT_a means of ΔT_0 and ΔT_L

$\dfrac{\Delta T_0}{\Delta T_L}$	1	1.2	1.5	1.7	2
$\dfrac{\Delta T_a}{\Delta T_{\ln}}$	1	1.0028	1.0137	1.023	1.04

$\Delta T_{\ln}$ = logarithmic mean temperature difference between the hot and cold fluids, defined by Eq. (12-9)

When A_t and U_m are available and the inlet and outlet temperatures of the hot and cold fluids are given, the total heat transfer rate Q for the heat exchanger can be readily calculated from Eq. (12-11).

Example 12-4. A counter-flow, shell-and-tube-type heat exchanger is to be used to cool water from $T_{h,in} = 22°C$ to $T_{h,out} = 6°C$, using brine entering at $T_{c,in} = -2°C$ and leaving at $T_{c,out} = 3°C$. The overall heat transfer coefficient is estimated to be $U_m = 500 \text{ W/m}^2 \cdot °C$. Calculate the heat transfer *surface area* for a design heat load of $Q = 10 \text{ kW}$.

Solution. The temperature profiles in the exchanger are shown in the accompanying figure. The logarithmic mean temperature difference becomes

$$\Delta T_{\ln} = \frac{\Delta T_0 - \Delta T_L}{\ln (\Delta T_0 / \Delta T_L)}$$

$$= \frac{19 - 8}{\ln (19/8)} = 12.7°C$$

The total heat transfer area is

$$A_t = \frac{Q_t}{U_m \Delta T_{\ln}}$$

$$= \frac{10,000}{(500)(12.7)} = 1.57 \text{ m}^2$$

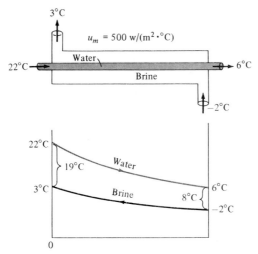

FIGURE EXAMPLE 12-4

Example 12-5. Engine oil at $T_i = 40°C$, flowing at a rate of $m = 0.05 \text{ kg/s}$, enters a $D_i = 2.5 \text{ cm}$ ID copper tube which is maintained at a uniform temperature $T_w = 100°C$ by steam condensing outside. Calculate the tube length required to have the outlet temperature of the oil $T_o = 80°C$.

Solution. The temperature profiles in the exchanger are shown in the accompanying figure. The physical properties of engine oil are taken at $(40 + 80)/2 = 60°C$ as

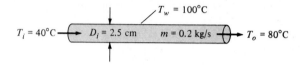

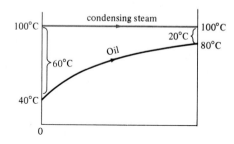

FIGURE EXAMPLE 12-5

$$c_p = 2047 \text{ J}/(\text{kg} \cdot {}^\circ\text{C}) \qquad \nu = 0.839 \times 10^{-4} \text{ m}^2/\text{s}$$

$$\rho = 864 \text{ kg}/\text{m}^3 \qquad k = 0.14 \text{ W}/(\text{m} \cdot {}^\circ\text{C})$$

$$\text{Pr} = 1050$$

The Reynolds number for flow inside the tube is

$$\text{Re} = \frac{4m}{\pi D(\nu\rho)}$$

$$= \frac{(4)(0.05)}{\pi(0.025)(0.839 \times 10^{-4})(864)} = 35.1$$

The flow is laminar. By assuming fully developed flow and a constant tube wall temperature, the Nusselt number for oil flow is given by

$$\text{Nu} = 3.66$$

and

$$h_i = 3.66 \frac{k}{D}$$

$$= 3.66 \frac{0.14}{0.025} = 20.5 \text{ W}/(\text{m}^2 \cdot {}^\circ\text{C})$$

Assuming that the thermal resistance of the copper tube is negligible, the overall heat transfer coefficient becomes

$$U = h_i = 20.5 \text{ W}/(\text{m}^2 \cdot {}^\circ\text{C})$$

The logarithmic mean temperature difference is

$$\Delta T_{\text{ln}} = \frac{\Delta T_0 - \Delta T_L}{\ln (\Delta T_0/\Delta T_L)}$$

$$= \frac{60 - 20}{\ln (60/20)} = 36.4 {}^\circ\text{C}$$

The tube length L is determined from the overall energy balance for the tube:

$$mc_p \, \Delta T_{\text{oil}} = UA_s \, \Delta T_{\text{ln}}$$

$$(0.05)(2047)(80 - 40) = (20.5)(\pi \times 0.025 \times L)(36.4)$$

This gives

$$L = 69.8 \text{ m}$$

12-5 CORRECTION FOR THE LMTD

The LMTD defined by Eq. (12-9) is strictly applicable for single-pass, non-cross-flow heat exchangers. For cross-flow and mutlipass arrangements, expressions for the effective temperature difference between the hot and cold fluids can be developed, but the resulting expressions are so complicated that they are not useful for practical purposes. Therefore, for such situations, it is customary to introduce a *correction factor F* so that the simple LMTD can be adjusted to represent the effective temperature difference ΔT_{corr} for cross-flow and multipass arrangements:

$$\Delta T_{\text{corr}} = F(\Delta T_{\text{ln}} \text{ for counter-flow}) \tag{12-12}$$

where ΔT_{ln} should be computed on the basis of counter-flow conditions; that is, ΔT_0 and ΔT_L in the definition of the LMTD given by Eq. (12-9) should be taken as (see Fig. 12-6e)

$$\Delta T_0 = T_{h,\text{out}} - T_{c,\text{in}} \tag{12-13a}$$

$$\Delta T_L = T_{h,\text{in}} - T_{c,\text{out}} \tag{12-13b}$$

where the subscripts c and h refer, respectively, to the cold and hot fluids. Figure 12-8 shows the correction factor F for some commonly used heat exchanger configurations. In these figures, the abscissa is a dimensionless ratio P, defined as

$$P = \frac{t_2 - t_1}{T_1 - t_1} \tag{12-14a}$$

where T refers to the *shell-side temperature*, t to the *tube-side temperature*, and the subscripts 1 and 2, respectively, to the *inlet* and *outlet* conditions. The parameter R appearing on the curves is defined as

$$R = \frac{T_1 - T_2}{t_2 - t_1} = \frac{(mc_p)_{\text{tube side}}}{(mc_p)_{\text{shell side}}} \tag{12-14b}$$

Note that the correction factors in Fig. 12-8 are applicable whether the hot fluid is in the shell side or the tube side. Correction-factor charts for several other flow arrangements are given by Bowman, Mueller, and Nagle.

Generally F is less than unity for cross-flow and multipass arrangements; it is unity for a true counter-flow heat exchanger. It represents the degree of departure of the true mean temperature difference from the LMTD for a counter-flow heat exchanger.

We note from Fig. 12-8 that the value of the parameter P ranges from 0 to 1; it represents the thermal effectiveness of the tube-side fluid. The value of R ranges from zero to infinity, with zero corresponding to pure vapor condensation on the shell side and infinity to evaporation on the tube side.

Thus, if the inlet and outlet temperatures of the hot and cold fluids are available, the simple LMTD based on the flow arrangement of Fig. 12-6e can be corrected as described above and the corrected effective temperature ΔT_{corr} can be calculated. Then the total heat transfer rate Q between the hot and cold fluids is calculated from Eq. (12-11) by replacing ΔT_{ln} by ΔT_{corr}.

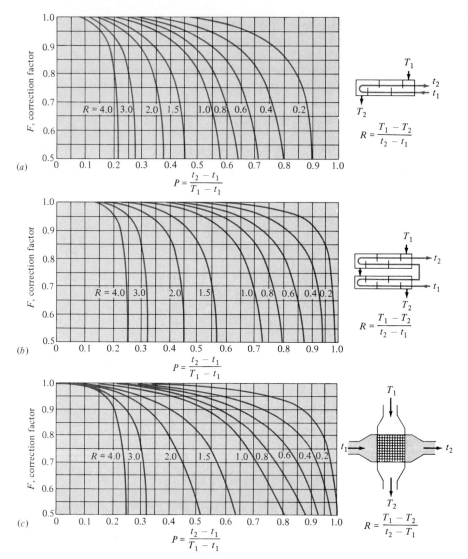

FIGURE 12-8
Correction factor F for computing $\Delta T_{corrected}$ for multipass and cross-flow exchanger. (a) one shell pass and two tube pass or multiple of two tube pass; (b) two shell pass and four tube pass or multiple of four tube pass; (c) single pass, cross-flow, both fluids unmixed. (From Bowman, Mueller, and Nagle.)

Example 12-6. A one-shell-pass, two-tube-pass heat exchanger with the flow arrangement shown in Fig. 12-8a is used to heat water flowing at a rate of $m_c = 1.5\,\text{kg/s}$ from $T_1 = 25°C$ to $T_2 = 80°C$ with pressurized water entering the tubes at $t_1 = 200°C$ and leaving it at $t_2 = 100°C$. The overall heat transfer coefficient is $U_m = 1250\,\text{W/(m}^2 \cdot °C)$. Calculate the heat transfer surface A required.

Solution. The temperature profiles in the exchanger are shown in the accompanying figure. The logarithmic mean temperature difference becomes

$$\Delta T_{\ln} = \frac{\Delta T_0 - \Delta T_L}{\ln(\Delta T_0 / \Delta T_L)}$$

$$= \frac{75 - 20}{\ln(75/20)} = 41.6°C$$

The parameters P and R are

$$P = \frac{t_2 - t_1}{T_1 - t_1}$$

$$= \frac{100 - 200}{25 - 200} = 0.57$$

$$R = \frac{T_1 - T_2}{t_2 - t_1}$$

$$= \frac{25 - 80}{100 - 200} = 0.55$$

Then, from Fig. 12-8a, the correction factor F is

$$F = 0.89$$

and the corrected temperature difference becomes

$$\Delta T_{corr} = F \Delta T_{\ln}$$

$$= (0.89)(41.6) = 37.02°C$$

The heat transfer area A is determined by an overall energy balance:

$$U_m A_s \Delta T_{corr} = m_c c_{p_c} (t_2 - t_1)$$

$$(1250)(A_s)(37.02) = (1.5)(4.18)(80 - 25)$$

which gives

$$A_s = 7.45 \text{ m}^2$$

FIGURE EXAMPLE 12-6

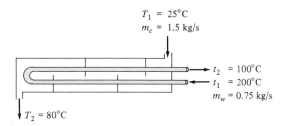

$T_1 = 25°C$
$m_c = 1.5$ kg/s

$t_2 = 100°C$
$t_1 = 200°C$
$m_w = 0.75$ kg/s

$T_2 = 80°C$

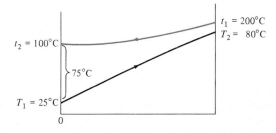

$t_2 = 100°C$

$75°C$

$T_1 = 25°C$

$t_1 = 200°C$
$T_2 = 80°C$

0

Example 12-7. A two-shell-pass, four-tube-pass heat exchanger with the flow arrangement shown in Fig. 12-8*b* is to be used to heat water with oil. Water enters the tubes at a flow rate of $m_c = 2$ kg/s and temperature $t_1 = 20°C$, and leaves at $t_2 = 80°C$. Oil enters the shell side at $T_1 = 140°C$ and leaves at $T_2 = 90°C$. Calculate the heat transfer area required for an overall heat transfer coefficient of $U_m = 300$ W/(m² · °C).

Solution. This is a multipass heat exchanger. Therefore, the LMTD based on a *counter-flow* arrangement should be corrected, using the correction factor given in Fig. 12-8*b*. The temperature distribution for the counter-flow arrangement is illustrated in the accompanying figure.

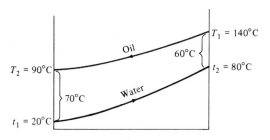

FIGURE EXAMPLE 12-7

The counterflow LMTD becomes

$$\Delta T_{\ln} = \frac{\Delta T_o - \Delta T_L}{\ln (\Delta T_0/\Delta T_L)}$$

$$= \frac{70 - 60}{\ln (70/60)} = 64.9°C$$

The parameters P and R are

$$P = \frac{t_2 - t_1}{T_1 - t_1}$$

$$= \frac{80 - 20}{140 - 20} = 0.5$$

$$R = \frac{T_1 - T_2}{t_2 - t_1}$$

$$= \frac{140 - 90}{80 - 20} = 0.83$$

Then, from Fig. 12-8, the correction factor F is

$$F = 0.97$$

The required heat transfer area A is determined by an overall energy balance:

$$U_m A_s (F \Delta T_{\ln}) = m_c c_{p_c} (t_2 - t_1)$$

$$(0.3)(A_s)(0.97 \times 64.9) = (2)(4.18)(80 - 20)$$

which gives

$$A_s = 26.6 \text{ m}^2$$

12-6 ε-NTU METHOD

Rating and sizing are two important problems encountered in the thermal analysis of heat exchangers. The rating problem is concerned with determining the transfer rate, the fluid outlet temperatures, and the pressure drops for an existing heat exchanger or one that is already sized; hence the heat transfer surface area and the flow passage dimensions are available. The sizing problem is concerned with determining the matrix dimensions needed to meet the specified heat transfer and pressure drop requirements. If we are not concerned with the pressure drop, the rating problem involves determination of the total heat transfer rate for an existing heat exchanger, and the sizing problem involves determination of the total heat transfer surface required to meet a specified heat transfer rate.

If the inlet and outlet temperatures of the hot and cold fluid and the overall heat transfer coefficient are specified, the LMTD method, with or without the correction, can be used to solve the rating or sizing problem.

In some situations only the inlet temperatures and the flow rates of the hot and cold fluids are given, and the overall heat transfer coefficient can be estimated. For such cases, the logarithmic mean temperature cannot be determined because the outlet temperatures are not known. Therefore, using the LMTD method for the thermal analysis of heat exchangers will involve tedious iterations in order to determine the proper value of LMTD which will satisfy the requirement that the heat transferred in the heat exchanger be equal to the heat carried out by the fluid.

In such situations the analysis can be simplified significantly by using the ε-NTU, or effectiveness method, developed originally by Kays and London.

In this method, effectiveness ε is defined as

$$\varepsilon = \frac{Q}{Q_{\text{max}}} = \frac{\text{actual heat transfer rate}}{\text{maximum possible heat transfer rate}}$$
from one stream to another

The maximum possible heat transfer rate Q_{max} is obtained with a counter-flow exchanger if the temperature change of the fluid with the minimum value of mc_p equals the difference in the inlet temperatures of the hot and cold fluids. Here we consider $(mc_p)_{\text{min}}$, because the energy given up by one fluid should equal that received by the other fluid. If we consider $(mc_p)_{\text{max}}$, then the other fluid should undergo a temperature change greater than the maximum available temperature difference; that is, ΔT for the other fluid should be greater than $T_{h,\text{in}} - T_{c,\text{in}}$. This is not possible. Given this consideration, Q_{max} is chosen as

$$Q_{\text{max}} = (mc_p)_{\text{min}}(T_{h,\text{in}} - T_{c,\text{in}}) \tag{12-15}$$

Then, given ε and Q_{max}, the actual heat transfer rate Q is

$$\boxed{Q = \varepsilon(mc_p)_{\text{min}}(T_{h,\text{in}} - T_{c,\text{in}})} \tag{12-16}$$

Here $(mc_p)_{\min}$ is the smaller of $m_h c_{ph}$ and $m_c c_{pc}$ for the hot and cold fluids; $T_{h,\text{in}}$ and $T_{c,\text{in}}$ are the inlet temperatures of the hot and cold fluids, respectively.

Clearly, if the effectiveness ε of the exchanger is known, Eq. (12-16) provides an explicit expression for the determination of Q through the exchanger.

A considerable amount of effort has been directed toward the development of simple explicit expressions for the effectiveness of various types of heat exchangers. For convenience in such developments, a dimensionless parameter called the *number of (heat) transfer units* (NTU) is defined as

$$\text{NTU} = AU_m/C_{\min} \tag{12-17}$$

where $C_{\min} = (mc_p)_{\min}$, that is, the smaller of mc_{ph} and mc_{pc}. The physical significance of NTU can be viewed as follows:

$$\text{NTU} = \frac{AU_m}{C_{\min}} = \frac{\text{heat capacity of exchanger, W/}^\circ\text{C}}{\text{heat capacity of flow, W/}^\circ\text{C}} \tag{12-18}$$

For simplicity in the notation, we adopt the abbreviation

For a specified value of $U_m/C_{\min}$, the NTU is a measure of the actual heat transfer area A, or the "physical size" of the exchanger. The higher the NTU, the larger the physical size.

Analytic expressions for ε-NTU relations have been developed for various types of heat exchangers, including parallel-flow, counter-flow, cross-flow, multipass, and many others. The reader should consult Kays and London for extensive documentation of effectiveness charts for various flow arrangements.

In Figs. 12-9 to 12-13 we present some effectiveness charts for typical flow arrangements. Also, Table 12-3 lists some explicit expressions for the effectiveness. In these figures, the physical significance of $(C_{\min}/C_{\max}) \to 0$ needs further clarification.

Consider the actual heat transfer rate Q through the heat exchanger, given by

$$Q = m_h c_{ph}(T_{h,\text{in}} - T_{h,\text{out}}) = m_c c_{pc}(T_{c,\text{out}} - T_{c,\text{in}}) \tag{12-19}$$

Then, from Eqs. (12-16) and (12-19), we write

$$\varepsilon = \frac{C_h(T_{h,\text{in}} - T_{h,\text{out}})}{C_{\min}(T_{h,\text{in}} - T_{c,\text{in}})} \tag{12-20}$$

or

$$\varepsilon = \frac{C_c(T_{c,\text{out}} - T_{c,\text{in}})}{C_{\min}(T_{h,\text{in}} - T_{c,\text{in}})} \tag{12-21}$$

where we define

$$C_h \equiv m_h c_{ph} \qquad C_c \equiv m_c c_{pc} \tag{12-22}$$

and $C_{\min} \equiv$ smaller and C_h and C_c. In the case of *condensers* and *boilers*, the fluid temperature on the boiling or condensing side remains essentially con-

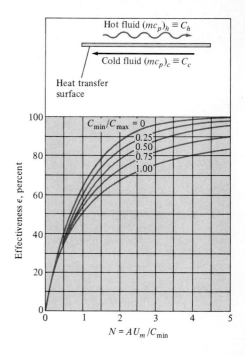

FIGURE 12-9
Effectiveness for a parallel-flow heat exchanger. (From Kays and London.)

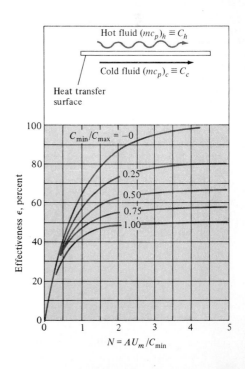

FIGURE 12-10
Effectiveness for a counterflow heat exchanger. (From Kays and London.)

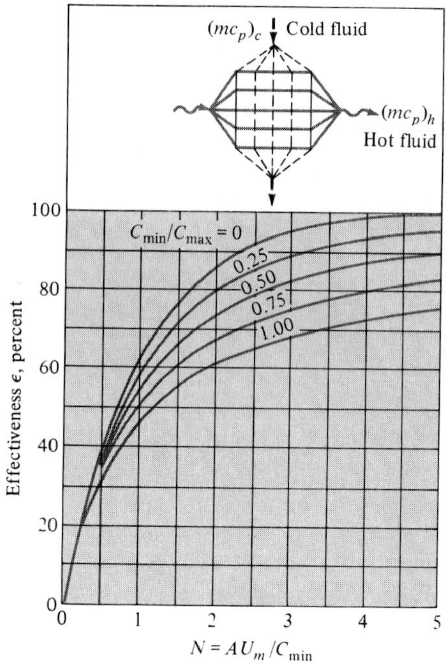

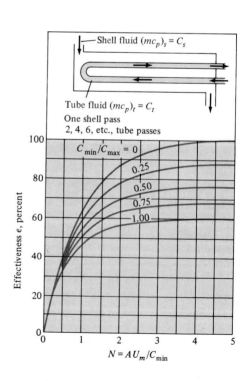

FIGURE 12-11
Effectiveness for a cross-flow heat exchanger, both fluids unmixed. (from Kays and London.)

FIGURE 12-12
Effectiveness for a single shell pass heat exchanger with two, four, six, etc. tube passes. (From Kays and London.)

$(mc_p)_c$ Cold fluid

$(mc_p)_h$ Hot fluid

Effectiveness ϵ, percent

$C_{min}/C_{max} = 0$

0.25
0.50
0.75
1.00

$N = AU_m/C_{min}$

Shell fluid $(mc_p)_s = C_s$

Tube fluid $(mc_p)_t = C_t$
One shell pass
2, 4, 6, etc., tube passes

Effectiveness ϵ, percent

$C_{min}/C_{max} = 0$

0.25
0.50
0.75
1.00

$N = AU_m/C_{min}$

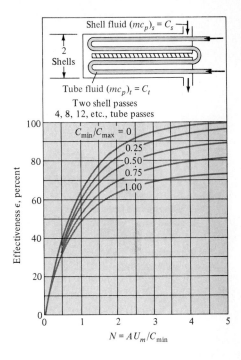

FIGURE 12-13

Effectiveness of a two shell pass heat exchanger with four, eight, twelve, etc. tube passes. (From Kays and London.)

TABLE 12-3
Heat exchanger effectiveness; $NTU = UA/C_{min}$, $C = C_{min}/C_{max}$

Flow arrangement	ε formula
Parallel flow	$\varepsilon = \dfrac{1 - \exp[-(NTU)(1 + C)]}{1 + C}$
Counter flow	$\varepsilon = \dfrac{1 - \exp[-(NTU)(1 - C)]}{1 - C\exp[-(NTU)(1 - C)]}$

Cross flow: Both fluids unmixed. Approximate formula is

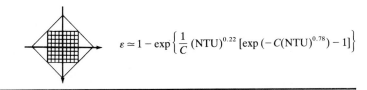

$$\varepsilon \simeq 1 - \exp\left\{\frac{1}{C}(NTU)^{0.22}[\exp(-C(NTU)^{0.78}) - 1]\right\}$$

stant. Now, if the effectiveness ε as defined by Eqs. (12-20) and (12-21) should remain finite while $T_{in} - T_{out}$ for the condensing or boiling side is practically zero, C_c or C_h on the phase change side should behave as an infinite specific heat capacity. Such a requirement implies that for a boiler or condenser we must have $C_{max} \to \infty$, and as a result,

$$C = \frac{C_{min}}{C_{max}} \to 0 \tag{12-23}$$

Therefore, the case $C \to 0$ *implies a boiler or a condenser*.

An examination of the results presented in Figs. 12-9 to 12-13 reveals that for $\varepsilon < 40$ percent, the capacity ratio $C = C_{min}/C_{max}$ does not have much effect on the effectiveness ε.

A counter-flow exchanger has the highest ε for specified values of NTU and C compared with that for other flow arrangements. Therefore, for a given NTU and C, a counter-flow arrangement yields maximum heat transfer performance.

Having established the framework for the definition and the determination of the heat exchanger effectiveness ε, we are now in a position to make use of the effectiveness concept in solving *rating* and *sizing* problems.

RATING PROBLEM. Suppose the inlet temperatures $T_{c,in}$ and $T_{h,in}$, the flow rates m_c and m_h, the physical properties of both fluids, the overall heat transfer coefficient U_m, and the total heat transfer area A are all given. The type of exchanger and its flow arrangement are specified. We wish to determine the total heat flow rate Q and the outlet temperatures $T_{h,out}$ and $T_{c,out}$. The calculation is as follows:

1. Calculate $C = C_{min}/C_{max}$ and NTU $= U_m A/C_{min}$ from the specified input data.
2. Knowing NTU and C, determine ε from the chart or from the equation for the specific geometry and flow arrangement.
3. Knowing ε, compute the total heat transfer rate Q from

$$Q = \varepsilon C_{min}(T_{h,in} - T_{c,in})$$

4. Calculate the outlet temperatures from

$$T_{h,out} = T_{h,in} - \frac{Q}{C_h}$$

$$T_{c,out} = T_{c,in} + \frac{Q}{C_c}$$

The preceding discussion of the ε-NTU method clearly illustrates that the rating problem when the outlet temperatures are not given can readily be solved with the ε-NTU method, but a tedious iteration procedure would be required to solve it with the LMTD method, and the convergence might not be easy.

SIZING PROBLEM. Suppose the inlet and outlet temperatures, the flow rate, the overall heat transfer coefficient, and the total heat transfer rate are given, and the flow arrangement is specified. We wish to determine the total heat transfer surface A. The procedure is as follows:

1. Knowing the outlet and inlet temperatures, calculate ε from Eqs. (12-20) and (12-21).
2. Calculate $C = C_{min}/C_{max}$.
3. Knowing ε and C, determine NTU from the appropriate ε-NTU chart.
4. Knowing NTU, calculate the heat transfer surface A from Eq. (12-17):

$$A = \frac{(\text{NTU})C_{min}}{U_m}$$

The use of the ε-NTU method is generally preferred for the design of compact heat exchangers for automotive, aircraft, air conditioning, and various industrial applications where the inlet temperatures of the hot and cold fluids are specified and the heat transfer rates are to be determined. In the process, power, and petrochemical industries, both the inlet and outlet temperatures of the hot and cold fluids are specified; hence the LMTD method is generally used.

Example 12-8. A counter-flow heat exchanger with the flow arrangement shown in Fig. 12-10 is used to cool mercury [$c_p = 1370 \text{ J/(kg} \cdot {}^{\circ}\text{C)}$] from $T_{h,in} = 110{}^{\circ}\text{C}$ to $T_{h,out} = 70{}^{\circ}\text{C}$ at a rate of $m_h = 1 \text{ kg/s}$, with water entering at $T_{c,in} = 30{}^{\circ}\text{C}$ at a rate of $m_c = 0.2 \text{ kg/s}$. The overall heat transfer coefficient is $U_m = 250 \text{ W/(m}^2 \cdot {}^{\circ}\text{C)}$. Calculate (*a*) the heat transfer *surface* required, and (*b*) the *exit temperature of the water*.

Solution. (*a*)   We determine the heat transfer surface area by using the effectiveness method:

$$C_h = m_h c_{ph} = (1)(1370) = 1370 \text{ W/}{}^{\circ}\text{C}$$

$$C_c = m_c c_{pc} = (0.2)(4180) = 836 \text{ W/}{}^{\circ}\text{C}$$

Therefore $C_{min} = C_c < C_h$, and

$$\frac{C_{min}}{C_{max}} = \frac{836}{1370} = 0.61$$

The total heat transfer rate Q is

$$Q = m_h c_{ph}(T_{h,in} - T_{h,out})$$

$$= (1)(1370)(110 - 70) = 54{,}800 \text{ W}$$

The effectiveness ε becomes

$$\varepsilon = \frac{Q}{Q_{max}} = \frac{Q}{C_{min}(T_{h,in} - T_{c,in})}$$

$$= \frac{54{,}800}{836(110 - 30)} = 0.82$$

From Fig. 12-10, we obtain

$$\text{NTU} = \frac{AU_m}{C_{\min}} = 2.7$$

which gives the area A as

$$A = \frac{(\text{NTU})C_{\min}}{U_m}$$

$$= \frac{(2.7)(836)}{250} = 9.03 \text{ m}^2$$

(*b*) The exist temperature of the water $T_{c,\text{out}}$ is determined from the energy balance:

$$Q = m_c c_{pc}(T_{c,\text{out}} - T_{c,\text{in}})$$

$$54{,}800 = (0.2)(4180)(T_{c,\text{out}} - 30)$$

which gives

$$T_{c,\text{out}} = 95.6°C$$

Note: Once we know the inlet and exit temperatures, this problem can easily be solved by LMTD, which gives a heat transfer area of $A = 8.75 \text{ m}^2$. Since LMTD doesn't involve reading values from a chart, it is more accurate.

Example 12-9. A cross-flow heat exchanger with the flow arrangement shown in Fig. 12-11 is to heat water with hot exhaust gas. The exhaust gas [$c_p = 1050 \text{ J/(kg} \cdot °C)$] enters at $T_{h,\text{in}} = 200°C$ and $m_h = 2.5 \text{ kg/s}$, while the water enters at $T_{c,\text{in}} = 30°C$ and $m_c = 1.5 \text{ kg/s}$. The overall heat transfer coefficient is $U_m = 150 \text{ W/(m}^2 \cdot °C)$, and the heat transfer surface is $A = 17.5 \text{ m}^2$. Calculate (*a*) the *total heat transfer rate*, and (*b*) the *outlet temperatures* of the water and exhaust gas.

Solution. (*a*) Since the exit temperatures are not known, the effectiveness method should be used to solve this problem.

$$C_h = m_h c_{ph} = (2.5)(1050) = 2625 \text{ W/°C}$$

$$C_c = m_c c_{pc} = (1.5)(4180) = 6270 \text{ W/°C}$$

Therefore $C_{\min} = C_h < C_c$, and

$$\frac{C_{\min}}{C_{\max}} = \frac{2625}{6270} = 0.42$$

$$\text{NTU} = \frac{AU_m}{C_{\min}} = \frac{17.5 \times 150}{2625} = 1$$

From Fig. 12-11, we obtain

$$\varepsilon = 0.48$$

The total heat transfer rate Q is

$$Q = \varepsilon Q_{\max} = \varepsilon C_{\min}(T_{h,\text{in}} - T_{c,\text{in}})$$

$$= (0.48)(2625)(200 - 30)$$

$$= 214.2 \text{ kW}$$

(*b*) The outlet temperature of the exhaust gas is determined as

$$Q = m_h c_{ph}(T_{h,\text{in}} - T_{h,\text{out}})$$
$$214{,}200 = 2625(200 - T_{h,\text{out}})$$

which gives

$$T_{h,\text{out}} = 118.4°C$$

The outlet temperature of the water is

$$Q = m_c c_{pc}(T_{c,\text{out}} - T_{c,\text{in}})$$
$$214{,}200 = 6270(T_{c,\text{out}} - 30)$$

which gives

$$T_{c,\text{out}} = 64.2°C$$

Example 12-10. A two-shell-pass, four-tube-pass heat exchanger with the flow arrangement shown in Fig. 12-13 is used to cool processed water flowing at a rate $m_h = 5$ kg/s from $t_1 = 75°C$ to $t_2 = 25°C$ on the tube side, with cold water entering the shell side at $T_1 = 10°C$ at a rate $m_c = 6$ kg/s. The overall heat transfer coefficient is $U_m = 750$ W/(m$^2 \cdot$°C). This heat exchanger is illustrated in the accompanying figure. Calculate (*a*) the heat transfer surface area, and (*b*) the outlet temperature of the coolant water.

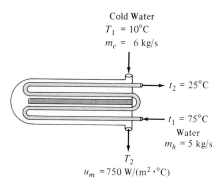

Cold Water
$T_1 = 10°C$
$m_c = 6$ kg/s

$t_2 = 25°C$

$t_1 = 75°C$
Water
$m_h = 5$ kg/s

T_2
$u_m = 750$ W/(m$^2 \cdot$°C)

FIGURE EXAMPLE 12-10

Solution.

$$C_h = m_h c_{ph} = (5)(4180) = 20{,}900 \text{ W/°C}$$
$$C_c = m_c c_{pc} = (6)(4180) = 25{,}080 \text{ W/°C}$$

Therefore $C_{\text{min}} = C_h < C_c$, and

$$\frac{C_{\text{min}}}{C_{\text{max}}} = \frac{20{,}900}{25{,}080} = 0.83$$

The total heat transfer rate Q is

$$Q = m_h c_{ph}(T_{h,\text{in}} - T_{h,\text{out}})$$
$$= (5)(4180)(75 - 25) = 1{,}045{,}000 \text{ W}$$

The effectiveness ε is

$$\varepsilon = \frac{Q}{Q_{max}} = \frac{Q}{C_{min}(T_{h,in} - T_{c,in})}$$

$$= \frac{1,045,000}{(20,900)(75 - 10)} = 0.77$$

From Fig. 12-13, we obtain

$$\text{NTU} = \frac{AU_m}{C_{min}} = 4.5$$

which gives

$$A = \frac{(\text{NTU})C_{min}}{U_m}$$

$$= \frac{(4.5)(20,900)}{750} = 125.4 \text{ m}^2$$

(*b*) The exit temperature of the coolant water $T_{c,out}$ is determined from the energy balance:

$$Q = m_c c_{pc}(T_{c,out} - T_{c,in})$$

$$= (6)(4180)(T_{c,out} - 10)$$

which gives

$$T_{c,out} = 51.7°C$$

Note: Once we know the inlet and exit temperatures, this problem can easily be solved by LMTD, which gives a heat transfer area of $A = 113.7 \text{ m}^2$. Since LMTD doesn't involve reading values from a chart, it is more accurate.

12-7 COMPACT HEAT EXCHANGERS

A heat exchanger that has a surface area density greater than about $700 \text{ m}^2/\text{m}^3$ is quite arbitrarily referred to as a compact heat exchanger. These heat exchangers are generally used for applications in which gas flows. Hence the heat transfer coefficient is low, and the smallness of weight and size is important. These heat exchangers are available in a wide variety of configurations of the heat transfer matrix, and their heat transfer and pressure drop characteristics have been studied extensively by Kays and London. Figure 12-14 shows typical heat transfer matrices for compact heat exchangers. Figure 12-14*a* shows a circular finned-tube array with fins on individual tubes, Fig. 12-14*b* shows a plain plate-fin matrix formed by corrugation, and Fig. 12-14*c* shows a finned flat-tube matrix.

The heat transfer and pressure drop characteristics of these configurations when they are used as compact heat exchangers are determined experimentally. For example, Fig. 12-15 shows typical heat transfer and friction factor data. Note that the principal dimensionless groups governing the experimental correlations include the Stanton, Prandtl, and Reynolds numbers,

$$\text{St} = \frac{h}{Gc_p} \qquad \text{Pr} = \frac{c_p\mu}{k} \qquad \text{Re} = \frac{GD_h}{\mu} \qquad (12\text{-}24)$$

Here G is the *mass velocity*, defined as

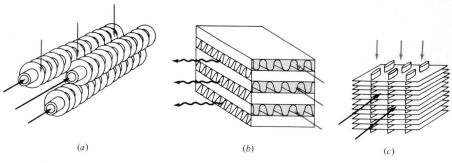

FIGURE 12-14
Typical heat transfer matrices for compact heat exchangers. (a) circular finned-tube matrix; (b) plain plate-fin matrix; (c) finned flat-tube matrix.

$$G = \frac{m}{A_{min}} \quad \text{kg/(m}^2 \cdot \text{s)} \tag{12-25}$$

where m = total mass flow rate of fluid in kilograms per second and A_{min} = minimum free-flow cross-sectional area in square meters, regardless of where this minimum occurs.

The hydraulic diameter D_h is defined as

FIGURE 12-15
Heat transfer and friction factor for low across circular finned-tube matrix. (From Kays and London.)

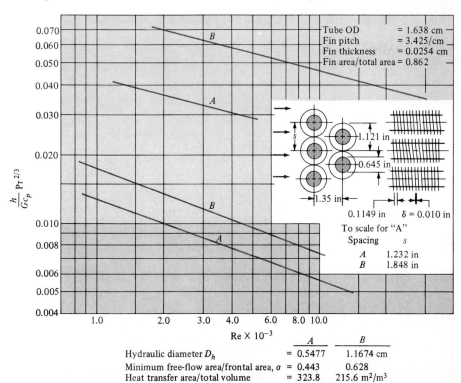

	A	B
Hydraulic diameter D_h	= 0.5477	1.1674 cm
Minimum free-flow area/frontal area, σ	= 0.443	0.628
Heat transfer area/total volume	= 323.8	215.6 m²/m³

$$D_h = 4 \frac{LA_{\min}}{A} \tag{12-26}$$

where A is the total heat transfer area and the quantity $LA_{\min}$ can be regarded as the minimum free-flow passage volume, since L is the flow length of the heat exchanger matrix.

Thus, once the heat transfer and the friction factor charts for a specified matrix are available and the Reynolds number Re for the flow is given, the heat transfer coefficient h and the friction factor f for flow across the matrix can be evaluated. Then the rating and sizing problems associated with the heat exchanger matrix can be performed using either the LMTD or the effectiveness method of analysis. We now describe the pressure drop analysis for compact heat exchangers.

As the fluid enters a heat exchanger, it experiences pressure losses owing to contraction resulting from an area change and an irreversible free expansion after a sudden contraction. As the fluid passes through the heat exchanger matrix (i.e., the core), it experiences pressure loss because of fluid friction. Also, depending on whether heating or cooling takes place, a pressure change results from flow acceleration or deceleration. Finally, as the fluid leaves the heat exchanger matrix, there are pressure losses associated with the area change and the flow separation. Therefore, in general, the pressure drop associated with flow through a compact heat exchanger matrix consists of three components: the *core friction*, the *core acceleration*, and the *entrance* and *exit losses*. Here we present the pressure drop analysis for finned-tube exchangers only.

Pressure Drop for Finned-Tube Exchangers

For heat exchangers with sufficiently long cores operating at low Reynolds numbers, the entrance and exit losses are considered negligible. Then, for flow normal to finned-tube banks, such as that illustrated in Fig. 12-14a, the entrance and exit losses are generally accounted for in the friction factor. Then the total pressure drop for flow across the tube bank becomes

$$\Delta P = \frac{G^2}{2\rho_i} \left[\underbrace{(1 + \sigma^2)\left(\frac{\rho_i}{\rho_o} - 1\right)}_{\substack{\text{Flow} \\ \text{acceleration}}} + \underbrace{f\left(\frac{A}{A_{\min}}\right)\left(\frac{\rho_i}{\rho_m}\right)}_{\substack{\text{Core} \\ \text{friction}}} \right] \tag{12-27}$$

where $\sigma = \dfrac{\text{minimum free-flow area}}{\text{frontal area}} = \dfrac{A_{\min}}{A_{\text{fr}}}$

$\dfrac{A}{A_{\min}} = \dfrac{\text{total heat transfer area}}{\text{minimum free-flow area}} = \dfrac{4L}{D_h}$

$G = \text{mass velocity, kg/(m}^2 \cdot \text{s)} = \dfrac{\rho u_\infty A_{\text{fr}}}{A_{\min}} = \dfrac{\rho u_\infty}{\sigma}$

$\rho_m = 2\left(\dfrac{1}{1/\rho_i + 1/\rho_o}\right)$

$\rho_i, \rho_o = \text{density at inlet and exit, respectively}$

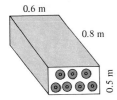

$P = 2$ atm
$T = 500$ K
$m = 12$ kg/s

0.6 m

0.8 m

0.5 m

$D_h = 0.5477$ cm

$\dfrac{A_{\min}}{A_{\text{ft}}} = 0.443$

FIGURE EXAMPLE 12-11

Example 12-11. Hot air at $P = 2$ atm pressure and 500 K with a mass flow rate $m = 12$ kg/s flows across a circular finned-tube matrix configuration like those shown in Figs. 12-14a and 12-15. The dimensions of the heat exchanger matrix are given in the accompanying figure. Calculate (a) the heat transfer coefficient, (b) the friction factor, and (c) the ratio of the pressure drop due to core friction to the inlet pressure.

Solution. The physical properties of hot air at 500 K and 2 atm are

$$\rho_a = 2 \times 0.7048 = 1.41 \text{ kg/m}^3 \qquad c_p = 1.0295 \text{ KJ/(kg} \cdot {}^\circ\text{C)}$$

$$\mu_a = 2.671 \times 10^{-5} \text{ kg/(m} \cdot \text{s)} \qquad \text{Pr}_a = 0.68$$

(a) Heat transfer coefficient:

$$G_a = \frac{m_a}{A_{\min}} = \frac{m_a}{\sigma_a A_{\text{fr}}}$$

$$= \frac{12}{(0.443)(0.8 \times 0.5)} = 67.72 \text{ kg/(m}^2 \cdot \text{s)}$$

$$\text{Re} = \frac{G_a D_h}{\mu_a}$$

$$= \frac{(67.72)(0.5477 \times 10^{-2})}{(2.671 \times 10^{-5})} = 13.9 \times 10^3$$

From Fig. 12-15 for Re = 13,900, we have

$$\frac{h_a \, \text{Pr}_a^{2/3}}{G_a c_{pa}} = 0.0052$$

Then

$$h_a = 0.0052 \, \frac{G_a c_{pa}}{\text{Pr}^{2/3}}$$

$$= 0.0052 \, \frac{(67.72)(1.0295 \times 10^3)}{(0.68)^{2/3}} = 469 \text{ W/(m}^2 \cdot {}^\circ\text{C)}$$

(b) From Fig. 12-15 for Re = 13,900, we have the friction factor:

$$f = 0.023$$

(c) The pressure drop across the heat exchanger due to core friction is determined from Eq. (12-27):

$$\frac{A}{A_{\min}} = \frac{4L}{D_h} = \frac{4(0.6)}{0.5477 \times 10^{-2}} = 438.2$$

$$\frac{\rho_i}{\rho_m} \cong 1$$

$$\Delta P_f = \frac{G^2}{2\rho_i} \left(f \, \frac{A}{A_{\min}} \, \frac{\rho_i}{\rho_m} \right)$$

$$= \frac{(67.72)^2}{(2 \times 1.41)} (0.023 \times 438.2 \times 1) = 16{,}390 \text{ N/m}^2$$

and the ratio of the pressure drop due to core friction to the inlet pressure is

$$\frac{\Delta P_f}{p} = \frac{16{,}390}{2 \times 1.0132 \times 10^5} = 0.081$$

PROBLEMS

Overall Heat Transfer Coefficient

12-1. Determine the overall heat transfer coefficient U_o based on the outer surface of a $D_i = 2.5$ cm, $D_o = 3.34$ cm steel pipe $[k = 54 \text{ W}/(\text{m} \cdot °\text{C})]$ for the following conditions: The inside and outside heat transfer coefficients are, respectively, $h_i = 1200 \text{ W}/(\text{m}^2 \cdot °\text{C})$ and $h_o = 2000 \text{ W}/(\text{m}^2 \cdot °\text{C})$; the fouling factors for the inside and outside surfaces are $F_i = F_o = 0.00018 \,(\text{m}^2 \cdot °\text{C})/\text{W}$.

12-2. Determine the overall heat transfer coefficient U_i based on the inside surface of a $D_i = 2.5$ cm, $D_o = 3.34$ cm brass tube $[k = 110 \text{ W}/(\text{m} \cdot °\text{C})]$ for the following conditions: The inside and outside heat transfer coefficients are, respectively, $h_i = 1200 \text{ W}/(\text{m}^2 \cdot °\text{C})$ and $h_o = 2000 \text{ W}/(\text{m}^2 \cdot °\text{C})$; the fouling factors for the inside and outside surfaces are $F_i = F_o = 0.00018 \,(\text{m}^2 \cdot °\text{C})/\text{W}$.

12-3. Determine the overall heat transfer coefficient U_o based on the outer surface of a $D_i = 2.5$ cm, $D_o = 3.34$ cm brass tube $[k = 110 \text{ W}/(\text{m}^2 \cdot °\text{C})]$ for the following conditions: The inside heat transfer coefficient is $h_i = 1000 \text{ W}/(\text{m}^2 \cdot °\text{C})$ and the fouling factors for the inside and outside surfaces are $F_i = F_o = 0.00018 \,(\text{m}^2 \cdot °\text{C})/\text{W}$. The outside heat transfer coefficients are (a) $h_o = 1500 \text{ W}/(\text{m}^2 \cdot °\text{C})$, (b) $h_o = 2000 \text{ W}/(\text{m}^2 \cdot °\text{C})$, and (c) $h_o = 2500 \text{ W}/(\text{m}^2 \cdot °\text{C})$.

12-4. Water at a mean temperature of $T_m = 80°\text{C}$ and a mean velocity of $U_m = 3 \text{ m/s}$ flows inside a 2.5-cm-ID thin-walled copper tube. Atmospheric air at $T_\infty = 20°\text{C}$ and a velocity of $U_\infty = 8 \text{ m/s}$ flows across the tube. Neglecting the tube wall resistance, calculate the overall heat transfer coefficient and the rate of heat loss per meter length of the tube.

12-5. Water at $T_i = 25°\text{C}$ and a velocity of $U_m = 1.5 \text{ m/s}$ enters a long brass condenser tube with $D_i = 1.34$ cm and $D_o = 1.58$ cm $[k = 110 \text{ W}/(\text{m} \cdot °\text{C})]$. The heat transfer coefficient for condensation at the outer surface of the tube is $h_o = 12,000 \text{ W}/\text{m}^2 \cdot °\text{C}$. Calculate the overall heat transfer coefficient U_o based on the outer surface.

  Answer: 3600 W/(m² · °C)

12-6. Engine oil at $T_i = 50°\text{C}$ and a mean velocity of $U_m = 0.25 \text{ m/s}$ enters a $D_i = 2.22$ cm, $t = 0.17$ cm thick brass horizontal tube $[k = 110 \text{ W}/(\text{m} \cdot °\text{C})]$. Heat is dissipated from the outer surface of the tube by free convection into ambient air at $T_\infty = 20°\text{C}$. Calculate the overall heat transfer coefficient U_o based on the tube outer surface.

  Answer: 4.9 W/(m² · °C)

12-7. Engine oil at $T_i = 50°\text{C}$ and a mean velocity of $U_m = 0.25 \text{ m/s}$ enters a steel pipe with $D_i = 1.34$ cm and $D_o = 1.58$ cm $[k = 54 \text{ W}/(\text{m} \cdot °\text{C})]$. Heat is dissipated from the horizontal outer surface of the pipe by free convection into ambient air at $T_\infty = 20°\text{C}$. Calculate the overall heat transfer coefficient U_o based on the tube outer surface.

12-8. Engine oil at a mean temperature $T_i = 80°\text{C}$ and mean velocity $U = 0.25 \text{ m/s}$ flows inside $D = 1$ cm ID, thin-walled, horizontal copper tubing. The outer surface of the tube dissipates heat by free convection into atmospheric air at a temperature $T_\infty = 25°\text{C}$. Calculate (a) the temperature of the tube wall, (b) the overall heat transfer coefficient, and (c) the heat loss per meter length of the tube.

Logarithmic Mean Temperature Difference Method

12-9. A counter-flow shell-and-tube heat exchanger is to be used to cool water from $T_{h,in} = 30°C$ to $T_{h,out} = 10°C$, using brine at $T_{c,in} = 1°C$ and $T_{c,out} = 5°C$. The overall heat transfer coefficient is estimated to be $U_m = 500$ W/(m$^2 \cdot$°C). Calculate the heat transfer surface area for a design heat load of $Q = 10$ kW.

12-10. Steam condenses at $T_h = 60°C$ on the shell side of a steam condenser while cooling water flows inside the tubes at a rate of $m_c = 3$ kg/s. The inlet and outlet temperatures of the water are $T_{c,in} = 20°C$ and $T_{c,out} = 50°C$, respectively. The overall heat transfer coefficient is $U_m = 2000$ W/(m$^2 \cdot$°C). Calculate the surface area required.
Answer: 8.7 m^2

12-11. A counter-flow shell-and-tube heat exchanger is to be used to heat water from $T_{c,in} = 10°C$ to $T_{c,out} = 70°C$, with oil $[C_p = 2100$ J/(kg$\cdot$°C)] entering at $T_{h,in} = 120°C$ and leaving at $T_{h,out} = 60°C$ at a rate of $m_h = 5$ kg/s. The overall heat transfer coefficient is $U_m = 300$ W/m$^2 \cdot$°C. Calculate the heat transfer surface required.

12-12. A counter-flow heat exchanger is to cool $m_h = 1$ kg/s of water from 65°C to 5°C by using a refrigerant $[C_{pc} = 920$ J/(kg$\cdot$°C)] entering at $T_{c,in} = -20°C$ and a flow rate of $m_c = 8$ kg/s. The overall heat transfer coefficient is $U_m = 1000$ W/(m$^2 \cdot$°C). Calculate the heat transfer area required.

12-13. Engine oil at 60°C, flowing at a rate $m = 0.5$ kg/s, enters a $D_i = 3$ cm ID copper tube which is maintained at a uniform temperature $T_w = 100°C$ by condensing steam outside. Calculate the tube length required if the outlet temperature is to be $T_o = 80°C$.

12-14. A counter-flow heat exchanger is to be used to heat $m_c = 3$ kg/s of water from $T_{c,in} = 20°C$ to $T_{c,out} = 80°C$, using hot exhaust gas $[c_p = 1000$ J/(kg$\cdot$°C)] entering at $T_{h,in} = 220°C$ and leaving at $T_{h,out} = 90°C$. The overall heat transfer coefficient is $U_m = 200$ W/(m$^2 \cdot$°C). Calculate the heat transfer surface required.

12-15. A single-pass, counter-flow, shell-and-tube heat exchanger is used to heat water from $T_{c,in} = 15°C$ to $T_{c,out} = 80°C$ at a rate $m_c = 5$ kg/s, using oil entering the shell side at $T_{h,in} = 140°C$ and leaving at $T_{h,out} = 90°C$. The overall heat transfer coefficient is $U_m = 400$ W/(m$^2 \cdot$°C). Calculate the heat transfer surface required.
Answer: 50.5 m^2

12-16. A counter-flow heat exchanger is to be designed to cool $m_h = 0.5$ kg/s of oil $[c_{0p} = 2000$ J/(kg$\cdot$°C)] from $T_{h,in} = 60°C$ to $T_{h,out} = 40°C$ with cooling water entering at $T_{c,in} = 20°C$ and leaving $T_{c,out} = 30°C$. The overall heat transfer coefficient is $U_m = 200$ W/(m$^2 \cdot$°C). Calculate the heat transfer surface required.
Answer: 4.05 m^2

12-17. The design heat load for an oil cooler is $Q = 500$ kW. Calculate the heat transfer surface A required if the inlet and outlet temperature differences are, respectively, $\Delta T_0 = 40°C$ and $\Delta T_L = 15°C$ and the average value of the overall heat transfer coefficient is $U_m = 500$ W/(m$^2 \cdot$°C).
Answer: 39.2 m^2

12-18. Engine oil at a mean temperature $T_i = 80°C$ and mean velocity $U = 0.2$ m/s flows inside $D = 1.9$ cm ID, thin-walled horizontal copper tubing. At the outer surface, atmospheric air at $T_\infty = 15°C$ and a velocity of $U_\infty = 5$ m/s flows across the tube. Neglecting the tube wall resistance, calculate the overall heat transfer coefficient and the rate of heat loss to the air per meter length of the tube.
Answer: 18.0 W/(m$^2 \cdot$°C), 69.8 W/m length

12-19. Water at a mean temperature $T_i = 80°C$ and a mean velocity $u = 0.15\,\text{m/s}$ flows inside a $D = 2.5\,\text{cm}$ ID, thin-walled horizontal copper tube. At the outer surface of the tube, heat is dissipated by free convection into atmospheric air at a temperature $T_\infty = 15°C$. Assuming that the tube wall resistance is negligible, calculate the overall heat transfer coefficient and the heat loss per meter length of the tube.

Correction for LMTD

12-20. A one-shell-pass, two-tube-pass heat exchanger with the flow arrangement shown in Fig. 12-8a is to heat water with enthylene glycol. Water enters the shell side at $T_1 = 30°C$ and leaves at $T_2 = 70°C$ with a flow rate of $m_c = 1\,\text{kg/s}$, while ethylene glycol enters the tube at $t_1 = 100°C$ and leaves at $t_2 = 60°C$. The overall heat transfer coefficient is $U_m = 200\,\text{W/(m}^2 \cdot °\text{C)}$. Calculate the heat transfer surface A required.
Answer: $55.7\,\text{m}^2$

12-21. A one-shell-pass, two-tube-pass heat exchanger is to be designed to heat $m_c = 1\,\text{kg/s}$ water entering the shell side at $T_{c,\text{in}} = 20°C$. The hot fluid [$c_p = 2000\,\text{J/(kg}\cdot°\text{C)}$] enters the tube at $T_{h,\text{in}} = 80°C$ with $m_h = 0.5\,\text{kg/s}$ and leaves the exchanger at $T_{h,\text{out}} = 30°C$. The overall heat transfer coefficient is $U_m = 300\,\text{W/(m}^2 \cdot °\text{C)}$. Calculate the total heat transfer area A required.

12-22. Air flowing at a rate $m_c = 1\,\text{kg/s}$ is to be heated from $T_{c,\text{in}} = 20°C$ to $T_{c,\text{out}} = 50°C$ with hot water entering the tubes at $T_{h,\text{in}} = 90°C$ and leaving at $T_{h,\text{out}} = 60°C$. A two-shell-pass, four-tube-pass heat exchanger with the flow arrangement shown in Fig. 12-8b is to be used. The overall heat transfer coefficient is $U_m = 400\,\text{W/(m}^2 \cdot °\text{C)}$. Determine the surface area required.
Answers: $1.91\,\text{m}^2$

12-23. A two-shell-pass, four-tube-pass heat exchanger with the flow arrangement shown in Fig. 12-8b is used to heat water with hot exhaust gases. Water enters the tubes at $t_1 = 50°C$ and leaves it at $t_2 = 125°C$ with a flow rate of $m_c = 10\,\text{kg/s}$, while the hot exhaust gas enters the shell side at $T_1 = 300°C$ and leaves at $T_2 = 125°C$. The total heat transfer surface is $A = 800\,\text{m}^2$. Calculate the overall heat transfer coefficient.
Answer: $35.6\,\text{W/(m}^2 \cdot °\text{C)}$

12-24. A two-shell-pass, four-tube-pass heat exchanger with the flow arrangement shown in Fig. 12-8b is to heat water with ethylene glycol. Water enters the shell side at $T_1 = 30°C$ and leaves at $T_2 = 70°C$ with a flow rate of $m_c = 1\,\text{kg/s}$, while ethylene glycol enters the tube at $t_1 = 100°C$ and leaves at $t_2 = 60°C$. The overall heat transfer coefficient is $U_m = 200\,\text{W/(m}^2 \cdot °\text{C)}$. Calculate the required heat transfer surface area.
Answer: $30.3\,\text{m}^2$

12-25. A cross-flow heat exchanger with the flow arrangement shown in Fig. 12-8c is to heat $m_c = 2\,\text{kg/s}$ of air flowing on the shell side from $T_1 = 10°C$ to $T_2 = 50°C$, with hot water entering the tube side at $t_1 = 80°C$ and leaving at $t_2 = 45°C$, as illustrated in the accompanying figure. The overall heat transfer coefficient is $U_m = 250\,\text{W/(m}^2 \cdot °\text{C)}$. Calculate the required heat transfer surface.
Answer: $11.3\,\text{m}^2$

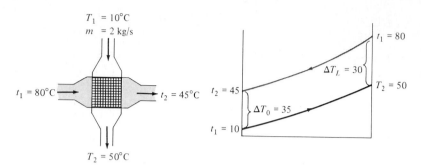

FIGURE PROBLEM 12-25

12-26. A single-pass cross-flow heat exchanger with the flow arrangement shown in Fig. 12-8c is used to heat water flowing at a rate $m_c = 0.5$ kg/s from $T_1 = 25°C$ to $T_2 = 80°C$ by using oil $[c_p = 2100$ J/(kg·°C)]. The oil enters the tubes at $t_1 = 175°C$ at a rate $m_h = 0.5$ kg/s. The overall heat transfer coefficient is $U_m = 300$ W/(m²·°C). Calculate the required heat transfer surface.

12-27. A cross-flow heat exchanger with the flow arrangement shown in Fig. 12-8c is to heat water with hot exhaust gas. The exhaust gas enters the exchanger at $T_{h,in} = 250°C$ and leaves at $T_{h,out} = 110°C$. The cold water enters at $T_{c,in} = 25°C$ and leaves at $T_{c,out} = 100°C$ with a flow rate of $m_c = 5$ kg/s. The overall heat transfer coefficient is estimated to be $U_m = 200$ W/(m²·°C). Calculate the required heat transfer surface.

12-28. A cross-flow heat exchanger with the flow arrangement shown in Fig. 12-8c is to heat air with hot water. The total heat transfer rate is 200 kW. The water enters the tubes at $t_1 = 85°C$ and leaves at $t_2 = 30°C$, while the air enters the shell side at $T_1 = 15°C$ and leaves at $T_2 = 50°C$. The overall heat transfer coefficient is $U_m = 75$ W/(m²·°C). Calculate the required heat transfer surface area.
Answer: 149.0 m²

ε-NTU Method

12-29. Hot chemical products $[c_{ph} = 2500$ J/(kg·°C)] at $T_{h,in} = 600°C$ and at a flow rate $m_h = 30$ kg/s are used to heat cold chemical products $[c_{pc} = 4200$ J/(kg·°C)] at $T_{c,in} = 100°C$ and $m_c = 30$ kg/s in the parallel-flow arrangement shown in Fig. 12-9. The total heat transfer surface is $A = 50$ m², and the overall heat transfer coefficient can be taken as $U_m = 1500$ W/(m²·°C). Calculate the outlet temperatures of the hot and cold chemical products for $m_h = m_c$.
Answers: 350°C, 249°C

12-30. Hot chemical products $[c_{ph} = 2500$ J/(kg·°C)] at $T_{h,in} = 600°C$ and at a flow rate $m_h = 30$ kg/s are used to heat cold chemical products $[c_{pc} = 4200$ J/(kg·°C)] at $T_{c,in} = 100°C$ and $m_c = 30$ kg/s in the counter-flow arrangement shown in Fig. 12-10. The total heat transfer surface is $A = 50$ cm², and the overall heat transfer coefficient can be taken as $U_m = 1500$ W/(m²·°C). Calculate the outlet temperatures of the cold and hot chemical products for $m_h = m_c$.

12-31. A counter-flow heat exchanger with the flow arrangement of Fig. 12-10 is to heat air to $T_{c,out} = 500°C$ with the exhaust gas from a turbine. Air enters the exchanger at $T_{c,in} = 300°C$ and $m_c = 4$ kg/s, while the exhaust gas enters at $T_{h,in} = 650°C$ and $m_h = 4$ kg/s. The overall heat transfer coefficient is $U_m = 80$ W/(m$^2 \cdot$°C). The specific heat for both air and the exhaust gas can be taken as $c_{ph} = c_{pc} = 1100$ J/(kg$\cdot$°C). Calculate the heat transfer surface A required and the outlet temperature of the exhaust gas.
Answers: 77 m^2, 450°C

12-32. A counter-flow heat exchanger with the flow arrangement shown in Fig. 12-10 is to heat cold fluid entering at $T_{c,in} = 30°C$ and $m_c c_{pc} = 15,000$ W/°C with hot fluid entering at $T_{h,in} = 120°C$ and $m_h c_{ph} = 10,000$ W/°C. The overall heat transfer coefficient is $U_m = 400$ W/(m$^2 \cdot$°C), and the total heat transfer surface is $A = 20$ m^2. Calculate the total heat transfer rate Q and the outlet temperatures of the hot and cold fluids.
Answers: 77°C, 58.7°C

12-33. A counter-flow exchanger is to heat air entering at $T_{c,in} = 400°C$ and $m_c = 6$ kg/s with exhaust gas entering at $T_{h,in} = 800°C$ and $m_h = 4$ kg/s. The overall heat transfer coefficient is $U_m = 100$ W/(m$^2 \cdot$°C), and the outlet temperature of the air is 551.5°C. the specific heat for both air and the exhaust gas may be taken as $c_{ph} = c_{pc} = 1100$ J/(kg$\cdot$°C). Calculate the heat transfer surface A required and the outlet temperature of the hot exhaust gas.

12-34. A counter-flow shell-and-tube heat exchanger with the flow arrangement shown in Fig. 12-10 is used to heat water with hot exhaust gases. The water enters at $T_{c,in} = 25°C$ and leaves at $T_{c,out} = 80°C$ at a rate $m_c = 2$ kg/s. Exhaust gases $[c_p = 1030$ J/(kg$\cdot$°C)] enter at $T_{h,in} = 175°C$ and leave at $T_{h,out} = 90°C$. The overall heat transfer coefficient is $U_m = 200$ W/(m$^2 \cdot$°C). Calculate the heat transfer surface area required.
Answer: 32.4 m^2

12-35. A cross-flow heat exchanger with the flow arrangement shown in Fig. 12-11 is to heat water with hot exhaust gas. The water enters the tubes at $T_{c,in} = 25°C$ and $m_c = 1$ kg/s, while the exhaust gas enters the exchanger at $T_{h,in} = 200°C$ and $m_h = 2$ kg/s. The total heat transfer surface is $A = 30$ m^2, and the overall heat transfer coefficient is $U_m = 120$ W/(m$^2 \cdot$°C). The specific heat for the exhaust gas may be taken as $c_{ph} = 1100$ J/(kg$\cdot$°C). Calculate the total heat transfer rate Q and the outlet temperatures $T_{c,out}$ and $T_{h,out}$ of the water and the exhaust gas.

12-36. A cross-flow heat exchanger with the flow arrangement shown in Fig. 12-11 has a heat transfer surface $A = 12$ m^2. It will be used to heat air entering at $T_{c,in} = 10°C$ and $m_c = 3$ kg/s with hot water entering at $T_{h,in} = 80°C$ and $m_h = 0.4$ kg/s. The overall heat transfer coefficient can be taken as $U_m = 300$ W/(m$^2 \cdot$°C). Calculate the total heat transfer rate Q and the outlet temperatures of the air and water.

12-37. A cross-flow heat exchanger with the flow arrangement shown in Fig. 12-11 is used to heat water with an engine oil. Water enters at $T_{c,in} = 30°C$ and leaves at $T_{c,out} = 85°C$ at a rate $m_c = 1.5$ kg/s, while the engine oil $[c_{ph} = 2300$ J/(kg$\cdot$°C)] enters at $T_{h,in} = 120°C$ at a flow rate $m_h = 3.5$ kg/s. The heat transfer surface is $A = 30$ m^2. Calculate the overall heat transfer coefficient U_m using the ε-NTU method.

12-38. A cross-flow heat exchanger with the flow arrangement shown in Fig. 12-11 is used to heat pressurized water to $T_{c,out} = 90°C$ with a hot exhaust gas. The hot gas [$c_{ph} = 1050 \, \text{J}/(\text{kg} \cdot {}°\text{C})$] enters the exchanger at $T_{h,\text{in}} = 300°\text{C}$ and a flow rate $m_h = 1 \, \text{kg}/\text{s}$, while the pressurized water enters at $T_{c,\text{in}} = 30°\text{C}$ and a flow rate $m_c = 0.25 \, \text{kg}/\text{s}$. The heat transfer surface area is $A = 3 \, \text{m}^2$. Calculate the overall heat transfer coefficient U_m, using the ε-NTU method.

12-39. A cross-flow heat exchanger with the flow arrangement shown in Fig. 12-11 is to heat water with hot exhaust gas. The exhaust gas [$c_p = 1100 \, \text{J}/(\text{kg} \cdot {}°\text{C})$] enters at $T_{h,\text{in}} = 300°\text{C}$ and $m_h = 3 \, \text{kg}/\text{s}$, while the water enters at $T_{c,\text{in}} = 20°\text{C}$ and $\dot{m}_c = 2 \, \text{kg}/\text{s}$. The overall heat transfer coefficient is $U_m = 100 \, \text{W}/(\text{m}^2 \cdot {}°\text{C})$, and the heat transfer surface is $A = 20 \, \text{m}^2$. Calculate (*a*) the total heat transfer rate and (*b*) the outlet temperatures of water and exhaust gas.

12-40. A single-shell-pass, two-tube-pass condenser with the flow arrangement shown in Fig. 12-12 is to condense steam at $T = 80°\text{C}$ with cooling water entering at $T_{c,\text{in}} = 20°\text{C}$ and $m_c = 500 \, \text{kg}/\text{s}$. The total heat transfer rate is given by $Q = 14,000 \, \text{kW}$. The overall heat transfer coefficient is $U_m = 4000 \, \text{W}/(\text{m}^2 \cdot {}°\text{C})$. Calculate the total heat transfer surface A and the outlet temperature $T_{c,\text{out}}$ of the cooling water.

12-41. A one-shell-pass, two-tube-pass heat exchanger with the flow arrangement shown in Fig. 12-12 is to heat water entering at $T_{c,\text{in}} = 40°\text{C}$ and leaving at $T_{c,\text{out}} = 140°\text{C}$, with $m_c = 2 \, \text{kg}/\text{s}$, with pressurized hot water entering at $T_{h,\text{in}} = 300°\text{C}$ and $m_h = 2 \, \text{kg}/\text{s}$. The overall heat transfer coefficient is $U_m = 1250 \, \text{W}/(\text{m}^2 \cdot {}°\text{C})$. Calculate the heat transfer surface A required and the outlet temperature of the hot water.

12-42. A single-shell-pass and two-tube-pass heat exchanger with the flow arrangement shown in Fig. 12-12 is used to heat water entering at $t_{c,\text{in}} = 15°\text{C}$ and $m_c = 2 \, \text{kg}/\text{s}$ with ethylene glycol [$c_p = 2600 \, \text{J}/(\text{kg} \cdot {}°\text{C})$] entering at $T_{h,\text{in}} = 85°\text{C}$ and $m_h = 1 \, \text{kg}/\text{s}$. The overall heat transfer coefficient is $U_m = 500 \, \text{W}/(\text{m}^2 \cdot {}°\text{C})$, and the heat transfer surface $A = 10 \, \text{m}^2$. Calculate the rate of heat transfer Q and the outlet temperatures of the water and ethylene glycol.

12-43. A two-shell-pass, four-tube-pass heat exchanger with the flow arrangement shown in Fig. 12-13 is to cool $m_h = 3 \, \text{kg}/\text{s}$ of oil [$c_{ph} = 2100 \, \text{J}/(\text{kg} \cdot {}°\text{C})$] from $T_{h,\text{in}} = 85°\text{C}$ to $T_{h,\text{out}} = 35°\text{C}$ with water [$c_{pc} = 4180 \, \text{J}/(\text{kg} \cdot {}°\text{C})$] entering the exchanger at $T_{c,\text{in}} = 14°\text{C}$ and $m_c = 2 \, \text{kg}/\text{s}$. The overall heat transfer coefficient is $U_m = 400 \, \text{W}/(\text{m}^2 \cdot {}°\text{C})$. Calculate the total heat transfer surface required.
Answer: 31.5 m^2

12-44. A two-shell-pass, four-tube-pass heat exchanger with the flow arrangement shown in Fig. 12-13 is to be used to heat $m_c = 1.2 \, \text{kg}/\text{s}$ water from $T_{c,\text{in}} = 20°\text{C}$ to $T_{c,\text{out}} = 80°\text{C}$ by using $m_h = 2.2 \, \text{kg}/\text{s}$ oil entering at $T_{h,\text{in}} = 160°\text{C}$. The overall heat transfer coefficient is $U_m = 300 \, \text{W}/(\text{m}^2 \cdot {}°\text{C})$, and the specific heat of oil is $c_{ph} = 2100 \, \text{J}/(\text{kg} \cdot {}°\text{C})$. Determine the heat transfer surface required.
Answer: 13.9 m^2

12-45. Air at 2 atm pressure and 500 K and with a velocity $u_\infty = 20 \, \text{m}/\text{s}$ flows across a compact heat exchanger matrix that has the configuration shown in Figs. 12-14*a* and 12-15 (case A). Calculate the heat transfer coefficient and the friction factor.
Answer: 441 W/(m$^2 \cdot {}°$C)

12-46. Air at atmospheric pressure and 600 K and at a mass flow rate $m = 10 \, \text{kg}/\text{s}$ flows across the circular finned-tube matrix configuration shown in Figs. 12-14*a* and 12-15 (case B). Calculate (*a*) the heat transfer coefficient, (*b*) the friction factor, and (*c*) the ratio of the pressure drop due to core friction to the inlet pressure.

REFERENCES

1. Afgan, N. H., and E. U. Schlunder: *Heat Exchangers*: *Design and Theory*, McGraw-Hill, New York, 1974.
2. Bowman, R. A., A. C. Mueller, and W. M. Nagle: "Mean Temperature Difference in Design," *Trans. ASME* **62**:283–294 (1940).
3. Fraas, A. P., and M. N. Ozisik: "Steam Generators for High Temperature Gas-Cooled Reactors," ORNL-3208, Oak Ridge National Laboratory, Oak Ridge, Tenn., April 1963.
4. Fraas, A. P., and M. N. Ozisik: *Heat Exchanger Design*, Wiley, New York, 1964.
5. Gardner, K. A.: "Efficiency of Extended Surfaces," *Trans. ASME* **67**:621–631 (1945).
6. *Heat Exchangers*, The Patterson-Kelley Co., East Stroudsburg, Pa., 1960.
7. Hyrnisak, W.: *Heat Exchangers*, Academic, New York, 1958.
8. Kakac, S., A. E. Bergles, and F. Mayinger (eds.): *Heat Exchangers*: *Thermal-Hydraulic Fundamentals and Design*, Hemisphere, Washington, D.C., 1981.
9. Kakac, S., R. K. Shah, and A. E. Bergles (eds.): *Low Reynolds Number Flow Heat Exchangers*, Hemisphere, Washington, D.C., 1982.
10. Kays, W. M., and A. L. London: *Compact Heat Exchangers*, 2d ed., McGraw-Hill, New York, 1964.
11. Kern, D. Q.: *Process Heat Transfer*, McGraw-Hill, New York, 1950.
12. Mueller, A. C.: "Heat Exchangers," in W. M. Rohsenow and J. P. Hartnett (eds.), *Handbook of Heat Transfer*, McGraw-Hill, New York, 1973, chap. 18.
13. Schlunder, E. U.: *Heat Exchanger Design Handbook*, Hemisphere, Washington, D.C., 1982.
14. Tubular Exchanger Manufacturers Association, *Standards*, TEMA, New York, 1959.
15. Walker, G.: *Industrial Heat Exchangers*, Hemisphere, Washington, D.C., 1982.

TABLE A-1
Conversion factors

1. Acceleration
   1 ft/s^2 = 0.3048 m/s^2
   1 m/s^2 = 3.2808 ft/s^2

2. Area
   1 in^2 = 6.4516 cm^2
   1 in^2 = 6.4516 × 10^{-4} m^2
   1 ft^2 = 929 cm^2
   1 ft^2 = 0.0929 m^2
   1 m^2 = 10.764 ft^2

3. Density
   1 lb/in^3 = 27.680 g/cm^3
   1 lb/in^3 = 27.680 × 10^3 kg/m^3
   1 lb/ft^3 = 16.019 kg/m^3
   1 kg/m^3 = 0.06243 lb/ft^3
   1 slug/ft^3 = 515.38 kg/m^3
   1 lb·mol/ft^3 = 16.019 kg·mol/m^3
   1 kg·mol/m^3 = 0.06243 lb·mol/ft^3

4. Diffusivity (heat, mass, momentum)
   1 ft^2/s = 0.0929 m^2/s
   1 ft^2/h = 0.2581 cm^2/s
   1 ft^2/h = 0.2581 × 10^{-4} m^2/s
   1 m^2/s = 10.7639 ft^2/s
   1 cm^2/s = 3.8745 ft^2/h

5. Energy, heat, power
   1 J = 1 W·s = 1 N·m
   1 J = 10^7 erg
   1 Btu = 1055.04 J
   1 Btu = 1055.04 W·s
   1 Btu = 1055.04 N·m
   1 Btu = 252 cal
   1 Btu = 0.252 kcal
   1 Btu = 778.161 ft·lbf
   1 Btu/h = 0.2931 W
   1 Btu/h = 0.2931 × 10^{-3} kW
   1 Btu/h = 3.93 × 10^{-4} hp
   1 cal = 4.1868 J (or W·s or N·m)
   1 cal = 3.968 × 10^{-3} Btu
   1 kcal = 3.968 Btu
   1 hp = 550 ft·lbf/s
   1 hp = 745.7 W = 745.7 N·m/s
   1 Wh = 3.413 Btu
   1 kWh = 3413 Btu

6. Heat capacity, heat per unit mass, specific heat
   1 Btu/(h·°F) = 0.5274 W/°C
   1 W/°C = 1.8961 Btu/(h·°F)
   1 Btu/lb = 2325.9 J/kg
   1 Btu/lb = 2.3259 kJ/kg

1 Btu/(lb · °F) = 4186.69 J/(kg · °C)
1 Btu/(lb · °F) = 4.18669 kJ/(kg · °C)
  [or J/(g · °C)]
1 Btu/(lb · °F) = 1 cal/(g · °C) = 1 kcal/(kg · °C)

7. Heat flux
1 Btu/(h · ft²) = 3.1537 W/m²
1 Btu/(h · ft²) = 3.1537 × 10⁻³ kW/m²
1 W/m² = 0.31709 Btu/(h · ft²)

8. Heat generation rate
1 Btu/(h · ft³) = 10.35 W/m³
1 Btu/(h · ft³) = 8.9 kcal/(h · m³)
1 W/m³ = 0.0966 Btu/(h · ft³)

9. Heat transfer coefficient
1 Btu/(h · ft² · °F) = 5.677 W/(m² · °C)
1 Btu/(h · ft² · °F) = 5.677 × 10⁻⁴ W/(cm² · °C)
1 W/(m² · °C) = 0.1761 Btu/(h · ft² · °F)
1 Btu/(h · ft² · °F) = 4.882 kcal/(h · m² · °C)

10. Length
1 Å = 10⁻⁸ cm
1 Å = 10⁻¹⁰ m
1 μm = 10⁻³ mm
1 μm = 10⁻⁴ cm
1 μm = 10⁻⁶ m
1 in = 2.54 cm
1 in = 2.54 × 10⁻² m
1 ft = 0.3048 m
1 m = 3.2808 ft
1 mi = 1609.34 m
1 mi = 5280 ft
1 light year = 9.46 × 10¹⁵ m

11. Mass
1 oz = 28.35 g
1 lb = 16 oz
1 lb = 453.6 g
1 lb = 0.4536 kg
1 kg = 2.2046 lb
1 g = 15.432 g
1 slug = 32.1739 lb
1 t (metric) = 1000 kg
1 t (metric) = 2205 lb
1 ton (short) = 2000 lb
1 ton (long) = 2240 lb

12. Mass flux
1 lb · mol/(ft² · h)
  = 1.3563 × 10⁻³ kg · mol/(m² · s)
1 kg · mol/(m² · s) = 737.3 lb · mol/(ft² · h)
1 lb/(ft² · h) = 1.3563 × 10⁻³ kg/(m² · s)
1 lb/(ft² · s) = 4.882 kg/(m² · s)
1 kg/(m² · s) = 737.3 lb/(ft² · h)
1 kg/(m² · s) = 0.2048 lb/(ft² · s)

13. Pressure, force
1 N = 1 kg · m/s²
1 N = 0.22481 lbf
1 N = 7.2333 pdl
1 N = 10⁵ dyn
1 lbf = 32.174 ft · lb/s²
1 lbf = 4.4482 N
1 lbf = 4.4482 kg · m/s²
1 lbf = 32.1739 pdl
1 lbf/in² ≡ 1 psi = 6894.76 N/m²
1 lbf/ft² = 47.880 N/m²
1 bar = 10⁵ N/m² = 10⁵ Pa
1 atm = 14.696 lbf/in²
1 atm = 2116.2 lbf/ft²
1 atm = 1.0132 × 10⁵ N/m²
1 atm = 1.0132 bar
1 Pa = 1 N/m²

14. Specific heat
1 Btu/(lb · °F) = 1 kcal /(kg · °C) = 1 cal/(g · °C)
1 Btu/(lb · °F) = 4186.69 J/(kg · °C)
  [or W · s/(kg · °C)]
1 Btu/(lb · °F) = 4.18669 J/(g · K)
  [or W · s/(g · °C)]
1 J/(g · °C) = 0.23885 Btu/(lb · °F)
  [cal/(g · °C) or kcal/(kg · °C)]

15. Speed
1 ft/s = 0.3048 m/s
1 m/s = 3.2808 ft/s
1 mi/h = 1.4667 ft/s
1 mi/h = 0.44704 m/s

16. Surface tension
1 lbf/ft = 14.5937 N/m
1 N/m = 0.068529 lbf/ft

17. Temperature
1 K = 1.8°R
$T(°F) = 1.8(K - 273) + 32$

$T(K) = \dfrac{1}{1.8}(°F - 32) + 273$

$T(°C) = \dfrac{1}{1.8}(°R - 492)$

$\Delta T(°C) = \Delta T(°F)/1.8$

18. Thermal conductivity
1 Btu/(h · ft · °F) = 1.7303 W/(m · °C)
1 Btu/(h · ft · °F) = 1.7303 × 10⁻² W/(cm · °C)
1 Btu/(h · ft · °F) = 0.4132 cal/(s · m · °C)
1 W/(m · °C) = 0.5779 Btu/(h · ft · °F)
1 W/(cm · °C) = 57.79 Btu/(h · ft · °F)

TABLE A-1 (*continued*)

<table>
<tr><td>

19. Thermal resistance
$1\ \mathrm{h\cdot°F/Btu} = 1.896°\mathrm{C/W}$
$1°\mathrm{C/W} = 0.528\ \mathrm{h\cdot°F/Btu}$

20. Viscosity
$1\ \mathrm{P} = 1\ \mathrm{g/(cm\cdot s)}$
$1\ \mathrm{P} = 10^2\ \mathrm{cP}$
$1\ \mathrm{P} = 241.9\ \mathrm{lb/(ft\cdot h)}$
$1\ \mathrm{cP} = 2.419\ \mathrm{lb/(ft\cdot h)}$
$1\ \mathrm{lb/(ft\cdot s)} = 1.4882\ \mathrm{kg/(m\cdot s)}$
$1\ \mathrm{lb/(ft\cdot s)} = 14.882\ \mathrm{P}$
$1\ \mathrm{lb/(ft\cdot s)} = 1488.2\ \mathrm{cP}$
$1\ \mathrm{lb/(ft\cdot h)} = 0.4134 \times 10^{-3}\ \mathrm{kg/(m\cdot s)}$

</td><td>

$1\ \mathrm{lb/(ft\cdot h)} = 0.4134 \times 10^{-2}\ \mathrm{P}$
$1\ \mathrm{lb/(ft\cdot h)} = 0.4134\ \mathrm{cP}$

21. Volume
$1\ \mathrm{in}^3 = 16.387\ \mathrm{cm}^3$
$1\ \mathrm{cm}^3 = 0.06102\ \mathrm{in}^3$
$1\ \mathrm{oz\ (U.S.\ fluid)} = 29.573\ \mathrm{cm}^3$
$1\ \mathrm{ft}^3 = 0.0283168\ \mathrm{m}^3$
$1\ \mathrm{ft}^3 = 28.3168\ \mathrm{liters}$
$1\ \mathrm{ft}^3 = 7.4805\ \mathrm{gal\ (U.S.)}$
$1\ \mathrm{m}^3 = 35.315\ \mathrm{ft}^3$
$1\ \mathrm{gal\ (U.S.)} = 3.7854\ \mathrm{liters}$
$1\ \mathrm{gal\ (U.S.)} = 3.7854 \times 10^{-3}\ \mathrm{m}^3$
$1\ \mathrm{gal\ (U.S.)} = 0.13368\ \mathrm{ft}^3$

</td></tr>
</table>

Constants

g_c = gravitational acceleration conversion factor	$= 32.1739\ \mathrm{ft\cdot lb/(lbf\cdot s^2)}$
	$= 4.1697 \times 10^8\ \mathrm{ft\cdot lb/(lbf\cdot h^2)}$
	$= 1\ \mathrm{g\cdot cm/(dyn\cdot s^2)}$
	$= 1\ \mathrm{kg\cdot m/(N\cdot s^2)}$
	$= 1\ \mathrm{lb\cdot ft/(pdl\cdot s^2)}$
	$= 1\ \mathrm{slug\cdot ft/(lbf\cdot s^2)}$
J = mechanical equivalent of heat	$= 778.16\ \mathrm{ft\cdot lbf/Btu}$
$\mathscr{R}$ = gas constant	$= 1544\ \mathrm{ft\cdot lbf/(lb\cdot mol\cdot °R)}$
	$= 0.730\ \mathrm{ft^3\cdot atm/(lb\cdot mol\cdot °R)}$
	$= 0.08205\ \mathrm{m^3\cdot atm/(kg\cdot mol\cdot K)}$
	$= 8.314\ \mathrm{J/(g\cdot mol\cdot K)}$
	$= 8.314\ \mathrm{N\cdot m/(g\cdot mol\cdot K)}$
	$= 8314\ \mathrm{N\cdot m/(kg\cdot mol\cdot K)}$
	$= 1.987\ \mathrm{cal/(g\cdot mol\cdot K)}$
σ = Stefan-Boltzmann constant	$= 0.1714 \times 10^{-8}\ \mathrm{Btu/(h\cdot ft^2\cdot °R^4)}$
	$= 5.6697 \times 10^{-8}\ \mathrm{W/(m^2\cdot K^4)}$

APPENDIX
B

PHYSICAL
PROPERTIES

TABLE B-1
Physical properties of gases at atmospheric pressure

T, K	ρ, kg/m^3	c_p, kJ/(kg · °C)	μ, kg/(m · s)	ν, m^2/s $\times 10^6$	k, W/(m · K)	α, m^2/s $\times 10^4$	Pr
Air							
100	3.6010	1.0266	0.6924×10^{-5}	1.923	0.009246	0.02501	0.770
150	2.3675	1.0099	1.0283	4.343	0.013735	0.05745	0.753
200	1.7684	1.0061	1.3289	7.490	0.01809	0.10165	0.739
250	1.4128	1.0053	1.488	10.53	0.02227	0.13161	0.722
300	1.1774	1.0057	1.983	16.84	0.02624	0.22160	0.708
350	0.9980	1.0090	2.075	20.76	0.03003	0.2983	0.697
400	0.8826	1.0140	2.286	25.90	0.03365	0.3760	0.689
450	0.7833	1.0207	2.484	31.71	0.03707	0.4222	0.683
500	0.7048	1.0295	2.671	37.90	0.04038	0.5564	0.680
550	0.6423	1.0392	2.848	44.34	0.04360	0.6532	0.680
600	0.5879	1.0551	3.018	51.34	0.04659	0.7512	0.680
650	0.5430	1.0635	3.177	58.51	0.04953	0.8578	0.682
700	0.5030	1.0752	3.332	66.25	0.05230	0.9672	0.684
750	0.4709	1.0856	3.481	73.91	0.05509	1.0774	0.686
800	0.4405	1.0978	3.625	82.29	0.05779	1.1951	0.689
850	0.4149	1.1095	3.765	90.75	0.06028	1.3097	0.692
900	0.3925	1.1212	3.899	99.3	0.06279	1.4271	0.696
950	0.3716	1.1321	4.023	108.2	0.06525	1.5510	0.699
1000	0.3524	1.1417	4.152	117.8	0.06752	1.6779	0.702
1100	0.3204	1.160	4.44	138.6	0.0732	1.969	0.704

TABLE B-1 (*continued*)

T, K	ρ, kg/m^3	c_p, kJ/(kg·°C)	μ, kg/(m·s)	ν, m^2/s × 10^6	k, W/(m·K)	α, m^2/s × 10^4	Pr
Air							
1200	0.2947	1.179	4.69	159.1	0.0782	2.251	0.707
1300	0.2707	1.197	4.93	182.1	0.0837	2.583	0.705
1400	0.2515	1.214	5.17	205.5	0.0891	2.920	0.705
1500	0.2355	1.230	5.40	229.1	0.0946	3.262	0.705
1600	0.2211	1.248	5.63	254.5	0.100	3.609	0.705
1700	0.2082	1.267	5.85	280.5	0.105	3.977	0.705
1800	0.1970	1.287	6.07	308.1	0.111	4.379	0.704
1900	0.1858	1.309	6.29	338.5	0.117	4.811	0.704
2000	0.1762	1.338	6.50	369.0	0.124	5.260	0.702
2100	0.1682	1.372	6.72	399.6	0.131	5.715	0.700
2200	0.1602	1.419	6.93	432.6	0.139	6.120	0.707
2300	0.1538	1.482	7.14	464.0	0.149	6.540	0.710
2400	0.1458	1.574	7.35	504.0	0.161	7.020	0.718
2500	0.1394	1.688	7.57	543.5	0.175	7.441	0.730
Helium							
33	1.4657	5.200	50.2	3.42	0.0353	0.04625	0.74
144	3.3799	5.200	125.5	37.11	0.0928	0.5275	0.70
200	0.2435	5.200	156.6	64.38	0.1177	0.9288	0.694
255	0.1906	5.200	181.7	95.50	0.1357	1.3675	0.70
366	0.13280	5.200	230.5	173.6	0.1691	2.449	0.71
477	0.10204	5.200	275.0	269.3	0.197	3.716	0.72
589	0.08282	5.200	311.3	375.8	0.225	5.215	0.72
700	0.07032	5.200	347.5	494.2	0.251	6.661	0.72
800	0.06023	5.200	381.7	634.1	0.275	8.774	0.72
900	0.05286	5.200	413.6	781.3	0.298	10.834	0.72
Carbon dioxide							
220	2.4733	0.783	11.105 × 10^{-6}	4.490	0.010805	0.05920	0.818
250	2.1657	0.804	12.590	5.813	0.012884	0.07401	0.793
300	1.7973	0.871	14.958	8.321	0.016572	0.10588	0.770
350	1.5362	0.900	17.205	11.19	0.02047	0.14808	0.755
400	1.3424	0.942	19.32	14.39	0.02461	0.19463	0.738
450	1.1918	0.980	21.34	17.90	0.02897	0.24813	0.721
500	1.0732	1.013	23.26	21.67	0.03352	0.3084	0.702
550	0.9739	1.047	25.08	25.74	0.03821	0.3750	0.685
600	0.8938	1.076	26.83	30.02	0.04311	0.4483	0.668
Carbon monoxide							
220	1.55363	1.0429	13.832 × 10^{-6}	8.903	0.01906	0.11760	0.758
250	1.3649	1.0425	15.40	11.28	0.02144	0.15063	0.750
300	1.13876	1.0421	17.843	15.67	0.02525	0.21280	0.737
350	0.97425	1.0434	20.09	20.62	0.02883	0.2836	0.728
400	0.85363	1.0484	22.19	25.99	0.03226	0.3605	0.722
450	0.75848	1.0551	24.18	31.88	0.0436	0.4439	0.718
500	0.68223	1.0635	26.06	38.19	0.03863	0.5324	0.718

TABLE B-1 (*continued*)

T, K	ρ, kg/m^3	c_p, kJ/(kg·°C)	μ, kg/(m·s)	ν, m^2/s × 10^6	k, W/(m·K)	α, m^2/s × 10^4	Pr
Carbon monoxide							
550	0.62024	1.0756	27.89	44.97	0.04162	0.6240	0.721
600	0.56850	1.0877	29.60	52.06	0.04446	0.7190	0.724
Ammonia, NH$_3$							
220	0.9304	2.198	7.255 × 10^{-6}	7.8	0.0171	0.2054	0.93
273	0.7929	2.177	9.353	11.8	0.0220	0.1308	0.90
323	0.6487	2.177	11.035	17.0	0.0270	0.1920	0.88
373	0.5590	2.236	12.886	23.0	0.0327	0.2619	0.87
423	0.4934	2.315	14.672	29.7	0.0391	0.3432	0.87
473	0.4405	2.395	16.49	37.4	0.0467	0.4421	0.84
Steam (H$_2$O vapor)							
380	0.5863	2.060	12.71 × 10^{-6}	21.6	0.0246	0.2036	1.060
400	0.5542	2.014	13.44	24.2	0.0261	0.2338	1.040
450	0.4902	1.980	15.25	31.1	0.0299	0.307	1.010
500	0.4405	1.985	17.04	38.6	0.0339	0.387	0.996
550	0.4005	1.997	18.84	47.0	0.0379	0.475	0.991
600	0.3652	2.026	20.67	56.6	0.0422	0.573	0.986
650	0.3380	2.056	22.47	64.4	0.0464	0.666	0.995
700	0.3140	2.085	24.26	77.2	0.0505	0.772	1.000
750	0.2931	2.119	26.04	88.8	0.0549	0.883	1.005
800	0.2739	2.152	27.86	102.0	0.0592	1.001	1.010
850	0.2579	2.186	29.69	115.2	0.0637	1.130	1.019
Hydrogen							
30	0.84722	10.840	1.606 × 10^{-6}	1.895	0.0228	0.02493	0.759
50	0.50955	10.501	2.516	4.880	0.0362	0.0676	0.721
100	0.24572	11.229	4.212	17.14	0.0665	0.2408	0.712
150	0.16371	12.602	5.595	34.18	0.0981	0.475	0.718
200	0.12270	13.540	6.813	55.53	0.1282	0.772	0.719
250	0.09819	14.059	7.919	80.64	0.1561	1.130	0.713
300	0.08185	14.314	8.963	109.5	0.182	1.554	0.706
350	0.07016	14.436	9.954	141.9	0.206	2.031	0.697
400	0.06135	14.491	10.864	177.1	0.228	2.568	0.690
450	0.05462	14.499	11.779	215.6	0.251	3.164	0.682
500	0.04918	14.507	12.636	257.0	0.272	3.817	0.675
550	0.04469	14.532	13.475	301.6	0.292	4.516	0.668
600	0.04085	14.537	14.285	349.7	0.315	5.306	0.664
700	0.03492	14.574	15.89	455.1	0.351	6.903	0.659
800	0.03060	14.675	17.40	569	0.384	8.563	0.664
900	0.02723	14.821	18.78	690	0.412	10.217	0.676
1000	0.02451	14.968	20.16	822	0.440	11.997	0.686
1100	0.02227	15.165	21.46	965	0.464	13.726	0.703
1200	0.02050	15.366	22.75	1107	0.488	15.484	0.715
1300	0.01890	15.575	24.08	1273	0.512	17.394	0.733
1333	0.01842	15.638	24.44	1328	0.519	18.013	0.736

TABLE B-1 (*continued*)

T, K	ρ, kg/m^3	c_p, kJ/(kg·°C)	μ, kg/(m·s)	ν, m^2/s × 10^6	k, W/(m·K)	α, m^2/s × 10^4	Pr
Oxygen
100	3.9918	0.9479	7.768 × 10^{-6}	1.946	0.00903	0.023876	0.815
150	2.6190	0.9178	11.490	4.387	0.01367	0.05688	0.773
200	1.9559	0.9131	14.850	7.593	0.01824	0.10214	0.745
250	1.5618	0.9157	17.87	11.45	0.02259	0.15794	0.725
300	1.3007	0.9203	20.63	15.86	0.02676	0.22353	0.709
350	1.1133	0.9291	23.16	20.80	0.03070	0.2968	0.702
400	0.9755	0.9420	25.54	26.18	0.03461	0.3768	0.695
450	0.8682	0.9567	27.77	31.99	0.03828	0.4609	0.694
500	0.7801	0.9722	29.91	38.34	0.04173	0.5502	0.697
550	0.7096	0.9881	31.97	45.05	0.04517	0.6441	0.700
600	0.6504	1.0044	33.92	52.15	0.04832	0.7399	0.704
Nitrogen
100	3.4808	1.0722	6.862 × 10^{-6}	1.971	0.009450	0.025319	0.786
200	1.7108	1.0429	12.947	7.568	0.01824	0.10224	0.747
300	1.1421	1.0408	17.84	15.63	0.02620	0.22044	0.713
400	0.8538	1.0459	21.98	25.74	0.03335	0.3734	0.691
500	0.6824	1.0555	25.70	37.66	0.03984	0.5530	0.684
600	0.5687	1.0756	29.11	51.19	0.04580	0.7486	0.686
700	0.4934	1.0969	32.13	65.13	0.05123	0.9466	0.691
800	0.4277	1.1225	34.84	81.46	0.05609	1.1685	0.700
900	0.3796	1.1464	37.49	91.06	0.06070	1.3946	0.711
1000	0.3412	1.1677	40.00	117.2	0.06475	1.6250	0.724
1100	0.3108	1.1857	42.28	136.0	0.06850	1.8591	0.736
1200	0.2851	1.2037	44.50	156.1	0.07184	2.0932	0.748

From E. R. G. Eckert and R. M. Drake, *Analysis of Heat Mass Transfer*, McGraw-Hill, New York, 1972.

TABLE B-2
Physical properties of saturated liquids

T, °C	ρ, kg/m^3	c_p, kJ/(kg·°C)	ν, m^2/s	k, W/(m·K)	α, m^2/s × 10^7	Pr	β, K^{-1}
Ammonia, NH$_3$							
−50	703.69	4.463	0.435 × 10^{-6}	0.547	1.742	2.60	
−40	691.68	4.467	0.406	0.547	1.775	2.28	
−30	679.34	4.476	0.387	0.549	1.801	2.15	
−20	666.69	4.509	0.381	0.547	1.819	2.09	
−10	653.55	4.564	0.378	0.543	1.825	2.07	
0	640.10	4.635	0.373	0.540	1.819	2.05	
10	626.16	4.714	0.368	0.531	1.801	2.04	
20	611.75	4.798	0.359	0.521	1.775	2.02	2.45 × 10^{-3}
30	596.37	4.890	0.349	0.507	1.742	2.01	
40	580.99	4.999	0.340	0.493	1.701	2.00	
50	564.33	5.116	0.330	0.476	1.654	1.99	
Carbon dioxide, CO$_2$							
−50	1,156.34	1.84	0.119 × 10^{-6}	0.0855	0.4021	2.96	
−40	1,117.77	1.88	0.118	0.1011	0.4810	2.46	
−30	1,076.76	1.97	0.117	0.1116	0.5272	2.22	
−20	1,032.39	2.05	0.115	0.1151	0.5445	2.12	
−10	983.38	2.18	0.113	0.1099	0.5133	2.20	
0	926.99	2.47	0.108	0.1045	0.4578	2.38	
10	860.03	3.14	0.101	0.0971	0.3608	2.80	
20	772.57	5.0	0.091	0.0872	0.2219	4.10	14.00 × 10^{-3}
30	597.81	36.4	0.080	0.0703	0.0279	28.7	
Dichlorodifluoromethane (Freon-12), CCl$_2$F$_2$							
−50	1,546.75	0.8750	0.310 × 10^{-6}	0.067	0.501	6.2	2.63 × 10^{-3}
−40	1,518.71	0.8847	0.279	0.069	0.514	5.4	
−30	1,489.56	0.8956	0.253	0.069	0.526	4.8	
−20	1,460.57	0.9073	0.235	0.071	0.539	4.4	
−10	1,429.49	0.9203	0.221	0.073	0.550	4.0	
0	1,397.45	0.9345	0.214	0.073	0.557	3.8	
10	1,364.30	0.9496	0.203	0.073	0.560	3.6	
20	1,330.18	0.9659	0.198	0.073	0.560	3.5	
30	1,295.10	0.9835	0.194	0.071	0.560	3.5	
40	1,257.13	1.0019	0.191	0.069	0.555	3.5	
50	1,215.96	1.0216	0.190	0.067	0.545	3.5	
Engine oil (unused)							
0	899.12	1.796	0.00428	0.147	0.911	47,100	
20	888.23	1.880	0.00090	0.145	0.872	10,400	0.70 × 10^{-3}
40	876.05	1.964	0.00024	0.144	0.834	2,870	
60	864.04	2.047	0.839 × 10^{-4}	0.140	0.800	1,050	
80	852.02	2.131	0.375	0.138	0.769	490	

TABLE B-2 (*continued*)

T, °C	$\rho,$ kg/m^3	$c_p,$ kJ/(kg·°C)	$\nu,$ m^2/s	$k,$ W/(m·K)	$\alpha,$ m^2/s × 10^7	Pr	β, K^{-1}
Engine oil (unused)							
100	840.01	2.219	0.203	0.137	0.738	276	
120	828.96	2.307	0.124	0.135	0.710	175	
140	816.94	2.395	0.080	0.133	0.686	116	
160	805.89	2.483	0.056	0.132	0.663	84	
Ethylene glycol, $C_2H_4(OH_2)$							
0	1,130.75	2.294	57.53 × 10^{-6}	0.242	0.934	615	
20	1,116.65	2.382	19.18	0.249	0.939	204	0.65 × 10^{-3}
40	1,101.43	2.474	8.69	0.256	0.939	93	
60	1,087.66	2.562	4.75	0.260	0.932	51	
80	1,077.56	2.650	2.98	0.261	0.921	32.4	
100	1,058.50	2.742	2.03	0.263	0.908	22.4	
Eutectic calcium chloride solution, 29.9 % $CaCl_2$							
−50	1,319.76	2.608	36.35 × 10^{-6}	0.402	1.166	312	
−40	1,314.96	2.6356	24.97	0.415	1.200	208	
−30	1,310.15	2.6611	17.18	0.429	1.234	139	
−20	1,305.51	2.688	11.04	0.445	1.267	87.1	
−10	1,300.70	2.713	6.96	0.459	1.300	53.6	
0	1,296.06	2.738	4.39	0.472	1.332	33.0	
10	1,291.41	2.763	3.35	0.485	1.363	24.6	
20	1,286.61	2.788	2.72	0.498	1.394	19.6	
30	1,281.96	2.814	2.27	0.511	1.419	16.0	
40	1,277.16	2.839	1.92	0.523	1.445	13.3	
50	1,272.51	2.868	1.65	0.535	1.468	11.3	
Glycerin, $C_3H_5(OH)_3$							
0	1,276.03	2.261	0.00831	0.282	0.983	84.7 × 10^3	
10	1,270.11	2.319	0.00300	0.284	0.965	31.0	
20	1,264.02	2.386	0.00118	0.286	0.947	12.5	0.50 × 10^{-3}
30	1,258.09	2.445	0.00050	0.286	0.929	5.38	
40	1,252.01	2.512	0.00022	0.286	0.914	2.45	
50	1,244.96	2.583	0.00015	0.287	0.893	1.63	
Mercury, Hg							
0	13,628.22	0.1403	0.124 × 10^{-6}	8.20	42.99	0.0288	
20	13,579.04	0.1394	0.114	8.69	46.06	0.0249	1.82 × 10^{-4}
50	13,505.84	0.1386	0.104	9.40	50.22	0.0207	
100	13,384.58	0.1373	0.0928	10.51	57.16	0.0162	
150	13,264.28	0.1365	0.0853	11.49	63.54	0.0134	

TABLE B-2 (*continued*)

T, °C	ρ, kg/m^3	c_p, kJ/(kg·°C)	ν, m^2/s	k, W/(m·K)	α, m^2/s × 10^7	Pr	β, K^{-1}
Air							
200	13,144.94	0.1360	0.0802	12.34	69.08	0.0116	
250	13,025.60	0.1357	0.0765	13.07	74.06	0.0103	
315.5	12,847	0.134	0.0673	14.02	8.15	0.0083	
Methyl chloride, CH$_3$Cl							
−50	1,052.58	1.4759	0.320 × 10^{-6}	0.215	1.388	2.31	
−40	1,033.35	1.4826	0.318	0.209	1.368	2.32	
−30	1,016.53	1.4922	0.314	0.202	1.337	2.35	
−20	999.39	1.5043	0.309	0.196	1.301	2.38	
−10	981.45	1.5194	0.306	0.187	1.257	2.43	
0	962.39	1.5378	0.302	0.178	1.213	2.49	
10	942.36	1.5600	0.297	0.171	1.166	2.55	
20	923.31	1.5860	0.293	0.163	1.112	2.63	
30	903.12	1.6161	0.288	0.154	1.058	2.72	
40	883.10	1.6504	0.281	0.144	0.996	2.83	
50	861.15	1.6890	0.274	0.133	0.921	2.97	
Sulfur dioxide, SO$_2$							
−50	1,560.84	1.3595	0.484 × 10^{-6}	0.242	1.141	4.24	
−40	1,536.81	1.3607	0.424	0.235	1.130	3.74	
−30	1,520.64	1.3616	0.371	0.230	1.117	3.31	
−20	1,488.60	1.3624	0.324	0.225	1.107	2.93	
−10	1,463.61	1.3628	0.288	0.218	1.097	2.62	
0	1,438.46	1.3636	0.257	0.211	1.081	2.38	
10	1,412.51	1.3645	0.232	0.204	1.066	2.18	
20	1,386.40	1.3653	0.210	0.199	1.050	2.00	1.94 × 10^{-3}
30	1,359.33	1.3662	0.190	0.192	1.035	1.83	
40	1,329.22	1.3674	0.173	0.185	1.019	1.70	
50	1,299.10	1.3683	0.162	0.177	0.999	1.61	
Water, H$_2$O							
0	1,002.28	4.2178	1.788 × 10^{-6}	0.552	1.308	13.6	
20	1,000.52	4.1818	1.006	0.597	1.430	7.02	0.18 × 10^{-3}
40	994.59	4.1784	0.658	0.628	1.512	4.34	
60	985.46	4.1843	0.478	0.651	1.554	3.02	
80	974.08	4.1964	0.364	0.668	1.636	2.22	
100	960.63	4.2161	0.294	0.680	1.680	1.74	
120	945.25	4.250	0.247	0.685	1.708	1.446	
140	928.27	4.283	0.214	0.684	1.724	1.241	
160	909.69	4.342	0.190	0.680	1.729	1.099	
180	889.03	4.417	0.173	0.675	1.724	1.004	

TABLE B-2 (*continued*)

T, °C	ρ, kg/m^3	c_p, kJ/(kg · °C)	ν, m^2/s	k, W/(m · K)	α, m^2/s $\times 10^7$	Pr	β, K^{-1}
Water, H_2O							
200	866.76	4.505	0.160	0.665	1.706	0.937	
220	842.41	4.610	0.150	0.652	1.680	0.891	
240	815.66	4.756	0.143	0.635	1.639	0.871	
260	785.87	4.949	0.137	0.611	1.577	0.874	
280.6	752.55	5.208	0.135	0.580	1.481	0.910	
300	714.26	5.728	0.135	0.540	1.324	1.019	

From E. R. G. Eckert and R. M. Drake, *Analysis of Heat Mass Transfer*, McGraw-Hill, New York, 1972.

TABLE B-3
Physical properties of liquid metals

Metal	Melting point, °C	Boiling point, °C	T, °C	ρ, kg/m³	c_p, kJ/(kg·°C)	$\mu \times 10^4$, kg/(m·s)	$\nu \times 10^6$, m²/s	k, W/(m·°C)	$\alpha \times 10^6$, m²/s	Pr
Bismuth	271	1477	315	10,011	0.144	16.2	0.160	16.4	11.25	0.0142
			538	9,739	0.155	11.0	0.113	15.6	10.34	0.0110
			760	9,467	0.165	7.9	0.083	15.6	9.98	0.0083
Lead	327	1737	371	10,540	0.159	2.40	0.023	16.1	9.61	0.024
			704	10,140	0.155	1.37	0.014	14.9	9.48	0.0143
Lithium	179	1317	204.4	509.2	4.365	5.416	1.1098	46.37	20.96	0.051
			315.6	498.8	4.270	4.465	0.8982	43.08	20.32	0.0443
			426.7	489.1	4.211	3.927	0.8053	38.24	18.65	0.0432
			537.8	476.3	4.171	3.473	0.7304	30.45	15.40	0.0476
Mercury	−38.9	357	−17.8	13,707.1	0.1415	18.334	0.1342	9.76	5.038	0.0266
			100	13,384.5	0.1373	12.420	0.0928	10.51	5.716	0.0162
			200	13,144.9	0.1570	10.541	0.0802	12.34	6.908	0.0116
Sodium	97.8	883	93.3	931.6	1.384	7.131	0.7689	84.96	56.29	0.0116
			204.4	907.5	1.339	4.521	0.5010	80.81	66.80	0.0075
			315.6	878.5	1.304	3.294	0.3766	75.78	66.47	0.00567
			426.7	852.8	1.277	2.522	0.2968	69.39	64.05	0.00464
			537.8	823.8	1.264	2.315	0.2821	64.37	62.09	0.00455
			648.9	790.0	1.261	1.964	0.2496	60.56	61.10	0.00408
			760.0	767.5	1.270	1.716	0.2245	56.58	58.34	0.00385
Potassium	63.9	760	426.7	741.7	0.766	2.108	0.2839	39.45	69.74	0.0041
			537.8	714.4	0.762	1.711	0.2400	36.51	67.39	0.0036
			648.9	690.3	0.766	1.463	0.2116	33.74	64.10	0.0033
			760.0	667.7	0.783	1.331	0.1987	31.15	59.86	0.0033
NaK (56% Na, 44% K)	−11.1	784	93.3	889.8	1.130	5.622	0.6347	25.78	27.76	0.0246
			204.4	865.6	1.089	3.803	0.4414	26.47	28.23	0.0155
			315.6	838.3	1.068	2.935	0.3515	27.17	30.50	0.0115
			426.7	814.2	1.051	2.150	0.2652	27.68	32.52	0.0081
			537.8	788.4	1.047	2.026	0.2581	27.68	33.71	0.0076
			648.9	759.5	1.051	1.695	0.2240	27.68	34.86	0.0064

TABLE B-4
Physical properties of metals

Metal	Melting point, °C	ρ, kg/m³	c_p, kJ/(kg·°C)	k, W/(m·°C)	α, m²/s × 10⁵	−100°C	0°C	100°C	200°C	300°C	400°C	600°C	800°C	1000°C
				Properties at 20°C					Thermal conductivity k, W/(m·°C)					
Aluminum														
Pure	660	2,707	0.896	204	8.418	215	202	206	215	228	249			
Al-Cu (Duralumin), 94–96% Al, 3–5% Cu, trace Mg		2,787	0.883	164	6.676	126	159	182	194					
Al-Si (Silumin, copper-bearing), 86.5% Al, 1% Cu		2,659	0.867	137	5.933	119	137	144	152	161				
Al-Si (Alusil), 78–80% Al, 20–22% Si		2,627	0.854	161	7.172	144	157	168	175	178				
Al-Mg-Si, 97% Al, 1% Mg, 1% Si, 1% Mn		2,707	0.892	177	7.311		175	189	204					
Beryllium	1277	1,850	1.825	200	5.92									
Bismuth	272	9,780	0.122	7.86	0.66									
Cadmium	321	8,650	0.231	96.8	4.84									
Copper														
Pure	1085	8,954	0.3831	386	11.234	407	386	379	374	369	363	353		
Aluminum bronze, 95% Cu, 5% Al		8,666	0.410	83	2.330									
Bronze, 75% Cu, 25% Sn		8,666	0.343	26	0.859									
Red brass, 85% Cu, 9% Sn, 6% Zn		8,714	0.385	61	1.804		59	71						
Brass, 70% Cu, 30% Zn		8,522	0.385	111	3.412	88	128	144	147	147				

TABLE B-4 (*continued*)

Metal	Melting point, °C	Properties at 20°C				Thermal conductivity k, W/(m·°C)								
		ρ, kg/m³	c_p, kJ/(kg·°C)	k, W/(m·°C)	α, m²/s × 10⁵	−100°C	0°C	100°C	200°C	300°C	400°C	600°C	800°C	1000°C
German silver, 62% Cu, 15% Ni, 22% Zn		8,618	0.394	24.9	0.733	19.2		31	40	45	48			
Constantan, 60% Cu, 40% Ni		8,922	0.410	22.7	0.612	21		22.2	26					
Iron														
Pure	1537	7,897	0.452	73	2.034	87	73	67	62	55	48	40	36	35
Wrought iron, 0.5% C		7,849	0.46	59	1.626		59	57	52	48	45	36	33	33
Steel (C max ≈ 1.5%):														
Carbon steel														
C ≈ 0.5%		7,833	0.465	54	1.474		55	52	48	45	42	35	31	29
1.0%		7,801	0.473	43	1.172		43	43	42	40	36	33	29	28
1.5%		7,753	0.486	36	0.970		36	36	36	35	33	31	28	28
Nickel steel														
Ni ≈ 0%		7,897	0.452	73	2.026									
20%		7,933	0.46	19	0.526									
40%		8,169	0.46	10	0.279									
80%		8,618	0.46	35	0.872									
Invar 36% Ni		8,137	0.46	10.7	0.286									
Chrome steel														
Cr = 0%		7,897	0.452	73	2.026	87	73	67	62	55	48	40	36	35
1%		7,865	0.46	61	1.665		62	55	52	47	42	36	33	33
5%		7,833	0.46	40	1.110		40	38	36	36	33	29	29	29
20%		7,689	0.46	22	0.635		22	22	22	22	24	24	26	29
Cr-Ni (chrome-nickel): 15% Cr, 10% Ni		7,865	0.46	19	0.527									

18% Cr, 8% Ni (V2A)		7,817	0.46	16.3	0.444		16.3	17	17	19	19	22	27	31
20% Cr, 15% Ni		7,833	0.46	15.1	0.415									
25% Cr, 20% Ni		7,865	0.46	12.8	0.361									
Tungsten steel W = 0%		7,897	0.452	73	2.026									
1%		7,913	0.448	66	1.858									
5%		8,073	0.435	54	1.525									
10%		8,314	0.419	48	1.391									
Lead	328	11,373	0.130	35	2.343	36.9	35.1	33.4	31.5	29.8				
Magnesium Pure	650	1,746	1.013	171	9.708	178	171	168	163	157				
Mg-Al (electrolytic) 6–8% Al, 1–2% Zn		1,810	1.00	66	3.605		52	62	74	83				
Molybdenum	2,621	10,220	0.251	123	4.790	138	125	118	114	111	109	106	102	99
Nickel Pure (99.9%)	1,455	8,906	0.4459	90	2.266	104	93	83	73	64	59			
Ni-Cr 90% Ni, 10% Cr		8,666	0.444	17	0.444		17.1	18.9	20.9	22.8	24.6			
80% Ni, 20% Cr		8,314	0.444	12.6	0.343		12.3	13.8	15.6	17.1	18.0			
Silver: Purest	962	10,524	0.2340	419	17.004	419	417	415	412					
Pure (99.9%)		10,525	0.2340	407	16.563	419	410	415	374	362	360			
Tin, pure	232	7,304	0.2265	64	3.884	74	65.9	59	57					
Tungsten	3,387	19,350	0.1344	163	6.271		166	151	142	133	126	112	76	
Uranium	1,133	19,070	0.116	27.6	1.25									
Zinc, pure	420	7,144	0.3843	112.2	4.106	114	112	109	106	100	93			

From E. R. G. Eckert and R. M. Drake, *Analysis of Heat Mass Transfer*, McGraw-Hill, New York, 1972.

TABLE B-5
Physical properties of insulating materials

Material	$T,$ °C	$k,$ W/(m·°C)	$\rho,$ kg/m³	$c_p,$ kJ/(kg·°C)	$\alpha,$ m²/s × 10⁷
Asbestos					
Loosely packed	−45	0.149			
	0	0.154	470–570	0.816	3.3–4
	100	0.161			
Asbestos-cement boards	20	0.74			
Sheets	51	0.166			
Felt, 40 laminations/in	38	0.057			
	150	0.069			
	260	0.083			
20 laminations/in	38	0.078			
	150	0.095			
	260	0.112			
Corrugated, 4 plies/in	38	0.087			
	93	0.100			
	150	0.119			
Balsam wool	32	0.04	35		
Board and slab					
Cellular glass	30	0.058	145	1.000	
Glass fiber, organic bonded	30	0.036	105	0.795	
Polystyrene, expanded extruded (R-12)	30	0.027	55	1.210	
Mineral fiberboard; roofing material	30	0.049	265		
Wood, shredded/cemented	30	0.087	350	1.590	
Cardboard, corrugated	—	0.064			
Celotex	32	0.048			
Corkboard	30	0.043	160		
Cork, regranulated	32	0.045	45–120	1.88	2–5.3
Ground	32	0.043	150		
Diatomaceous earth (Sil-o-cel)	0	0.061	320		
Felt, hair	30	0.036	130–200		
Wool	30	0.052	330		
Fiber, insulating board	20	0.048	240		
Glass wool	23	0.038	24	0.7	22.6
Insulex, dry	32	0.064			
Kapok	30	0.035			
Loose fill					
Cork, granulated	30	0.045	160	—	
Diatomaceous silica, coarse powder	30	0.069	350	—	
Diatomaceous silica, fine powder	30	0.091	400	—	
	30	0.052	200	—	
	30	0.061	275	—	
Glass fiber, poured or blown	30	0.043	16	0.835	
Vermiculite, flakes	30	0.068	80	0.835	
		0.063	160		

TABLE B-5 (*continued*)

Material	T, °C	k, W/(m·°C)	ρ, kg/m^3	c_p, kJ/(kg·°C)	α, m^2/s × 10^7
Formed/Foamed-in-Place					
Mineral wool granules with asbestos/					
inorganic binders, sprayed	30	0.046	190		
Polyvinyl acetate cork mastic;					
sprayed or troweled	30	0.100			
Urethane, two-part mixture; rigid foam	30	0.026	70	1.045	
Magnesia, 85%	38	0.067	270		
	93	0.071			
	150	0.074			
	204	0.080			
Rock wool, 10 lb/ft^3	32	0.040	160		
Loosely packed	150	0.067	64		
	260	0.087			
Sawdust	23	0.059			
Silica aerogel	32	0.024	140		
Wood shavings	23	0.059			

From A.I. Brown and S.M. Macro, *Introduction to Heat Transfer*, 3d ed., McGraw-Hill, New York, 1958; *International Critical Tables*, McGraw-Hill, New York, 1926–1930.

TABLE B-6
Physical properties of nonmetals

Material	T, °C	k, W/(m·°C)	ρ, kg/m³	c_p, kJ/(kg·°C)	α, m²/s × 10⁷
Asphalt	20–55	0.74–0.76			
Brick					
Building brick, common	20	0.69	1600	0.84	5.2
Face		1.32	2000		
Carborundum brick	600	18.5			
	1400	11.1			
Chrome brick	200	2.32	3000	0.84	9.2
	550	2.47			9.8
	900	1.99			7.9
Diatomaceous earth, molded					
and fired	200	0.24			
	870	0.31			
Fireclay brick, burned 1330°C	500	1.04	2000	0.96	5.4
	800	1.07			
	1100	1.09			
Burned 1450°C	500	1.28	2300	0.96	5.8
	800	1.37			
	1100	1.40			
Missouri	200	1.00	2600	0.96	4.0
	600	1.47			
	1400	1.77			
Magnesite	200	3.81		1.13	
	650	2.77			
	1200	1.90			
Clay	30	1.3	1460	0.88	
Cement, portland	23	0.29	1500		
Mortar	23	1.16			
Coal, anthracite	30	0.26	1200–1500	1.26	
Powdered	30	0.116	737	1.30	
Concrete, cinder	23	0.76			
Stone 1-2-4 mix	20	1.37	1900–2300	0.88	8.2–6.8
Cotton	20	0.06	80	1.30	
Glass, window	20	0.78 (avg)	2700	0.84	3.4
Corosilicate	30–75	1.09	2200	–	
Plate (soda lime)	30	1.4	2500	0.75	
Pyrex	30	1.4	2225	0.835	
Paper	30	0.011	930	1.340	
Paraffin	30	0.020	900	2.890	
Plaster, gypsum	20	0.48	1440	0.84	4.0
Metal lath	20	0.47			
Wood lath	20	0.28			
Rubber, vulcanized					
Soft	30	0.012	1100	2.010	
Hard	30	0.013	1190	—	

TABLE B-6 (*continued*)

Material	$T,$ °C	$k,$ W/(m · °C)	$\rho,$ kg/m^3	$c_p,$ kJ/(kg · °C)	$\alpha,$ m^2/s × 10^7
Sand	30	0.027	1515	0.800	
Stone					
Granite		1.73–3.98	2640	0.82	8–18
Limestone	100–300	1.26–1.33	2500	0.90	5.6–5.9
Marble		2.07–2.94	2500–2700	0.80	10–13.6
Sandstone	40	1.83	2160–2300	0.71	11.2–11.9
Teflon	30	0.35	2200	—	
Tissue, human skin	30	0.37	—	—	
Fat layer	30	0.20	—	—	
Muscle	30	0.41	—	—	
Wood (across the grain)					
Balsa	30	0.055	140		
Cypress	30	0.097	460		
Fir	23	0.11	420	2.72	0.96
Maple or oak	30	0.166	540	2.4	1.28
Yellow pine	23	0.147	640	2.8	0.82
White pine	30	0.112	430		

TABLE B-7
Abbreviated saturated steam table

Temperature T, °C	Pressure P, kPa	Enthalpy of evaporation h_{fg}, kJ/kg	Liquid density ρ_l, kg/m³	Vapor density ρ_v, kg/m³
0.01	0.6113	2501.3	1000.0	0.0049
5	0.8721	2489.6	1000.0	0.0068
10	1.2276	2477.7	1000.0	0.0094
15	1.7051	2465.9	999.0	0.0129
20	2.339	2454.1	998.0	0.0173
25	3.169	2442.3	997.0	0.0231
30	4.246	2430.5	996.0	0.0304
35	5.628	2418.6	994.0	0.0397
40	7.384	2406.7	992.1	0.0512
45	9.593	2394.8	990.1	0.0655
50	12.349	2382.7	988.1	0.0831
55	15.758	2370.7	985.2	0.1045
60	19.940	2358.5	983.3	0.1304
65	25.03	2346.2	980.4	0.1614
70	31.19	2333.8	977.5	0.1983
75	38.58	2321.4	974.7	0.2421
80	47.39	2308.8	971.8	0.2935
85	57.83	2296.0	968.1	0.3536
90	70.14	2283.2	965.3	0.4235
95	84.55	2270.2	961.5	0.5045
100	101.35	2257.0	957.9	0.598
110	143.27	2230.2	950.6	0.826
120	198.53	2202.6	943.4	1.121
130	270.1	2174.2	934.6	1.496
140	361.3	2144.7	925.9	1.965

Temperature T, °C	Pressure P, MPa	Enthalpy of evaporation h_{fg}, kJ/kg	Liquid density ρ_l, kg/m³	Vapor density ρ_v, kg/m³
150	0.4758	2114.3	916.6	2.549
160	0.6178	2082.6	907.4	3.256
170	0.7917	2049.5	897.7	4.119
180	1.0021	2015.0	887.3	5.153
190	1.2544	1978.8	876.4	6.388
200	1.5538	1940.7	864.3	7.852
210	1.9062	1900.7	852.5	9.578
220	2.318	1858.5	840.3	11.602
230	2.795	1813.8	827.1	13.970
240	3.344	1766.5	813.7	16.734
250	3.973	1716.2	799.4	19.948
260	4.688	1662.5	783.7	23.691
270	5.499	1605.2	768.1	28.058
280	6.412	1543.6	750.8	33.145
290	7.436	1477.1	732.1	39.108
300	8.581	1404.9	712.3	46.147
310	9.856	1326.0	691.1	54.50
320	11.274	1238.6	667.1	64.57
330	12.845	1140.6	640.6	76.95
340	14.586	1027.9	610.5	92.62
350	16.513	893.4	574.7	113.47
360	18.651	720.5	528.3	143.99
370	21.03	441.6	451.9	203.05
374.14	22.09	0	316.96	316.96

Adapted from Joseph H. Keenan, Frederick G. Keyes, Philip G. Hill, and Joan G. Moore, *Steam Tables*, Wiley, New York, 1969.

TABLE C-1
Normal emissivity of surfaces

Surface	T, °C	ε_n	Surface	T, °C	ε_n
Metals			Metals		
Aluminum			Lead (*continued*)		
Highly polished, plate	200–600	0.038–0.06	Gray, oxidized	23	0.28
Bright, foil	21	0.04	Oxidized at 200°C	200	0.63
Heavily oxidized	100–500	0.20–0.33	Nickel		
Brass			Electrolytic	37–260	0.04–0.06
Highly polished	250–360	0.028–0.031	Pure, polished	260	0.07
Dull plate	50–350	0.22	Oxidized at 600°C	260–540	0.37–0.48
Oxidized	200–500	0.60	Platinum		
Chromium, polished	37–1100	0.08–0.40	Plate, polished	260–540	0.06–0.10
Copper			Filament	26–1225	0.04–0.19
Polished, electrolytic	80	0.018	Silver		
Polished	37–260	0.04–0.05	Polished	37–625	0.02–0.03
Calorized	37–260	0.18	Stainless steel		
Black oxidized	37	0.78	**Polished**	23	0.17
Gold, polished	37–260	0.02	**Cleaned**	23	0.21–0.39
Iron			Tin		
Polished	425–1025	0.14–0.38	Polished	37	0.05
Oxidized	100	0.74	Tungsten		
Cast iron, oxidized at 600°C	200–600	0.64–0.78	Filament	3300	0.39
Cast plate, smooth	22	0.80	Zinc		
Cast plate, rough	22	0.82	Polished	225–325	0.05–0.06
Lead			Oxidized at 400°C	400	0.11
Pure, polished	260	0.08	Galvanized	23-27	0.23–0.28

TABLE C-1 (*continued*)

Surface	T, °C	ε_n	Surface	T, °C	ε_n
Nonmetals			Nonmetals		
Alumina			Ice (*continued*)		
(85–99.5% Al_2O_3),			Rough crystals	0	0.985
effect of mean			Marble, light gray, polished	22	0.93
grain size			Mica	37	0.75
10 μm	1000–1560	0.30–0.18	Paints		
50 μm	1000–1560	0.39–0.28	Aluminum 10%,		
100 μm	1000–1560	0.50–0.40	lacquer 22%	100	0.52
Asbestos			Aluminum 26%,		
Paper	37	0.93	lacquer 27%	100	0.30
Board	37	0.96	Other aluminum paints	100	0.27
Brick			Lacquer, white	100	0.925
Magnesite, refractory	1000	0.38	Lacquer, black matte	80	0.97
Red, rough	21	0.93	Oil paints, all colors	100	0.92–0.96
Gray, glazed	1100	0.75	Oil paints	20	0.89–0.97
Silica	540	0.80	Paper		
Carbon			Ordinary	20–95	0.80–0.92
Filament	1050–1400	0.526	Tar	20	0.93
Candle soot	95–270	0.952	Porcelain, glazed	22	0.92
Lampblack	20	0.93–0.967	Quartz		
Ceramic			Glass, 1.98 mm thick	280	0.90
Earthenware, glazed	20	0.90	Glass, 1.98 mm thick	840	0.41
Earthenware, matte	20	0.93	Glass, 6.88 mm thick	280	0.93
Porcelain	22	0.92	Glass, 6.88 mm thick	840	0.47
Refractory, black	93	0.94	Rubber		
Clay, fired	70	0.91	Hard	23	0.94
Concrete, rough	37	0.94	Soft, gray	23	0.86
Glass			Soil	37	0.93–0.96
Smooth	22	0.94	Water, deep	0–100	0.96
Pyrex, lead, and soda	260–530	0.95–0.85	Wood	20	0.80–0.90
Ice					
Smooth	0	0.97			

TABLE C-2
Solar absorptivity of surfaces (receiving surface at room temperature)

Surface	α	Surface	α
Metals		**Nonmetals**	
Aluminum		Asphalt	
Polished	0.10	Pavement	0.85
Anodized	0.14	Pavement free from dust	0.93
Foil	0.15	Brick	
Brass		White glazed	0.26
Polished	0.3–0.5	Red	0.70–0.77
Dull	0.4–0.65	Concrete	
Chromium, electroplated	0.41	Uncolored	0.65
Copper		Brown	0.85
Highly polished	0.18	Black	0.91
Clean	0.25	Earth, plowed field	0.75
Tarnished by exposure	0.64	Granite	0.45
Gold	0.21	Grass	0.75–0.8
Iron		Gravel	0.29
Matte, oxidized	0.96	Leaves, green	0.71–0.79
Lead roofing, old	0.77	Magnesium oxide (MgO)	0.15
Nickel		Marble	
Highly polished	0.15	White	0.44
Polished	0.36	Ground, unpolished	0.47
Oxidized	0.79	Paints	
Platinum, bright	0.31	Oil, white lead	0.24–0.26
Silver		Oil, light cream	0.30
Highly polished	0.07	Oil, light green	0.50
Polished	0.13	Aluminum	0.55
Stainless steel, type 301		Oil, light gray	0.75
Polished	0.37	Oil, black on galvanized iron	0.90
Clean	0.52	Paper	
Zinc		White	0.28
Highly polished	0.34	Sand	0.76
Polished	0.55	Sawdust	0.75
		Slate	
		Silver gray	0.79
		Blue gray	0.85
		Dark gray	0.90
		Snow, clean	0.2–0.35
		Soot, coal	0.95
		Zinc oxide	0.15

TABLE D

Error function $\mathrm{erf}(\xi)$ $\qquad \left(\mathrm{erf}(\xi) = \dfrac{2}{\sqrt{\pi}} \displaystyle\int_0^{\xi} e^{-y^2}\, dy \qquad \mathrm{erf}(\infty) = 1 \right)$

ξ	$\mathrm{erf}(\xi)$	ξ	$\mathrm{erf}(\xi)$	ξ	$\mathrm{erf}(\xi)$	ξ	$\mathrm{erf}(\xi)$	ξ	$\mathrm{erf}(\xi)$	ξ	$\mathrm{erf}(\xi)$
0.00	0.00000	0.76	0.71754	1.52	0.96841	(*continued*)					
0.02	0.02256	0.78	0.73001	1.54	0.97059						
0.04	0.04511	0.80	0.74210	1.56	0.97263	0.40	0.42839	1.16	0.89910	1.92	0.99338
0.06	0.06762	0.82	0.75381	1.58	0.97455	0.42	0.44749	1.18	0.90484	1.94	0.99392
0.08	0.09008	0.84	0.76514	1.60	0.97636	0.44	0.46622	1.20	0.91031	1.96	0.99443
0.10	0.11246	0.86	0.77610	1.62	0.97804	0.46	0.48466	1.22	0.91553	1.98	0.99489
0.12	0.13476	0.88	0.78669	1.64	0.97962	0.48	0.50275	1.24	0.92050	2.00	0.99532
0.14	0.15695	0.90	0.79691	1.66	0.98110	0.50	0.52050	1.26	0.92524	2.10	0.99702
0.16	0.17901	0.92	0.80677	1.68	0.98249	0.52	0.53790	1.28	0.92973	2.20	0.99813
0.18	0.20094	0.94	0.81627	1.70	0.98379	0.54	0.55494	1.30	0.93401	2.30	0.99885
0.20	0.22270	0.96	0.82542	1.72	0.98500	0.56	0.57162	1.32	0.93806	2.40	0.99931
0.22	0.24430	0.98	0.83423	1.74	0.98613	0.58	0.58792	1.34	0.94191	2.50	0.99959
0.24	0.26570	1.00	0.84270	1.76	0.98719	0.60	0.60386	1.36	0.94556	2.60	0.999764
0.26	0.28690	1.02	0.85084	1.78	0.98817	0.62	0.61941	1.38	0.94902	2.70	0.999866
0.28	0.30788	1.04	0.85865	1.80	0.98909	0.64	0.63459	1.40	0.95228	2.80	0.999925
0.30	0.32863	1.06	0.86614	1.82	0.98994	0.66	0.64938	1.42	0.95538	2.90	0.999959
0.32	0.34913	1.08	0.87333	1.84	0.99074	0.68	0.66278	1.44	0.95830	3.00	0.999978
0.34	0.36936	1.10	0.88020	1.86	0.99147	0.70	0.67780	1.46	0.96105	3.20	0.999994
0.36	0.38933	1.12	0.88079	1.88	0.99216	0.72	0.69143	1.48	0.96365	3.40	0.999998
0.38	0.40901	1.14	0.89308	1.90	0.99279	0.74	0.70468	1.50	0.96610	3.60	1.000000

INDEX

431